Principles of Environmental Management

Second Edition

Principles of Environmental Management

The Greening of Business

Rogene A. Buchholz

Joseph A. Butt, S.J., College of Business Administration
Loyola University of New Orleans

Prentice Hall, Upper Saddle River, New Jersey 07458

Acquisitions Editor: *Lisamarie Brassini*
Associate Editor: *John Larkin*
Editorial Assistant: *Dawn Marie Reisner*
Editor-in-Chief: *James Boyd*
Marketing Manager: *Tamara Wederbrand*
Production Editor: *Maureen Wilson*
Managing Editor: *Dee Josephson*
Manufacturing Buyer: *Diane Peirano*
Manufacturing Supervisor: *Arnold Vila*
Manufacturing Manager: *Vincent Scelta*
Composition: *Omegatype Typography, Inc.*
Copyeditor: *Nancy Marcello*
Cover art: *Private Spaces XI, acrylic on board, 1996, Pamela Moore, New York, NY*
Cover design: *Bruce Kenselaar*

A Simon & Schuster Company
Upper Saddle River, New Jersey 07458

Library of Congress Cataloging-in-Publication Data

Buchholz, Rogene A.
Principles of environmental management : the greening of business
/ Rogene A. Buchholz—2nd ed.
p. cm.
Includes bibliographical references and index.
ISBN 0-13-684895-8
1. Environmental policy. 2. Environmental management.
3. Industrial management—Environmental aspects. 4. Social
responsibility of business. I. Title.
HC79.E5B83 1998
363.7—dc21 97-32583
CIP

Prentice-Hall International (UK) Limited, *London*
Prentice-Hall of Australia Pty. Limited, *Sydney*
Prentice-Hall Canada, Inc., *Toronto*
Prentice-Hall Hispanoamericana, S.A., *Mexico*
Prentice-Hall of India Private Limited, *New Delhi*
Prentice-Hall of Japan, Inc., *Tokyo*
Simon & Schuster Asia Pte. Ltd., *Singapore*
Editora Prentice-Hall do Brasil, Ltda., *Rio de Janeiro*

Printed in the United States of America

10 9 8 7 6 5 4 3 2 1

This book is dedicated to my wife Sandy,
whose interest in the environment and nature
and whose support of all my efforts to learn
about this subject and relate it to business
and management has been inspirational.

Contents

PART III: RESPONSES TO ENVIRONMENTAL PROBLEMS

Preface

Several years ago, the National Wildlife Federation (NWF) conducted an informal telephone survey of several instructors at schools of business and management around the country. These instructors, for the most part, taught courses in Business and Society, Business and Public Policy, or Strategic Management. The purpose of this survey was to determine if anything of substance regarding the natural environment and resource management was being taught at these schools.

The general conclusion of this survey was that not much environmental material was being taught by faculty in these schools. None of their courses dealt very thoroughly with environmental concepts and issues, and no course, either required or elective, exclusively dealt with environmental concerns. NWF saw this as a serious problem, so it decided to do something to promote the teaching of environmental concerns in schools of business and management.

The NWF formed an Outreach Committee by asking about half a dozen members of the Corporate Conservation Council (CCC) to serve on this committee. The CCC is composed of about 15 senior-level executives who are responsible for environmental affairs at some of the country's major corporations. The Outreach Committee was thus a subset of the CCC and included executives from 3M, Dow Chemical, Du Pont, and other major corporations.

The purpose of the Outreach Committee was to develop a plan to promote the inclusion of environmental materials in courses and programs where future managers of corporations are trained. It was believed that students of business and management should be sensitized to environmental issues and be made aware of the environmental consequences of managerial actions. The Outreach Committee planned to develop environmental materials such as cases that would be made available to faculty that wanted to use them, and promote the development of pilot courses in the environment at a few key schools around the country.

To accomplish this task, the Outreach Committee involved three faculty members from different schools to work on putting together the actual material. These three faculty started writing cases about environmental issues and putting together bibliographic and conceptual materials. The corporations represented on the Outreach Committee supported these efforts and made resources available to promote the development of these environmental materials. The author of this textbook was one of these three faculty, and thus this textbook is an outgrowth of this involvement.

The first edition of this book was the first comprehensive text devoted exclusively to environmental issues and responses to those issues by business corporations. Since that time, more books dealing with environmental concerns have appeared and more environmental courses are being taught in schools of business and management around the country and the world. Journals and academic organizations have been formed dealing with environmental issues and business corporations. Thus concern about the natural environment and business responses to these concerns have continued to grow as the natural environment has received increasing attention.

Environmental issues such as global warming, ozone depletion, deforestation, and species decimation are in the news constantly. Traditional issues such as air pollution, water pollution, pesticide use, and cleanup of hazardous waste dumps are also of continuing concern. Other issues such as wetlands protection and protection of endangered species have caused a great deal of controversy at the highest levels of government. These concerns make a book of this nature extremely important in introducing business and management students to such issues and dealing with the nature of corporate responses.

The book is organized into three major sections. The first section deals with the development of the environment as a major concern, and provides some background material related to ecology, environmental ethics, and environmental management from the standpoint of regulation and public policy. The purpose of this section is to provide a framework for the analysis of environmental issues and to deal with some fundamental concepts that may help in understanding more concrete problems. It is important that students have some background in these areas before dealing with concrete environmental issues of concern to business.

The second section deals with specific environmental issues starting with such macro issues as global warming and ozone depletion. The section then moves into more traditional environmental issues such as air pollution, water pollution, pesticide use, control of toxic substances, and solid and hazardous waste disposal. The section then deals with deforestation and species decimation and coastal erosion and wetlands protection, which are some of the most controversial issues in the country and the world. Each of these chapters discusses the nature of the problem, presents a history of legislation as well as current legislation and regulations that are pertinent to business, and discusses problems with current efforts. These chapters also try to link the issues to managerial practice.

Finally, the last section deals with approaches to these problems from the standpoint of business and society. The chapter on management theory discusses some of the problems apparent in traditional approaches with respect to the natural environment and presents some recent work that attempts to integrate the natural environment into theoretical approaches to management and the business organization. The next chapter discusses the greening of business with respect to manufacturing, marketing, strategy, and communication, and discusses the nature of a sustainable corporation. The last chapter in this section talks about strategies for the society as a whole related to the concept of sustainable development.

This second edition has been thoroughly updated with new material, and many of the chapters have been substantially reorganized because of new developments. A new chapter has been added on management theory as stated above because of the importance of this subject and the recent work that has been done in developing theo-

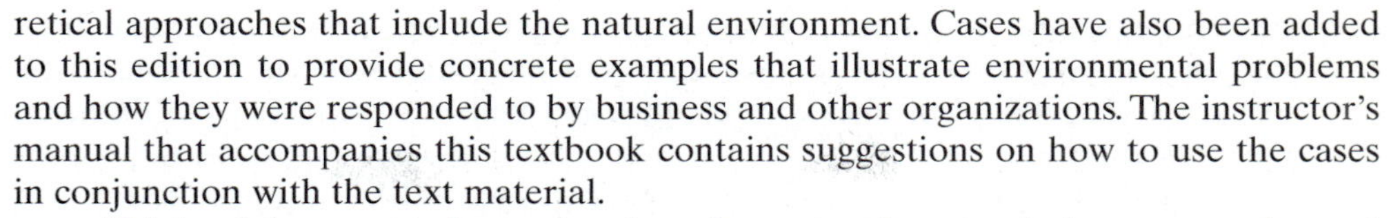

retical approaches that include the natural environment. Cases have also been added to this edition to provide concrete examples that illustrate environmental problems and how they were responded to by business and other organizations. The instructor's manual that accompanies this textbook contains suggestions on how to use the cases in conjunction with the text material.

This book is meant to be used as the primary text in a stand-alone course devoted to environmental concerns and business management. But it could also serve as a supplementary text in business and society and business policy and strategy courses where instructors want to go into some depth regarding environmental issues and their significance for management and corporations. The book should also have a market in the trade for corporations that have an emphasis on environmental concerns in their management training programs. Many activities of this sort are going on in corporations, especially those that have a major impact on the environment and that are seriously affected by environmental regulations.

There are many people who were extremely helpful in the writing of this book. First of all, Alison Reeves, the senior editor at Prentice Hall with whom I worked on the first edition, must be thanked for her immediate interest in this book and continuing support during the time it was being initially developed. This was something of a risky venture, because it was not clear how much of a market would eventually develop for a book of this nature. Yet she never hesitated in supporting the work and getting her management to commit to its publication. The same support was received for the second edition, particularly from Lisamarie Brassini and Natalie Anderson, who are the new associate editor and editor, respectively, in the management group at Prentice Hall.

Equally important are the people I worked with on the NWF project. My academic colleagues on this project, Jim Post at Boston University and Alfred Marcus at the University of Minnesota, provided a great deal of motivation and knowledge with respect to environmental interests. All the members of the CCC deserve credit for their support of this project, but a few must be singled out because of their direct and sustained contact with the project. Dr. Robert P. Bringer, staff vice-president, Environmental Engineering and Pollution Control of 3M Corporation; Robert L. Dostal, director of Safety, Environmental Affairs, Security and Loss Prevention of Dow Chemical; and Dr. J. R. Cooper, director, Environmental Affairs Division of Du Pont, were particularly helpful and supportive of the NWF project. And finally, the NWF staff, in particular Barbara Haas, director of the CCC, and Mark Haveman, project coordinator on the staff of the CCC, devoted a great deal of time and effort to the project. It was my involvement in this project that acquainted me with the environment in enough depth to be able to write this textbook.

Many other colleagues in the Social Issues in Management Division and the Organizations and the Natural Environment (ONE) Interest Group of the Academy of Management could also be mentioned, as we have had many interesting discussions of environmental issues over the years and I have learned a great deal from them. The support I received from the dean and faculty at Loyola University was very essential in developing a course in environmental management that enabled me to develop teaching materials that found their way into this textbook. The response of the students to the course has been extremely gratifying, and their interest in the material provided a great deal of motivation to write this textbook. And finally, my wife, to whom this book is dedicated, and I have enjoyed nature together on numerous hikes in mountains and canyons in many parts of the world and have had some interesting discussions about environmental problems in conjunction with our work in environmental and business ethics.

To all of these people and many others, I owe a debt of gratitude for providing me with many interesting opportunities to learn more about our world and the environment in which we live. If the human race is to have any chance of survival, and I am sincerely talking about a survival issue here, we must become scholars of the environment and adopt an interest in ecology and nature. Rather than just walk through nature as if it weren't there, we must become attuned to our natural environment and come to appreciate it for itself, and not just for the services it provides for human beings. Perhaps this could be called a reverence for nature, but whatever it is, some kind of environmental consciousness is essential if we are to cope successfully with the environmental problems we have created. Hopefully, this book may make a small contribution to the solution of these problems.

Rogene A. Buchholz
Joseph A. Butt, S.J., College of Business Administration
Loyola University of New Orleans

CHAPTER 1

Development of Environmental Concerns

There are some fundamental changes going on in American society and throughout the world that are impacting institutions and lifestyles as never before. These changes have to do with our relationship to the environment, and are being made in order to cope with some serious problems having to do with the depletion of the ozone layer, global warming, deforestation, species decimation, coastal erosion, wetlands protection, acid rain, water pollution, solid and hazardous waste disposal, toxic air emissions, and several other environmental problems of similar magnitude. Most if not all of these problems have become serious enough to require massive expenditures and changes in industrial processes that are unprecedented. Companies are phasing out highly useful products and are spending millions to reduce their environmental impact.

Events of the past decade have accelerated a steadily growing concern for the environment. Substantial attention has been given to the discovery of the ozone hole over the Antarctic and the link between depletion of the ozone layer and chlorofluorocarbons (CFCs), which are now believed to be the major culprit in destroying this layer. Subsequently, an international agreement was signed to reduce the production of these substances and accelerate the search for substitutes. The hot and dry summer of 1988 was confirmation enough for some scientists and policy makers to conclude that the greenhouse effect was for real, and was changing our climate in ways that are only beginning to be understood. The hot summer of 1995, during which hundreds of people in places like Chicago and Milwaukee as well as thousands of cattle and other animals died because of the heat, only heightened these concerns.

The tenth anniversary of Earth Day celebrated in the spring of 1990 and the Earth Summit held in Rio de Janeiro in 1992 brought with them a new awareness of environmental problems and a new sense of urgency that enabled people to be mobilized for environmental causes all over the world. Public opinion polls in the United States revealed a steady and widespread growth in public concern for environmental quality, even throughout the 1980s and 1990s when conservatives in the administration and in Congress did their best to cut back on environmental expenditures. Data from a variety of polls have indicated that the American public is increasingly aware of and concerned about the gravity of environmental problems and wants the federal government to become more actively involved in their solution. Many people believe tougher laws and regulations are needed and feel more money should be spent on environmental protection. They also say they are more willing than ever before to pay for environmental cleanup and believe economic growth must be sacrificed in order to protect the environment.[1]

Some writers suggest that environmental consciousness has "trickled down" from its core of relatively affluent and well-educated supporters to the American population in general because of four factors: (1) The less well off are being visibly affected by environmental problems far more than they are suffering due to environmental protection measures; (2) environmentalists have become more equity conscious, and through their adoption of the sustainable growth logic of the appropriate technology movement, have largely cast off charges of antigrowth obstructionism; (3) the environmental movement has managed to mobilize informational and political resources successfully and has avoided displacement of its goals even during years when the administration and Congress were anti-environment; and (4) the youthful core of supporters of the early 1970s have matured, largely retaining their environmental ideals, and have resolved claims of conflict between environment and equity by arguing that the two goals are in many cases consistent.[2]

Growing recognition exists among corporate leaders that environmental issues are here to stay, and that environmental protection must be considered a normal part of doing business. It is becoming increasingly clear that positive corporate environmental performance cannot be adequately addressed through either the enunciation of environmental policies by top management or the efforts of environmental staff specialists, but must be institutionalized by being incorporated into the fundamental responsibilities of line management. This task will require increased environmental awareness and sophistication on the part of managers throughout the organization and the development of new skills and strategies to respond to environmental problems.

While environmental concerns have been on the public agenda for over two decades, during which a good deal of environmental legislation has been passed and implemented, what is different about today's world is that the environment has become a survival issue rather than a quality-of-life issue as it was in the 1960s and 1970s. More and more business executives are identifying the environment as the issue that will most affect their companies as we move into the twenty-first century. And more and more scientists are suggesting that if we don't come to grips with environmental problems in this decade, irreversible processes will have been set in motion that will inevitably lead to serious environmental degradation all over the globe and widespread human suffering.

THE WORLD AT RISK

What are some of these environmental problems and how serious are they? The latter question is the subject of much debate. But at least we have a fairly good knowledge of what the problems are even if we are not sure of their exact causes in all instances or what to do about them. Perhaps most serious are the so-called global problems. These problems are global because they affect everyone and every country in the world to some degree. The two problems that are generally considered to be in this category include global warming and ozone depletion. These problems are truly global in nature and require something of an international solution.

The most controversial of these two problems is the global warming phenomenon. It is a fact that the buildup of infared absorbing trace gases such as carbon dioxide, methane, CFCs, and nitrous oxide have mounted dramatically over the past few decades primarily, but not solely, from industrial processes and products. Carbon dioxide is produced during the burning of fossil fuels in electric-generating plants, automobiles, and other such sources. Methane comes from cattle, and the huge feedlots that have sprung up around the country to serve our needs for beef are a major contributor to the problem. CFCs are released into the air when old automobiles or refrigerators are scrapped. At the same time this buildup has occurred, carbon dioxide–absorbing resources, such as the world's rain forests, are being cut down at an alarming rate. So the buildup of trace gases continues.

The 1980s were the the warmest decade on record, with 1988, 1987, and 1981 being the warmest years in that order.[3] The new decade started out with a bang, as 1990 was the warmest year in more than a century of record keeping.[4] Then temperatures stopped rising for a few years because of the explosion in the Philippines of Mount Pinatubo, which poured enough debris into the air to mask the effects of global warming. When this debris began to dissipate, warming continued with record heat during the summer of 1995 in the Midwest and other parts of the country.

While there is some dispute about this warming, most scientists seem to agree that some warming has taken place. The real controversy surrounds the question of linkage: Is there a link between climate change and the buildup of greenhouse gases or is the warming we have experienced due to other causes? The consequences of global warming, should it continue and result in several degrees of additional warming as some scientists are predicting, are severe. Many coastal cities would be flooded if the polar ice caps melted as a result of warming and the sea level rose significantly, and people would have to either build sea walls to keep out the water or move to another location.[5]

The critical policy question is what steps do we take at present. Do we take steps to limit carbon dioxide emissions based on what we know now and accept the limitations of our computer models, or do we wait for more evidence and develop more sophisticated models that have greater predictive power? If we wait, the risk is that things may rapidly deteriorate and be more costly to try and correct somewhere down the road, and may even become irreversible. But if we act now and limit emissions of carbon dioxide by requiring utilities to install expensive scrubbers, we may find that money was wasted if carbon dioxide should prove not to have been a major culprit.

The situation with ozone depletion is much less uncertain. The ozone layer in the stratosphere protects the earth and its people from exposure to excessive ultraviolet radiation, which can cause skin cancer and cataracts, reduce crop yields, deplete marine fisheries, and interfere with the process of photosynthesis, which of course is essential to all life on the planet. Two scientists at the University of California at Irvine had theorized already in 1974 that CFCs were the culprit. While the various CFC compounds proved to be remarkable products, useful in air-conditioning, insulation, and as a solvent in chip making, as well as a host of other uses because of their chemical inertness, this very stability meant that they did not break down in the lower reaches of the atmosphere, but drifted up to the ozone layer where they finally reacted with the ozone molecules to break them apart.

The use of CFC in aerosol spray cans was banned in this country on the basis of this theory, but no other action was taken, and the use of CFCs continued to grow worldwide. There was a great deal of debate, as there currently is about the greenhouse effect, as to whether CFCs were the real culprit and how serious a problem ozone depletion really was in the larger scheme of things. This debate ended for all practical purposes in the spring of 1987 when scientists discovered that ozone concentration was down 50 percent over the South Pole and had actually disappeared entirely in some places. Subsequently, scientists began to gather evidence that the ozone layer around the globe may be eroding much faster than predicted.[6]

These findings galvanized the industrial nations of the world to take action and put aside the debate. These nations signed the Montreal Protocol in 1987 agreeing to phase out CFC production over a period of years, allowing developing countries a reprieve from these restrictions in order to improve their economies. DuPont, the world's largest producer of CFCs, agreed to eliminate production of CFCs entirely ahead of schedule in order to protect the ozone layer. Based on new findings about the rapid deterioration of the ozone layer, many scientists believed the treaty was outdated and needed to be strengthened, which it was a few years later. But even if all companies were to stop production of CFCs today, there are six or seven years' worth of CFCs already in the atmosphere that will eventually drift up to the ozone layer to do their damage and expose us to more ultraviolet radiation.

There are two other problems that are more regional in nature but have global implications. Destruction of the tropical rain forests is one such problem. The tropical rain forests are scattered in an uneven green belt that lies roughly between the Tropic of Cancer and the Tropic of Capricorn. Rain forests grow in regions where at least four inches of rain falls monthly, where the mean monthly temperature exceeds 75 degrees F, and where frost never occurs. Not all the rain forests in the world are tropical; witness the Olympic Peninsula of Washington and the Tongass National Forest in Alaska. But these forests are much less diverse and contain far fewer species of trees and other forms of life than the tropical rain forests.[7]

The tropical rain forests cover less than 5 percent of the earth's surface, but are home to perhaps half of all the earth's species. Some scientists estimate that the actual number of species of insects in tropical forests might be between 30 million and 80 million. Fewer than half a million tropical species of any kind have been cataloged. There may be as many as 200 species of trees in a single acre of tropical rain forest. There are only about 400 species of trees in all of temperate North America. A single square mile of Amazonian Ecuador or Brazil may be home to more than 1,500 kinds of butterflies, while only about 750 occur in all the United States and Canada.[8]

Every year, some 50,000 square miles of tropical forest are lost to logging and field clearing, and an additional 60,000 square miles are seriously disrupted. This means that about 1 percent of the world's tropical forests are disappearing each year, and if current practices continue, all tropical forests will disappear in the next century.[9] About half of the original rain forest is already gone, with the remainder covering an area about the size of the 48 contiguous states. During the next 20 to 30 years this area will be reduced to scattered remnants except for major patches in the Amazon region and the interior of Africa. Clearing these forests for settlement is counterproductive because the soil is not rich enough to support long-term farming or cattle ranching. The nutrients in the soil are exhausted after a few years, and thus new land must be cleared.

Because of this destruction, biological extinction is occurring at a more rapid rate than at any time since the demise of the dinosaurs 66 million years ago.[10]

Acid rain is a phenomenon caused by sulfur dioxide (SO_2) emissions that affects some regions of the world much more than others. Acid rain is believed to place severe stress on many ecosystems in reducing the size and diversity of fish populations and playing a role in forest damage. Even though the National Acid Precipitation Assessment Program questioned the linkage between SO_2 emissions and the severity of the problem,[11] the new Clean Air Act, passed in the early 1990s, contained provisions for drastic reductions in these emissions and established an emissions trading system for utilities that would allow them to buy and sell emissions credits in a market arrangement. It was hoped that this system would provide better incentives for utilities to reduce SO_2 emissions than traditional regulation.

More traditional problems such as air pollution have seemingly become more serious despite the best efforts of government and industry to deal with them. Some 60 cities still violate one or more of the air pollution standards in existence. Although lead emissions are down by about 87 percent over the past few years, smog seems to be an intractable problem. Cities like Los Angeles are proposing drastic measures such as eliminating all backyard barbecuing and requiring cars to burn alternative fuels like methanol. The new Clean Air Act requires new controls on industrial installations releasing smog-forming chemicals and may cost as much as $25 billion a year when all its requirements go into effect toward the end of the decade.[12]

Regarding water pollution, many coastal towns along the Atlantic and Gulf of Mexico have had to close beaches during the summer months because of pollution. Groundwater is being contaminated by underground storage tanks, fertilizers, and pesticides, hazardous waste sites, and other sources, threatening 50 percent of the nation's drinking water for half the population. Wetlands are being destroyed at the rate of between 350,000 and 500,000 acres per year, much of that destruction taking place in the state of Louisiana, which has more wetlands that any other state in the Union. Because of this destruction of wetlands along the Gulf Coast, the state is experiencing coastal erosion at an alarming rate, which is a cause of grave concern.[13]

Approximately 158 million tons of municipal solid waste is generated each year in the United States, but many municipal landfills are close to overflowing and almost 70 percent are expected to reach capacity in 15 years. Municipalities are having trouble opening new landfills, however, because of the not in my backyard (NIMBY) effect. With respect to hazardous waste, approximately 30,000 potentially contaminated sites that may pose a threat to human health or the environment have been identified nationwide. But cleanup of these sites is proceeding slowly because of the problem of identifying what is in these sites and because of legal conflict over who is responsible.[14]

Finally, airborne toxic substances are a relatively new problem of significant proportions. Under Title III of the Superfund Amendments and Reauthorization Act (SARA) of 1986, also known as the Emergency Planning and Community Right-to-Know Act, facilities that manufacture, process, or use any of 309 designated chemicals in greater than specified amounts, must report routine releases of those chemicals. The EPA is required to make information from these reports available to the public. This Toxics Release Inventory, as it is called, is designed to assist citizen groups, local health officials, state environmental managers, and the EPA to identify and control toxic chemical problems.[15]

The nation's first inventory of toxic releases showed that in 1987, industry released 2.4 billion pounds of toxic substances into the air we breathe. The chemical industry headed the list with 886.6 million pounds of toxic releases. Emissions in eight states exceeded 100 million pounds. These emissions included 235 million pounds of carcinogens such as benzene and formaldehyde, and 527 million pounds of such neurotoxins as toluene and trichloroethlene. The EPA estimated that air toxins cause more than 2,000 cases of cancer each year based on only 20 chemicals, not the 239 that were included in the survey of toxic releases. Industry argues that these chemicals become so diluted in the air that they are innocuous. But the EPA says that living near chemical plants poses a cancer risk greater than the national average.[16]

Information from Eastern Europe and the former Soviet Union as these countries have opened up to the outside world show that these traditional problems are even worse than in the Western world. In Russia, for example, a quarter of the nation's drinking water is unsafe and all the major rivers are polluted. Dangerous pesticides are stockpiled around the country, posing risks comparable to those of chemical weapons. Life expectancy is steadily dropping, and infant mortality, which is already twice that of other industrialized countries, is expected to rise by another 25 percent.[17]

Seventy percent of the rivers in what was formerly Czechoslovakia are badly polluted. A third of the rivers and 9,000 lakes in East Germany are biologically dead. Eighty percent of Romania's river water is unpotable. In Hungary, some 1.3 billion cubic meters of untreated sewage is discharged into the country's surface waters each year. Half of Poland's cities and 35 percent of its industries do not treat their waste. In 1988, the Soviet Union could adequately treat only 30 percent of its sewage.[18]

Though data are scarce, hazardous wastes appear to have been indiscriminately dumped on land throughout the region. Some 15,000 hazardous dump sites are said to be awaiting evaluation in former East German territory. In the Soviet Union, more than half of nearly 6,000 official landfills do not meet sanitary regulations.[19] Similar conditions are believed to exist in other Eastern European nations with respect to the disposal of hazardous wastes.

Because these environmental factors are superimposed on a more general health-care crisis, sorting out the precise causes and consequences of health problems is impossible. Yet in the dirtiest areas of what was formerly Czechoslovakia, life expectancy is as much as five years less than in relatively clean parts of the country. In industrialized Poland, people die three to four years earlier than in the rest of the country. In Halle, East Germany, a center of the industrial chemistry industry, people can expect to live five years less than other East Germans. Shortened life expectancies, soaring cancer rates, and a host of other maladies have been recorded in highly polluted regions of Eastern Europe.[20]

CHANGES IN THE ENVIRONMENT

These problems pose serious challenges to our planet and our way of life, particularly for those of us who live in advanced industrial societies. Such problems have made many of us step back and take a look at traditional ways we have thought about the environment, how we have understood human beings in relation to the environment with which we are surrounded. The physical environment includes air, water, and land, without which life as we know it would be impossible. This environment provides a number of services that human beings cannot do without. Chief among them is provision of a habitat in which plant and animal as well as human life can survive. If this habitat is seriously degraded, plant and animal life will be adversely affected. And if plant and animal life are adversely affected, so will human life be affected, as we are all connected in a web of life such that if any part is seriously degraded, all parts of the whole are affected.

The physical environment is also called upon to provide resources that are used in the production process, whatever form that process might take, to produce goods and services for the members of society. Some of these resources are nonrenewable and are thus able to be completely exhausted after having been used for many years. Others, such as timber, are renewable, but conscious effort is generally needed to replace those renewable resources that are used. This replacement usually does not happen automatically, at least not fast enough to support a growing population. The physical environment is also used as a place to dispose of waste material that results from the production of goods and services as well as from their consumption. Problems arise when this waste material overwhelms the absorption capacity of the environment and serious degradation is the result.

Pollution of the physical environment interferes with its ability to provide a habitat in which life can survive and flourish. The ability of the physical environment to serve as a gigantic waste disposal facility depends on its dilutive capacity. Pollution occurs when the waste discharged into the environment exceeds its dilutive capacity—when air can no longer dilute the wastes dumped into it without air quality being adversely affected; water can no longer absorb the wastes dumped into it without some fundamental change taking place in the quality of the water; and land cannot absorb any more waste material without producing harmful effects that relate to land usage itself or drinking water supplies.

The amount of damage that results to a particular medium (air, water, land) varies by the type of pollutant, the amount of pollutant disposed of, and the distance from the source of pollution. These damages, however, alter the quality of the environment and render it, to some degree, unfit to provide its normal services. Thus the air can become harmful for human beings to breathe, water unfit to drink, and land unfit to live on because toxic wastes that begin seeping to the surface pose a threat to human health.

Before the advent of pollution control legislation, air, water, and land were treated as free goods available to anyone for dumping wastes. This caused no problem when the population was sparse, factories small, and products few in number compared to today. The environment's dilutive capacity was rarely exceeded and was perceived as infinite in its ability to absorb waste. Changes in society, however, began to cause serious pollution problems. The following factors were critical in this transformation.

Population growth and concentration: More people means more manufactured goods and services to provide for their needs, which in turn means more waste material to be discharged into the environment. The concentration of people in urban areas compounds the problem. Eventually the dilutive capacity of the air, water, and land in major industrial centers becomes greatly exceeded and a serious pollution problem results.

Rising affluence: As real income increases, people are able to buy and consume more goods and services, throw them away more quickly to buy something better, travel more miles per year using various forms of transportation, and expand their use of energy. In the process, much more waste material is generated for the society as a whole.

Technological change: Changes in technology have expanded the variety of products available for consumption, increased their quantity through increases in productivity, made products and packaging more complex, and raised the rate of obsolescence through rapid innovation. All of this has added to the waste disposal problem. In addition, the toxicity of many materials was initially unknown or not given much concern, with the result that procedures for the abatement of these pollution problems have lagged far behind the technology of manufacture.

Increased expectations and awareness: As society became more affluent, it could give attention to higher-order needs. Thus expectations for a higher quality of life have increased, and the physical environment is viewed as an important component of the overall quality of life. One cannot fully enjoy the goods and services that are available in a hostile or unsafe environment. In addition, the people's awareness of the harmful effects of pollution increased due to mounting scientific evidence, journalistic expose, and the attention given environmental problems by the media.

These forces combined about the mid-1960s to give birth to an environmental movement that developed very quickly. Many of the energies that had gone into the civil rights movement were channeled into the environmental movement as the former matured. The result was a major public policy effort to control pollution and correct for the deficiencies of the market system in controlling the amount and types of waste being discharged into the environment. These efforts have had a major impact on business and consumers alike and caused attitudinal and behavioral changes throughout society.

CHANGING CONSCIOUSNESS

Environmental problems pose a challenge to human self-understanding and the place of humans in the universe. The traditional view of humans and their relationship to nature has been dualistic, that idea that humans stand over nature and are somehow apart from nature. The task of humans has been to conquer nature, to take dominion over the animals and the natural world as some religious doctrines have emphasized, to gain more and more power over nature. This dualistic view led to an objectification of nature and allowed us to manipulate nature to our advantage and exploit it for our own purposes.

This view is being challenged by those who advocate that humans must instead see themselves as a part of nature, and through education about ecology must come to see themselves as but one link in the great chain of being. Only by adopting this perspective, it is argued, can humans see nature properly and understand what must be done to promote survival of the planet and the human race. Nature is both subject and object, as are human beings themselves, and we humans must learn to cooperate and live in harmony with nature rather than dominate it and use it solely for our own advantage.

Several aspects of this dualistic view of nature are proving to be problematic. While once nature indeed may have been able to take care of itself, today nature is defined by human activity. While we once may have thought of nature as a collection of forces that humans reacted to as they fought for survival, it is nature's survival that is now threatened by human activity.[21] Human beings determine what lives and dies as far as plants and animals are concerned. We determine the amount of ultraviolet radiation we will be exposed to through our willingness to curb production and use of ozone-destroying chemicals. And if there is anything to global warming, we also determine the kind of climate we will live in and what kind of threats from rises in sea levels have to be responded to in future years.

Another aspect of the dualistic view of nature is the belief that nature exists to serve human purposes, what has been called an anthropomorphic view of nature. This term simply refers to the human-centered way we have traditionally approached nature. We manipulate nature to serve our own purposes and our own sense of progress. Nature must be "developed" and has no value in its natural state. It only has value as it is shaped to serve some human purpose. Resources in their natural state have no utility, so they are taken out of the ground and processed by industrial systems to make something useful that can be sold to consumers. Only then does nature have any use or value and become part of the system where economic wealth is measured.

The system of national accounting we use in this country and other developed countries to measure economic progress incorporates the depreciation of plant and equipment but not the depletion of natural capital. The principal measure of economic progress is gross national product or its equivalent, but this measure does not take into account the depletion of nonrenewable and renewable resources and thus produces a misleading sense of national economic health. If all the environmental consequences of economic activity from resource depletion to numerous forms of environmental damage were included in this measure, real economic progress would be much less than conventional measures indicate. Nations should apply some kind of an ecological deflator if they are to measure real progress in human and social welfare.[22] But again, these measures reflect the dualism that exists in the world with respect to nature, as nature doesn't matter and impacts on nature need not be included in our measures of progress.

Environmental costs are not reflected in the price of a good or service, producing the tragedy of the commons, a term that refers to the things in society we share in common.[23] Before the advent of pollution control regulations, the rational executive found that his or her share of the costs of the waste material discharged into the commons, in this case the air, water, and land that we all shared, was less than the cost of purifying the wastes before releasing them into the commons. There was no incentive for a manufacturer to reduce wastes when the commons was treated as a free good available to all for the dumping of wastes. The problem is that the commons eventually became ruined when its carrying capacity was exceeded and air and water pollution became so bad as to threaten human health and the existence of many species of fish and other wildlife. Thus modern industrial societies took steps to regulate the use of the commons in the interests of society as a whole. This regulation has largely been accomplished on a national level, but with respect to global problems like ozone depletion, an international treaty had to be formulated to protect the ozone layer.

A final aspect of our dualistic approach to nature is that we tend to think in linear fashion with respect to our industrial systems. Modern industrial processes have been built to take resources out of the ground, combine them with labor and capital to produce something useful to serve consumer needs and wants, and then dispose of whatever waste is left over in some convenient place. Hence toxic chemical waste was simply set out on the back dock of factories to be hauled off by some waste hauler, and no one cared enough to keep any records as to where and how this waste was being disposed of or what potential threats it might eventually pose to human health and the environment. This process reflected the out-of-sight, out-of-mind kind of thinking.

This linear design assumes that resources are inexhaustible and that new sources of supply will continually be discovered, and that bottomless sinks exist in which to dispose of our waste materials. Both of these assumptions are now being questioned, as resources are being depleted and it is either not feasible from an economic standpoint or too environmentally costly to open up some new sources of supply. And bottomless sinks do not exist, as landfills are being closed all over the country and it is becoming more and more difficult to open new ones because of public opposition. The linear design again reflects a dualism, as nature is cyclical and consists of closed loops to recycle nutrients and other materials. Manufacturers are beginning to realize the value of emulating nature and recycling waste material rather than disposing of it with traditional methods.

There is thus a new consciousness emerging that will change the way we think about the environment and the way we do things. This change is beginning to impact our motivational patterns and institutions in major ways that have profound and long-lasting impacts. Change is being forced upon us by a rapidly deteriorating environment that is under attack all over the globe. Humans have no choice but to recognize themselves as a part of nature, because human existence depends on the ability to draw sustenance from a finite natural world. And the continuance of the human race depends on the ability to abstain from destroying the natural systems that sustain the world.

STAGES OF ENVIRONMENTAL CONCERN

There have been several changes in the United States in the past century relative to perspectives on nature that are reflected in the policies and practices of governments and corporations. The concerns of society have changed over the years with respect to its understanding of the environment and how human beings relate to the natural world in which they find themselves. We have not been totally ignorant of our environment and have recognized that the environment is affected by human activities and sometimes needs to be given consideration because of the impact of those activities. Four such periods in the United States can be identified that represent distinctive concerns about the environment that resulted in different policy approaches and practices to address these concerns (Exhibit 1.1).

Conservation

The first approach we took was exemplified in the conservation movement that began in the early years of this century. During the frontier days, we recklessly exploited our resources by cutting down trees as fast as possible, plowing up grassland on a vast scale, and destroying our wildlife. Several species were made extinct and others, including the buffalo, were brought to near extinction. It was recognized at that time that such wanton exploitation could not continue, and that we must take steps to conserve our resources for future use and not deplete them needlessly.

The conservation movement developed as an attempt to restrain the reckless exploitation of forests and wildlife that characterized the pioneer state of social devel-

EXHIBIT 1.1 Stages of Environmental Concern

Stage	*Principle or Focus*	*Ethic*
Conservation	Use resources wisely and do not deplete them needlessly. Emphasized efficient development and use of natural resources.	Instrumental view of nature in that nature has utility only as it serves human purposes.
Preservation	Certain areas of the country are to be preserved in their natural state and closed to development.	Nature has intrinsic value in its own right apart from the services it provides for human beings.
Protection	Focused on pollution control and dangers to human health	Human centered
Sustainability	Concerned with global problems, sustainable growth, and equity considerations	Eco-centered

opment. This movement curbed the destructive environmental impacts of individuals and corporations who exploited nature for profit without regard for the larger social good or the welfare of future generations. It emphasized that resources should be used wisely and that consideration should be given to a sustainable society. This movement began to get a glimpse of natural limits to resource exploitation that would require different norms of conduct for the society to become sustainable on a long-term basis.[24]

The conservation movement thus promoted the wise and efficient use of resources. During this era, the national park ideal began where we set aside areas of the country that had a particular scenic value. The idea was to conserve these areas for the enjoyment of citizens by limiting development to those things that were necessary to promote and encourage visitation. The essence of the conservation approach was rational planning to promote efficient development and use of natural resources. Resource management was at the heart of the conservation movement.[25] Such management should be accomplished in the long-term interests of the citizens of the country and future generations still to come.

The ethic behind the conservation movement, however, was the idea that nature is instrumental, that nature has value only for human purposes, whether nature is used to provide resources for human use or whether certain beautiful areas of the country are set aside for human enjoyment. Treating nature as instrumental involves the view that nature has utility only as it is used to provide something for human use, whether that use be the extraction of materials to make something useful or the preservation of mountain beauty for human enjoyment. The timber or mineral executive reduces nature to a commodity, something to be taken out and made into something useful. The tourist seeking scenic beauty reduces nature to pleasing images, enjoyed and then taken home on film or preserved in mental images. This approach does not recognize nature as a living system of which our human lives are part, and on which our lives and all lives depend.

Preservation

After several years of a conservation approach, it began to be recognized by many environmentalists that conservation of resources was not quite what they were after. Conservation still promoted development and exploitation of resources even though it emphasized a more intelligent and efficient use of those resources. What many environmentalists wanted was some approach that prevented further development and exploitation of any sort, whether enlightened or unenlightened. They began to recognize that conservation and preservation were not the same thing, and that the latter approach was based on different assumptions about nature and the relationship of humans to nature.

The Wilderness Act of 1964 ushered in a new stage where nature was recognized as having value in its own right that was independent of its potential use for human purposes. Certain areas of the country were set aside to be preserved in their natural state and closed to resource development through a permanent wilderness designation. It came to be believed that land and wildlife could be truly conserved only by leaving them in their natural state and eliminating human presence as much as possible. The Wilderness Act recognizes a wilderness "as an area where the earth and community of life are untrammeled by man, where man himself is a visitor who does not remain."[26]

This kind of thinking is also found in the Endangered Species Act where certain animals are protected for their own sakes, regardless of the effect on human beings. This act has had an impact in Louisiana and Texas with the controversy over the use of turtle excluder devices (TEDs) to protect the Kemp Ridley turtle from being drowned in shrimp nets, and in the Pacific Northwest in the controversy over the northern spotted owl and the continued logging of old-growth forests. The important values in this movement are ecological, which means that natural systems should be allowed to operate as free from human interference as possible.

Designating an area as wilderness has become a way to stop economic activity and prevent economic development in some areas of the country. Human activity is considered to be bad in these areas and natural conditions are believed to be good. Some supporters of this approach treat wilderness as a semi-sacred place that should be preserved and placed beyond humanity's intrusion. They consider that human beings can only be truly free in wilderness. Society enslaves people and only in a state of nature does humanity live in a state of fulfillment. Wilderness areas must be preserved so that people can seek a temporary release from civilization.

The preservation movement sees nature as having intrinsic value in its own right apart from the services it can provide for human beings. Treating nature as having intrinsic value results in more respect for animal rights in testing procedures and food production and the usage of animal pelts for fur coats and other clothing. Some ideas have also come out of this movement that would give trees a kind of legal standing so that they can be protected explicitly by environmental groups who now have to use the Endangered Species Act to stop logging. Thus preservation has come to mean something quite different from conservation, resulting in different policies and practices.

Environmental Protection

Modern-day concern about the environment began in the 1960s, when there was something of a social revolution taking place in this country. Concern about the environment took its place alongside civil rights, consumer protection, safety and health, and a host of other social issues that were on the public agenda at that time in our history. From society's point of view, concern about the environment stemmed from a desire to improve the quality of life in our society and protect human health. Many scholars made a linkage between the environment and consumerism, for example, arguing that people could not appropriately enjoy the products they were buying in the marketplace in an environment that was deteriorating where the air was unfit to breathe and the water risky to drink. This concern for the quality of life and health was reflected in all of the major social issues of those days, but was particularly evident in concern about the environment.

> Fifteen or so years ago, pollution and ecology were two terms rarely found in the lexicon of business. Today environmental survival and pollution abatement are major topics of the times and receive prominent exposure in the literature of business and economics. If any one issue provided the initial sustenance for social responsibility proponents, that issue was the effect of business operations and practices on the physical environment. Probably more words have been written on this subject than on most others of a business and social problems context.[27]

The environmental movement that began in the 1960s was initially concerned about air and water pollution, and was sparked by Rachel Carson's book *Silent Spring,* which pointed out the problem increasing pesticide usage was posing for our society. Warning of the dangers of unrestricted use of pesticides, she brought together the findings of toxicology, ecology, and epidemiology in a form accessible to politicians and the general public. She discussed the bioaccumulation of fat-soluble insecticides in the fatty tissues of fish and the birds that eat fish, the natural resistance of surviving insects to these toxins, the natural dispersion of the toxins far from the source of the substance, and the biochemical interaction of toxins in the human body without human permission or awareness. Weaving together scientific, moral, and political arguments, she combined scientific knowledge about the environment with the need for political action.[28]

As a result of Carson's writing and other concerns expressed in society, people began to be educated about the environment and what services it provides for human beings. The central value of environmentalism is respect for the laws of nature. Ecology is believed to be more fundamental than human wants and needs. The love of nature and recognition of natural limits leads to humility about the place of the human species in the ecosystem. Environmentalists see that the earth is a commons and the solutions to many environmental problems must be undertaken on a global scale.

The results of this movement were to increase the awareness of environmental problems throughout society and institutionalize environmental concerns in business and government through a host of legislation and regulation pertaining to environmental problems. Indeed, the primary result of this concern about the environment in the 1960s and 1970s was a host of new laws at the federal, state, and local levels to deal with environmental concerns and the establishment of new agencies such as the Environmental Protection Agency (EPA) to administer the laws and make sure the business community in particular was in compliance. The focus of this agency, and indeed of the movement itself, was on protection of the environment from serious degradation and harm from human and industrial activities and protection of human health.

Several new metaphors were developed in these early days of concern about the environment and several ideas were written about relative to the root of the problem. Garrett Hardin, for example, introduced the notion about the tragedy of the commons in getting at the root of environmental problems. Imagine a pasture that was common land and open to all herdsmen who wanted to use it for grazing. As rational beings, each herdsman would try to keep as many cattle as possible on the commons and maximize their gain. Since each herdsman receives all the proceeds from the sale of an additional animal, there is an incentive to keep adding animals. The additional overgrazing created by one more animal is shared by all the herdsmen and thus there is a net positive utility to each herdsman from adding additional animals.[29]

The problem is that each herdsman sharing the commons reaches this same conclusion and continues to add animals to his herd. Each is locked into a system that compels him to increase his herd without limit in a world that is limited. Eventually, the carrying capacity of the commons is exceeded, and the commons is unable to support any more animals because of overgrazing and is ruined. The pursuit of self-interest in a society that believes in freedom of the commons brings eventual ruin to all who want to use the commons. Decisions that are reached individually, do not, in fact, work out to be the best decisions for the entire society.

This analysis has application to the pollution problem our society has experienced. The rational executive finds that his share of the costs of the wastes discharged into the commons, in this case air and water, is less than the cost of purifying the wastes before releasing them. There is no incentive for a manufacturer to reduce wastes when the commons is treated as a free good available to all for the dumping of wastes. But the commons is eventually ruined when its carrying capacity is exceeded and air and water pollution become so bad as to threaten human health and the existence of many species of fish and other wildlife. Thus society has to take steps to regulate use of the commons in the interests of society as a whole.

Other writers developed the notion of the earth as a spaceship, and used this metaphor to argue for policies that were frugal instead of wasteful in order to ensure survival in the limited world in which we live. We were encouraged to think of the earth as a spaceship floating in a vast universe where everything we needed to survive was more or less self-contained. There were no external inputs or outputs that we could depend upon to help solve our problems. Just as on a spaceship, we needed to think about conserving our use of resources and recycle our waste as much as possible because eventually we would run out of places to store waste. The spaceship earth concept was used to get people to think holistically and accept the idea of limitations on human activity.[30]

Barry Commoner offered a somewhat more optimistic assessment of the human prospect. He believed the economy could continue to grow, and that the standard of living and jobs could increase while environmental quality could be improved. In the United States, he argued, neither the increase in population nor in affluence could account for the very large increase in environmental pollution. The real culprit, he argued, was changes in the technology of production that had been introduced over the past several decades. Natural products were replaced by synthetic products such as detergents, synthetic fibers, and plastics. These products have been the real problem as far as pollution is concerned. And these changes in the technology of production have also brought about a serious decline in the efficiency with which resources and capital have been used in our society.[31]

The production system is governed almost exclusively by economic considerations. Profit maximization governs the design of the system of production and therefore the fate of ecosystems. Technology is a social institution that reflects to a large degree the governing aims of the society in which it develops. In this country, technology is used to enhance what capitalists want, namely, profit maximization and domination of the market rather than the welfare of the people. The solution, according to Commoner, is some form of democratic socialism where technology is used for society's benefit rather than private gain. Society must develop an investment policy that is under social rather than private control. The key to any solution to our environmental problems is social governance of the choice of production technologies through democratic control of investment decisions.

Finally, Paul Ehrlich argued that the root cause of most environmental damage was excessive growth in human population. Either humanity must change its ways of reproducing or mass starvation was inevitable. Neither technological breakthroughs nor social adjustments, other than an end to population increase, were adequate. In *The Population Bomb,* he was absolutely explicit about the cause of the coming tragedy, and recommended luxury taxes on cribs, diapers, and other children's goods, and proposed other economic disincentives to reproduction. Population control for developing countries must involve even more drastic measures than for advanced countries. Food aid should be given only to those countries that have an aggressive population control policy and clear hope of obtaining food self-sufficiency.[32]

Taken as a whole, these writings pointed out the complexity of the environmental problem. Technology, affluence, overpopulation, and use of the commons all contribute to the problem. But no single cause is dominant, and it would be unduly optimistic to suggest that stabilization or correction of only one of these factors might be sufficient. In the final analysis, there is probably not a single principal cause of ecological damage. Solution to environmental problems requires a multifaceted approach that is global in nature. No single nation can solve environmental problems all by itself, and focusing on only one facet of the problem does not recognize the interrelatedness of natural and social phenomena.

Sustainability

The current environmental movement is worldwide in nature and views all environmental problems as in some sense global rather than as simply regional and local. All environmental problems are interrelated, reflecting the nature of ecology itself. As mentioned before, the so-called global problems such as global warming and the depletion of the ozone layer threaten the entire planet and require international cooperation for their solution. But all environmental problems are in some sense global rather than just regional or local in nature. It is difficult to talk about air pollution in one country and efforts being made to reduce it without talking about other countries' problems. The same is true of water pollution and waste disposal problems. While public policy measures can be implemented by individual countries to deal with these problems, these problems really do not respect the boundaries of nations or localities, are fundamentally global in nature, and in many cases require global solutions.

Current environmentalism transcends the old ideologies and has become something of a new ideology itself, cutting across liberal-conservative lines, and affecting both socialistic and capitalistic systems. People of different political persuasions all over the world have been able to unite behind environmental causes and all countries of whatever ideology have their share of environmental problems. The new environmentalism challenges old ways of thinking and of organizing reality and calls for new paradigms and intellectual constructs that are more comprehensive and less reductionist in nature. Instead of age-old battles between capitalism and socialism and conservative and liberal ideologies, we are now challenged to transcend these ways of thinking and focus on a more comprehensive and inclusive view of reality, where humans are a part of nature and have to take environmental effects into account in all their activities.

This new approach to the environment may also provide a useful base from which to make individual life choices, from which to take collective political action, and from which to decide a surprisingly broad range of public policy issues. Some believe that environmentalism now has the potential to become the first original ideological perspective to develop since the middle of the nineteenth century. Such an ideology could help to halt or slow the expansionism inherent in both capitalist and socialist systems, which tend to seek ever bigger economies well past the point where greater economic activity is either sustainable or desirable. This ideology questions whether expansion beyond a reasonable level is a net benefit at all, regardless of the manner in which those benefits are distributed.[33]

Several aspects to the current environmental context make it quite different from previous stages of environmental concern (Exhibit 1.2). The overall approach with regard to environmental problems is one of sustainability. The sustainable approach to the environment recognizes resource limitations and questions the wisdom of a continued emphasis on economic growth. But instead of emphasizing conservation or preservation, or talking about limits to growth as did an earlier movement, the banner of the new environmental movement is sustainable growth or sustainable development.

Only a few short years ago, there was intense debate about the limits to economic growth in the world in general, and in the advanced industrial nations in particular. The first Club of Rome's study emphasized resource shortages, pollution problems, and population pressures in the industrialized nations and the world.[34] With an impressive array of computer graphs and statistics, the study proceeded to show that even under the best of assumptions, the limits to growth on this planet would be reached sometime within the next hundred years. The most probable result would be a rather sudden and uncontrollable decline in both population and industrial capacity. If this danger were recognized, it would be possible to alter these growth trends and establish a condition of ecological and economic stability that could be sustained far into the future.

This study was followed by the second Club of Rome study, which made many of the same points and recommendations.[35] Then came the Global 2000 Report with equally pessimistic conclusions. These predictions became all too real with the oil embargo in the mid-1970s that caused long gasoline lines in the United States and brought

EXHIBIT 1.2 Current Environmental Context

- Sustainability
- Survival
- Global Problems
- Equity
- Eco-Centered

home to every American our vulnerability regarding energy resources. There was a great deal of emphasis placed on the search for alternative sources of energy in the hopes of reducing our dependence on foreign sources of oil and gaining some degree of energy independence. The government of the United States proposed an $88 billion Synfuels Corporation to promote the search for and development of new sources and forms of energy.

These efforts came to naught, however, with the election of the Reagan administration. Talk about limits to growth came to an end except perhaps in some isolated corners of academia. Instead the emphasis was on opportunity and the unlimited potential of technology and the human spirit. The debate about supply-side economics shifted concern from the redistribution of an existing set of resources in a zero-sum type of situation, to expanding the size of the pie and lifting the boats of everyone, rich and poor alike, through uninterrupted economic growth. Investment, growth, creativity, and entrepreneurship were hallmarks of the Reagan administration, which harbored an unbounded optimism in the future of America and the spirit of the American people.

It is no mystery that young people in the United States supported Reagan in record numbers. Common sense would indicate that young people just starting their careers and families want to hear about opportunity and don't want to hear about limits to growth and shrinking opportunities, particularly from middle-aged people who have made their mark in life and accumulated their share of the world's goods. Limits to growth only appeal to those who already have enough wealth to live comfortably and want to prevent further growth from threatening their lifestyle. The limits to growth movement was something of an elitist concern, and it did a great deal of harm to the environmental movement of earlier years in getting it labeled as antigrowth and obstructionist to those on the lower rungs of the economic ladder.

Sustainable growth, however, has a much better chance of being accepted and implemented in public and corporate policy. This concept is concerned with finding paths of social, economic, and political progress that meet the needs of the present without compromising the ability of future generations to meet their own needs. It reflects a change of values in regard to managing our resources in such a way that equity matters, and thus has an appeal to people at all levels of development.[36] People and nations at early stages of economic development obviously don't want to see resources depleted before they have had their share, and must be concerned about growth that is sustainable for many years to come.

Current environmental concerns also emphasize survival of the planet and its human occupants. Concern about the environment is not just a quality-of-life issue as it was in the 1960s and 1970s but has now become a survival issue for many people. The human race should be considered to be an endangered species, as we are intimately connected with our natural environment and cannot grow and flourish in a deteriorating environment. Nature is not just something to be exploited for our own purposes, but it provides a habitat in which we live and move and have our being. As this habitat goes, so go humans. If we destroy our habitat, human life will no longer survive.

What brings up the survival issue more than anything else are the so-called global problems such as global warming, depletion of the ozone layer, deforestation, and species decimation. These problems are different from the more traditional problems such as air and water pollution, pesticides, toxic substances, and waste disposal that were the focus of earlier decades. No one can escape these problems by moving to some pristine area as they affect the entire planet and every human being is in some way impacted by them. Climate change affects everyone on the planet as does increased exposure to ultraviolet rays because of depletion of the ozone layer. Global problems of this nature are comprehensive and require international cooperation for their solution.

Any attempt to create a sustainable society on a worldwide basis must take into account the inequities that presently exist between countries. The poor people of the world not only suffer disproportionately from environmental damage that is caused by the industrial nations that are better off economically, they have become a major cause of ecological decline themselves as they have been pushed onto marginal land by population growth and inequitable land development patterns. Economic deprivation and environmental degradation reinforce one another to form a downward spiral that is difficult to arrest, much less turn around. But poverty must be dealt with in order for a sustainable world to be developed.[37]

Poverty drives ecological deterioration when people in desperate situations overexploit their resource base and sacrifice the future to salvage what they can out of the present. Ecological decline, in turn, perpetuates poverty, as degraded ecosystems offer diminishing yields to their inhabitants, thus setting into motion a downward spiral of economic deprivation and ecological degradation. This poverty trap appears on every continent and the net effect is universal. The poor are usually concentrated in fragile regions where the land is least productive and tenure least secure. This geographic concentration of poverty is driven in part by heightened population growth rates that poverty itself brings. Thus failure to launch an assault on poverty that includes these elements will guarantee the continued destruction of much of our shared biosphere.[38]

Concern with sustainability also involves changing the focus of concern from protection of human health to a concern with broader ecosystems. The EPA, for example, has been admonished for paying too little attention to natural ecosystems over the 20-year course of its history. The agency has considered the protection of public health to be its primary mission and has been less concerned about risks posed to ecosystems. This lack of concern reflects society's views as expressed in legislation, as ecological degradation is viewed as a less serious problem because it is subtle, long-term, and cumulative. But natural ecosystems such as forests, wetlands, and oceans are extremely valuable, and the EPA was thus asked to correct this imbalance that currently exists in national environmental policy.[39]

Responding to human health risks and largely ignoring risks to ecosystems is inappropriate because in the real world there is little distinction between them, and there is no doubt that over time the quality of human life declines as the quality of natural ecosystems declines. Ecological degradation either directly or indirectly degrades human health as well as the economy. As the extent and quality of saltwater estuaries decline, both human health and local economies are adversely affected. As soils erode, forests, farmlands, and waterways can become less productive. And while the loss of species may not be noticed immediately, over time the decline in genetic diversity has implications for the future health of the human race as a whole.[40]

Human health and welfare in the final analysis rest upon the life support systems and natural resources provided by healthy ecosystems. Human beings are part of an interconnected and interdependent global ecosystem, and change in one part of the system affects other parts, often in an unexpected manner. Thus it has been recommended that when the EPA compares the risks posed by different environmental problems in order to set priorities for action, the risks posed to ecological systems must be an important part of its consideration. The EPA's priorities for action should reflect an appropriate balance between ecological, human health, and welfare concerns, and the agency should communicate to the public that it considers ecological risks to be just as serious as risks to human health and welfare, because of the inherent value of ecological systems and their links to human health and welfare.[41]

The EPA is thus being asked to take a more eco-centered approach, and find value in ecosystems as such, rather than simply focusing on human health and welfare. Such a suggestion is consistent with shifting the ethical focus of environmental concerns from a human-centered approach to more of an ecological approach, and attaining something of a balance between the two approaches. Ecosystems need to be protected for their own sakes, but ecosystems have a direct connection with human well-being, as human life cannot flourish in a degraded environment that is not capable of providing the services necessary to sustain it. It is well to remember that such connections exist in nature, and policy mechanisms and outcomes should reflect this fundamental nature of our environment.

THE END OF NATURE

The End of Nature is the title of a book by Bill McKibben that captures these concerns and provides a new way of viewing the present situation. This book's basic thesis is that nature as we have known it in the past in its pure form no longer exists. Human beings have conquered nature as the entire natural world now bears the stamp of humanity, as we have left our imprint on nature everywhere and have altered it beyond recognition in some cases. We have made nature a creation of our own, and have lost the otherness that once belonged to the natural world. The natural world is so affected by human technology that it is more and more becoming one of our own creations and thus is no longer the autonomous nature in which we sought refuge from human civilization.[42]

> In our times . . . human cunning has mastered the deep mysteries of the earth at a level far beyond the capacities of earlier peoples. We can break the mountains apart; we can drain the rivers and flood the valleys. We can turn the most luxuriant forests into throwaway paper products. We can tear apart the great grass cover of the western plains and pour toxic chemicals into the soil and pesticides onto the fields until the soil is dead and blows away in the wind. We can pollute the air with acids, the rivers with sewage, and the seas with oil—all this in a kind of intoxication with our power for devastation at an order of magnitude beyond all reckoning. . . . Our managerial skills are measured by the competence manifested in accelerating this process.[43]

What this view suggests is that the world has crossed a threshold with respect to the environment that can never be recrossed. We cannot return to a simpler age, something like the small-is-beautiful idea where humans would tread much more lightly on the environment and be less disruptive of nature. We have no choice in these matters, because our science and technology have taken us too far to turn back to a past age that may never have existed. Such notions of a return to a pristine past where nature was less affected by human activities are romantic and unrealistic. Human activities alter natural processes far greater than anyone can imagine, and nature has been subjugated and reconfigured according to human needs and desires.

Many are legitimately concerned that nature will be crowded out by such human interference, and oppose the idea that humans should exercise their dominion over nature for the sake of material progress. They have a sense of loss because nature's independence is being destroyed. For much of history, humans beings have not experienced nature as kind and gentle, but as harsh and dangerous, and therefore humans have felt compelled to subordinate nature in order to protect themselves. But humility toward nature is what is now being advocated, that human beings should neither control nor dictate to nature, but must learn to live in harmony with nature and take a responsibility for it that has thus far largely been avoided.

The earth has finite resources and a fragile environment, all of which gives us a responsibility to manage the human use of planet earth. We must develop new technologies and strategies that are environmentally sensitive and we must be more responsible in our use of resources. Not to take these kinds of steps and change our way of thinking will most assuredly lead to environmental degradation on a scale that far surpasses anything we have experienced in our lifetime. Nature no longer can take care of itself, and one of the most fundamental assumptions we have made about nature, namely that it can take care of itself without any conscious thought given to the impacts human activities have on the environment, must be discarded. We simply cannot proceed as we have in the past to exploit nature and not worry about the environmental consequences of our activities. The leaders of business and industry, as well as government and educational institutions, are beginning to think in terms of managing nature, managing planet earth, which basically means taking responsibility for nature to assure the survival of the world.

Managing nature involves making value judgments regarding the kind of planet we want. While science at least attempts to tell us what kind of planet we can have or are likely to have if certain trends continue, what we want is a value judgment. Value judgments include the answer to such questions as the following: How much species diversity should be maintained? How much of nature and what natural resources do we wish to leave for our children? Should the size or growth rate of the human population be curtailed to protect the global environment? How much climate change is acceptable? How much poverty is acceptable throughout the world? Science can tell us something about the broad patterns of global transformation taking place, but value questions about the pace and direction of those patterns have to be answered through political and economic systems.

In order to answer those value questions, we need better education about ecology to understand nature and the impacts we have on nature. We need new measurement systems to quantify these impacts and clarify the nature of the decisions we face relative to resource usage and environmental degradation. This way of thinking provides quite a challenge to all of us, but we simply must think in terms of taking responsibility for our actions and managing nature in the interests of the entire world and in the interests of future generations. We have to make conscious choices about the kind of world we want for ourselves and for our children.

Individuals and institutions have begun to respond to an increased awareness of global environmental change by altering their values, beliefs, and actions. Many steps have been taken to respond to environmental problems and begin to manage our relations with the environment in a more responsible manner. However, we must respond as a global species, pooling knowledge, coordinating our actions, and sharing what the planet has to offer. Only in adopting a global perspective do we have any realistic prospect for managing the planet's transformation along pathways of sustainable development.[44] The environment is important to the entire human race, not just one or a few countries. The earth is home to everyone, and everyone is affected by radical changes in the environment.

Efforts to manage the sustainable development of the earth must have three specific objectives according to one author: (1) to disseminate the knowledge and the means necessary to control human population growth, (2) to facilitate sufficiently vigorous economic growth and equitable distribution of its benefits to meet the basic needs of the human population in this and subsequent generations, and (3) to structure the growth in ways that keep its enormous potential for environmental transformation within safe limits yet to be determined.[45] The greatest responsibility and greatest immediate potential for the design of sustainable development strategies may be in the high-income and high-density regions of the industrialized world.[46]

Through a gradual awakening, people are beginning to develop a new perception of humanity's relationship to the earth's natural systems. People are crossing perceptual thresholds without necessarily even being aware of it, and new ways of thinking are emerging. Such changes are necessary to respond to environmental problems effectively and in time to save the world from irreversible destruction. There is a growing sense of the world's interdependence and connectedness, and an understanding that progress is an illusion if it destroys the conditions for life to thrive on earth. The leaders of industrial and developing countries alike recognize their common interest in and responsibility for participating in sustainable development. Looming threats to the world's climate and undermining of other global commons may soon make the transition to stronger international solutions inevitable.

Questions for Discussion

1. Do you agree that the world is at risk? In what ways? How would you categorize the risks mentioned in the chapter? Which are most severe and need immediate attention?
2. Is the distinction between global problems and more traditional environmental problems valid in your opinion? What would be a more meaningful distinction? What, if any, significant difference exists between these two groups of environmental problems?
3. Is socialism any better than capitalism in terms of its environmental record? Why or why not? What implications does your answer have for the future of socialistic systems?
4. What is pollution? Why does it occur? What factors were critical in making pollution a serious problem in the 1960s, and getting pollution on the national agenda?
5. What changes are necessary in human consciousness in order to develop an appropriate understanding of humans and their relationship to the environment? What implications does this change have for environmental policy and practices?
6. What is conservation? What ethic is behind this movement? What major conservation efforts have been attained? Is conservation important today? In what ways?
7. What is an intrinsic value? How does this ethical approach reflect itself in the preservation movement? Would you advocate this approach be extended? In what ways?

8. What was Rachel Carson's book about? What does the title *Silent Spring* mean? What effects did her book have on society? What has been the result of the environmental movement that began in the 1960s and continued into future years?
9. Describe the tragedy of the commons. List all the commons you can think of in our society. Does Hardin's metaphor apply to these commons? Does his analysis help you to understand what happens to these commons?
10. What was Commoner's assessment of the environmental problem? Do you agree or disagree with his analysis? Would his solution work given what we know about the world today?
11. In what ways does the current environmental movement differ from earlier environmental movements? What does it mean to say that the new environmentalism has become something of an ideology in itself? Will this kind of environmental concern continue for very long, or is it something of a fad in your opinion?
12. What is sustainable growth? How does this concept differ from limits to growth? Why did the latter fade from public consciousness? Is sustainable growth possible?
13. What does Bill McKibben mean by the end of nature? Do you agree or disagree with his thesis? What are the implications of his thesis for the future of the planet?
14. What does it mean to manage planet earth? How can science help us in this regard? What kind of value judgments have to be made? What kind of institutional changes are necessary to make this kind of approach a reality?

Endnotes

1. David Kirkpatrick, "Environmentalism: The New Crusade," *Fortune,* February 12, 1990, pp. 44–55. See also "Americans Speak Out," *National Wildlife,* April–May 1995, pp. 34–37.
2. Denton E. Morrison and Riley E. Dunlap, "Environmentalism and Elitism: A Conceptual and Empirical Analysis," *Environmental Management,* 10, no. 5 (1986), 581–589.
3. Stephen H. Schneider, "The Changing Climate," *Scientific American,* 261, no. 3 (September 1989), 72.
4. "Hot Times," *Time,* January 21, 1991, p. 65.
5. See Jodi Jacobson, "Holding Back the Sea," *State of the World 1990,* Linda Starke, ed. (New York: W. W. Norton, 1990), pp. 79–97.
6. See Cynthia Pollock Shea, "Protecting the Ozone Layer," *State of the World 1989,* Linda Starke, ed. (New York: W. W. Norton, 1989, pp. 77–96.
7. Peter H. Raven, "Endangered Realm," *The Emerald Realm: Earth's Precious Rain Forests,* Martha E. Christian, ed. (Washington, D.C.: National Geographic Society, 1990), p. 10.
8. Ibid.
9. Peter Ward, *The End of Evolution* (New York: Bantam Books, 1995), p. 256.
10. Raven, "Endangered Realm," p. 24.
11. S. Fred Singer, "The Answers on Acid Rain Fall on Deaf Ears," *Wall Street Journal,* March 6, 1990, p. A20.
12. See Michael D. Lemonick, "Forecast: Clearer Skies," *Time,* November 5, 1990, p. 33.
13. United States Environmental Protection Agency, *Environmental Progress and Challenges: EPA's Update* (Washington, D.C.: U.S. Government Printing Office, 1988), p. 44.
14. Ibid., pp. 78–109.
15. Ibid., p. 124.
16. Sharon Begley, "Is Breathing Hazardous to Your Health?" *Newsweek,* April 3, 1989, p. 25.
17. Kathy Lally, "Russia Levels with People on Pollution," *Times-Picayune,* October 8, 1992, p. A25. See also James S. Robbins, "Cleaning Up After Socialism: Depolluting the USSR," *Liberty,* 4, no. 6 (July 1991), 27–32.
18. Hillary E. French, *Green Revolutions: Environmental Reconstruction in Eastern Europe and the Soviet Union* (Washington, D.C.: Worldwatch Institute, 1990), p. 17.

19. Ibid., pp. 19–20.
20. Ibid., pp. 21–22. See also Linnet Myers, "Communist Legacy: A Poisoned Earth," *Chicago Tribune,* June 1, 1992, p. 1-1.
21. Peter A. A. Berle, "How Do We Define Nature?" *Audubon,* May 1991, p. 6.
22. Lester R. Brown, "The Illusion of Progress," *State of the World 1990,* Linda Starke, ed., pp. 7–9.
23. See Garrett Hardin, "The Tragedy of the Commons," *Science,* 162, no. 1 (December 13, 1968), 1243–1248.
24. John Rodman, "Four Forms of Ecological Consciousness," *Ethics and the Environment,* Donald Scherer and Thomas Atteg, eds. (Englewood Cliffs, NJ: Prentice Hall, 1983), p. 84.
25. William Tucker, *Progress and Privilege: America in the Age of Environmentalism* (Garden City, NY: Anchor Press, 1982), pp. 42–45.
26. Ibid., p. 129.
27. Arthur Elkins and Dennis W. Callaghan, *A Managerial Odyssey: Problems in Business and Its Environment,* 2nd ed. (Reading, MA: Addison-Wesley, 1978), p. 173.
28. Robert C. Pahlke, *Environmentalism and the Future of Progressive Politics* (New Haven: Yale University Press, 1989), pp. 28–32.
29. Hardin, "The Tragedy of the Commons."
30. Kenneth E. Boulding, "The Economics of the Coming Spaceship Earth," *Environmental Quality in a Growing Economy,* H. Jarrett, ed. (Baltimore, MD: Johns Hopkins Press, 1966).
31. Barry Commoner, "Economic Growth and Environmental Quality: How to Have Both," *Social Policy,* Summer 1985, pp. 18–26.
32. Paul Ehrlich, *The Population Bomb* (New York: Ballantine Books, 1971).
33. Pahlke, *Environmentalism,* pp. 3–7.
34. Donella H. Meadows, Dennis L. Meadows, Jorgen Randers, and William W. Behrens III, *The Limits to Growth: A Report for the Club of Rome's Project on the Predicament of Mankind* (New York: Universe Books, 1972).
35. Mihajlo D. Mesarovic, *Mankind at the Turning Point: The Second Report to the Club of Rome* (New York: Dutton, 1974).
36. William C. Clark, "Managing Planet Earth," *Scientific American,* 261, no. 3 (September, 1989), 48.
37. Alan B. During, "Ending Poverty," *State of the World 1990* (New York: W. W. Norton, 1990), pp. 135–36.
38. Ibid., p. 146.
39. United States Environmental Protection Agency, *Reducing Risk: Setting Priorities and Strategies for Environmental Protection* (Washington D.C.: U.S. Government Printing Office, 1990), p. 9.
40. Ibid.
41. Ibid., p. 17.
42. Bill McKibben, *The End of Nature* (New York: Random House, 1989).
43. Thomas Berry, *The Dream of the Earth* (San Francisco: Sierra Club Books, 1988), p. 7.
44. Clark, "Managing Planet Earth," p. 47.
45. Ibid., p. 49.
46. Ibid., p. 53.

Suggested Reading

Aptaker, Lewis. *Environmental Disasters in Global Perspective.* New York: Macmillan, 1993.

Bernards, Neal, ed. *Environmental Crisis: Opposing Viewpoints.* Boston: Greenhaven, 1991.

Boulding, Kenneth E. "The Economics of the Coming Spaceship Earth," *Environmental Quality in a Growing Economy,* H. Jarrett, ed. Baltimore, MD: Johns Hopkins Press, 1966.

Brundtland, G. H. *Our Common Future: World Commission on Environment and Development.* New York: Oxford University Press, 1987.

Dashefsky, H. Steven. *Environmental Literacy: Everything You Need to Know about Saving Our Planet.* New York: Random House, 1993.

Ehrlich, Paul. *The Population Bomb.* New York: Ballantine Books, 1971.

Fox, Stephen. *John Muir and His Legacy: The American Conservation Movement.* Boston: Little, Brown, 1981.

French, Hillary E. *Green Revolutions: Environmental Reconstruction in Eastern Europe and the Soviet Union.* Washington, D.C.: Worldwatch Institute, 1990.

Goudie, Andrew S. *Environmental Change.* New York: Oxford University Press, 1992.

Hartzog, George B., Jr. *Battling for the National Parks.* Mt. Kisco, NY: Moyer Bell, 1988.

Hays, Samuel P. *Beauty, Health, and Permanence: Environmental Politics in the United States, 1955–1985.* Cambridge, MA: Cambridge University Press, 1987.

Kempton, Willet M., et al. *Environmental Values in American Culture.* Cambridge: MIT Press, 1995.

Leopold, Aldo. *A Sand County Almanac.* New York: Oxford University Press, 1949.

McKibben, Bill. *The End of Nature.* New York: Random House, 1989.

Meadows, Donella H., Dennis L. Meadows, Jorgen Randers, and William W. Behrens III. *The Limits to Growth: A Report for the Club of Rome's Project on the Predicament of Mankind.* New York: Universe Books, 1972.

Mesarovic, Mihajlo D. *Mankind at the Turning Point: The Second Report to the Club of Rome.* New York: Dutton, 1974.

Moeller, Dade. *Environmental Health.* Cambridge: Harvard University Press, 1992.

Nash, Roderick. *Wilderness and the American Mind,* 3rd ed. New Haven, CT: Yale University Press, 1982.

Owens, Susan, and Peter L. Owens. *Environment, Resources, and Conservation.* Cambridge: Cambridge University Press, 1991.

Pahlke, Robert C. *Environmentalism and the Future of Progressive Politics.* New Haven: Yale University Press, 1989.

Petulla, Joseph M. *American Environmental History,* 2nd ed. Columbus, OH: Merrill, 1988.

Rodman, John. "Four Forms of Ecological Consciousness," *Ethics and the Environment,* Donald Scherer and Thomas Atteg, eds. Englewood Cliffs, NJ: Prentice Hall, 1983.

Speth, James Gustave. *Environmental Pollution: A Long-Term Perspective.* Washington, D.C.: World Resources Institute, 1988.

Tucker, William. *Progress and Privilege: America in the Age of Environmentalism.* Garden City, NY: Anchor Press, 1982.

Waite, Donald E. *Environmental Health Hazards.* Williamston, MI: Environmental Health Consultants, 1994.

Wolbarst, Anthony B., ed. *Environment in Peril.* Washington, D.C.: Smithsonian, 1992.

CHAPTER 2

Concepts and Principles of Ecology

The subject of ecology may not strike many business school students as being of immediate relevance to their concerns about the environment, but this chapter assumes that such a view is far from appropriate. Just as it is important for students in schools of business and management to have some knowledge of economics so they can understand how the economy works and the role business plays in the economy, so it is important for them to have some understanding of the way in which nature works in order to understand the impacts business activities have on the environment. This chapter is meant to provide business school students with at least a rudimentary understanding of the way in which nature works and acquaint them with some basic concepts and principles of ecology that may help in understanding the nature of environmental problems and what can and can't be done about them.

The lack of basic ecological knowledge and understanding means that many policy makers and business executives are not very well equipped to make decisions about activities that have environmental impacts. They simply don't understand what impacts these decisions may have on the environment and what alternatives exist that may have less severe environmental consequences. Ecological education is not done very well, if it is done at all, in our educational system, perhaps reflecting our perception that nature will take care of itself and we can go about our business of pursuing more and more economic growth irrespective of impacts on the environment. The concept of sustainable development suggests that such a course can lead to environmental and economic disaster, if economic growth is undermining the very environmental conditions that make that growth possible.

> With better ecological understanding, media and the public would have known that the spotted owl, though the legal focus of the fracas, is but a single species that scientists use as an indicator of the health and diversity of old-growth forests. Once the system is damaged so much that owls can't survive, other species—and ultimately the entire forest as a living system—may be affected beyond repair. Reporters and readers could pick up on the fact that a tree farm, though it might have as many trees as a forest, is radically different, and cannot be considered an adequate replacement for the complex old-growth ecosystem. Lacking both this specific information and the long-term perspective of biological science, decisionmakers often favor economists' and business managers' short-term, bottom-line-for-the-current-quarter judgments.[1]

The science of ecology has become more important as industrial societies have matured and as our knowledge of the world has become more complex. Even in primitive societies, ecology was important as human beings needed to have some knowledge of the environment, the forces of nature, and the plants and animals around them in order to survive. But as population increased and as human beings expanded their power to alter the environment, it has become more important than ever for people to have a better knowledge of the environment and of what they are doing to themselves and the planet on which they depend for survival. Ecology is the science that provides that knowledge.

As a distinct field of biology, ecology dates from about 1900, but only in past decades has the word become part of the general vocabulary. During the 1960s when social problems were brought to the nation's attention, the word was introduced to the general public as more and more attention was given to the physical environment and the pollution problems that were appearing all over the world that needed attention. People gradually became aware of the importance of the environmental sciences in providing an understanding of the environment and giving us tools to maintain the quality of human civilization. Ecology rapidly became extremely relevant to every person on earth.

Plants, microorganisms, and animals are not isolated entities that have no relation to each other. They are parts of a vast complex of natural machinery that ecology seeks to understand. They are related elements in a system that operates in a manner that can be examined by scientific means and described in a way that increases our knowledge of ourselves and the world in which we live.[2] Increasing our understanding of how nature is assembled and maintains itself is one of the most difficult and exciting challenges for ecologists. Comprehending how natural communities are organized and how their diversity is maintained is essential for their preservation, as animal and plant communities all over the world are under assault from ever-expanding human populations and activities.[3]

Ecology is historical and comparative in its approach as well as holistic in its outlook. Most sciences are reductionist in their approach in that they reduce nature to its components and seek to understand the structure and function of each of these components. Consequently, knowledge has become fragmented as different fields of science have developed and have become more and more specialized. Most scientists do not necessarily have to be holistic and understand the relationship of the parts to the whole, but can concentrate on the parts without dealing with interrelationships or the relation of the parts to the whole.

Ecologists, on the other hand, seek to understand the rationale for the present makeup of nature and the operation of the entire mechanism, as well as the history of its construction.[4] Ecologists seek a comprehensive understanding of the environment and how all of its components relate to one another. Everything is related to everything else in the environment, and these interrelationships are just as important as the components of the environment themselves. Ecology is thus a synthetic science where the whole is greater than the sum of the parts. When the whole is reduced to its parts, the quality of the whole is likely to disappear. The interrelationships between the parts can be studied separately from the breakdown and analysis of the parts themselves.

THE REALM OF ECOLOGY

The word "ecology" comes from the Greek *oikos,* and means "house" or "place to live." Taken literally, ecology refers to the study of organisms "at home" in their natural habitat.[5] Ecology is concerned with the biology of groups of organisms and with functional processes on the land, in the oceans and freshwater, and in the air. Ecology can also be defined as the study of the structure and function of nature. Ecologists seek to explain or understand nature, which is a search for knowledge in the pure scientific tradition. They also try to predict what will happen to organisms, populations, or communities under a particular set of circumstances.[6]

For more practical purposes, ecology can be considered as the study of organisms and their environment and all the external conditions and factors, living and nonliving (biotic and abiotic), that affect an organism. It is the science of the interrelations between living organisms and their environment, and deals with the interactions of plants and animals in natural systems.[7] The key words here are "interrelations" and "interactions" as the interrelations and interactions are as important as the individual organisms themselves.

Ecology is a science that proceeds at five levels: (1) the individual organism; (2) the population (consisting of individuals of the same species); (3) the community (consisting of a greater or lesser number of populations; (4) the ecosystem where the concept of abiotic components such as temperature, nutrients, and moisture are introduced that interact with the three previous levels; and (5) the biome level, which is a classification of ecosystems into general categories that contain similar types of organisms. The largest biome is, of course, planet earth itself.

At the level of the *organism,* ecology deals with how individuals are affected by (and how they affect) their environment. Is acid rain killing spruce trees and, if so, what is the mechanism of damage? Is the greenhouse effect for real and what are the implications for human life as the earth heats up further? At the level of *population,* ecology deals with the presence or absence of particular species, with their abundance or rarity, and with trends and fluctuations in their numbers. To understand population changes, the changes happening to individuals making up the population must be analyzed. *Community* ecology deals with the composition or structure of communities, and with the pathways followed by energy, nutrients, and other chemicals as they pass through communities.[8] Communities are not constant but are constantly changing because of interactions among the populations and because of disturbances caused by climactic and geological events as well as by human activities.

The concept of an *ecosystem* is the most fundamental concept in the field of ecology. The ecosystem is the basic functional unit in ecology since the concept includes organisms, populations, and communities, each influencing the properties of the other (Exhibit 2.1). The concept of the ecosystem is broad as its main function in ecological thought is to emphasize obligatory relationships, interdependence, and causal relationships. The parts of an ecosystem are operationally inseparable from the whole, and thus systems analysis techniques are especially appropriate for understanding how an ecosystem works. A pond, a lake, a tract of forest are convenient units of analysis for such purposes. An entity may be considered an ecosystem as long as the major

EXHIBIT 2.1 The Realm of Ecology

Biome	A classification of ecosystems into general categories that contain similar types of organisms
Ecosystem	The community of organisms and populations interacting with one another and with the chemical and physical factors making up their environment
Community	Populations of different plants and animals living and interacting in an area at a particular time
Population	Group of individual organisms of the same species living within a particular area
Organism	Any form of life including all plants and animals

components of the system are present and operate together to achieve some sort of functional stability, if only for a short period of time.[9]

> All human beings and human activities are imbedded in and dependent upon the ecosystems of our planet. Ecosystems are the machinery of nature, the machinery that supports our lives. Without the services provided by natural ecosystems, civilization would collapse and human life would not be possible. . . . An ecosystem consists of the physical environment and all the organisms in a given area, together with the network of interactions of these organisms with that physical environment and with each other.[10]

Ecosystems consist of various living and nonliving components (Figure 2.1). The major types of organisms that make up the living or biotic components of an ecosystem are classified as producers, consumers, and decomposers. Producers are organisms that can manufacture the organic compounds they require as sources of energy and nutrients. Consumers get the nutrients and energy they need by feeding either directly or indirectly on producers. Decomposers such as fungi and bacteria break down dead organic plant and animal matter into simpler inorganic compounds. The nonliving or abiotic components of an ecosystem include various physical and chemical factors, the former being factors such as precipitation, wind, and temperature, and the latter being nutrient elements and compounds that are required in large or small amounts for the survival, growth, and reproduction of organisms.[11]

Ecosystems are capable of self-maintenance and self-regulation as are their component populations and organisms. There is a natural tendency for ecosystems to resist change and to remain in state of equilibrium. Thus cybernetics, or the science of controls, has an important application in terms of understanding how ecosystems maintain themselves. Since human activities tend to disrupt the natural functioning of control mechanisms or substitute artificial mechanisms for natural ones, it is important to have some understanding of how an ecosystem works and how the balance of forces in an ecosystem is likely to be upset by human projects.

An ecosystem is not a static concept, however, even in the absence of human intervention. Ecosystems do develop and succeed each other; in other words, they are in a constant state of evolution. According to one ecologist, ecological succession may be

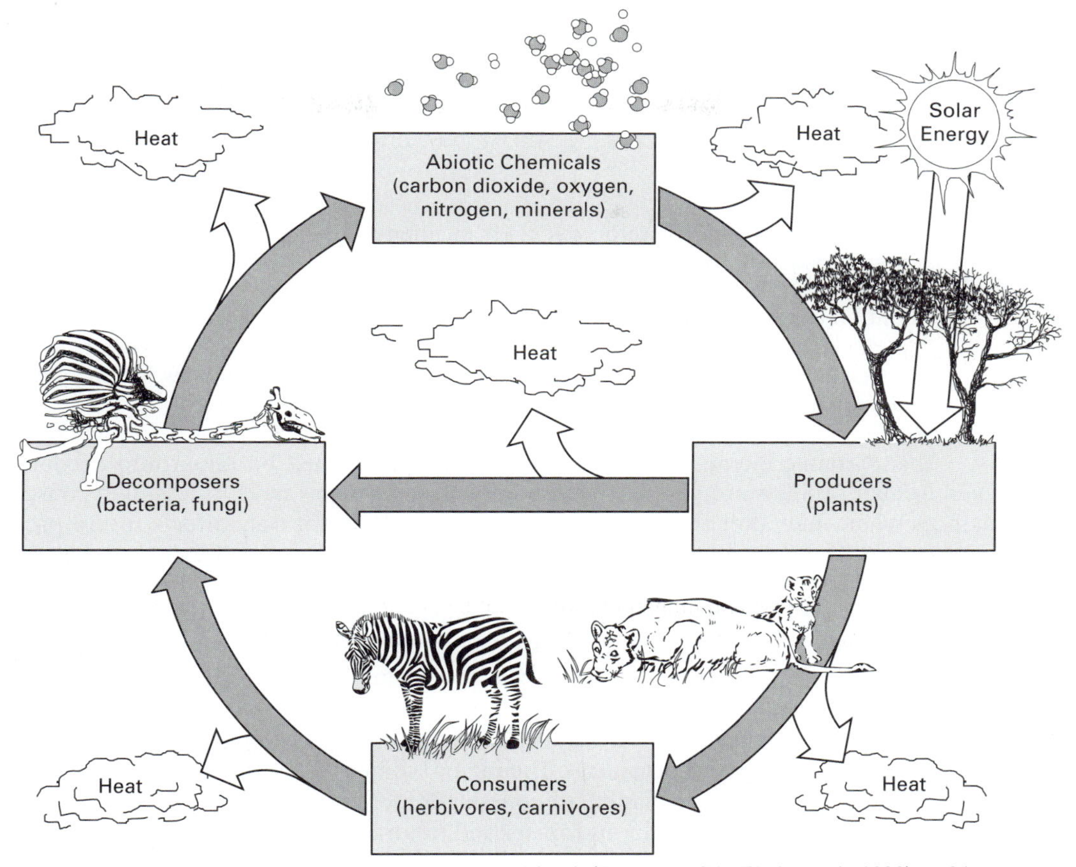

Source: G. Tyler Miller, *Living in the Environment,* 6th ed. (Belmont, CA: Wadsworth, 1990), p. 84.

FIGURE 2.1 Major Structural Components of an Ecosystem

defined in terms of three parameters: (1) Ecological succession is an orderly process of community development that involves changes of species and community structures and processes over time, but these changes are reasonably directional and predictable; (2) such succession results from modification of the physical environment by the community and is thus community controlled even though the physical environment determines the pattern, rate of change, and often sets limits as to how far development can proceed; and (3) succession culminates in another stabilized ecosystem where some kind of equilibrium can be maintained over a period of time.[12]

When human intervention is introduced, changes can be produced that are unnatural and life threatening. Since the industrial revolution, humans in so-called advanced societies have generally been preoccupied with obtaining as much production from the land mass available to them as possible. Intensive farming and forestry, for example, have been practiced to achieve high rates of production of readily harvestable products with little standing crop left to accumulate on the landscape. This goal of maximum production often conflicts with nature's strategy of maximum protection (trying to achieve maximum support of complex ecological structures), which often characterizes ecological development. Thus there is a conflict between humans and nature that must be recognized if rational land-use and resource extraction policies are to be established.[13]

> Many essential life-cycle resources, not to mention recreational and esthetic needs, are best provided for man by the less "productive" landscapes. In other words, the landscape is not just a supply depot but is also the oikos—the home—in which we must live. Until recently mankind has more or less taken for granted the gas-exchange, water-purification, nutrient-cycling, and other protective functions of self-maintaining ecosystems, that is, until his numbers and his environmental manipulations became great enough to affect regional and global balances.[14]

This dynamic element of ecosystem development must be taken into account when human projects are planned. Oftentimes these projects have unintended consequences when they alter the ecosystem in a manner that not only affects future productivity of the environment but also affects the ability of the ecosystem to sustain human as well as animal life. The ecosystem has multiple functions and therefore land-use policies as well as industrial development must recognize these multiple functions. The productive orientation of industrial societies must be balanced with the protective orientation of many natural processes in order to promote the welfare of the entire biological community.

Finally, ecosystems can be classified into more general categories called *biomes* that contain similar types of organisms. The major biomes within the most general type of biome, which is the entire planet, are grasslands, deserts, temperate forests, tropical forests, coniferous forests, deciduous forests, and tundra. The differences between these major biomes in various parts of the world are based mainly on differences in average temperature and average precipitation. Thus climate changes because of global warming, to be discussed later, can have major impacts on the planet and its fundamental structure.

IMPORTANT CONCEPTS

The concept of *cycles* is important in ecology, as the chemical elements, including those that are essential to human life, tend to circulate in the biosphere in characteristic paths from the environment to organisms and back to the environment. Of the 90 odd elements that are known to occur in nature, some 30 to 40 are required by human organisms, and the movement of these elements is sometimes designated as nutrient cycling. Some of these elements, such as carbon, hydrogen, oxygen, and nitrogen, are needed in large quantities, while others are needed in small or, even in some cases, minute quantities.[15]

The global cycling of carbon dioxide (CO_2) illustrates the concept of nutrient cycling (Figure 2.2). Carbon dioxide is released into the atmosphere by humans as they take in oxygen that is essential to human life and release carbon dioxide as a waste product. Carbon dioxide is also released from the soil because of agricultural activities,

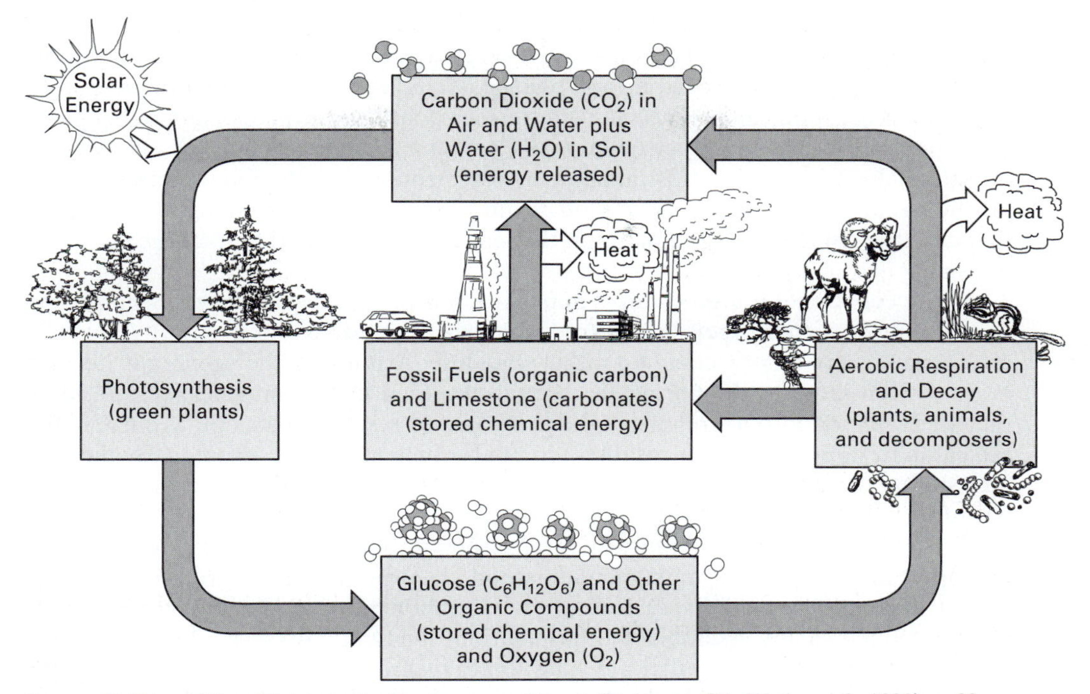

Source: G. Tyler Miller, *Living in the Environment,* 6th ed. (Belmont, CA: Wadsworth, 1990), p. 93. © Wadsworth Publishing Co.

FIGURE 2.2 Simplified Diagram of the Carbon Cycle

especially frequent plowing of large land areas. This carbon dioxide is used by plants in the process of photosynthesis, which releases oxygen into the atmosphere. Carbon dioxide is also absorbed by the carbonate system of seawater. Human activities on a global scale can upset this balance and interfere with the recycling process. Release of carbon dioxide into the air through the increased burning of fossil fuels and a subsequent decrease in the removal capacity of the green belt through the destruction of the rain forests in South and Central America can have an effect on the atmosphere and on world climate.

Corporate managers need to understand the nature of these cyclical processes in ecology in order to understand what effect major industrial processes can have on the ecosystem upon which we depend for our very existence. Besides the carbon cycle, there is also a nitrogen cycle, a phosphorus cycle, and a sulfur cycle. All of the major macro environmental problems such as the greenhouse effect, acid rain, and depletion of the ozone layer of the atmosphere have cyclical components. There is another aspect to this notion of a cycle in ecology that relates to conservation of natural resources. As some nonrenewable resources become more and more scarce, the concept of recycling must become a major goal for society. This concept is especially important in relation to those resources that are essential for human life and for which no effective and readily available substitutes can be found.

The concept of a *food chain* is also an important concept in terms of understanding the environment. The transfer of food energy from its source in plants through a series of organisms where eating and being eaten is repeated a number of times is referred to as the food chain. Since at each transfer point a large proportion of the available energy is lost, the number of steps or links in a sequence is limited. The concept of a food chain is more or less familiar to most people, since human beings occupy a position at or near the end of most food chains. Human beings, for example, eat big fish that eat little fish, that eat zooplankton that eat phytoplankton that fixes the energy of the sun.[16]

The concept of a food chain is important for another reason beyond simply understanding how energy gets transferred from one organism to another. Pollutants also are transferred in this process, and rather than lose their effect, they generally become more concentrated as they progress through the food chain. Thus pesticide residues that wash off soil into a stream or lake can come to reside in fish, and as smaller fish are eaten by larger fish, these residues tend to become more concentrated. By the time humans eat these fish, the concentrations of some pollutants can already have reached harmful levels to human health.

The concept of a *community* mentioned previously deserves to be highlighted as an important concept that must be understood by corporate managers, especially in a society that emphasizes individualism as a key component of its ideology. A community can be defined as any assemblage of populations living in a prescribed area or physical habitat that has characteristics in addition to its individual and population components. Major communities are those of sufficient size and organizational completeness that they are relatively independent of inputs and outputs from adjacent communities. Minor communities are relatively dependent on neighboring aggregations.[17]

The community concept is important because it emphasizes the fact that diverse organisms usually live together in an orderly manner and are not just haphazardly strewn over the earth as independent beings. There is an interdependence factor built into the notion of community that is important to recognize. Communities just don't happen because of ecological accidents, but they evolve out of mutual needs and patterns of dependence where an ecological balance can be maintained and organisms can be provided for their needs and interests. Abrupt changes in the physical environment can disrupt communities and cause organisms to become extinct if they can't adapt to the changes.

Organisms are thus not self-contained units that are independent of their surroundings, and the statement "As the community goes so goes the organism" is something that must be remembered. Often the best way to control a particular organism, whether it is desired to encourage or discourage its development, is to modify the community in which it exists, rather than mount a direct attack on the organism itself. For example, mosquitoes can often be controlled more efficiently by modifying the entire aquatic community in which they develop, such as lowering water levels in marshes and swamps, than by attempting to poison the organisms directly.[18]

Human welfare, like that of the mosquito, depends on the nature of the communities and ecosystems in which humans live and work out their existence. Industrial processes and products that alter essential elements of the community or create communities where the balance of ecological factors are changed, are ultimately going to affect humans for good or ill in some fashion. It is well to have some idea of how humans are going to be affected before these changes are introduced and have irreversible effects or at least cause problems that are going to be very expensive to reverse.

Carrying capacity is another important concept that deserves to be highlighted. Every ecosystem does have limits in terms of the size of various populations that it can support, whether we are talking about human beings or animal populations. Every species or organism has certain needs that the community must provide in order for it to survive and continue to exist. But if any population gets too large in relation to its community, the ecosystem is overloaded and cannot provide the basic needs to every organism. Human beings need space, clean air, water, food, and other essentials in order to survive and maintain a quality existence, but if the human population gets too large relative to its environment, the carrying capacity of that ecosystem may be overtaxed, and human welfare may be affected adversely.

The same concept applies to essential elements of the environment such as air and water. Every medium has a certain ability to absorb waste material without serious harm done to the quality of that medium. Thus air, for example, can absorb a certain amount of waste material without serious harm done to its quality. But if the carrying capacity of that air is exceeded, the air starts to become fouled by certain pollutants and the quality of the air is affected. Its natural dilutive capacity is violated and human health is affected as a result because of exposure to harmful pollutants.

There are various limits on ability of organisms to spread from one type of ecosystem to another and from one part of an ecosystem to another. Every species and each individual organism has a range of tolerance to variations in chemical and physical factors in its environment. Some organisms survive only within narrow temperature ranges. Others need a certain level of precipitation to survive. Individual organisms within a species may have slightly different tolerance ranges because of differences in their genetic makeup.[19]

Thus the existence, abundance, and distribution of a species in an ecosystem are determined by the levels of one or more physical or chemical factors in the environment. *The range of tolerance* includes an optimum within which a species can thrive and operate most efficiently, levels on each side of the optimum where the physical and chemical factors are either below or above the optimum and only a smaller population size can be supported, and lower and upper limits of tolerance where no organisms of a particular species can survive (Figure 2.3).

A related concept is that of *limiting factors,* a very important concept in ecological literature. Organisms depend for their success, and indeed their very existence, on a complex set of conditions. Any one condition that approaches or exceeds the limits of tolerance is said to be a limiting condition or a limiting factor. Under steady-state conditions, an organism can be said to be no stronger than the weakest link in its chain of requirements. But under dynamic conditions, organisms may substitute a closely related substance for one that is required but is deficient in the environment. Or they may be able to alter the conditions in which they are living so as to reduce their requirements.[20]

Something that is in short supply may thus be a limiting factor, but something that is in oversupply may also provide limits. Factors such as heat, light, and water in greater amounts than required also inhibit an organism's development. Organisms thus have an ecological minimum and maximum. The range in between this minimum and maximum represents the limits of tolerance. Organisms with a wide range of tolerance for

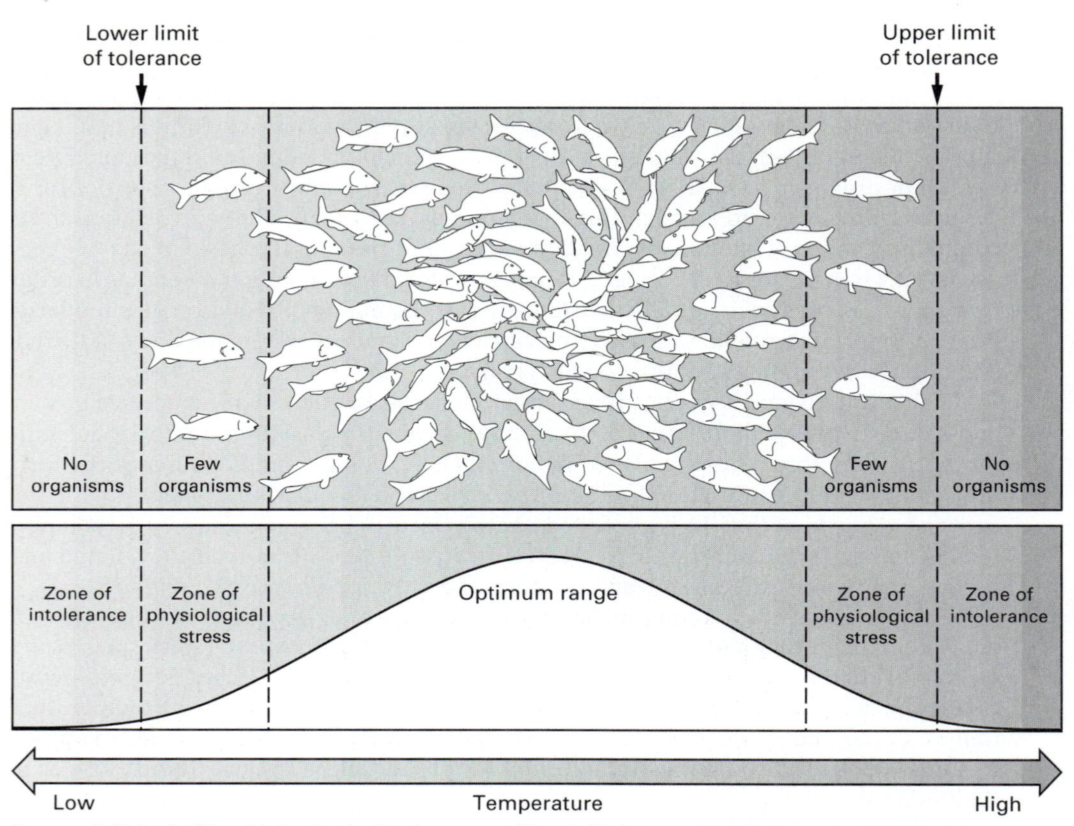

Source: G. Tyler Miller, *Living in the Environment,* 6th ed. (Belmont, CA: Wadsworth, 1990), p. 85.

FIGURE 2.3 Range of Tolerance to Temperature for a Population of Organisms

all factors are likely to be most widely distributed, but some organisms may have a wide range of tolerance for one factor and a narrow range for another; thus their habitats are more restricted.[21]

Primary attention should be given to those factors that are believed to be "operationally significant" to the organism at some time during its life cycle. If policy makers are concerned about the environmental impact of a particular project, they should focus on those environmental conditions most likely to be critical or limiting to the organism that will be affected by the project. The purpose of an environmental analysis, such as preparation of an environmental impact statement, should be (1) to discover by observation, analysis, and experiment which factors are "operationally significant" to the organisms under consideration; and (2) to determine how these factors affect the individual, population, or community. Focusing on these objectives gives a decision maker a better chance of predicting with reasonable accuracy the environmental effects of disturbances or proposed environmental alterations.[22]

Many organisms can change their tolerance of physical factors such as preciptation if they are exposed to gradually changing conditions. This adaption is a useful protective device, but it can also be dangerous as organisms come closer to their *limits of tolerance*. Without any warning, the next small change in the environment can trigger a *threshold effect* where a harmful or even fatal reaction occurs, kind of a straw-that-breaks-the-camel's-back type of effect. This threshold effect partly explains why many environmental problems seem to arise suddenly, when in actuality they have been building up for a long time.[23] The effects of global warming, for example, may not be immediately apparent, but are building up over time until a threshold is reached when it is too late to prevent serious damage.

The *habitat* of an organism is simply the place where an organism lives and where we would go to find it in the total nature of things. Some organisms are of such a nature that they need specialized habitats and thus to find them means one must search out habitats that have the characteristics they need to survive. The Endangered Species Act, for example, does not just address the organism and its right to exist, but also has implications for the habitat in which that organism lives and moves and has its being. Thus environmentalists hope to preserve old-growth forests in the western part of the United States by using the act to preserve the habitat for a certain kind of owl that is on the endangered species list. The owl cannot be preserved without preserving its habitat. The term "habitat" may also refer to the place occupied by an entire community.[24]

The four major habitats are the atmosphere, marine and estuarine habitats, freshwater habitats, and terrestrial habitats. The atmosphere is a habitat that is under severe threat from modern industrial processes. The ozone layer that protects humans from ultraviolet rays of the sun is being depleted, and some industrial nations have taken steps to limit the production and use of certain compounds that are causing the depletion. The greenhouse effect has to do with the increased amounts of carbon dioxide dumped into the air and the apparent warming of the earth that results. Using the atmosphere as a place in which to dispose of waste material is now causing some serious global environmental problems that need to be addressed. Pollutants dumped into the air prevent it from providing the kind of habitat that plant and animal life need in order to survive.

The oceans of the world have been a source of food for centuries, and provide a rich habitat for many thousands of species of fish and other marine animals and plants. The role the oceans play in controlling the earth's atmosphere and climates is better understood than in the past as science discovers more about the importance of this habitat. The oceans are also a source of minerals as more and more deposits of minerals are found that are useful in industrial processes. They are also affected by the production of oil, of course, as offshore drilling technology was developed to tap vast pools of oil underneath the ocean bottom. And they may even provide living space for future generations who want to live under the sea to escape the crowded conditions on earth. Humans must consider the oceans an integral part of their total life support system, and not as an inert supply depot that provides resources for the taking or as a vast waste disposal system that is impossible to pollute.[25]

An estuary has been defined as a semi-enclosed coastal body of water that has a free connection with the open sea. It is thus strongly affected by tidal action, and consists of a mixture of seawater and freshwater from land drainage. The mouths of rivers, coastal bays, tidal marshes, and bodies of water behind offshore beaches are examples of estuaries. Many different kinds of seafood are found in estuaries, and many commercial and sport fisheries are dependent on the preservation of these habitats. When they become fouled with pollutants, these fisheries are threatened with extinction. Estuaries also function as places where waterfowl and other birds find a habitat either in a permanent sense or only as a transition to a more permanent place during their migration.[26]

Freshwater habitats include standing water such as lakes and ponds and running water such as streams or rivers. While freshwater habitats occupy a relatively small portion of the earth's surface as compared to other habitats, their importance is far greater because they are the cheapest and most convenient source of water for domestic and industrial needs, and freshwater ecosystems provide the cheapest and most convenient waste disposal systems for the majority of human communities. Because of abuses regarding usage of this resource, the scarcity of freshwater often becomes the limiting factor to the growth of human communities.[27]

Finally, land masses are where the majority of human beings live along with a rich diversity of plant and animal life. Humans derive most of the resources necessary for the production of goods from the land and depend on the land for most of the food necessary for their existence. The terrestrial habitat is thus of utmost importance in the survival and growth of the human species, and yet land usage often follows destructive patterns because of the prevailing productive orientation toward land. The destruction of the Amazon rain forest is a good example of destroying an essential resource that to some has more value in the short run when it is cleared for farming because it is not seen as productive in its natural state. To an ecologist, nothing could make less sense and be further from the truth.

The concept of *ecological niche* is different from that of habitat in that the former is a more inclusive term that includes more than the physical space occupied by an organism. The concept of niche refers to the organism's functional role in the community or its status in terms of its activities, its rate of metabolism and growth, its effect on other organisms with which it comes into contact, and the extent it modifies or is capable of modifying important operations in the ecosystem. By way of analogy, it could be said that the habitat is the organism's address and the niche is its profession or role in the community. If we wish to become acquainted with some individual in our human community, we would first want to know his or her address in order to find them. But to really know the person, we would want to know something about his or her occupation, interests, associates, and the part he or she plays in community life in general.[28]

So it is with all organisms. They have an address where they can be found. But then we also need to know something about the particular niche or role they play in the larger scheme of things. They may perform some essential role in the ecosystem that cannot be replaced, and thus if the organism is destroyed, the ecosystem may be altered in some fundamental way. This concept is important in relation to the Amazon rain forest, where the habitat of certain plants and animals that are a form of unique species is rapidly being destroyed even before scientists have had a chance to study them and determine the unique niche they play in nature and the important role they may play with respect to the welfare of human beings.

The concept of a *biological clock* refers to natural rhythms or cycles that are in the nature of physiological mechanisms for measuring time in some fashion. There are two theories regarding the biological clock: (1) the endogenous timer hypothesis where the organism has some internal mechanism that can measure time without environmental clues; and (2) the external timer hypothesis where the internal clock is timed by external signals from the environment, such as seasonal changes, or changes in light,

EXHIBIT 2.2 Important Concepts of Ecology

Cycles	The circulation of the chemical elements in the biosphere from the environment to organisms and back to the environment
Food Chain	The transfer of food energy from its source in plants through a series of organisms where eating and being eaten is repeated a number of times
Community	Any assemblage of populations living in a prescribed area or physical habitat that has characteristics in addition to its individual and population components
Carrying Capacity	Maximum population of a particular species that a given habitat can support over a given period of time
Range of Tolerance	The ability of species and organisms to respond to variations of physical and chemical factors in its environment
Limiting Factors	Single factor that limits the growth, abundance, and distribution of the population of a particular organism in an ecosystem
Threshold Effect	A harmful or fatal reaction triggered when the limits of an organism's tolerance have been exceeded
Habitat	The place where an organism lives. The four major habitats are the atmosphere, marine and estuarine, freshwater, and terrestrial
Ecological Niche	An organism's functional role in the community or its status. Habitat is the organism's address; niche refers its role in the community
Biological Clock	Natural rhythms or cycles that are in the nature of physiological mechanisms for measuring time in some fashion

temperature, tides, and similar factors. With regard to human beings, for example, many depend on an alarm clock to wake them up in the morning, but others seem to have an internal clock that wakes them up without an external source. This clock seems to work even in instances where there is a change in the time involved, as when it is desired to rise earlier than normal in order to catch an early flight.[29]

These persistent rhythms have been found in a wide variety of organisms. The migratory habits of birds, for example, depends on some kind of biological clock that tells them when it is necessary to head north or south, depending on the time of year. Other organisms know when it is time to shed their skin or make other adjustments in regard to changing seasons. From the point of view of human activities, particularly in regard to the prevalence of external cues, if the environment is altered and these cues are no longer present or are changed in some fashion, the organism will have to adapt or may be made extinct if it can't adapt to these changes. Thus the total ecosystem may be affected by a change in some element that alters the biological rhythms of organisms in that ecosystem.

PRINCIPLES OF ECOLOGY

The science of ecology deals with principles the same as any other science, and part of the scientific process consists of the discovery of new principles that describe how nature works so that we can have a better understanding of the environment in which we live. Many such principles can be found in the ecological literature, and only a few such principles that form the chapters of a recent book on ecology will be presented in this chapter. The purpose is to acquaint the student with some important principles of ecology that have application to the practice of management (Exhibit 2.3).

1. *The Distribution of Species Is Limited by Barriers and Unfavorable Environments:* Every plant and animal seems to have a restricted distribution on earth. Barriers to dispersal of species—land, water, mountains—have set the broad limits to distribution of species on a global scale. The movement of species by humans from one continent to another in the last two hundred years has produced problems of one sort or another, providing evidence that ecological ignorance of natural dispersal can cause long-range problems. Newly introduced species have produced environmental damage and have destroyed native organisms.

Some ecologists recommend that no one should be allowed to introduce any exotic animal or plant from one continent to another unless rigorous research has been done to show that the organism will not do damage to the environment or other native organisms. It is recommended that we stop the ecologically naive practice of assuming that newly introduced species are harmless additions. While there may be some short-run commercial benefits to be gained from the introduction of new species, the long-run consequences can be disastrous.[30]

2. *No Population Increases Without Limit:* A population is a group of interbreeding organisms belonging to the same species. Two processes add plants or animals to a population: the addition of new organisms through births or seed production and the addition through movement into a population called immigration. There are also two processes that remove organisms from a population: deaths and emigration or movement out of a population. If it is desired to find out why a population either increased or decreased, it must be determined which of these four processes changed to allow the increase or decrease[31] (Figure 2.4).

There are four components of the environment that act to change births, deaths, immigration, and emigration: (1) weather, (2) food, (3) other organisms, and (4) a place in which to live.[32] A change in any of these components will have an effect on the population and components must be considered when human projects alter the environment in some significant manner. Predictions must be made as to what populations will

EXHIBIT 2.3 Principles of Ecology

1. The Distribution of Species Is Limited by Barriers and Unfavorable Environments
2. No Population Increases Without Limit
3. Good and Poor Places Exist for Every Species
4. Overexploited Populations Can Collapse
5. Communities Can Rebound from Disturbances
6. Communities Can Exist in Several Stable Configurations
7. Keystone Species May Be Essential to a Community
8. Natural Systems Recycle Essential Materials
9. Climates Change—Communities Change
10. Natural Systems Are Products of Evolution

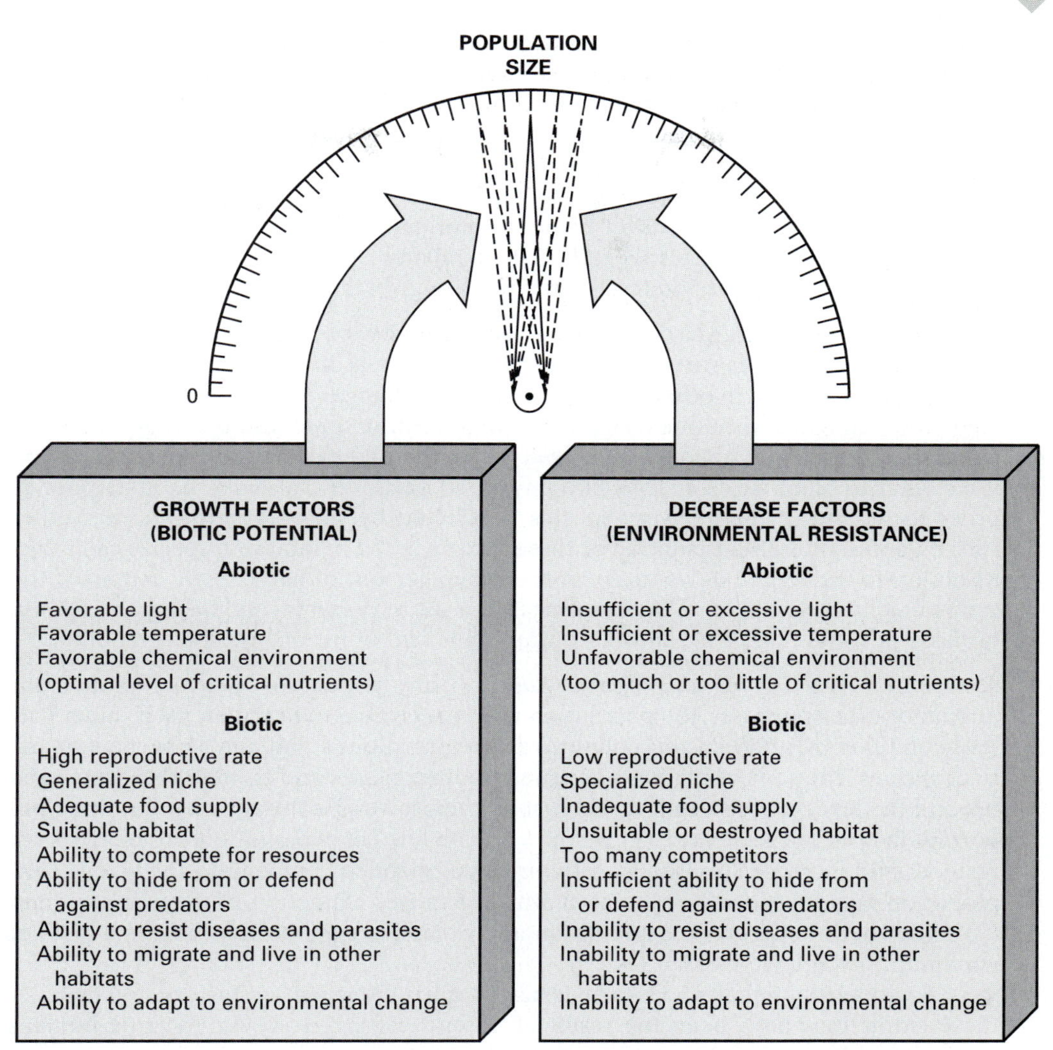

Source: G. Tyler Miller, *Living in the Environment,* 6th ed. (Belmont, CA: Wadsworth, 1990), p. 136. © Wadsworth Publishing Co.

FIGURE 2.4 Determinants of Population Size

be affected and whether they will increase or decrease because of changes in the environment. Ignorance of these effects can cause some serious long-range problems.

This principle also holds true for human beings. For most of human history, the human population rose and fell because of the combined effect of starvation, disease, climactic catastrophes, and self-inflicted losses because of wars and other human events. But over the last several centuries, the human population has been on a growth spiral because of improvements in agriculture, public health, housing, and a reduction of warfare. This growth has been extremely rapid during the last century.

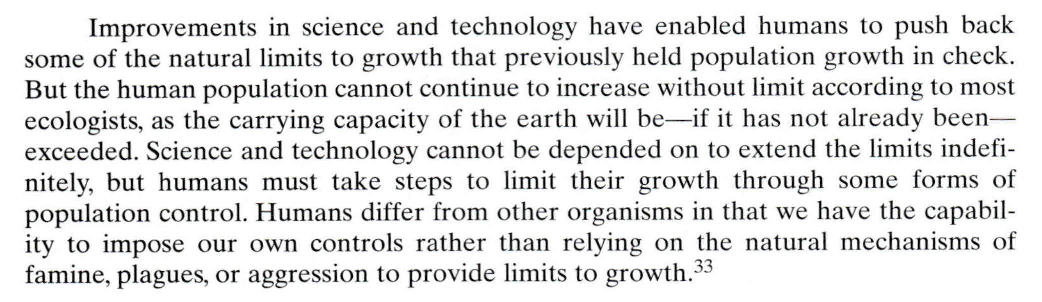

Improvements in science and technology have enabled humans to push back some of the natural limits to growth that previously held population growth in check. But the human population cannot continue to increase without limit according to most ecologists, as the carrying capacity of the earth will be—if it has not already been—exceeded. Science and technology cannot be depended on to extend the limits indefinitely, but humans must take steps to limit their growth through some forms of population control. Humans differ from other organisms in that we have the capability to impose our own controls rather than relying on the natural mechanisms of famine, plagues, or aggression to provide limits to growth.[33]

3. *Good and Poor Places Exist for Every Species:* Every species is dependent on a good habitat in order for it to survive and grow. This is the reason some species are found in some habitats and not in others. Human beings, for example, can't survive in extremely cold or hot climates without altering the environment in some fashion. In order to protect a species, we need to know what kind of a habitat it needs in order to survive. And if we want to eliminate a species like unwanted pests, we can make habitats poor in order to get rid of them. Every species is affected by the weather, nutrients, other species, and shelter. Manipulation of these factors in the habitat of a species can affect its ability to thrive and develop. By proper management of habitats, we can assist the growth and development of those species we want to preserve, and hasten the demise of those pests and other predators we want to get rid of for one reason or another.[34]

4. *Overexploited Populations Can Collapse:* If any population is overexploited for human or other purposes, the population may not be able to maintain itself under current conditions. Buffalo in this country were overexploited and almost became an extinct species. The same thing has happened to other plants and animals that were used beyond the level at which they could sustain themselves. Ecologists mention three important factors in this regard: (1) Below a certain level of exploitation, populations are resilient and increase survival, growth, or recruitment to compensate for loss; (2) exploitation may be raised to a point at which it causes extinction of the resource; and (3) somewhere between no exploitation and excessive exploitation there is a level of maximum sustainable yield.[35]

Extinctions of plants and animals have occurred throughout history, and many of these extinctions have been the result of natural causes. However, the role humans have played in the extinction of species is being recognized as more important than previously expected. Even in prehistoric times, it is suspected that humans played an important role in the demise of some species of animals because of overkill. And certainly in modern times where technology has given humans greater powers of destruction and the means to alter habitats on a scale not known before, the role that humans play in the extinction of species because of overexploitation is terribly important. The exploitation of populations can be managed if we know the conditions necessary for continuance of the population and take steps to maintain a balance between human needs and the needs of the population itself.

5. *Communities Can Rebound from Disturbances:* Populations of plants and animals exist in a matrix of other plants and animals that is called a community. When these communities are disturbed by something like an oil spill, the community can rebound and reach a new level of stability, but it most likely will not return to its original con-

figuration, especially if the disturbance is large enough. One of the tasks of ecologists who study communities is to find the limits of stability of natural communities and see how they rebound from disturbances of one kind or another.

Oil spills like the one in Prince William Sound have many far-reaching and long-lasting effects on biological communities. Crude oil contains many substances that are toxic to marine animals and plants and may stay in the water for some length of time posing a long-term danger. A more immediate disturbance is that birds and sea animals become coated with the oil and are killed as a result. Organisms accumulate hydrocarbons from oil in their tissues, and if these hydrocarbons are not broken down, they may reach high concentrations. If the species is edible, it eventually may pose a cancer threat to humans. Thus oil spills have multiple effects on marine organisms and the community in which they exist (Figure 2.5).

The recovery time varies depending on the climactic zone and other factors, but eventually the community will recover from an oil spill or other disturbance. However, this ability of communities to recover from major disturbances must not be translated into a license to disturb natural communities or allow us to be complacent and not take the proper precautions to prevent oil spills from occurring or being ready to contain them if they should happen. It is recommended that we should treat natural communities the same way we treat our bodies. We have all recovered from various diseases, but we do not go about needlessly exposing ourselves to these diseases. We try to avoid them if possible, and so should we try to avoid disturbances to natural communities.[36]

FIGURE 2.5 Effects of Oil Pollution on Marine Organisms

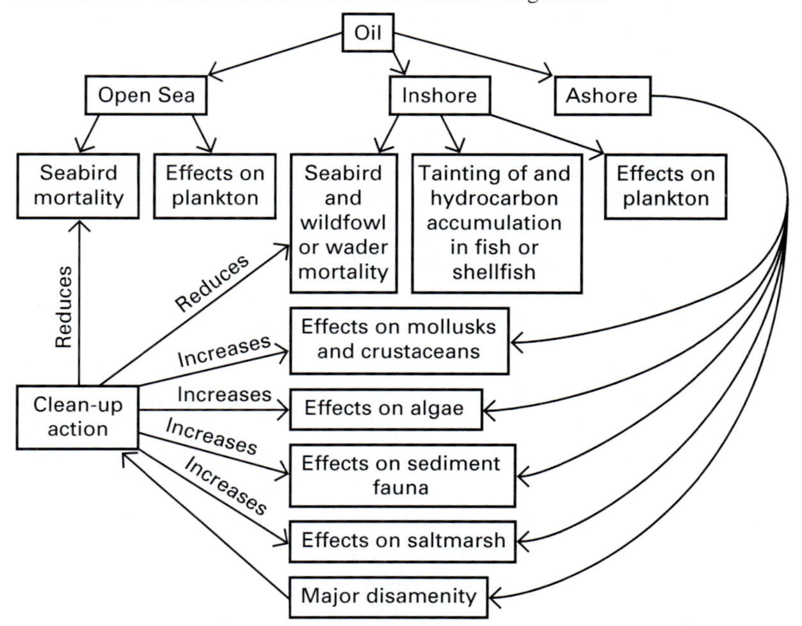

Source: From *The Message of Ecology* by Charles J. Krebs. Copyright © 1988 by Harper & Row Publishers, Inc. Reprinted by permission of Addison Wesley Educational Publishers, Inc.

6. *Communities Can Exist in Several Stable Configurations:* Sometimes a biological community can rebound from a disturbance and return to its original configuration, but if the disturbance is large enough or is permanent in nature, the community may shift to a new configuration. The fish community of Lake Erie, for example, has moved to a new configuration because of uncontrolled overfishing, erosion and nutrient pollution, introduced species, stream destruction, dams and shoreline changes, and toxic chemicals and pesticides.[37] While communities can thus exist in several stable configurations, humans will typically find some of these configurations more desirable than others.

The simple belief that biological communities will return to their pristine configuration once a disturbance is removed is a myth, as they may not revert to their original condition when the stress is removed. The promise of Exxon to return Prince William Sound to its original condition was a bit unrealistic from an ecological point of view and somewhat arrogant public relations material. We need to find out more about what forces can move communities from one configuration to another. At the present time, we introduce major disturbances into biological communities and then act surprised by the results of our disturbances. If we can discover the rule by which communities change, we would be in a better position to restore them when they are damaged. Biological communities are resilient, but it is not easy or cheap to return them to a desirable state, and they will not necessarily return to their original state naturally.[38]

7. *Keystone Species May Be Essential to a Community:* This principle is important when it comes to determining what species belong on rare and endangered species lists. Biological communities contain hundreds of plants and animals and the important question to ask is whether they are of equal importance in a community. Species are important if their removal causes the community to change in some significant manner either by gaining or losing some additional species. The questions to be asked concern how replaceable a species is and how much its loss reverberates to other species in a community.[39]

Every community has one or two dominant species at various levels in that they are the most abundant or contain the most biomass. But these dominant species, according to some ecologists, are not necessarily essential to a community. The essential species are called keystone species in that their activities determine the structure of the entire community, and those species often turn out to be unexpected. If communities are controlled by competition, the removal of one or a few species may not be noticed. However, if a community is controlled by a keystone species, the removal of this species can have dramatic consequences.[40]

8. *Natural Systems Recycle Essential Materials:* Nature consists of various kinds of recycling mechanisms that conserve essential resources. Gaseous elements such as nitrogen, oxygen, and hydrogen all circulate in global cycles and long-distance transfers are common. The oxygen we breath is part of a large pool or reservoir that has had many different sources. Solid elements circulate in local cycles where there are no mechanisms for long-distance transfers. Nutrient cycles have been studied in many biological systems, particularly those that can and have been exploited by human activities.[41]

Human activities can disrupt these nutrient cycles and enough nutrients can be lost so as to cause a downgrading of a particular ecological system. But ecosystems can also be affected by nutrient additions as is the case in the eutrophication of lakes and the problem that acid rain causes for large bodies of water as well as for trees and other

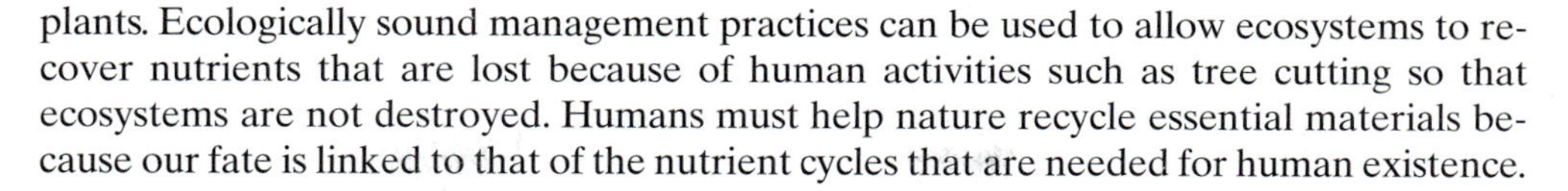

plants. Ecologically sound management practices can be used to allow ecosystems to recover nutrients that are lost because of human activities such as tree cutting so that ecosystems are not destroyed. Humans must help nature recycle essential materials because our fate is linked to that of the nutrient cycles that are needed for human existence.

9. *Climates Change—Communities Change:* Climates change on all time scales and these changes have dramatic implications for all populations, including human populations, that exist on earth. Not all the changes that occur in biological communities are due to changes in climate, but an examination of the changes in human communities that have taken place throughout history shows that climate plays a major role in these changes. While climates change because of natural causes, human activities have come to play a larger role as their scale and extent has increased. The greenhouse effect is one such example, and if, indeed, humans are changing the climate in various parts of the world because of increased industrial activity, communities have no choice but to change accordingly. This principle must be recognized as these global environmental problems become more important.

10. *Natural Systems Are Products of Evolution:* According to some ecologists, one of the greatest and most persistent errors of biological thought is to regard each species as a constant and fixed type in which all organisms are genetically equal. Supposedly, Charles Darwin was one of the first scientists to deal with this mistake and recognize cumulative change in the characteristics of organisms over many generations. While the concept of evolution had been elucidated before Darwin, his contribution was to recognize that the mechanism for evolution is natural selection.[42]

Natural selection operates over time to produce organisms that are adapted to their environment. Favorable attributes give higher reproductive rates or better survival rates, thus individuals with these favorable attributes are selected because they live and survive longer, and these attributes thus become more common in the population over several generations. As humans change environments, they are also thus going to change the evolutionary patterns in ways they do not understand, and organisms are going to adapt to the new environment and the process of natural selection is going to favor some attributes over others because of changes in the environment.

The evolutionary process is the net result of ecology in action where change is inevitable. Since Darwin recognized the principle of natural selection, biologists have continually discovered how much natural selection has changed organisms to adapt to the local physical and biological environment. But there is still much to learn about the evolutionary process and how human activities affect this process. "We are one species in a world whose biological heritage is as irreplaceable as our own cultural heritage. We are the custodians of this biological heritage and our goals must be to study it, to understand it, to enjoy it, and to pass it to our children undiminished."[43]

These and other concepts and principles of ecology are important for business managers and students to know something about. A narrow outlook on the part of business managers and students with respect to business activities is likely to lead to environmental disaster. By focusing on economic concerns and ignoring environmental impacts, the environment is left to fend for itself, and under the pressure of more and more demands from human activities, is becoming seriously degraded throughout the world. Education is one important component of a strategy to counter this narrow outlook, education about important ecological concepts and principles. By taking these concepts and principles into account, business managers can develop policies that are consistent with environmental preservation and learn to live in harmony with nature instead of undermining the conditions for continued economic development.

Questions for Discussion

1. What is the science or study of ecology? How does ecology differ from other sciences? What is the focus of its concern? When did the science of ecology begin?
2. Distinguish between an ecosystem, community, population, and organism. How do these four components of ecology relate to each other and to the whole of ecology?
3. What is an ecosystem? Why is it important for business executives to have some knowledge about the ways in which an ecosystem functions? Describe a major impact that a business activity has had on an ecosystem. What was done to mitigate the impact?
4. Describe the concept of cycles and their importance to the study of ecology. What relevance does the concept of cycles have to environmental problems that concern business?
5. Why is the concept of a food chain important? Describe how pesticide residues, for example, can be concentrated as they move up the food chain. What implications does this concentration have for human beings?
6. Distinguish between carrying capacity and limiting factors. What implications do both these concepts have for economic and population growth? How do they relate to a concept such as sustainable growth?
7. What is a habitat? Describe the four major habitats. How do habitats relate to the Endangered Species Act and the controversy over old-growth forests? What are the important habitats in your area and how are they being protected?
8. What is the importance of an ecological niche? How does this concept differ from a habitat? What is the importance of this concept to economic development, particularly in countries that have vast rain forests that are being cut down?
9. What is the notion of a biological clock? Do humans have a biological clock inside them? Give evidence for your opinion. How can economic development activities interfere with the functioning of this internal mechanism?
10. Review the ten ecological principles described in the chapter. Think of their relevance to business activities. Which are most important, in your opinion, for business to consider? Are there other ecological principles that you are aware of that should be considered?

Endnotes

1. Joseph S. Levine, "Your Child Can Learn Science," *World Monitor,* 4, no. 2 (February 1991), 44.
2. Paul R. Ehrlich, *The Machinery of Nature* (New York: Simon & Schuster, 1986), p. 13.
3. Ibid., p. 236.
4. Ibid., p. 226.
5. Eugene P. Odum, *Fundamentals of Ecology,* 3rd ed. (Philadelphia: Saunders, 1971), p. 3.
6. Michael Begon, John L. Harper, and Colin R. Townsend, *Ecology: Individuals, Populations, and Communities* (Sunderland, MA: Sinauer Associates, Inc., 1986), p. xi.
7. Odum, *Fundamentals,* p. 3.
8. Begon, Harper, and Townsend, *Ecology,* p. x.
9. Odum, *Fundamentals,* p. 9.
10. Ehrlich, *Machinery,* p. 239.

11. G. Tyler Miller, Jr., *Living in the Environment,* 6th ed. (Belmont, CA: Wadsworth, 1990), pp. 80–83.
12. Odum, *Fundamentals,* p. 251.
13. Ibid., p. 267.
14. Ibid.
15. Ibid., p. 86.
16. Ibid., p. 63.
17. Ibid., p. 140.
18. Ibid., pp. 140–141.
19. Miller, *Living in the Environment,* pp. 84–85.
20. Eugene P. Odum, *Basic Ecology* (New York: Saunders College Publishing, 1983), pp. 221–222.
21. Ibid., p. 223.
22. Ibid., p. 225.
23. Miller, *Living in the Environment,* p. 86.
24. Odum, *Fundamentals,* pp. 234–235.
25. Ibid., p. 324.
26. Ibid., p. 352.
27. Ibid., p. 295.
28. Ibid., pp. 234–235.
29. Ibid., p. 245.
30. Charles J. Krebs, *The Message of Ecology* (New York: Harper & Row, 1988), pp. 15–16.
31. Ibid., pp. 17–18.
32. Ibid., p. 18.
33. Ibid., p. 29.
34. Ibid., pp. 31–43.
35. Ibid., p. 48.
36. Ibid., pp. 76–77.
37. Ibid., p. 89.
38. Ibid., p. 97.
39. Ibid.
40. Ibid., p. 106.
41. Ibid., pp. 113–114.
42. Ibid., p. 147.
43. Ibid., p. 167.

Suggested Reading

Andow, David A., ed. *Ecological Interactions.* New York: Westview, 1996.

Cherrett, J. M., ed. *Ecological Concepts: The Contributions of Ecology to an Understanding of the Natural World.* New York: Blackwell, 1989.

Costanza, Robert, ed. *Ecological Economics.* New York: Columbia University Press, 1991.

Ehrlich, Paul R. *The Machinery of Nature.* New York: Simon & Schuster, 1986.

Hayward, Tim. *Ecological Thought: An Introduction.* New York: Blackwell, 1995.

Howell, Dorothy J. *Ecology for Environmental Professionals.* New York: Greenwood, 1993.

Krebs, Charles J. *The Message of Ecology.* New York: Harper & Row, 1988.

Morgan, Sally. *Ecology and Environment: The Cycles of Life.* New York: Oxford University Press, 1993.

Odum, Eugene P., ed. *Ecology and Our Endangered Life-Support Systems,* 2nd rev. ed. New York: Sinauer, 1993.

Pickett, Steward T., et al. *Ecological Understanding: The Nature of Theory and the Theory of Nature.* Washington, D.C.: Academic Press, 1994.

Ricklefs, Robert E. *Ecology,* 3rd ed. Boston: W. H. Freeman, 1995.

Roaema, J., and J. A. Verkleij, eds. *Ecological Responses to Environmental Stresses.* Netherlands: Kluwer Academic, 1991.

Woodley, Stephen, et al., ed. *Ecological Integrity and the Management of Ecosystems.* New York: St. Lucie Press, 1993.

CHAPTER 3

Social Responsibility and Environmental Ethics

Making value judgments about the environment and the kind of world in which we want to live not only involves some knowledge about ecology and the ecological consequences of our actions; such value judgments also involve ethical considerations. These ethical considerations are the subject of a field of endeavor called environmental ethics, a form of applied ethics that deals with environmental issues and concerns from an ethical perspective. Environmental ethics is a field in its own right separate from the field of business ethics with its own set of scholars and body of literature, resting on different theories and conceptual foundations.[1]

Environmental ethics involves critical thinking with respect to policies of the private and public sectors that have been developed in response to environmental problems, and questions the assumptions upon which those policies are based. An ethical approach to these issues provides a means of questioning traditional approaches and advocates new ways of thinking about the environment. Questions can be asked about the responsibilities humans have for the environment and how these responsibilities ought to be reflected in the policies adopted by the government and private companies as well as in the habits of the population as a whole.

The traditional approach of Western societies toward nature is to objectify nature and see nature as existing to serve human purposes. Nature is there to be manipulated to serve human interests and has no interests of its own that deserve to be respected. Trees have no value in and of themselves; they are valuable only as they are cut down and processed into lumber that can then be used to build houses. Scenic areas of the country have been set aside to be used for human enjoyment as national parks, but they have no particular value in and of themselves. The desert is considered a wasteland that needs to be developed rather than as a particular kind of ecosystem with its own value and its own beauty that deserves protection from the developers.

This approach to nature is anthropocentric in that right and wrong are determined by human interest and the promotion of human welfare as the ultimate objective. This view has sometimes been called a humanistic ethic as concern for the environment is of a lower priority than a concern for humans, who are helped or hurt by the conditions of their environmental surroundings.[2] This approach is human centered and does not hold that nature has intrinsic value. Nature has value only because people value it, and environmental preservation is good only because it is good for human beings.

It has been argued that only persons are moral agents, and thus only humans have moral status and can pursue moral claims. Something that has no conscious sentiency and so can experience no pleasure or pain, joy or suffering, does not deserve moral respect. [3] Why should a tree not be cut down if no harm comes to any human being from destruction of that part of nature? Why do certain species need to be preserved if they have no use for human beings? Nothing can have intrinsic value except the activities, experiences, and lives of conscious, sentient beings. The ecological community has no intrinsic value over and above that contained in the lives of its human members.[4]

Ethics in general has been defined as a concern with actions and practices that are directed to improving the welfare of people. Ethicists explore the concepts and language that are used to direct such actions and practices to enhance human welfare, and are concerned with clarifying what constitutes human welfare and the kind of conduct necessary to promote such welfare.[5] Ethics is the quest for an understanding of what constitutes a good life and a concern for creating the conditions for humans to attain that good life.[6] It is the study of what is good and right for human beings and what goals people ought to pursue and what actions they ought to perform. Ethics concerns itself with human conduct, meaning human activity that is done knowingly and, to a large extent, willingly.[7]

All of these definitions are profoundly anthropocentric in that they say nothing about the welfare of nature as such and say nothing about certain rights nature or animals have that may interfere with the human quest for the good life. It may be good for humans to use animals extensively in testing substances for toxic effects, but what about the rights of animals to be protected from such abuse? Human beings knowingly cut down the rain forest in the quest for a better life without thinking or caring about the ecological function of that forest. By limiting ethical concerns to human beings, nature is shut out of serious ethical consideration and is subject to exploitation and abuse in the interests of promoting human welfare.

The traditional approach to nature serves the interests of economic society, and ethical concerns are largely supportive of a materialistic approach to human welfare. Nature is there to be exploited and used for human and materialistic purposes. The corporation is seen as an economic institution with economic responsibilities to produce goods and services as efficiently as possible. Its ethical responsibilities are largely tied up with an efficient use of resources for purposes of promoting economic growth, which is believed to enhance human welfare. Concern for nature, if it exists at all, is limited to conservation of resources coupled with a concern to minimize pollution. But there has been little by way of respect for the rights of nature or a concern to view nature from an ecological rather than an economic perspective. An economic ethic is at the base of the traditional view of the corporation and poses a problem when trying to incorporate environmental concerns into corporate behavior.

AN ECONOMIC ETHIC

The traditional view of the firm and its ethical responsibilities assumes that there is and can be no divergence between the operation of a successful business organization and ethical behavior on the part of the organization. Such a divergence cannot exist because ethical responsibilities are defined in terms of marketplace performance. What is considered to be ethical is exactly the same as what is considered to be good business. The ethical notion that forms the basis of this view is the principle of economizing. A business organization is formed to provide goods and services that people in a society are willing to buy at prices they can afford. In order to do this successfully, business must economize in the use of resources—combine resources efficiently—so that it can earn profits to continue in business and expand into new markets.

The ethical performance of business is thus tied up with marketplace performance. If a business organization is successful and earns a satisfactory level of profits, this means that the business has economized in the use of resources, assuming that competition exists in the markets it is serving. The business has produced something people want to buy, and has done so in such a way that it has met the competition. Successful performance in the marketplace is ethical behavior, and there is no divergence between being ethical and being successful in the marketplace. Successful business performance and acceptable ethical behavior are believed to be one and the same thing.

This view is espoused by economists like Milton Friedman who in response to the notion of social responsibility argued that the social responsibility of business is to increase its profits.[8] In other words, the social and ethical responsibilities of business are exhausted in terms of marketplace performance. As long as business performs its economizing function well, it has fulfilled its social and ethical responsibilities and nothing more need be said. Thus the traditional view of ethics and business subsumes ethics under marketplace performance and does not necessitate any conscious ethical considerations of business's responsibilities to society or to the environment other than successful economic performance. Ethics is totally captured by the notion of economizing, which can be promoted by the development of principles related to an efficient combination of resources.

Within this system, management had no motivation to pay attention to environmental matters. Companies disposed of their wastes as efficiently as possible, which in the case of hazardous wastes meant setting barrels of such waste on the back dock to be picked up and disposed of by a waste hauler. But no records were kept of where this waste was eventually deposited and whether it was disposed of safely. The corporation, as well as society, operated with an out-of-sight, out-of-mind mentality. With respect to air pollution, air was considered to be a free good available to dump wastes into at will with no concern for air quality. The same was true for disposal of wastes into rivers and streams as well as lakes. They were there for disposal purposes at no cost to the corporation.

CORPORATE SOCIAL RESPONSIBILITY

While not everyone adhered to this view and accepted the notion that business was solely an economic institution with only economic responsibilities, it does seem that this view of ethics and business was the prevailing view in our society since its inception. And as long as the system worked well enough for most people, there were not likely to be any serious questions raised about the ethical behavior of corporations outside of the marketplace context. It was the concern with social responsibilities that began to raise serious questions about this view of ethics and business. The problems that social responsibility advocates addressed, such as pollution and unsafe workplaces, were in large part created by the drive for efficiency in the marketplace. Thus it began to be argued that there was a divergence between the performance of business in the marketplace and its performance as far as the social impacts of its behavior were concerned.

People began to believe that cleaning up pollution, providing safer workplaces, producing products that were safe to use, promoting equal opportunity, and attempting to eliminate poverty in our society had something to do with promoting human welfare and creating the "good life" in our society. Yet business was causing some of these problems and perpetuating others in its quest for an efficient allocation of resources. For example, by economizing in the use of resources and disposing of its waste material as cheaply as possible, business was causing some serious pollution problems regarding air quality and poisoning of drinking water. By always hiring the best-qualified person for a job opening and not having some kind of an affirmative action program, business was helping to perpetuate the effects of discrimination against minorities and women.

It was at these points of intersection between the economic performance of business and changing social values of society that ethical questions began to arise. Business increasingly came to be viewed as a social as well as an economic institution with social impacts that management needed to consider. Social responsibility advocates strongly argued that management needed to take the social impacts of business into account when developing policies and strategies, and much effort was devoted to convincing management to take its social responsibilities seriously. A great deal of research was done to help management redesign corporate organizations and develop policies and practices that would enable corporations to respond to the social expectations of society and measure their social performance.

The deficiencies of the traditional view of ethics and business began to be exposed. It became clear that there were many points of divergence between good business performance and what society expected of its business organizations in terms of ethical behavior. An economic ethic doesn't include the social aspects of corporate activities and encourage management to pay attention to the social impacts of corporate operations. Thus it provides no means or rationale for management to internalize the social costs of production, which includes environmental damages, and take these costs into account in its decisions. The problem facing management theorists who accept the fact that a divergence often exists between ethical behavior and marketplace performance is how to get management to pay attention to these social or external costs of production.

One way to do this is simply to argue that good ethics is good business by making the case that ethical behavior in the broadest sense will lead to success in the marketplace. Ethical behavior of the kind envisioned by social responsibility advocates will be rewarded by increased profits, improved performance on the stock market, and other relevant measures of business success. Ethical considerations are not exhausted by economizing in the use of resources and deserve conscious reflection and attention. But by choosing to be ethical in all aspects of business operations and following such principles in its decisions, management will be economically successful as well as ethical in its behavior.

The social responsibility advocates tried to make this argument in convincing management to take the notion seriously. They made various arguments based on the notion of long-run self-interest, that by being socially responsible business was taking account of its long-range health and survival. Business could not remain a healthy and viable organization in a deteriorating society. Thus it made sense for business to devote some of its resources to helping solve some of the most serious social problems of society, whether it be environmental deterioration, discrimination, or poverty, because business could function better in a society where most of its members shared in a high standard of living and enjoyed an improved quality of life.

Other arguments had to do with gaining a better public image, that by being socially responsible business organizations could improve their image in society and in this way gain more customers and provide more of an incentive for investors to put their money in the company. There are many examples of companies that have tried to present themselves as concerned about public health and the environment through their advertising and contributions program. Finally, other arguments had to do with the avoidance of government regulation, that by being socially responsible and effectively responding to changing social expectations, business might be able to eliminate the necessity of onerous government regulations that would affect its profits and other aspects of performance.

These arguments were never very convincing because they were not based on a solid moral philosophy about the nature of the corporation and its management but were more in the nature of moralizing about certain aspects of business behavior. It now seems that social responsibility was more of a doctrine than a serious theory of the corporation. Scholars and executives who advocated social responsibility seemed to do so more as an article of faith than as a theoretical paradigm that could bid for serious attention and begin to compete with the economic theory as an alternative description of the ethical responsibilities of corporations. The bottom line of corporate organizations as well as for the nation as a whole remained economic in nature.

There are several reasons for this development, not the least of which is the difficulty of implementing social responsibility in a competitive context. Being socially responsible usually costs money. Pollution control equipment is expensive to buy and operate. Ventilation equipment to take toxic fumes out the workplace is expensive. Proper disposal of toxic wastes in landfills can be very costly and time-consuming. These efforts cut into profits, and in a competitive system, companies that go very far in this direction will simply price themselves out of the market. This is a fact of life for companies operating in a free enterprise system that the social responsibility advocates never took seriously.

> . . . every business . . . is, in effect, "trapped" in the business system it has helped to create. It is incapable, as an individual unit, of transcending that system . . . the dream of the socially responsible corporation that, replicated over and over again can transform our society is illusory. . . . Because their aggregate power is not unified, not truly collective, not organized, they [corporations] have no way, even if they wished, of redirecting that power to meet the most pressing needs of society. . . . Such redirection could only occur through the intermediate agency of government rewriting the rules under which all corporations operate.[9]

Management has to be concerned about the economic performance of the organization. It cannot set aside these requirements to pursue social objectives that conflict with economic performance and expect to remain in business for very long. When there is a choice to be made between an ethical ought and a technical must, something business must do to remain a viable organization within the system, it seems clear which path most managements will follow. Technical business matters are the ultimate values—a technical business necessity is a must that always takes precedence over an ethical ought that would be nice to implement but is simply not practical under most business conditions.[10]

It could be argued that social responsibility theory and principles cannot provide answers to the problems of finance, personnel, production, and general management decision making. The businessperson's role is defined largely, though not exclusively, in terms of private gain and profit, and to suggest that this can be set aside for adherence to a set of social responsibilities, however well-intentioned, that may conflict with that role is startlingly naive and romantic. The businessperson is locked into a going system of values and ethics that largely determine the actions to be taken. There is little question that at any given time individuals who are active within an institution are subject in large measure to its prevailing characteristics.

Social responsibility doctrines proved to be amorphous, fuzzy, and provided no clear guidelines for managerial behavior. The critics of social responsibility were right when they exposed the difficulty of providing a legitimate basis for social action on the part of corporate managers. Social responsibility advocates provided no sound moral basis for managerial social action other than some impossible-to-measure notions of enlightened self-interest or creation of a better corporate image, worthy goals perhaps, but certainly difficult to implement in a competitive context. Thus concern for the environment, as well as other social issues, had no solid basis in either theory or practice.

PUBLIC POLICY AND SOCIAL RESPONSIVENESS

Some scholars turned to public policy as an alternative to social responsibility because of these frustrations. They found it difficult to teach students anything substantive about social responsibility that might be useful to their careers as managers of corporate organizations. And during the 1970s, most, if not all, social issues became public policy matters as more and more legislation was passed and regulations were issued dealing with environmental protection, workplace safety and health, equal opportunity, consumer protection, and other social concerns. Events seemed to be overtaking corporate managers, even those who made their best efforts to be socially responsible. These efforts did not stem the tide of government involvement in more and more aspects of corporate behavior.

The public policy approach seemed to offer several advantages over the notion of social responsibility as a theoretical underpinning or framework for an expanded notion of corporate responsibilities. When regulations were issued, these effectively operationalized the social responsibilities of management in great detail. These regulations sometimes specified what kind of wood could be used in ladders, what standards had to be met with regard to specific pollutants, what kind of waste disposal methods would pass muster with federal agencies, and similar concerns. The public policy approach took the institutional context of business into account and provided a legitimacy for socially responsible actions on the part of management, as government, acting on behalf of its citizens, had a legitimate right grounded in democratic theory, to provide guidelines for managers and shape corporate behavior to correspond more closely with societal expectations.

There did not seem to be, at least on the surface, a need for a theoretical underpinning for public policy at least as it affected corporations, as business has a moral obligation to obey the law as a good citizen. Failure to do so subjects the corporation to all sorts of penalties and other problems in society. Thus the social responsibility of business is not only to perform well in the marketplace and meet its economic objectives but also to follow the directives of society at large as expressed in and through the public policy process. The public policy process and marketplace are both sources of guidelines for managerial behavior.[11]

> Society can choose to allocate its resources any way it wants and on the basis of any criteria it deems relevant. If society wants to enhance the quality of air and water, it can choose to allocate resources for the production of these goods and put constraints on business in the form of standards. . . . These nonmarket decisions are made by those who participate in the public policy process and represent their views of what is best for themselves and society as a whole. . . . It is up to the body politic to determine which market outcomes are and are not appropriate. If market outcomes are not to be taken as normative, a form of regulation which requires public participation is the only alternative. The social responsibility of business is not operational and certainly not to be trusted. When business acts contrary to the normal pressures of the marketplace, only public policy can replace the dictates of the market.[12]

Other scholars turned to corporate social responsiveness as an alternative, and avoided the frustration of dealing with social responsibility by focusing on the manner in which corporations were responding to societal expectations. This effort was more pragmatic and management oriented and less philosophical in its approach. The attempt of scholars proceeding in this direction is to discover patterns of responsiveness that will help in understanding how corporations have coped with a changing social environment, and to try and explain differences in responses between corporations. This research was largely directed to help corporations respond more effectively to social problems by identifying key variables within the organization that were determinative of the response pattern.

> Corporate social responsiveness refers to the capacity of a corporation to respond to social pressures. The literal act of responding, or of achieving a generally responsive posture, to society is the focus of corporate social responsiveness. . . . One searches the organization for mechanisms, procedures, arrangements, and behavioral patterns, that, taken collectively, would mark the organization as more or less capable of responding to social pressures. It then becomes evident that organizational design and managerial competence play important roles in how extensively and how well a company responds to social demands and needs.[13]

Many scholars dealing with social responsiveness have utilized the stakeholder concept to describe the various constituencies to which corporations have to respond and prescribe corporate responsibilities. The stakeholder model is a useful tool to analyze and describe the various relationships a corporation has to its main constituents in society, but it is by no means a serious theoretical attempt to provide a new paradigm that would even begin to replace the economic paradigm. Stakeholder relations can be analyzed in economic terms, and often are, and while a manager might have to balance the interests of various stakeholder groups in order to resolve a problem, he or she does so in an economic context. In order to stay in business and continue to make a profit, business may have to respond to certain stakeholder interests.

Both of these approaches, public policy and social responsiveness, have been criticized on the basis of ignoring the deeper value issues involved in corporate responsibility. The criticism has been made that scholars in both camps tried to ignore deeper value issues by focusing on nonnormative concerns. The advocates of social responsiveness urged corporations to avoid philosophic questions of social responsibility and concentrate on more pragmatic matters related to responding effectively to social pressures. They shunned normative questions by attempting to conduct value-free inquiry into the corporate response processes and developing various techniques such as social forecasting and issues management that could improve the ability of corporations to respond to social concerns.[14]

The same is true for advocates of the public policy approach, as they have been criticized for failing to acknowledge how thoroughly saturated the public process is with value-laden phenomena. There was a hope that by using the public policy approach, scholars could escape the subjectivity and vagueness of corporate social responsibility philosophizing and substitute a more objective and value-neutral basis for measuring and judging business social performance. If business adhered to the standards of performance expressed in the law and existing public policy, then it could be judged as being socially responsive to the changing expectations of society. But when one digs beneath the surface of the public policy approach, one finds it is plagued by the same kinds of ethical dilemmas that plagued earlier attempts to deal with the social responsibilities of business. Public policy is, in the final analysis, all about values and value conflicts, and public policy solutions to social problems are built on some conception of the good life that has to do with the promotion of human welfare by corporate activities.[15]

Neither of these approaches has advanced an understanding of the normative dimension related to corporate behavior, and both have contributed little, if anything, to the development of an alternative theory of the corporation for supporting environmental responsibility. The economic theory of the corporation is largely intact as social responsiveness and public policy concerned themselves with mostly peripheral matters as far as the primary responsibilities of the corporation are concerned. Responses to social concerns are made within the established framework of the traditional enterprise where economizing is dominant over social values. And public policy responses are often shaped to correspond with the dominant economic value system that determines corporate behavior. All of the normative questions about corporate social and environmental responsibilities were still on the table and largely unanswered.[16]

THE NEED FOR AN ENVIRONMENTAL ETHIC

The normative question that is unanswered has to do with the nature of these attempts to broaden the ethical responsibilities of business. All of the previously described attempts to develop broader notions of the good society and prescribe a corresponding set of social responsibilities for corporations suffer from the same problem. They are profoundly anthropocentric, that is to say human centered. This characteristic leaves them much less than satisfactory in a world besieged by environmental problems on a scale that was not imagined possible just a few years ago when environmental problems were seen as a quality-of-life issue, not the survival issue they are today. All of these efforts to deal with corporate responsibility and redefine the good life reflect the arrogance of Western culture with respect to its approach to nature.

While the social responsibility issues that developed in the 1960s did include environmental concerns, these concerns were largely human centered. The public policy measures passed in the 1960s and 1970s were largely based on the protection of human health, not on a concern for the protection of the environment for its own sake. The typical stakeholder map includes stockholders, creditors, employees, consumers, government, and so forth, but it never includes plants and animals or nature in general, all of which have a significant stake in corporate activities. The social contract model, which is often used to support ethical responsibilities, is a contract between human beings. The natural world is never included in the bargain.

With regard to environmental concerns and the development of an adequate theoretical support for the environmental responsibilities of corporations, scholars simply must take a broader ecological perspective and abandon an anthropological perspective. Human beings are not the center of the universe nor should they be the sole center of concern. They are but one species in a world populated by millions of species. Human beings could not survive without the existence of these species, whereas other plant and animal species could survive quite well without human interference. Humans stand at the end of the food chain and thus stand to suffer the most from poisons introduced into the chain and will be without sustenance if any part of the food chain is broken. Humans are extremely vulnerable and dependent on the environment for survival and cannot exist without the services and resources the environment provides.

Economic growth cannot take place without the appropriate environmental conditions to support such growth. The notion that policy makers have to make a tradeoff between economic growth and environmental protection in decisions about public and corporate policy no longer makes sense. The two goals are consistent with each other. The environment must be protected and enhanced for economic growth to take place. If the environment is destroyed, as is taking place in the Amazon rain forest, for example, economic growth will eventually come to a halt as resources are exhausted. All we will be left with is a ravaged earth that cannot support human life, and perhaps not any form of life at all. Economic growth that undermines the conditions for that growth is not sustainable. This awareness must sink in to Western consciousness and must become a part of ethical thinking for any kind of reasonable theories to be developed that would offer a significant challenge to the dominant economic paradigm. Such a task is no longer merely an interesting intellectual exercise; it is necessary to the survival of planet earth and all the life forms that exist on earth.

Since the corporation now has additional responsibilities largely because of government regulation that attempts to make business respond to environmental concerns, it may be fruitful to try and discover something called an environmental ethic that can provide theoretical support for this process. The fact is that despite several years of discussion about social responsibilities, environmental concerns have become matters of public policy in the sense that laws have been passed and regulations issued to make business respond to environmental problems. Voluntary responses out of a sense of social responsibility have not been relied on to get the job done. But there is a need for an environmental ethic to support these efforts so they are less subject to political changes. Such an ethic has implications for business and government in terms of environmental responsibilities and practices.

CURRENT DIRECTIONS IN ENVIRONMENTAL ETHICS

As environmental problems impinge themselves on the public mind, intellectual disciplines such as philosophy are struggling to develop new environmental paradigms that overcome the traditional approach to the environment that has pervaded industrial societies. Sometimes called the greening of philosophy, new thinking in philosophy is attempting to provide alternatives to the dualism between humans and nature that has undergirded traditional approaches to the environment, approaches in which the environment is seen as something separate from humans to be dominated and manipulated for human purposes. This traditional paradigm was consistent with institutional practices in which environmental impacts were largely ignored with a "nature can take care of itself" attitude. Now that institutional behavior is changing, there is a need for new philosophical paradigms to provide a rationale for these changes and to indicate a path for future changes. Some believe that this task is vital to human survival on a planet that is being systematically destroyed by traditional ways of thinking and acting.

> The upshot is that the dominant ethical systems of our times, those clustered as the Western ethic and other kindred human chauvinistic systems, are far less defensible, and less satisfactory, than has been commonly assumed, and lack an adequate and nonarbitrary basis. Furthermore, alternative theories are far less incoherent than is commonly claimed, especially by philosophers. Yet although there are viable alternatives to the Dominion thesis, the natural world is rapidly being preempted in favor of human chauvinism—and of what it ideologically underwrites, the modern economic-industrial superstructure—by the elimination or overexploitation of those things that are not considered of sufficient instrumental value for human beings. Witness the impoverishment of the nonhuman world, the assaults being made on tropical rainforests, surviving temperate wildernesses, wild animals, the oceans, to list only a few of the victims of man's assault on the natural world. Observe also the associated measures to bring primitive or recalcitrant peoples into the Western consumer society and the spread of human-chauvinistic value systems. The time is fast approaching when questions raised by an environmental ethic will cease to involve live options. As things stand at present, however, the ethical issues generated by the preemptions—especially given the weakness and inadequacy of the ideological and value-theoretical basis on which the damaging chauvinistic transformation of the world is premised and the viability of alternative environmental ethics—are not merely of theoretical interest but are among the most important and urgent questions of our times, and perhaps the most important questions that human beings, whose individual or group self-interest is the source of most environmental problems, have ever asked themselves.[17]

Treating nature as instrumental places the natural world in a utilitarian position with relationship to human beings. The material world has value only to the extent it can serve humans. Such an approach promotes an unhealthy separation between humans and the rest of nature. Humans must also ask how they can serve nature and recognize a mutual interdependence between human life and life in the natural world. Human life, in fact, is part of the natural world, and the dualism and individualism characteristic of Western thought is no longer functional. Such thinking leads to policies and practices that undermine the conditions for supporting human life and activities by destroying the natural world on which we all depend.

Some philosophers have responded to this challenge by devoting their efforts to developing new notions about environmental ethics and extending moral consideration to nature through new philosophical systems. Thus far, the greening of philosophy seems to be taking two paths, one path that deals with moral extensionism and eligibility, and the other path with biocentrism or deep ecology. Both these approaches are attempting to bring nature into the moral realm and make all or parts of nature an active moral agent that deserves moral consideration in our actions and practices.

Moral Extensionism and Eligibility

The concept of rights has come into standard usage in terms of extending moral consideration to entities and has changed a host of institutional practices that have impacts on human beings. The civil rights movement extended the notion of rights to blacks and other minorities, and had as a goal the extension and implementation of basic rights afforded to Americans of any race or color. Equal rights dealt with the same problems with regard to women, and were used to press for equal treatment for women in all aspects of society. The right to a safe and hazard-free workplace found its way into legislation and regulation regarding safety and health in the workplace. Something called a consumer bill of rights dealt with product safety and other aspects of the marketplace that needed attention. The concept of human rights has thus been extended into many aspects of life in our society and has been used as the basis of much legislation and regulation that has changed the nature of institutional behavior to respect those rights.

The question is whether this notion of rights can also be extended to the natural world or at least some of its components and whether this extension can help to deal with environmental problems in an effective manner. Where does the ethical cutoff fall with regard to moral eligibility? What aspects of nature should be brought into the moral realm and thereby given moral consideration? The philosophical approach to extending rights to nature has taken place around the issue of moral extensionism and eligibility.

Many philosophers extend ethics only so far as animals on the grounds that animals are sentient beings in that they are able to suffer and feel pain. But more radical thinkers widened the circle to include all life such as plants. Still others see no reason to draw a moral boundary at the edge of life and argued for ethical considerations for rocks, soil, water, air, and biophysical processes that constitute ecosystems. Some are even led to the conclusion that the universe has rights superior to those of its most precocious life form.[18]

Peter Singer argues that the view holding that the effects of our actions on nonhuman animals have no intrinsic moral significance is arbitrary and morally indefensible. He makes an analogy between the way we now treat animals with the way we used to treat black slaves. The white slaveowners limited their moral concern to the white race and did not regard the suffering of a black slave as having the same moral significance as the suffering of a white person. Thus the black could be treated inhumanly with no moral compulsion. This way of thinking and treating blacks is now called racism, but we could just as well substitute the word "specieism" in regard to the manner in which animals are treated. The logic of racism and the logic of specieism are the same.[19]

Just as our concern about equal treatment of blacks through legislation and regulation moved us to a different level of moral consciousness, so too will treating of animals as beings who have interests and can suffer and therefore deserve moral consideration move us to a different level of moral consciousness (see Figure 3.1). This level may involve the stopping of certain practices such as using animals for testing purposes and subjecting them to slow and agonizing deaths. It may also involve stopping the practice of raising animals in crowded conditions solely for the purpose of human consumption. The decision to avoid specieism of this kind will be difficult, but no more difficult, states Singer, than it would have been for a white Southerner to go against the traditions of his society and free his slaves.[20]

The creatures in Singer's moral community have to possess nervous systems of sufficient sophistication to feel pain; that is, they have to be sentient beings. Ethics ends

FIGURE 3.1 The Expanding Concept of Rights

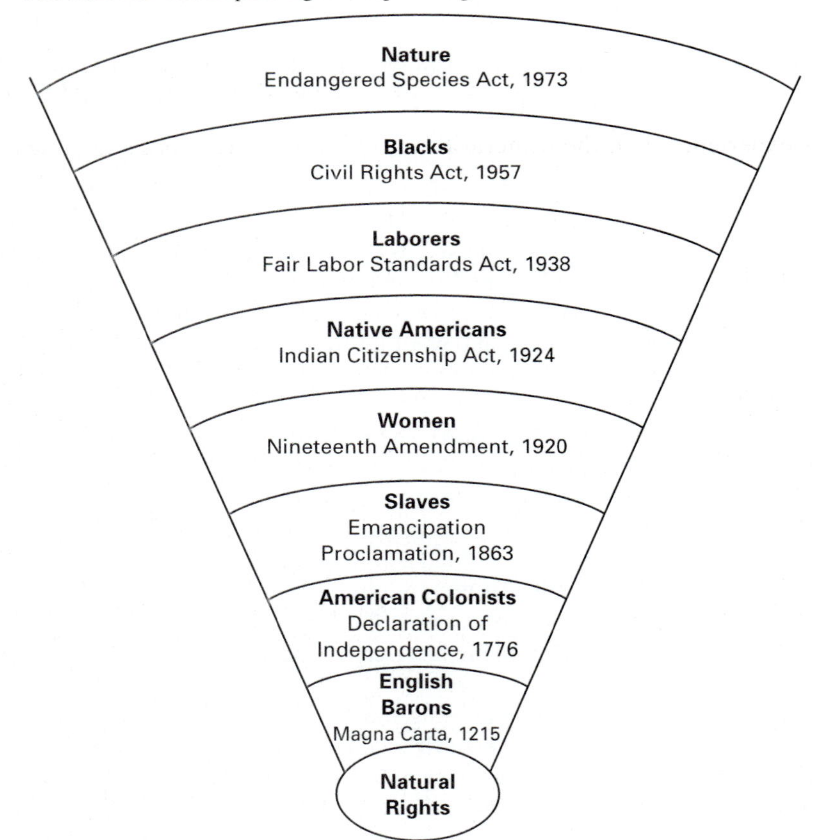

Source: From Nash, Roderick. *The Rights of Nature: A History of Environmental Ethics.* © 1989. (Madison: The University of Wisconsin Press.) Reprinted by permission of The University of Wisconsin Press.

at the boundary of sentience. A tree or a mountain or a rock being kicked does not feel anything and therefore does not possess any interests or rights. Since they cannot be harmed by human action they have no place in ethical discourse. There is nothing we can do that matters to them and thus they are not deserving of moral consideration.[21]

Other philosophers, such as Joel Fineberg, also limited their moral concerns to animals. Fineberg excluded plants from the rights community on the grounds that they had insufficient "cognitive equipment" to be aware of their wants, needs, and interests. He also denied rights to incurable "human vegetables" and using the same logic disqualified certain species from moral consideration. Protection of rare and endangered species became protection of humans to enjoy and benefit from them. Even less deserving of rights were mere things.[22] While many philosophers now find these requirements too limiting, Singer and Fineberg at least deserve credit for helping liberate moral philosophy from its fixation on human beings.

Scholars such as Christopher Stone pushed the boundaries of moral eligibility further to include other aspects of the natural world. Stone saw no logical or legal reason to draw any ethical boundaries whatsoever. Why should the moral community end with humans or even animals? While this idea may seem absurd to many people, so did the extension of certain rights to women and blacks at one point in our history. The extension of rights in this manner would help environmentalists better protect the environment and also reflects the view that nature needs to be preserved for its own sake and not just for the interests of human beings.[23]

Stone's experience with the Mineral King case in the late 1960s stimulated him to write his landmark essay with the title, "Should Trees Have Standing?" which made a case for the extension of moral concern to the plant community. Mineral King was a beautiful valley in the Southern Sierras that was the subject of a proposal by Walt Disney Enterprises for development into a massive ski resort. The Sierra Club saw itself as a longtime guardian of this region and tried to stop the development. But the U.S. Court of Appeals of California ruled that since the Sierra Club was not itself injured it had no standing or legal reason to sue against the development. But something was going to be injured, Stone reasoned, and the courts should be receptive to its need for protection. Thus he argued in his essay that society should give legal rights to forests, oceans, rivers, and other so-called natural objects in the environment, and indeed, to the natural environment as a whole.[24]

The question of moral extensionism and eligibility also comes into play when we talk about the obligations we have to future generations. There is probably a general agreement that it would be wrong to use all of the earth's resources and to contaminate the environment that we pass on to our children. But at what point do we draw the line? How many resources do we leave to future generations and in what condition shall we leave the environment? Do future generations in any sense have any rights to the resources we are presently using and do they have a right to a clean environment? Since they are not yet alive, they cannot lay claim to a livable environment and do not seem, at least on the surface, to have any interests in our present activities.

But, as Joel Fineberg argues, whatever future human beings turn out to be like, they will have interests that we can affect, for good or ill, in the present. The interests that these future generations will have do need to be protected from irresponsible invasions of those interests by present generations. Present generations have some responsibility to save something for the future, so their offspring can enjoy some of the amenities that we presently enjoy. Rather than focus solely on the rights and interests of present individuals, Fineberg argues that we have obligations to consider the good of the continuing human community.[25]

Richard DeGeorge, on the other hand, believes that future persons, either individually or as a class, do not presently have the right to existing resources. Future generations or individuals have rights only to what is available when they come into existence—when their future rights become actual and present. It is only when a being actually exists that it has needs and wants and interests. It then has a right only to the kind of treatment or to the goods available to it at the time of its conception. It cannot have a reasonable claim to what is not available. However, to argue that future generations do not have present rights does not mean that present generations do not have some obligations to try and provide certain kinds of environments and leave open as many possibilities as feasible for future generations. But the needs of the present and already existing people take precedence over consideration of the needs of future generations.[26]

Toward a Biocentric Ethic

Kenneth Goodpaster argues that the extension of rights beyond certain limits is not necessarily the best way to deal with moral growth and social change with respect to the environment. The last thing we need, he states, is a liberation movement with respect to trees, animals, rivers, and other objects in nature. The mere enlargement of the class of morally considerable beings is an inadequate substitute for a genuine environmental ethic. The extension of rights to other objects or to future generations does not deal with deeper philosophical questions about human interests and environmental concerns. Moral considerations should be extended to systems as well as individuals. Societies need to be understood in an ecological context and it is this larger whole that is the bearer of value. An environmental ethic, while paying its respects to individualism and humanism, must break free of them and deal with the way the universe is operating.[27]

John Rodman, a political theorist at California's Claremont Graduate School, protests the whole notion of extending human-type rights to nonhumans, because this action categorizes them as "inferior human beings" and "legal incompetents" who need human guardianship. This was the same kind of mistake that some white liberals made in the 1960s with regard to blacks. Instead we should respect animals and everything else in nature "for having their own existence, their own character and potentialities, their own forms of excellence, their own integrity, their own grandeur." Instead of giving nature rights or legal standing within the present political and economic order, Rodman urged environmentalists to become more radical and change the order. All forms of domestication must end along with the entire institutional framework associated with owning land and using it in one's own interests.[28]

Another philosopher from the University of Wisconsin, J. Baird Callicott, an admirer of Aldo Leopold's land ethic, declared that the animal liberation movement was not even allied with environmental ethics, as it emphasized the rights of individual organisms. The land ethic, on the other hand, was holistic and had as its highest objective the good of the community as a whole. The animal rights advocates simply added individual animals to the category of rights holders, whereas "ethical holism" calculated right and wrong in reference not to individuals but to the whole biotic community. The whole, in other words, carried more ethical weight than any of its component parts. Oceans and lakes, mountains, forests, and wetlands are assigned a greater value than individual animals who might happen to reside there.[29]

Thus biocentric ethics or deep ecology leads the more radical philosophers to devalue individual life relative to the integrity, diversity, and continuation of the ecosystem as a whole. This approach offended many proponents of animal rights, and presumably those who advocated rights for plants and other aspects of the natural community, to say nothing of those whose moral community ended with human society. According to Roderick Nash, this perspective on environmental ethics has created entirely new definitions of what liberty and justice mean on planet earth and an evolution of ethics to be ever more inclusive (see Figure 3.2). This approach recognizes that there can be no individual welfare or liberty apart from the ecological matrix in which individual life must exist. "A biocentric ethical philosophy could be interpreted as extending the esteem in which individual lives were traditionally held to the biophysical matrix that created and sustained those lives."[30]

This new approach to the environment holds that some natural objects and ecosystems have intrinsic value and are morally considerable in their own right apart

FIGURE 3.2 The Evolution of Ethics

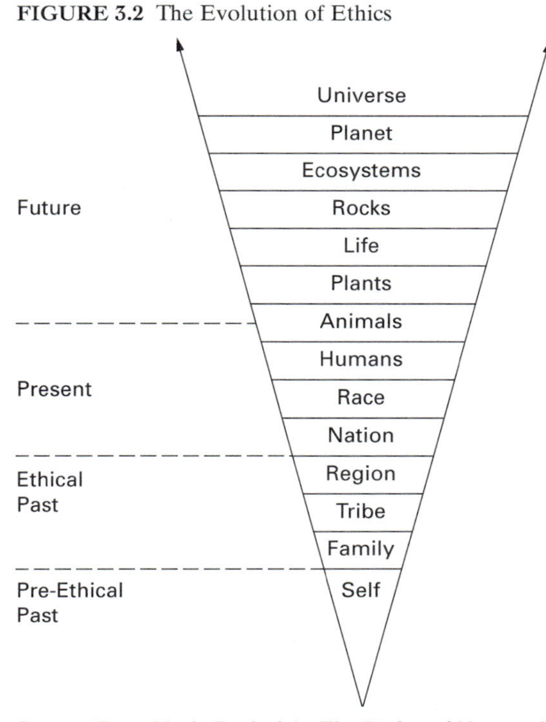

Source: From Nash, Roderick. *The Rights of Nature: A History of Environmental Ethics.* © 1989. (Madison: The University of Wisconsin Press.) Reprinted by permission of The University of Wisconsin Press.

from human interests. Nature is not simply a function of human interests, but has value in and of itself apart from human interests. The naturalistic ethic respects each life form and sees it as part of a larger whole. All life is a sacred thing and we must not be careless about species that are irreplaceable. Particular individuals come and go, but nature continues indefinitely, and humans must come to understand their place in nature. Each life form is constrained to flourish in a larger community, and moral concern for the whole biological community is the only kind of an environmental ethic that makes sense and preserves the integrity of the entire ecosystem.[31]

Nature itself is a source of values, it is argued, including the value we have as humans, since we are a part of nature. The concept of value includes far more than a simplistic human-interest satisfaction. Value is a multifaceted idea with structures that are rooted in natural sources.[32] Value is not just a human product. When humans recognize values outside themselves, this does not result in dehumanizing of the self or a reversion to beastly levels of existence. On the contrary, it is argued, human consciousness is increased when we praise and respect the values found in the natural world, and this recognition results in a further spiritualizing of humans.[33] Thus this school of thought holds that there are natural values that are intrinsic to the natural object itself apart from humans and their particular valuing activities. Values are found in nature as well as in humans. Humans do not simply bestow value on nature, as nature also conveys value to humans.[34]

Such an ethic was advocated several years ago by Aldo Leopold, a wildlife management expert in the state of Wisconsin, who in 1948 talked about a land ethic as a different approach to the natural world. Such a land ethic changes the role of human from conquerer of nature to a plain member and citizen of nature. We abuse land, said Leopold, because we regard it as a commodity belonging to us. When we see land as a community to which we belong, we may begin to use it with love and respect. His most widely quoted precept with regard to land usage is that "a land-use decision is right when it tends to preserve the integrity, stability, and beauty of the biotic community. It is wrong when it tends otherwise."[35]

This approach calls us to a new kind of relationship with the earth, as it does not involve simply measuring water pollution, for example, and taking steps to reverse this pollution; it is a matter of coming to know water through being aware of it in a new way, as a fellow citizen of our earth community. This approach involves (1) the notion that other members of the earth community deserve respect or moral consideration of their own simply because they are there and not just because they are useful to humans, and (2) the notion that a consciously developed relationship with these fellow natural beings is essential to understanding what the ethics of respect demands.[36]

The world of nature is not to be defined in terms of commodities that are capable of producing wealth for humans who manage them in their own interests. All things in the biosphere are believed to have an equal right to live and reach their own individual forms of self-realization. Instead of a hierarchial ordering of entities in descending order from God through humans to animals, plants, and rocks, where the lower creatures are under the higher ones and are ruled by them, nature is seen as a web of interactive and interdependent life that is ruled by its own natural processes. These processes must be understood if we are to work in harmony with nature and preserve the conditions for our own continued existence. Protecting the rain forest is not just a matter of someone from the outside trying to preserve the rain forest as something apart from human existence; it is a matter of seeing oneself as part of the rain forest and acting to protect oneself from extinction.[37]

As stated by Thomas Berry, "Any diminishment of the natural world diminishes our imagination, our emotions, our sensitivities, and our intellectual perception, as well as our spirituality."[38] Human beings are integral with the entire earth and even with the universe as the larger community we belong to by the very nature of our existence. But most of us do not live within this perspective and are not in intimate communion with the natural world. We have become autistic and do not hear the voices of nature. We have been blindly pursuing more and more economic growth and all the while replacing nature with our own view of reality, and in the process have destroyed much that is good and beautiful.

> The mountains and valleys, the rivers and the sea, the birds and other animals, this multitude of beings that compose the natural world no longer share in our lives and we have ceased to share in their lives except as natural resources to be plundered for their economic value. . . . We might wonder how it was that we let this fascinating world be taken from us to be replaced with the grime of our cities disintegrating the very stones of our buildings, as well as increasing the physical and emotional stress under which we live. It was, of course, the illusion of a better life, foisted on us largely through hypnotic advertising and the promise of economic enrichment.[39]

Biocentrism or deep ecology thus accords nature ethical status that is at least equal to that of human beings. From the perspective of the ecosystem, the difference is between thinking that people have a right to a healthy ecosystem or thinking that the ecosystem itself possesses intrinsic or inherent value.[40] Deep ecologists argue for a biocentric perspective and a holistic environmental ethic regarding nature. Human beings are to step back into the natural community as a member and not the master. The philosophy of conservation for Holmes Ralston, for example, was comparable to arguing for better care for slaves on plantations. The whole system was unethical, not just how people operated within the system. In Ralston's view, nothing matters except the liberation of nature from the system of human dominance and exploitation. This process involves a reconstruction of the entire human relationship with the natural world.[41]

The heart of deep ecology is the idea that identity of the individual is indistinguishable from the identity of the whole. The sense of self-realization in deep ecology goes beyond the modern Western sense of the self as an isolated ego striving for hedonistic gratification. Self in a new sense is experienced as integrated with the whole of nature. Human self-interest and interest of the ecosystem are one and the same. There is a fundamental interconnectedness of all things and all events that must be taken into consideration in our thinking and practices.[42]

On close examination, both these approaches prove useful in understanding the relationship between humans and nature. But these approaches treat the environment differently and make different assumptions about the locus of moral consideration. Moral extensionism and eligibility uses the vehicle of rights to extend moral concern to more and more aspects of nature, but these rights are bestowed by human beings; they are not intrinsic to nature itself. Biocentrism or deep ecology assumes nature already has intrinsic value that needs to be recognized by liberating nature from the system in which it is currently trapped. Further, while moral extensionism and eligibility stress the individual to the exclusion of the whole, biocentrism or deep ecology subordinates the individual to the good of the whole.

NEW DIRECTIONS IN ENVIRONMENTAL ETHICS

Pragmatic philosophy offers another way of understanding the environment that shows the possibility of overcoming or undercutting some of these problems inherent in current thinking.[43] Within this philosophical approach, humans and their environment—organic and inorganic—take on an inherently relational aspect. To speak of organism and environment in isolation from each other is never true to the situation, for no organism can exist in isolation from an environment, and an environment is what it is in relation to an organism. The properties attributed to the environment belong to it in the context of that interaction. What we have is interaction as an indivisible whole, and it is only within such an interactional context that humans and their relationship to the environment can be properly understood.

The concept of value in pragmatic philosophy is not something subjective, housed either as a content of mind or in any other sense within the organism, but neither is it something "there" in an independently ordered universe. Objects and situations, as they emerge in human experience, possess qualities that are as real in their emergence as the processes within which they emerge. Value or valuing experiences in relation to these qualities are traits of nature; novel emergents in the context of organism-environment interaction.[44] Thus value is neither subjective nor objective, it is neither extrinsic nor intrinsic, but it emerges in the interaction of the organism and its environment.

Humans are concrete organisms emmeshed in an environment with which they are continuous. Human development is ecologically connected with its biological as well as its cultural world.[45] Growth involves precisely this deepening and expansion of perspective to include ever-widening horizons of the cultural and natural worlds to which we are inseparably bound. This receives its most intense form in Dewey's understanding of experiencing the world religiously as a way of relating one's self with the universe as the totality of conditions with which the self is connected.[46] This unity can be neither apprehended in knowledge nor realized in reflection, for it involves such a totality not as a literal content of the intellect, but as an imaginative extension of the self, not an intellectual grasp, but a deepened attunement to the universe.[47]

Such an experience brings about a change in moral consciousness, not just a change in the intellect alone. It allows one to "rise above" the divisiveness we impose through arbitrary and illusory in-group, out-group distinctions by "delving beneath" to the sense of the possibilities of a deep-seated harmonizing of the self with the totality of the conditions to which it relates. And, this ultimately involves the entire universe, for the emphasis on continuity reveals that at no time can we separate our developing selves from any part of the universe and claim that it is irrelevant. Indeed, while environmentalists may seek to describe "objective" relationships among interacting individuals—human, nonhuman, organic, and inorganic—that make up the biosphere, yet the properties attributed to the individuals are not possessed by them independently of the interactions in which they exhibit themselves.

Nature cannot be dehumanized, nor can humans be denaturalized. Humans exist within and are part of nature, and any part of nature provides a conceivable relational context for the emergence of value. The understanding of "human interests," of what is valuable for human enrichment, has to be expanded not just in terms of long range versus short range and conceivable versus actual, but in terms of a greatly extended notion of human interest or human welfare. Further, to increase the experience of value is not to increase something subjective or within us, but to increase the value ladenness of relational contexts within nature. Dewey's understanding of experiencing the world religiously provides the ultimate context within which such an ethics must be located. While every situation or context is in some sense unique, no situation or context is outside the reaches of moral concern.

Such an ethics cannot be called anthropocentrism. True, only humans can evaluate; and, without evaluation as a judgment concerning what best serves the diversity of valuings, the valuable could not emerge. Further, humans can speak of nonhuman types of experience only analogically in reference to their own. But, though the concept of the valuable emerges only through judgments involving human intelligence, value emerges—either positively or negatively—at any level of environmental interaction involving sentient organisms. While the value level emergent in organism-environment contexts increases with the increased capacity of the organism to experience in conscious and self-conscious ways, as long as there are sentient organisms experiencing, value is an emergent contextual property of situations.

As James stresses, as moral agents we are forbidden "to be forward in pronouncing on the meaningfulness of forms of existence other than our own." We are commanded to "tolerate, respect, and indulge those whom we see harmlessly interested and happy in their own ways, however unintelligible these may be to us. Hands off: neither the whole of truth nor the whole of good is revealed to any single observer."[48] Though some may question the claim that a distinction in levels of value emergence can be made, when push comes to shove, when all the abstract arguments are made, is it not the case that claims of the valuable must be seen in light of its promotion of or irrepressible harm to human welfare, actual or potential? Does anyone really think that the preservation of the spotted owl and the preservation of the AIDS virus have equal moral claim?

This evaluation of the relative merits of the spotted owl and the AIDS virus in terms of their promotion of or harm to human welfare cannot be pushed into a position of anthropocentrism. The attempt to do so comes from a failure to adequately cut beneath the either/or of anthropocentrism/biocentrism. In fact, both/and is closer to the position intended, but even this is inadequate, for it fails to capture the radical conceptual shift that, in making the conjunction, changes the original extremes of the positions that are brought together through pragmatic philosophy. There is no "all or none" involved. It is not the case that all value is such only in relation to humans.

Yet, neither is it the case that all value has equal claim irrespective of its relation to the welfare of humans. Value is an emergent contextual property of situations as long as and whenever there are sentient organisms experiencing, yet the value level emergent in organism-environment contexts increases with the increased capacity of the organism to experience in conscious and self-conscious ways. The biological egalitarianism of biocentrism can perhaps be thought consistently, but it cannot be maintained in practice. Surely one is not willing to move from the theoretical egalitarianism of humans and the AIDS virus to an implementation of such theory in practice.

This view does mean that humans can ignore the value contexts of sentient organisms within nature. To do so is not to evaluate in terms of conflicting claims, but to exploit through egocentric disregard for the valuings of other organisms. We must make judgments that provide protection for the welfare of humans, yet such judgments must consider the value-laden contexts involving other sentient organisms to the largest degree consistent with this goal. And, while this position does not allow for the emergence of value in nonsentient contexts, such contexts are ultimately included in moral deliberation. For, is it really possible to envision any aspect of nature, any relational context in nature, anything in nature that cannot provide a conceivable experiential context for sentient organisms?

The problem is not that environments are ultimately valuable in their actual or potential relational contexts of emergent value, but that valuings and the valuable environments that allow for them are taken far too narrowly. At no point can one draw the line between human welfare and the welfare of the environment of which it is a part and with which it is intertwined in an ongoing interactive process. Here many would object that to value nonsentient nature in terms of its potentiality for yielding valuing experiences is to say that it has merely instrumental value, and if nature is merely an instrument, then no real environmental ethic is possible. Yet, within the above framework, the entire debate concerning instrumental versus intrinsic value is problematic from the beginning.

Everything that can conceivably enter into experience has the potential for being an intrinsic relational aspect of the context within which value emerges, and any value, as well as any aspect of the context within which it emerges, involves consequences and is therefore instrumental in bringing about something further. Thus, Dewey holds that no means-end distinction can be made, but rather there is an ongoing continuity in which the character of the means enters into the quality of the end, which in turn becomes a means to something further.[49]

Moreover, evaluations grow, gain novel direction and novel contexts in the resolution of conflicting and novel interests, and it is with choice and creative resolution in these problematic contexts that morality is concerned. If everything has intrinsic value, then decision becomes somewhat arbitrary. If, for example, every tree has its own intrinsic value and the right to exist, irrespective of its potential for valuing experiences, how can we choose which trees to cut down? Yet, common sense tells us we cannot "save" them all. Arguments must be made, and the literature itself shows that arguments are made in terms of the potential for valuing experiences and, ultimately, when hard choices must be made, for the valuing experiences of humans.[50]

It may be further questioned as to whether the ideal of "fully attained growth" in the union of self and universe merges into an ecocentrism in which value is given to the system rather than to the individual. Here again, these alternatives do not hold within the pragmatic framework. Sometimes the system is more important, sometimes the individual, and this is dependent on the contexts in which meaningful moral situations emerge and the conflicting claims at stake. Further, no absolute break can be made between the individual and the system, for each is inextricably linked with the other and gains its significance in terms of the other. The whole notion of an isolated individual is an abstraction, for diversity and continuity have been seen to be inextricably interrelated. Neither individuals nor whole systems are the bearers of value, but rather value emerges in the interactions of individuals, and wholes gain their value through the interactions of individuals, while the value of individuals cannot be understood in isolation from the relationships that constitute their ongoing development.

While this view cannot tell us what position to take on specific issues, it gives a directive for understanding what is at issue, for making intelligent choices, and for engaging in reasoned debate on the issues. What is needed for an environmental ethics is the development of the reorganizing and ordering capabilities of creative intelligence, the imaginative grasp of authentic possibilities, the vitality of motivation, and a deepened attunement to the sense of concrete human existence in its richness, diversity, and multiple types of interrelatedness with the natural environments in which it is embedded. The resulting deep-seated harmonizing can bring about the change in moral consciousness needed for the implementation of an environmental ethic.

Questions for Discussion

1. Describe an anthropocentric approach to nature. Do you agree that this approach is characteristic of Western industrial societies? Why or why not? What does it mean to say that this approach treats nature as having only instrumental value?
2. What is an economic ethic? Is this ethic consistent with an instrumental approach to nature? What role does nature play in such an economic outlook? What is the goal of a business organization in this view?
3. Are ethics and marketplace performance consistent with each other? Where do ethics and the marketplace performance begin to diverge from each other? What problems was society addressing through the concept of social responsibility?
4. What are the difficulties of using the social responsibility concept to support a broader notion of corporate responsibility to society? Why is it difficult to implement social responsibility on a scale large enough to effectively solve pollution problems?
5. What is the relation between economic growth and environmental problems? Is a trade-off necessary between these two goals? Or is further economic growth dependent on a certain level of environmental protection?
6. With what issues are moral extensionism and eligibility concerned? On what basis are rights extended to animals? Should rights also be extended to the plant community? What about rocks and other inanimate objects?
7. Do future generations have any rights with respect to the resources we are using and the pollution we are producing? What do philosophers think about this issue? What is a common-sense view with respect to this issue?
8. What is a biocentric ethic? How does this approach to environmental ethics differ from extending rights to nature? How does intrinsic value differ from instrumental value? What is the approach of deep ecology to environmental ethics?
9. What are the important components of a relational view between the organism and its environment? How does this view deal with the problem of intrinsic versus instrumental value? How do values come about and where do they reside?
10. What does it mean to say that nature cannot be dehumanized nor can humans be denaturalized? How can humans make value judgments that take nature into account? How can we make morally justifiable distinctions between different parts of nature? Can we make justifiable decisions to use parts of nature for our own purposes?

Endnotes

1. For an extended discussion of the separation of the two fields of business ethics and environmental ethics, see Rogene A. Buchholz and Sandra B. Rosenthal, "Business Ethics/Environmental Ethics: If Ever the Twain Shall Meet," paper presented at the IABS 7th Annual Conference, Santa Fe, New Mexico, March 21–24, 1956.
2. Holmes Rolston III, "Just Environmental Business," *Just Business: New Introductory Essays in Business Ethics,* Tom Regan, ed. (New York: Random House, 1984), pp. 325–343.
3. W. K. Frankena, "Ethics and the Environment," *Ethics and Problems of the 21st Century,* K. E. Goodpaster and K. M. Sayre, eds. (Notre Dame, IN: University of Notre Dame Press, 1979), p. 11.
4. Ibid., p. 17.
5. Charles W. Powers and David Vogel, *Ethics in the Education of Business Managers* (Hastings-on-Hudson, NY: Hastings Center, 1980), p. 1.
6. Robert C. Solomon and Kristine R. Hanson, *Above the Bottom Line: An Introduction to Business Ethics* (New York: Harcourt Brace Jovanovich, 1983), p. 9.
7. Richard T. DeGeorge, *Business Ethics,* 2nd ed. (New York: Macmillan, 1986), p. 15.
8. Milton Friedman, "The Social Responsibility of Business Is to Increase Its Profits," *New York Times Magazine,* September 13, 1970, pp. 122–126.
9. Neil W. Chamberlain, *The Limits of Corporate Responsibility* (New York: Basic Books, 1973), pp. 4, 6.
10. See Benjamin and Sylvia Selekman, *Power and Morality in a Business Society* (New York: McGraw-Hill, 1956).
11. See Lee E. Preston and James E. Post, *Private Management and Public Policy* (Englewood Cliffs, NJ: Prentice Hall, 1975), pp. 12–13.
12. Rogene A. Buchholz, "An Alternative to Social Responsibility," *MSU Business Topics,* Summer 1977, pp. 12–16.
13. William C. Frederick, "From CSR1 to CSR2: The Maturing of Business and Society Thought," Graduate School of Business, University of Pittsburgh, 1978, Working Paper No. 279, p. 1.
14. William C. Frederick, "Toward CSR3: Why Ethical Analysis is Indispensable and Unavoidable in Corporate Affairs," *California Management Review,* XXVII, no. 2 (Winter 1986), 131.
15. Ibid., p. 133.
16. Ibid.
17. R. and V. Routley, "Against the Inevitability of Human Chauvinism," *Ethics and Problems of the 21st Century,* K. E. Goodpaster and K. M. Sayre, eds. (Notre Dame, IN: University of Notre Dame Press, 1979), p. 57.
18. Roderick Frazier Nash, *The Rights of Nature: A History of Environmental Ethics* (Madison, WI: University of Wisconsin Press, 1989), p. 125.
19. Peter Singer, "The Place of Nonhumans in Environmental Issues," *Moral Issues in Business,* 4th ed., William Shaw and Vincent Barry, eds. (Belmont, CA: Wadsworth, 1989), p. 471.
20. Ibid., p. 474.
21. Nash, *The Rights of Nature,* pp. 140–141.
22. Ibid., p. 126.
23. Christopher D. Stone, "Should Trees Have Standing?—Toward Legal Rights for Natural Objects," *Moral Issues in Business,* 4th ed., Shaw and Barry, eds., pp. 475–479.
24. Nash, *The Rights of Nature,* pp. 128–129.
25. "The Environment," *Moral Issues in Business,* 4th ed., Shaw and Barry, eds., pp. 452–453.
26. Richard T. DeGeorge, "The Environment, Rights, and Future Generations," *Ethics and Problems of the 21st Century,* Goodpaster and Sayre, eds., pp. 93–105.

27. K. E. Goodpaster, "From Egoism to Environmentalism," *Ethics and Problems of the 21st Century,* Goodpaster and Sayre, eds., pp. 21–33.
28. Nash, *The Rights of Nature,* p. 152.
29. Ibid., p. 153.
30. Ibid., p. 160.
31. Rolston, "Just Environmental Business," pp. 325–343.
32. Holmes Rolston III, *Philosophy Gone Wild: Essays in Environmental Ethics* (Buffalo, NY: Prometheus Books, 1987), p. 121.
33. Ibid., p. 141.
34. Ibid., pp. 103–104.
35. Nash, *The Rights of Nature,* p. 71.
36. Sara Ebenreck, "An Earth Care Ethics," *The Catholic World: Caring for the Endangered Earth,* 233, no. 1396 (July–August 1990), 156.
37. Ibid., p. 157.
38. Thomas Berry, "Spirituality and Ecology," *The Catholic World: Caring for the Endangered Earth,* 233, no. 1396 (July–August 1990), 159.
39. Ibid., p. 161.
40. Nash, *The Rights of Nature,* p. 10.
41. Ibid., p. 150.
42. Ibid., p. 151.
43. Pragmatism in this context refers to a philosophical school of thought, rather than the American "practical" approach to which it so often refers, a movement that incorporates the writings of its five major contributors: Charles Peirce, William James, John Dewey, G. H. Mead, and C. I. Lewis.
44. The claim by Anthony Weston that pragmatism is a form of subjectivism is misplaced. "Beyond Intrinsic Value: Pragmatism in Environmental Ethics," *Environmental Ethics,* Vol. 7 no. 4 (Winter 1985), p. 321.
45. Bob Pepperman Taylor's objections to Dewey as an environmentalist stem from an ongoing illicit abstraction both of the social, cultural, and biological dimensions of the human in Dewey's philosophy from concrete human existence, and of aesthetic sensibility from the very fiber of human life. See "John Dewey and Environmental Thought: A Response to Chaloupka," *Environmental Ethics* Vol. 12 (Summer 1990).
46. It should perhaps be pointed out here that this is quite different from theistic beliefs, which often foster environmental indifference.
47. And thus, James holds that the broadest forms of moral commitment are held by those who appreciate the religious dimension of existence.
48. "On a Certain Blindness in Human Beings," in *Talks to Teachers* (New York: W. W. Norton, 1958). In "American Pragmatism Reconsidered: William James' Ecological Ethic," *Environmental Ethics,* Vol. 14 (Summer 1992), Robert Fuller argues for a possible panpsychism in James such that even inorganic matter has sentience and thus engages in valuing in at least some rudimentary fashion. This interpretation of James or this kind of justification for concern with inorganic nature is unacceptable in the present context.
49. See the debate between Weston, "Beyond Intrinsic Value," and Eric Katz, "Searching for Intrinsic Value: Pragmatism and Despair" in *Environmental Ethics,* vol. 9, no. 3 (Fall 1987), pp. 231–241.
50. Thus, for example, old-growth forest is valuable in that it has the potential for yielding valuing experiences for individuals. But here problematic situations emerge. For, the old-growth forest, as cut down for lumber, has the potential for yielding valuing experiences for humans as they desire more housing. The old-growth forest, as a forest, has the potential for providing valuing experiences for individuals as they experience the joys of the outdoors. Further,

in these and various other value dimensions of the old-growth forest, its potential for the production of valuing experiences extends not just to actual valuings, or even to the valuings of actual individuals, but to its potential for the production of valuing experiences into an indefinite future. These potentialities for future valuing are not something that can be excluded from the present problematic context, for these potentialities to be affected are not in the future; they are there within the present context, to be affected by our present decisions.

Suggested Reading

Bormann, F. Herbert, and Stephen R. Kellert, eds. *Ecology, Economics, Ethics: The Broken Circle.* New Haven: Yale University Press, 1992.

Botzler, Richard G. *Environmental Ethics: Divergence and Convergence.* New York: McGraw-Hill, 1993.

Brennan, Andrew. *Thinking About Nature: An Investigation of Nature, Value, and Ecology.* Athens: University of Georgia Press, 1988.

Cahn, Robert. *Footprints on the Planet: A Search for an Environmental Ethic.* New York: Universe Books, 1978.

Callicott, J. Baird. *In Defense of the Land Ethic: Essays in Environmental Philosophy.* Albany NY: SUNY Press, 1988.

Chamberlain, Neil W. *The Limits of Corporate Responsibility.* New York: Basic Books, 1973.

Cooper, David E., and Joy A. Palmer, ed. *Environment in Question: Ethics and Global Issues.* New York: Routledge, 1992.

Daly, Herman E., ed. *Economics, Ecology, and Ethics.* San Francisco: W. H. Freeman, 1980.

DesJardins, Joseph R. *Environmental Ethics: An Introduction to Environmental Philosophy.* Cincinnati: Thomson International Press, 1993.

Devall, Bill, and George Sessions. *Deep Ecology: Living as if Nature Mattered.* Salt Lake City: Gibbs M. Smith, 1985.

Elliott, Robert, ed. *Environmental Ethics.* New York: Oxford University Press, 1993.

Gaard, Grets. *Ecofeminism: Women, Animals, Nature.* Philadelphia: Temple University Press, 1993.

Hardin, Garret. *Exploring New Ethics for Survival,* 2nd ed. New York: Viking Press, 1978.

Hargrove, Eugene C. *Foundations of Environmental Ethics.* Englewood Cliffs, NJ: Prentice Hall, 1989.

Leopold, Aldo. *A Sand County Almanac.* New York: Oxford University Press, 1949.

Nash, Roderick Frazier. *The Rights of Nature: A History of Environmental Ethics.* Madison, WI: University of Wisconsin Press, 1989.

Partridge, Ernest, ed. *Responsibilities for Future Generations: Environmental Ethics.* Buffalo, NY: Prometheus Books, 1981.

Passmore, John. *Man's Responsibility for Nature: Ecological Problems and Western Traditions.* New York: Scribner, 1980.

Pepper, David. *Eco-Socialism: From Deep Ecology to Social Justice.* New York: Routledge, 1993.

Pojman, Louis P., ed. *Environmental Ethics: Readings in Theory and Application.* New York: Jones and Bartlett, 1994.

Preston, Lee E., and James E. Post. *Private Management and Public Policy.* Englewood Cliffs, NJ: Prentice Hall, 1975.

Regan, Tom, ed. *Just Business: New Introductory Essays in Business Ethics.* New York: Random House, 1984.

Regan, Tom. *The Case for Animal Rights.* Berkeley: University of California Press, 1983.

Rolston, Holmes, III. *Environmental Ethics: Duties to and Values in the Natural World.* Philadelphia: Temple University Press, 1988.

Rolston, Holmes, III. *Philosophy Gone Wild: Essays in Environmental Ethics.* Buffalo, NY: Prometheus Books, 1987.

Singer, Peter. *Animal Liberation.* New York: New York Review Books, 1975.

Taylor, Paul W. *Respect for Nature: A Theory of Environmental Ethics.* Lawrenceville, NJ: Princeton University Press, 1986.

VanDeVeer, Donald, and Christine Pierce. *Environmental Ethics and Policy Book: Philosophy, Ecology, and Economics.* Cincinnati: Thomson International Press, 1994.

Westra, Laura. *Environmental Proposal for Ethics: The Principle of Integrity.* New York: Rowman, 1993.

CHAPTER 4

The Policy-Making Process

The environment in some sense can be considered a commons that provides resources and services as well as a habitat for everyone on the planet. While the concept of private property does apply to land and in some cases to water, with respect to controlling pollution, this concept does not work very well for all who are affected. Pollution does not respect the boundaries of private property and affects many people who do not own property. Environmental problems such as pollution are often considered to be externalities, which generally means that a third party who is not involved in a marketplace transaction is unintentionally harmed.

The consenting parties to the transaction are able to damage the third party without compensating it, thus the exchange does not adequately reflect the true costs to society. If a river is polluted and the fish are unsafe to eat and the water unsafe to swim in because of transactions between producers and consumers who have an interest in the lowest prices possible and want to dump environmental costs onto someone else, third parties who may not have been involved in any transactions with the producers have to pay the costs in that they may not be able to fish in the river anymore and swimming may be prohibited because it is unsafe.

There is no way the value of the environment or any of its components can be determined through a market process, since there is nothing to be exchanged. People cannot take a piece of dirty air, for example, and exchange it for a piece of clean air on the market, at least given the current state of technology. Some call this inability of market systems to respond to environmental pollution and degradation market failure, but to use this term is not entirely accurate. Market systems were not designed to factor in environmental costs, and it is not fair to blame the system for not doing something for which it was not designed. Property rights are not appropriately assigned with regard to the environment and nature often lacks a discrete owner to look after its interests. The rights of nature can be violated by market exchanges, and as a common property resource, nature can be overused and degraded as it is subject to the tragedy of the commons.

Environmental degradation and pollution are external to normal market processes and are not taken into account in the price mechanism unless these costs are determined by some other process and imposed on the market system. The market system by itself cannot determine the value of the environment and determine the price of clean air or the value of preserving a particular piece of wilderness area in its natural state. These decisions have to be made through some other process and imposed on the market system in order for these values to be internalized and reflect themselves in market decisions. The ecological functions of environmental entities have no value as far as the market is concerned. It is only their economic utility or instrumental value that is of importance in a market economy.

Market systems evolved to serve human needs and wants; they were not constructed to protect the environment. The environment is treated as a source of raw materials to be used in the production process and as a bottomless sink in which to dispose of waste materials. The environment has no value in and of itself, but it is only worth something as it can be used to serve some human purpose such as enhancing living standards through the creation of more and more economic wealth. Market systems are limited in their scope and cannot be used to determine the value of the environment or any of its components.

A tree in the Amazon rain forest, for example, has no value as far as its ecological function is concerned. It has value only in Western market societies as it is cut down to be used for lumber, or left standing because it may be economically beneficial as far as providing nuts and fruits is concerned. Or, perhaps, economic decision makers can be convinced that the tree has value because it is a rare species that may eventually have some kind of medicinal value in providing a cure for some disease. But its ecological value in terms of providing the world with a carbon sink and contributing to the diversity of plant life in that part of the world has no economic value. From a strictly economic point of view, the tree has no value in its natural state and must be cut down and made into something useful or left standing as long as it is seen as productive in some other sense.

The same is true of wetlands, which cover vast areas of some parts of the world, both along coastal areas and inland. These wetlands perform valuable ecological functions that we are only now beginning to understand, yet from an economic point of view they have value only because they contain resources that can be exploited or can be filled in to provide land area for residential or commercial development. Most people probably consider wetland areas as wasteland and fail to appreciate the valuable ecological functions they perform in terms of providing a habitat for fish and other forms of wildlife and acting as water reservoirs to prevent flooding, just to mention a few such functions.

The belief that markets alone will bring about as utopian a social order as we may achieve should be abandoned as far as environmental policy is concerned. Concepts of resource and welfare economics are largely obsolete and irrelevant for environmental purposes. Other concepts and processes must be looked to in order to set priorities in solving environmental and social problems. The air we breathe and the water we drink and the land we live on are in some sense common to humankind. The question is how this commons is going to be managed and to what ends and purposes. Who is going to determine environmental policy and what institution is going to implement the policy that has been formulated?

The policy-making process with respect to the environment in our society includes many different groups and individuals who have various degrees of influence in terms of managing the commons. The term *public policy* is often used to refer to this policy-making process and may be useful in understanding how the commons is managed in our society and how society decides on environmental policy with respect to corporations. The general goals of public policy are determined through a political process in which citizens participate constrained only by rights of the kind protected by the constitution.[1] Over the last several decades, public policy has become an ever more important determinant of corporate behavior, as market outcomes have been increasingly altered through the public policy process. This is especially true with respect to environmental goals of the larger society.

THE NATURE OF PUBLIC POLICY

There are many concepts related to public policy that can be discussed in order to better understand how the commons is managed. Regarding the concept of public policy itself, Anderson, Brady, and Bullock state that a useful definition of public policy will describe public policy as a pattern of governmental activity on some topic or matter that has a purpose or goal. Public policy is purposeful, goal-oriented behavior, rather than random or chance behavior, that is formulated and implemented in order to deal with a public problem. Public policy consists of courses of action, according to these authors, rather than separate, discrete decisions or actions performed by government officials. Furthermore, public policy refers to what governments actually do, not what they say they will do or intend to do with respect to some public problem.[2] With these criteria in mind, the authors offer the following as their definition of public policy.

> A goal directed or purposeful course of action followed by an actor or set of actors in an attempt to deal with a public problem. This definition focuses on what is done, as distinct from what is intended, and it distinguishes policy from decisions. Public policies are developed by governmental institutions and officials through the political process (or politics). They are distinct from other kinds of policies because they result from the actions of legitimate authorities in a political system.[3]

Theodore J. Lowi defines public policy as a government's expressed intention, which is sometimes called purpose or mission. Lowi further points out that a public policy is usually backed by a sanction, which is a reward or punishment to encourage obedience to the policy. Governments have many different sanctions or techniques of control to assure that their policies are followed.[4] Thomas R. Dye defines public policy as whatever governments choose to do or not to do. Dye argues that public policy must include all actions of government and not just stated intentions of either government or government officials. He also points outs that public policy must include what government chooses not to do, as government inaction with respect to particular issues can have as great an impact on society as government action.[5]

Preston and Post offer a much different definition of public policy. They refer to policy, first of all, as principles guiding action, and emphasize that this definition stresses the idea of generality, by referring to principles rather than specific rules, programs, practices, or the actions themselves, and also emphasize activity or behavior as opposed to passive adherence.[6] Public policy, then, refers to the principles that guide action relating to society as a whole. These principles may be made explicit in law and other formal acts of governmental bodies, but Preston and Post are quick to point out that a narrow and legalistic interpretation of the term "public policy" should be avoided. Policies can be implemented without formal articulation of individual actions and decisions. These are called implicit policies by Preston and Post.[7]

The first few definitions are unnecessarily restrictive. Government need not engage in a formal action for public policies to be put into effect. The Preston and Post definition, however, confuses principles and action. Principles can guide action, but the principles themselves are not necessarily the policy. Policy does more appropriately refer to a specific course of action with respect to a problem, but not to the principles that guide the action. Public policy involves choices related to the allocation of scarce resources to achieve goals and objectives. But public policy makers cannot ride off in all directions at once and must make choices among contending allocations of scarce resources. These choices represent courses of action taken with respect to particular problems.

Thus public policy is a specific course of action taken collectively by society or by a legitimate representative of society, addressing a specific problem of public concern, that reflects the interests of society or particular segments of society. This definition emphasizes a course of action rather than principles. It does not restrict such action to government, it refers to the collective nature of such action, and does not claim that each and every public policy represents the interests of society as a whole. Enough interests have to be represented, however, so that the policy is supported and can be implemented effectively.

The public policy agenda is that collection of topics and issues with respect to which public policy may be formulated.[8] There are many problems and concerns that various people in society would like to have acted on, but only those important enough to receive serious attention from policy makers comprise the public policy agenda. Such an agenda does not exist in concrete form but is found in the collective judgment of society, actions and concerns of interest groups, legislation introduced into Congress, cases being considered by the Supreme Court, and similar activities. The manner in which problems in our society get on the public policy agenda is complex and involves many sets of actors.

THE PUBLIC POLICY PROCESS

The specific course of action that is eventually taken with respect to a problem is decided through the public policy process. The term "public policy process" refers to the various processes by which public policy is formed. There is no one single process by which public policy is made in our country.[9] It is made by means of a complex, subtle, and not always formal process. Many agents who do not show up on any formal organization chart of government nevertheless influence the outcome of the public policy process.[10] The public policy process refers to all the various methods by which public policy is made in our society. Formulation of public policy is not limited to formal acts of government, but can be achieved by interest groups that bring issues to public attention and attempt to influence public opinion as well as government.

Even when public policy is formalized by government, however, there still is no single process involved. Public policy can be made through legislation passed by

Congress, regulations created by federal agencies and published in the Federal Register, executive orders issued by the President, or decisions handed down by the Supreme Court. The process of making public policy begins in the society as problems and issues are defined. These issues may find their way into formal institutions for some policy decisions and then are returned to the society again for implementation.[11]

Most public policy that affects business and the environment, however, is the result of formal government action, particularly at the federal level. Interest group pressure and public opinion eventually translate into some kind of legislation and/or regulation on most environmental issues that prescribe a specific form of business behavior. The public policy process allows citizens to express their shared values regarding health, well-being, safety, and respect and reverence for nature, and translates these values into specific policies with regard to some aspect of the environment.

Values are assigned to particular entities in the public policy process and decisions are made about the allocation of resources through a political process. The business of the political process is to pursue the common or public interest of the community, which is separate from the aggregate private interests of individuals as defined by efficient markets. The political process is a complex amalgam of power and influence that involves many actors pursuing different interests who try to persuade and influence others in order to achieve their objectives.

Politics has often been called the art of the possible, meaning a balancing of interests is necessary to resolve conflicts between interests in order to arrive at a common course of action. People usually have to be willing to give up something of what they want in order to reach agreement among all the members of a group. The usual outcome of the political process reflects the principle that no one gets everything of what he or she wants and yet everyone has to get something in order to satisfy oneself that the objective is worth pursuing. Thus compromise and negotiation are necessary skills to participate effectively in the political process.

The function of a political process is to organize individual effort to achieve some kind of collective goal or objective that individuals or private groups find difficult, if not impossible, to achieve by themselves. But people have many different ideas about what ought to be done with respect to environmental or other public policy problems and what goals should be accomplished. The political system manages such conflict by (1) establishing rules of the game for participants in the system, (2) arranging compromises and balancing interests of the various participants, (3) enacting compromises in the form of public policy measures, and (4) enforcing these public policies.[12] The outcome of the political process depends on how much power and influence one has, how skillful one is at compromising and negotiating, and the variety and strength of other interests involved. Decisions can be made by vote where the majority rules, by building a consensus, or by exercising raw power and coercing other members of a group to agree with your course of action.

People pursue their own interests through the political process based on the values they hold relative to the objectives being sought collectively. But these values cannot be expressed directly or precisely, particularly in a representative democracy. Individual preferences are rarely matched because of the need for compromise, and the outcome is highly uncertain because of the complex interactions that take place between all the parties to a transaction. Yet resources for the attainment of public policy objectives are allocated through the political process that combines individual preferences into common objectives and courses of action.

The reason public policy decisions have to be made through a political process is the nature of the goods and services that are provided through the public policy process. These goods and services can appropriately be referred to as public goods and services as distinguished from the private goods and services pursued in the market system. Just as in the market system, these public goods and services are provided to meet the demands of people for these goods and services as expressed through the political system.

Public goods and services are indivisible in the sense that the quantity produced cannot be divided into individual units to be purchased by people according to their individual preferences. For all practical purposes, one cannot, for example, buy a piece of clean air to carry around and breathe wherever one goes. Nor can one buy a share of national defense over which one would have control. This indivisibility gives these goods their public character because if people are to have public goods and services at all, they must enjoy roughly the same amount.[13] No one owns these goods and services individually—they are collectively owned in a sense and private property rights do not apply. Thus there is nothing to be exchanged and the values people have in regard to these goods and services and decisions about them cannot be made through the exchange process.

One might argue, however, that even though public goods and services have these characteristics, they could still be provided through the market system rather than the public policy process. Suppose, for example, the market offered a consumer the following choice: two automobiles in a dealer's showroom are identical in all respects, even as to gas mileage. The only difference is that one car has pollution control equipment to reduce emissions of pollutants from the exhaust while the other car has no such equipment. The car with the pollution control equipment sells for $500 more than the other.

If a person values clean air, it could be argued that he or she would choose the more expensive car to reduce air pollution. However, such a decision would be totally irrational from a strictly self-interest point of view. The impact that one car out of all the millions on the road will have on air pollution is infinitesimal—it cannot even be measured. Thus there is no relationship in this kind of a decision between costs and benefits—one would, in effect, be getting nothing for one's money unless one could assume that many other people would make the same decision. Such actions, however, assume a common value for clean air that doesn't exist. Thus the market never offers consumers this kind of choice. Automobile manufacturers know that pollution control equipment won't sell in the absence of federally mandated standards.

There is another side to the coin, however. If enough people in a given area did buy the more expensive car so that the air was significantly cleaner, there would be a powerful incentive for others to be free riders. Again, the impact of any one car would not alter the character of the air over a region. One would be tempted to buy the polluting car for a cheaper price and be a free rider by enjoying the same amount of clean air as everyone else and not paying a cent for its provision.

Because of these characteristics of human behavior and the nature of public goods and services, the market system will not work to provide them for a society that wants them. When goods are indivisible among large numbers of people, the individual consumer's actions as expressed in the market will not lead to the provision of these goods.[14] Society must register its desire for public goods and services through the political process because the bilateral exchanges facilitated by the market are insufficiently inclusive.[15] Only through the political process can compromises be reached that will resolve the value conflicts that are inevitable in relation to public goods and services.

Value conflicts are more pronounced in the public policy process because of the existence of a diverse value system. There is no underlying value system into which other values can be translated, no common denominator by which to assess trade-offs and make decisions about resource allocation. What is the overall objective, for example, of clean air and water, safe disposal of hazardous waste, or preservation of endangered species? One could say that all these goods and services are meant to improve the quality of life for all members of society. But if this is the objective, what kind of common value measure underlies all these goods and services so that benefits can be assessed in relation to costs, and trade-offs analyzed in view of this common objective of improving the quality of life?

The costs of pollution control equipment, for example, can be determined in economic terms. The benefits this equipment provides should be positive in improving health by reducing the amount of harmful pollutants people have to breathe and improving the asthetic dimension by making the air smell better. Safety may also be enhanced through an improvement of visibility for aircraft. The difficulty lies in translating all these diverse benefits into economic terms so that a direct comparison with costs can be made.

What is the price tag for the lives saved by avoiding future diseases that may be caused by pollution? What is the economic value of having three more years added on to one's life span because of living in a cleaner environment? What is the value of reducing the probability that children will be born with abnormalities because of toxic substances in the environment? What is the appropriate value of being able to see the mountains from one's house in Los Angeles and enjoy whatever benefits this view provides?[16]

The difficulty of expressing all these intangibles in economic terms so that people's preferences are matched should be apparent. But in spite of these difficulties, insurance agents, legal experts, scientists, and agency administrators routinely assign values to human life, ranging from a few dollars to many millions of dollars, depending on the methods used to calculate these values. One of the most precise ways of calculating the value of a human life is to break down the body into its chemical elements. Some experts have determined that the value of a human life on this basis is about $8.37, which has increased $1.09 in six years because of inflation.[17] Obviously, such a method is not acceptable for public policy purposes.

There are at least five ways of determining the value of a human life: (1) calculating the present value of estimated future earnings that are foregone due to premature death, (2) calculating the present value of the losses others experience because of a person's death, (3) examining the value placed on an individual life by presently established social policies and practices, (4) the "willingness to pay" method where people are asked how much they would be willing to pay to reduce the probability of their death by a certain amount, and (5) looking at the compensation people accept as wage premiums for dangerous jobs or hazardous occupations.[18]

The diversity of economic valuation that results from these techniques is not surprising. Such diversity renders the use of analytical techniques such as those described above highly questionable. There seems to be no way to force a translation of the diversity into a common value system that is acceptable, realistic, and appropriate. Should more money be spent on reducing the emissions from coke ovens than on improving highway safety? How much money should be spent on cleaning up existing dump sites for hazardous wastes? For these kinds of public policy questions, the political process seems to be a reasonable way to aggregate the diversity of people's values to make a decision about a course of action when there is no common value system to use for more rational calculations.

The universal motivating principle in the public policy process is the public interest rather than self-interest. This principle is invoked by those who make decisions about public policy. Elected public officials often claim to be acting in the interests of the nation as a whole or of their state or congressional district. Public interest groups also claim to be devoted to the general or national welfare. These claims make a certain degree of sense. When politicians have to make a decision about the provision of some public good or service, they cannot claim to be acting in the self-interest of everyone in their constituency. When goods and services are indivisible across large numbers of people, it is impossible for individual preferences to be matched. Nor can public policy makers claim to be acting in their own self-interest, as such a claim is not politically acceptable. Some more general principle such as the public interest has to be invoked to justify the action.

The definition of the public interest, however, is problematical. The term can have at least four meanings.[19] The public interest can refer to the aggregation, weighing, and balancing of a number of special interests. In this view the public interest results through the free and open competition of interested parties who have to compromise their differences to arrive at a common course of action. The public interest is the sum total of all the private interests in the community that are balanced for the common good. This definition allows for a diversity of interests.

The public interest can also refer to a common or universal interest that all or at least most of the members of a society share. A decision is in the public interest if it serves the ends of the whole public rather than those of some sector of the public, if it incorporates all of the interests and concepts of value that are generally accepted in our society. Such a definition assumes a great deal of commonality as to basic wants and needs of the people who comprise a society.

There is also an idealist perspective as to the meaning of the public interest. Such a definition judges alternative courses of action in relation to some absolute standard of value, which in many cases exists independently of the preferences of individual citizens. The public interest is more than the sum of private interests; it is something distinct and apart from basic needs and wants of human beings. Such a definition has a transcendent character and refers to such abstractions as "intelligent goodwill" or "elevated aspirations" or "the ultimate reality" that human beings should strive to attain. The difficulty with this definition is finding someone with a God-like character who can define these abstractions in an acceptable manner.

Another definition of the public interest focuses on the process by which decisions are made rather than the specification of some ideal outcome. This definition involves the acceptance of some process, such as majority rule, to resolve differences among people. If the rules of the game have been strictly followed, which in a democratic setting means that interested parties have had ample opportunity to express their views, then the outcome of the process has to be in the public interest by definition.

These definitions all have their problems, making an acceptable definition as difficult to arrive at as a specific public policy itself. Most public policies undoubtedly reflect all of these definitions in some manner. Before leaving this subject, one additional caveat must be mentioned. Those in a position of power and influence in society to shape public policy can never really escape their own self-interest and legitimately claim to be acting solely in the public interest or general good of society, however it is defined. Politicians want to get reelected and will vote for those goods and services they believe have an appeal to the majority of their constituency. Public interest groups want to extend their power and influence in society, and might more appropriately be called special interest groups. Thus the definition of the public interest can never be entirely divorced from the self-interest of those making the decisions.[20]

Whatever definition of the public interest is invoked, resources are allocated in the public policy process by a group of decision makers in the public policy process who have been most active and influential in arriving at a common course of action. They are the ones who consciously allocate resources for the production of public goods and services they believe the public wants; that is, those goods and services they believe serve the public interest. If they make the wrong decisions and do not adequately serve the public interest, however it may be defined, they can be held accountable and removed from their position of power and influence.

The average person simply plays the role of citizen by voting for a representative of his or her choice, contributing money to a campaign, writing elected public officials on particular issues, and similar measures. Joining large social movements such as the environmental movement is another way for the average person to exercise political influence. Widespread support for issues such as this has an effect on the voting of elected public officials. Finally, people can join environmental groups or support them with contributions and fulfill a political role in this fashion. Most citizens, however, simply elect others to engage in the business of governing in the public interest and go about their daily tasks with a minimum of political participation.

However, citizens are supposedly sovereign over the public policy process as consumers are supposedly sovereign over the market system. The vote is the ultimate power that citizens have in a democratic system. A public official can be voted out of office if he or she does not perform as the majority of citizens in his or her constituency would like. The citizens can then vote someone else into office whom they believe will make decisions about allocation of resources for production of public goods and services that are more consistent with the citizens' preferences as a whole. In the interim period between votes, citizens can express their preferences and try to influence the outcome of the public policy process either individually, through contact with public officials, or collectively, through interest groups.

There are two problems with this notion of citizen sovereignty that need to be mentioned. One concerns the idea of manipulation, as candidates for elected public office are advertised and packaged as are products. In recent years, television advertising has been used more and more in political campaigns. Are citizens thus being manipulated by these promotional techniques and voting for an image created on television rather than for an individual whom they have little or no chance to know? Has citizen sovereignty been rendered obsolete by the packaging of candidates to appeal to the prejudices of people with little consideration given to the merits of issues important to the election?

Another problem with citizen sovereignty is the bad reputation that the average citizen has with regard to participation in the political process. Voter turnouts are often very low in many elections. Most of those who do vote probably know little about the candidates and the issues at stake in the election. Most people are not interested in public issues much of the time, particularly those that do not affect them directly. Taking such an interest means spending time on political concerns that might be more profitably devoted to the family or to leisure activities. Most citizens do not derive primary satisfactions from political participation, and unlike the marketplace, they do not have to participate to fulfill their basic needs and wants. The cost of participation in public affairs seems greater than the return. People who do not participate thus sacrifice their sovereignty and power to the minority in the society who do have a strong interest in political life and choose to actively participate in the formulation of public policy for the society as a whole.[21]

Thus the major institutional force operative in the public policy process is government, primarily the federal government and to a lesser extent state and local government. Government is the principal institution involved in formulating public policy with respect to environmental concerns that shapes the behavior of business organizations. Many policies do not become public policies until they are adopted, implemented, and enforced by some governmental institution. Government lends legitimacy to policies by making them legal obligations that command the loyalty of citizens. Government policies extend to all people in a society while the policies of other groups or organizations such as business reach only a segment of society. Government also monopolizes the legitimate use of force in seeing to it that public policies are followed by those who are affected. Only government can legitimately imprison violators of its policies.[22]

THE ADMINISTRATIVE PROCESS

To implement laws created by Congress, administrative agencies have been created in many areas of concern. The agencies that have the most direct impact on business are the regulatory agencies. The first such agency was the Interstate Commerce Commission (ICC) created in 1887 to deal with the railroad problem. Then followed other such regulatory agencies dealing with specific industries such as communications, transportation, and financial institutions. In the 1960s and 1970s, Congress enacted hundreds of laws dealing with the environment, civil rights, consumer issues, and other social matters, and created many new agencies to implement this legislation. This new type of regulation has come to be called social regulation to distinguish it from the earlier type of regulation that dealt with a specific industry. Figure 4.1 presents an historical perspective to agency growth, showing the growth of traditional industry regulation in the New Deal era, and the surge of social regulation that is of more recent vintage.

An administrative agency has been defined as "a governmental body other than a court or legislature which takes action that affects the rights of private parties."[23] These agencies may be called boards, agencies, administrative departments, and so forth, but in the regulatory area they are often referred to as commissions. The State Governmental Affairs Committee defined a regulatory commission as "one which (1) has decision making authority, (2) establishes standards or guidelines conferring benefits and imposing restrictions on business conduct, (3) operates principally in the

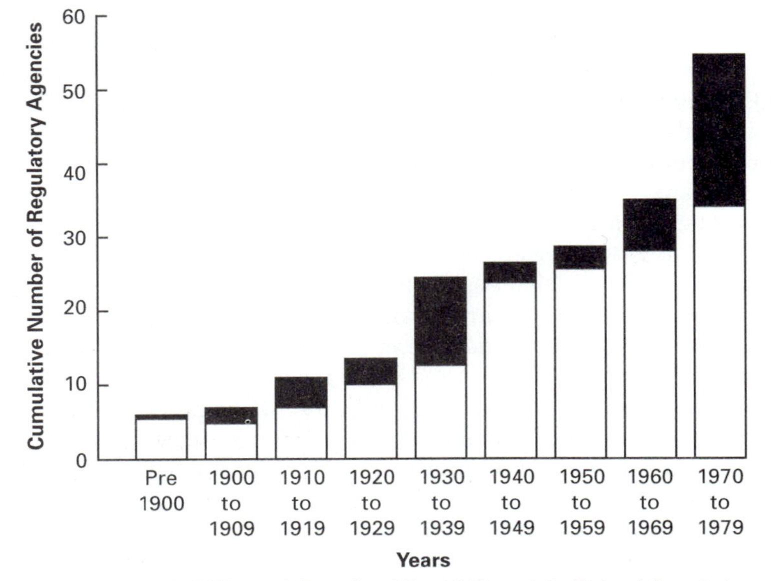

Source: Kenneth Chilton, *A Decade of Rapid Growth in Federal Regulation* (St. Louis, MO: Washington University, Center for the Study of American Business, 1979), p. 5. Reprinted with permission.

FIGURE 4.1 A Historical Perspective to Agency Growth

sphere of domestic business activity, (4) has its head and/or members appointed by the president . . . [generally subject to Senate confirmation], and (5) has its legal procedures governed by the Administrative Procedures Act."[24]

These regulatory commissions have specialized functions to implement governmental policy in specifically defined fields. Congress cannot immerse itself in all the details of each activity regulated or pass legislation that mandates specific forms of business behavior. Thus it passes laws that are broad in scope and more or less sets general goals to be accomplished. The task of implementing these laws is given to the regulatory agencies that are largely composed of so-called experts in areas like safety and health or the environment. Congress, for example, gives the Environmental Protection Agency (EPA) the power to set standards to improve air and water quality, but Congress does not specify what kind of standards should be established and for what substances. It is up to the EPA to determine these standards based on its expertise. In this manner, the EPA and other regulatory agencies make public policy.

Congress creates an administrative agency by passing a statute that specifies the name, composition, and powers of the agency. This statute is called the enabling legislation for the agency. Thus the agencies are theoretically a creature of Congress and accountable to Congress for agency activities. Congress can amend the enabling legislation to change agency behavior. Each House has oversight committees that review the work of the agencies, hold hearings, and propose amendments to the enabling legislation. Congress can also control agency activities through the appropriations process by attaching riders forbidding the agency to spend any money on particular cases.

Agencies are also subject to specific statutes that govern their activities. The Administrative Procedures Act (APA) passed in 1946 specifies formal procedures with which agencies must comply and establishes standards and prerequisites for judicial review of agency action. Agency actions that are going to significantly affect the environment are subject to the National Environmental Policy Act (NEPA) that under certain conditions requires the development of an environmental impact statement before undertaking the action. Finally, the Freedom of Information Act (FOIA) and Government in the Sunshine Act require, with certain exceptions, that agency documents be publicly available and that agency proceedings be open to the public.[25]

Judicial review of agency action is important because many of the regulations issued by agencies that are opposed by business wind up in the courts. Despite congressional oversight, the primary task of assuring that agencies comply with congressional dictates has fallen on the courts. The courts may overturn an agency's action for any of the following reasons: (1) The agency failed to comply with the procedures specified in its enabling legislation, the APA, NEPA, or FOIA; (2) the agency's action conflicts with its enabling legislation and therefore exceeds the scope of its authority; (3) the agency's decision is premised on an erroneous interpretation of the law; (4) the agency's action conflicts with the Constitution; or (5) the agency erred in the substance of its action. The last reason has to do with standards of evidence to support an agency's findings and the consideration of all relevant factors in a decision.[26]

There are two general types of regulatory agencies. Some agencies are independent in the sense that they are not located within a department of the executive branch of government. Since they are not part of the legislative or judicial branch either, a fourth branch of government seems to have emerged that combines the functions of the other three in the making, interpreting, and implementing of legislation. In creating these agencies and making them independent, Congress sought to fashion them into an arm of the legislative branch and insulate them from presidential control. But many Presidents have considered these commissions to be adjuncts of the executive branch and have argued that they should be able to coordinate and direct the independent agencies.[27]

Other agencies are located within the executive branch in one of the cabinet departments. These agencies include the Food and Drug Administration (FDA) as part of the Department of Health and Human Services, the Antitrust Division of the Department of Justice, the Labor-Management Services Administration and OSHA in the Department of Labor, and the National Highway Traffic Safety Administration (NHTSA) in the Department of Transportation. Even here, however, there is some question whether these agencies are subject to presidential influence and guidance or whether they are free to use the regulatory authority granted them by Congress. Some believe that the President's power to appoint and dismiss cabinet officers carries with it an implicit authority to direct actions by regulatory agencies within the executive departments. Others argue that these agencies may accept White House advice but that ultimately they are as independent as the separate regulatory commissions.[28]

Regulatory activities may be pursued in a number of ways, but the rule-making procedure is generally preferred by the EPA and other social regulatory agencies. Rule making is the process of promulgating rules, and results in regulations of greater certainty and consistency, and allows for broader input from the public. The APA definition of a rule is "an agency statement of general or particular applicability and future effect designed to complement, interpret, or prescribe law or policy."[29] Thus rule making is the enactment of regulations that will generally be applicable at some future time.

Under the rule-making process, an agency must first publish a proposed regulation in the Federal Register. The Federal Register is a legal newspaper in which the executive branch of the U.S. government publishes regulations, orders, and other documents of government agencies. It was created by Congress for the government to communicate with the public about the administration's actions on a daily basis. This procedure provides an opportunity for public comment. Any interested individual or organization concerned with a pending regulation may comment on it directly in writing or orally at a hearing within a certain comment period.

The Federal Register gives detailed instructions on how, when, and where a viewpoint can be expressed. After the agency receives and considers the comments, it may publish a final version of the regulation in the Federal Register or discontinue the rule-making procedure. If a final regulation is published, the agency must also include a summary and discussion of the major comments it received during the comment period. The final regulation may take effect no sooner than 30 days following its publication. After a final rule has been adopted by the agency, it is also published in the Code of Federal Regulations. The code contains all the rules and regulations that any given agency has passed over the course of its existence.

Government agencies are very important actors in the public policy process. They combine the functions of the legislative branch in making administrative law, the executive branch through enforcing agency actions, and the judicial branch in ajudicating disputes. The administrative process has grown because of the need for specialized application of the laws Congress passes. The Congress did not wish to increase executive power by giving these functions to the President, and thus created administrative agencies as an alternative. These agencies are subject to control by the other three branches through congressional oversight, the presidential power of appointment and issuance of executive orders, and judicial review. But they also have shown a great deal of autonomy at times in formulating public policy. Thus business is surprised because a law passed by Congress often turns out to be quite different than anticipated when implemented by the agencies.

> Administrative agencies wield power because they constitute mobilizations of resources that can be used to allocate political values. They develop distinctive institutional points of view on what policies are deemed in the public interest. They push unabashedly within political arenas to advance these viewpoints. Moreover, agencies are supported by external political groups as well as opposed by them, and thus they engage fully in the political conflict that inevitably envelopes those possessing power.[30]

This political conflict limits the power agencies can exercise in the government. There are several features in the bureaucratic environment, some of which have been already mentioned, that limit the power of agencies. Because of the separation of powers concept, Congress and the executive branch have some ability to control agency behavior. The press makes the exposure of agency scandal a principal objective. The Freedom of Information Act requires agencies to open themselves to public scrutiny. The power of public interest lobbies and built-in mechanisms of client advocacy also provide limits on agency behavior. Statutory encouragement and protection of whistle blowing is also a factor in the environment. Finally, the institutionalization of citizen participation in decision making provides another check on agency power. This system of checks and balances leads one author to comment that bureaucratic power in the United States is "probably more inhibited than in any other country on earth."[31]

ADMINISTRATIVE STRUCTURE FOR ENVIRONMENTAL POLICY

An administrative structure has been created at the federal level to implement laws related to the environment. Actually, when Congress passes environmental legislation, it is often quite broad so that it needs a great deal of interpretation. Regulations issued by agencies provide this interpretation and are quite specific regarding requirements and procedures to be followed. This process is something of a lawmaking function in and of itself. While the Congress passes statutory legislation regarding the environment, it is up to the agencies to implement them and in the process create what is called administrative law. Several federal agencies are responsible for attaining environmental objectives and developing policies to deal with environmental problems.

Environmental Protection Agency

The main federal agency involved in protecting and enhancing the physical environment and one with much sweeping authority over various aspects of pollution control is the Environmental Protection Agency (EPA). Its responsibilities involve the development and implementation of programs that range across the whole domain of environmental management. The EPA began on July 9, 1970, when President Nixon sent a reorganization plan to Congress that took the various units dealing with the environment from existing departments and agencies and relocated them in a single, new independent agency. The plan became effective on December 2, 1970, when the EPA officially opened its doors.

The EPA was formed from 15 separate components of 5 executive departments and independent agencies. Programs related to air pollution control, solid waste management, radiation, and drinking water were transferred to the EPA from the Department of Health, Education, and Welfare (now the Department of Health and Human Services). The water pollution control program was transferred from the Department of Interior and the authority to register and regulate pesticides from the Department of Agriculture. The responsibility to set tolerance levels for pesticides in food was transferred from the Food and Drug Administration and a pesticide research program came from the Department of Interior. Finally, the responsibility for setting environmental radiation protection standards came from the old Atomic Energy Commission.

The EPA now has responsibility for pollution control in seven areas of the environment: air, water, solid and hazardous waste, pesticides, toxic substances, radiation, and noise. Its general responsibilities in these areas include (1) the establishment and enforcement of standards, (2) monitoring of pollution in the environment, (3) conducting research into environmental problems and holding demonstrations when appropriate, and (4) assisting state and local governments in their efforts to control pollution (see Exhibit 4.1). The EPA is headquartered in Washington, D.C., with regional offices and laboratories located throughout the country (see Figures 4.2 and 4.3).

EXHIBIT 4.1 ENVIRONMENTAL PROTECTION AGENCY
401 M Street SW, Washington, D.C. 20460

Purpose: To protect and enhance the physical environment.

Regulatory Activity: In cooperation with state and local governments, the agency controls pollution through regulation, surveillance, and enforcement in eight areas: air, water quality, solid waste, pesticides, toxic substances, drinking water, radiation, and noise. Its activities in each area include development of: (1) national programs and technical policies; (2) national emission standards and effluent guidelines; (3) rules and procedures for industry reporting, registration and certification programs; and (4) ambient air standards. EPA issues permits to industrial dischargers of pollutants and for disposal of industrial waste; sets standards which limit the amount of radioactivity in the environment; reviews proposals for new nuclear facilities; evaluates and regulates new chemicals and chemicals with new uses; and establishes and monitors tolerance levels for pesticides occurring in or on foods.

Established: 1970

Legislative Authority:

Enabling Legislation: Reorganization Plan No. 3 of 1970, effective December 2, 1970
The EPA is responsible for the enforcement of the following acts:

- Water Quality Improvement Act of 1970 (84 Stat. 94)
- Clean Air Act Amendments of 1970 (84 Stat. 1676)
- Federal Water Pollution Control Act Amendments of 1972 (86 Stat. 819)
- Federal Insecticide, Fungicide, and Rodenticide Act of 1972 (86 Stat. 975)
- Marine Protection, Research, and Sanctuaries Act of 1972 (86 Stat. 1052)
- Noise Control Act of 1972 (86 Stat. 1234)
- Provisions of the Energy Supply and Environmental Coordination Act of 1974 (88 Stat. 246)
- Safe Drinking Water Act of 1974 (88 Stat. 1661)
- Resource Conservation and Recovery Act of 1976 (90 Stat. 95)
- Toxic Substances Control Act of 1976 (90 Stat. 2005)
- Clean Air Act Amendments of 1977 (91 Stat. 685)
- Clean Water Act of 1977 (91 Stat. 1566)

Organization: This independent agency, located within the Executive branch, is headed by an administrator.

Budgets and Staffing
(Fiscal Years 1970–1994)

	70	*75*	*80*	*85*	*90*	*91*	*92*	*93*	*94*
Budget ($ millions)	582	1614	1897	2042	3175	3378	3642	3450	3391
Staff	4093	10440	13045	12590	15587	16241	16874	18131	18439

Note: Budget figures exclude construction grants.

From Ronald J. Penoyer, *Directory of Federal Regulatory Agencies,* 2nd ed. (St. Louis, MO: Washington University Center for the Study of American Business, 1980), p. 27. Reprinted with permission.

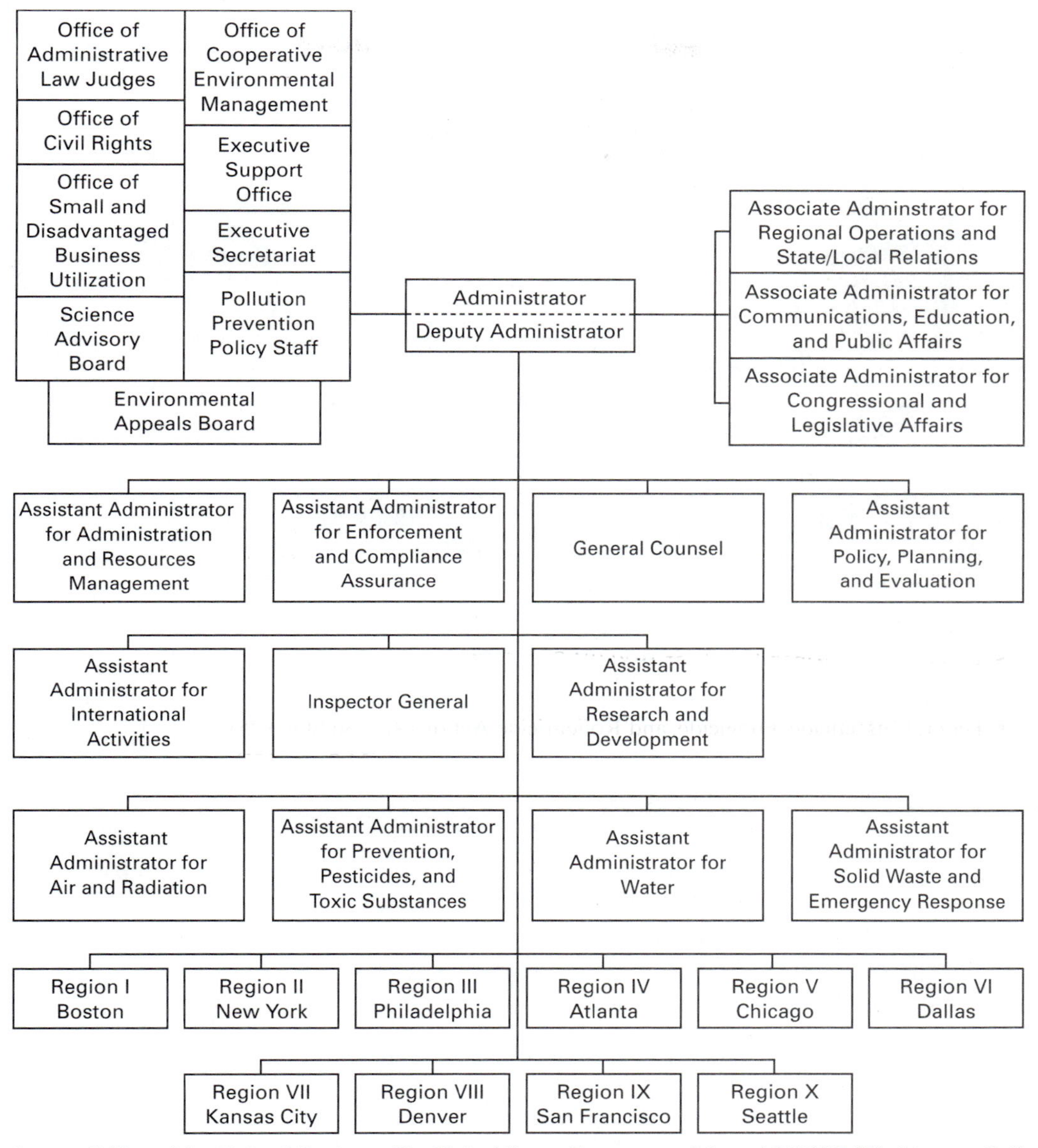

Source: Office of the Federal Register, *The United States Government Manual 1996/97* (Washington, D.C.: U.S. Government Printing Office, 1996.

FIGURE 4.2 United States Environmental Protection Agency Organization Chart

The research arm of the EPA is the Office of Research and Development (ORD), which is responsible for a national research program that pursues technological controls of all forms of pollution. This office directly supervises the research activities of the EPA's national laboratories and gives technical policy direction to those laboratories that support the program responsibilities of the regional offices. The general functions

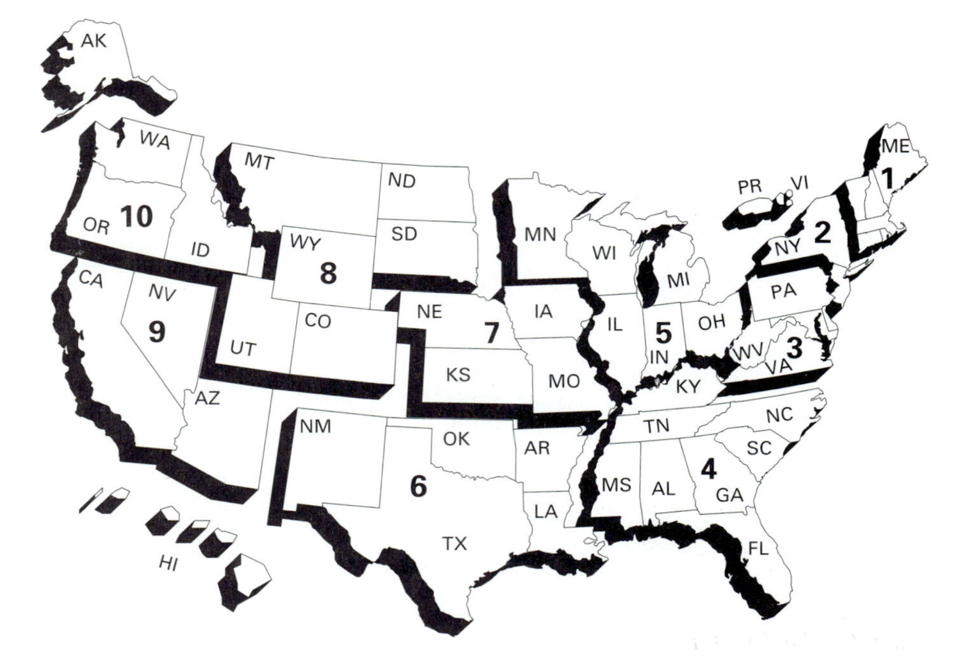

Office of Public Affairs
US EPA Region 1
JFK Federal Building
Boston, MA 02203
(617) 565-3424

Office of External Programs
US EPA Region 2
26 Federal Plaza
New York, NY 10278
(212) 264-2515

Office of Public Affairs
US EPA Region 3
841 Chestnut Street
Philadelphia, PA 19107
(215) 597-9370

Office of Public Affairs
US EPA Region 4
345 Courtland Street NE
Atlanta, GA 30365
(404) 347-3004

Office of Public Affairs
US EPA Region 5
230 S. Dearborn Street
Chicago, IL 60604
(312) 353-2072

Office of Public Affairs
US EPA Region 6
1445 Ross Avenue
Dallas TX, 75202
(214) 655-2200

Office of Public Affairs
US EPA Region 7
726 Minnesota Avenue
Kansas City, KS 66101
(913) 551-7003

Office of Public Affairs
US EPA Region 8
999 18th Street
Denver, CO 80202
(303) 293-1692

Office of Public Affairs
US EPA Region 9
75 Hawthorne Stret
San Francisco, CA 94105
(415) 744-1020

Office of Public Affairs
US EPA Region 10
1200 Sixth Avenue
Seattle, WA 98101
(415) 744-1020
1-800-424-4EPA

Source: Preserving Our Future Today: Your Guide to the U.S. Environmental Protection Agency (Washington, D.C.: EPA, 1991), p. 31.

FIGURE 4.3 United States Environmental Protection Agency Regional Organization

of the office include management of selected demonstration programs, planning for agency environmental quality monitoring programs, coordination of agency monitoring efforts with those of other federal agencies, the states, and other public bodies, and dissemination of agency research, development, and demonstration results.[32]

The Office of Legal and Enforcement Council acts as the EPA's law firm and is responsible for carrying out all of its legal responsibilities and activities. The enforcement philosophy of the EPA is to encourage voluntary compliance by communities and private industry with environmental laws, and encourage state and local governments to perform enforcement actions when needed. If these efforts fail, the EPA is authorized to enforce the law through inspection procedures with respect to all aspects of its responsibilities and criminal investigation units with regard to hazardous waste disposal.

Council on Environmental Quality

This council is composed of three members appointed by the President with the advice and consent of the Senate. The President designates one of the members as chair. The responsibilities of the council include (1) recommending to the President national policies that further environmental quality; (2) performing a continuing analysis of changes or trends in the national environment; (3) reviewing and appraising programs of the federal government to determine their contributions to sound environmental policy; (4) conducting studies, research, and analyses relating to ecological systems and environmental quality; (5) assisting the President in the preparation of the annual environmental quality report to the Congress; and (6) overseeing implementation of the National Environmental Policy Act. The chair also serves as chair of the President's Commission on Environmental Quality.

White House Office on Environmental Policy

This office was established in 1993 to supplement the work of the Council on Environmental Quality. Under the Clinton administration, the new office was expected to give environmental issues more prominence in the administration's deliberations on economic and national policy matters, and have broader influence and a more focused mandate to coordinate environmental policy. The head of the office, for example, was slated to sit in on meetings of the Domestic Policy Council and the National Security Council to represent environmental concerns.[33]

Forest Service

Located in the Department of Agriculture, the Forest Service's mission is to achieve quality land management under the sustainable multiple-use management concept to meet the diverse needs of people with respect to usage of forest resources. The service manages 155 national forests, 20 national grasslands, and 8 land utilization projects on over 191 million acres in 44 States, the Virgin Islands, and Puerto Rico. The nation's need for wood and paper products is supposed to be balanced with other benefits that national forests and grasslands provide such as recreation, wildlife habitat, livestock forage, and water supplies. The service advocates a conservation ethic in promoting the health, productivity, diversity, and beauty of forests and associated lands. It also performs basic and applied research to develop the scientific information and technology needed to protect, manage, use, and sustain the natural resources of the nation's forests and rangelands.[34]

Fish and Wildlife Service

The United States Fish and Wildlife Service is located in the Department of the Interior, and is responsible for migratory birds, endangered species, certain marine mammals, and inland sport fisheries. Its mission is to conserve, protect, and enhance fish and wildlife for the continuing benefit of the American people. The service strives to foster an environmental stewardship ethic based on ecological principles and scientific knowledge of wildlife, works with the states to improve the conservation and management of the nation's fish and wildlife resources, and administers a national program providing opportunities to the American public to understand, appreciate, and wisely use these resources. The service is composed of a headquarters office in Washington, D.C., 7 regional offices, and a variety of field units and installations including more than 500 national wildlife refuges and 166 waterfowl production areas totaling more than 92 million acres, 78 national fish hatcheries, and a nationwide network of wildlife law enforcement agents.[35]

National Park Service

Also located in the Department of the Interior, the National Park Service is dedicated to conserving in an unimpaired state the natural and cultural resources and values of the National Park System for the enjoyment, education, and inspiration of this and future generations. There are more than 365 units in the National Park System including national parks and monuments, scenic parkways, preserves, trails, riverways, seashores, lakeshores, recreation areas, and historic sites associated with important movements, events, and personalities in American history. The service also administers the Wild and Scenic Rivers System, the National Trails System, and the National Register of Historic Places.[36]

Office of Surface Mining Reclamation and Enforcement

The primary goal of this office is to assist states in operating a nationwide program that protects society and the environment from the adverse effects of coal mining, while ensuring that surface mining can be done without permanent damage to land and water resources. This office is located in the Department of the Interior. Since most coal mining states have assumed primary responsibility for regulating coal mining and reclamation activities within their borders, the main objectives of the office are to oversee mining regulatory and abandoned mine reclamation programs in states with primary responsibility, to assist states in meeting the objectives of the Surface Mining Control and Reclamation Act of 1977, and to regulate mining and reclamation activities in those states choosing not to assume primary responsibility.[37]

Bureau of Land Management

This bureau, also located in the Department of the Interior, is responsible for management of more than 270 million acres of public lands located primarily in the West and Alaska with small scattered parcels in other states. The bureau is supposed to develop land-use plans for these areas with public involvement to provide orderly use and development while maintaining and enhancing the quality of the environment. Under certain conditions, the bureau makes land available for sale to individuals, organizations, local governments, and other federal agencies when such transfer is in the public interest. Lands may be leased to government agencies and to nonprofit organizations for certain purposes. The bureau also has minerals management responsibilities for public lands under its jurisdiction.[38]

POLICY TOOLS

There are various tools or methods that can be utilized in regulating the environment that have different impacts and implications for business organizations. The traditional way that regulation is accomplished is through the issuance of rules or standards, as mentioned earlier. This method has also been called command and control regulation. The EPA issues standards that apply to air, water, and other aspects of the environment, and then enforces these standards through inspections and fines for those organizations out of compliance. There are in general two kinds of standards: health standards based on some scientific analysis of what it takes to protect human beings, and technology-based standards related to the kind of technology used to control pollution. The former is used to control air pollution, for example, and the latter is used by and large for surface water pollution.

There are a great many problems with this type of regulation. Many critics argue that command and control regulation is too costly and inflexible. Without some way of weighing the costs in relation to the benefits provided, a good deal of money can be spent without accomplishing much in the way of reducing pollution and protecting human health. This type of regulation also does not give business much flexibility to devise its own methods of reducing pollution that are tailored to its unique situation and may be much less costly. Such regulations also focus largely on "end-of-pipe" solutions; that is, on cleaning up pollution after it has been produced, rather than on providing incentives for business to reduce the waste it produces. For all of these reasons and more, many alternative tools or methods of controlling pollution have been proposed or implemented.

Benefit-Cost Analysis While currently being touted as something new in relation to regulation, benefit-cost analysis has actually been in effect with respect to the EPA since the early 1980s when then President Reagan issued an executive order for benefit-cost analysis that applied to all agencies within the executive branch. The purpose of this analysis was to require an assessment of the potential benefits and costs of each new major regulatory procedure. Agencies were required to choose regulatory goals and set priorities to maximize the benefits to society and choose the most cost-efficient means among legally available options. The implementation of this process did slow down the issuance of regulations during the Reagan years, but was less effective during later administrations.

On the surface, such a procedure makes a great deal of sense, as it is obvious that regulations should not be issued where the costs to business exceed the benefits that are going to be provided for society. But when it comes to actual implementation of the procedure, many difficulties present themselves. The procedure is subject to all of the problems associated with a utilitarian approach to decision making. It is difficult to quantify all the benefits and costs as mentioned earlier in this chapter, making estimates of benefits and costs open to question. Benefit-cost analysis is thus something of a quasi-scientific procedure, as estimates of benefits and costs may vary by factors of a hundred or even a thousand in different studies. The procedure also does not take into account how the benefits and costs are distributed. All the benefits could go to a few

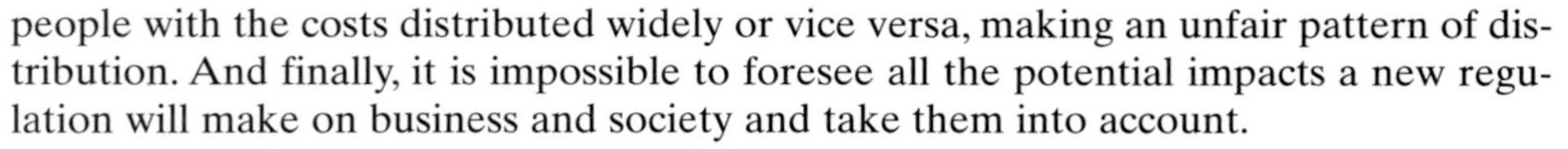

people with the costs distributed widely or vice versa, making an unfair pattern of distribution. And finally, it is impossible to foresee all the potential impacts a new regulation will make on business and society and take them into account.

The procedure is also antidemocratic in nature. It removes decision making with respect to regulation from the democratic process where people can comment on regulations, and places the final decision in the hands of so-called analytical experts. The process creates a bureaucracy that thwarts democracy as well as delays good regulations that provide incentives for business to lower the cost of compliance.[39] It spite of these difficulties, however, Congress wants to pass a law making this procedure mandatory for all regulatory agencies. And administrations can use executive orders, as did former President Reagan, to impose benefit-cost analysis on agencies within the executive branch.

Performance-Based Standards Rather than issue standards related to health or technology, and then mandating design specifications or specific technologies to meet these standards, the EPA has been encouraged to issue performance standards. These performance standards would set certain goals with respect to pollution control, but would allow business the flexibility to meet these goals in the best way they can, using whatever methods are cheapest and quickest for each facility. Use of these standards is supposed to provide greater economic incentives and encourage innovation. Demonstration projects have been established to provide business the flexibility to achieve environmental goals beyond what the law requires while requiring accountability for performance.[40]

Pollution Prevention Another trend that has appeared in recent years is the emphasis on pollution prevention rather than just treating pollution after it has been produced. While various ways to treat waste to protect the environment have been developed over the years, there is a growing realization that avoiding waste altogether, wherever possible, is an even better approach. Traditional end-of-pipe technology offers only a partial solution to the problem of pollution and fails to protect the environment completely. The Pollution Prevention Act of 1990 focuses industry, government, and public attention on reducing the amount of pollution through changes in production, operations, and usage of raw materials. Manufacturers are encouraged to modify equipment and processes, redesign products, substitute raw materials, and make improvements in management techniques, training, and inventory control to produce less pollution.

Risk-Based Management Until recently, the EPA made no attempt to rank environmental problems in relative order of importance. Its regulatory agenda was dictated by a series of separate laws each dealing with a separate problem, and the agency was committed to enforcing all the laws under its jurisdiction. Separate offices were established to implement specific laws, which meant that environmental problems tended to be viewed separately as each program office was concerned primarily with those problems it was mandated to remediate. Questions related to the relative seriousness or urgency of these problems were unasked, and there was little correlation between the relative resources dedicated to different environmental problems and the relative risks posed by those problems.[41]

As the 1990s began, however, an effort was begun to direct more of the EPA's funds and efforts to the most serious environmental problems facing the nation. The agency is trying to identify the most significant environmental risks by improving its scientific data collection activities and analytical methodologies that support the assessment, comparison, and reduction of different environmental risks. These determinations are then implemented through a strategic planning process based on comparative assessments of relative risk posed to society. Program offices have developed four-year strategic plans that are geared toward dealing with high-risk problems. The EPA's budget also reflects these risk-based priorities.[42]

Emissions Trading Trading of emission credits with respect to pollution has been advocated by economists and policy makers for some years as an alternative to command and control regulations. Such a procedure would utilize the power of markets to help resolve environmental problems by allowing business to buy and sell credits in the open market, where the value of the credits would be set by supply and demand conditions. It is argued that such a market-based approach would arrive at a much more realistic value for pollution control activities than other methods. Business would also have more flexibility to make decisions with regard to cleaning up its pollution and would have better incentives than a command and control structure provides.

The market-based approach that has the greatest visibility at the present time and is being implemented on a nationwide scale is the sulfur dioxide allowance trading program that was created as part of the new Clean Air Act to deal with the problem of acid rain. The acid rain program was set up to reduce SO_2 emission from utilities by 10 million tons per year from 1980 levels by the year 2000, leaving the choice of technology largely to the polluters, which are mostly electric utilities. Utilities were allocated allowances based on their historic fuel use and the desired level of emissions with each allowance being equal to one ton of sulfur dioxide. At the end of each year a utility must hold allowances equal to its yearly emissions as allowed under the standards. Those utilities that are cleaner than the standards can sell allowances to those utilities that do not meet the standards. Each utility can decide whether it is better to install pollution control technology or buy allowances if it is not in compliance.

After several years of operation with allowances being traded on the Chicago Board of Exchange, one study has found a relatively low level of participation by utilities in this program. The important variables related to the level of participation included (1) management uncertainty over the treatment of allowances by its public utility commission, (2) the relative cost of allowances compared to other sulfur dioxide control strategies, (3) public opinion in the utility's region that might be strongly against business buying the right to pollute, (4) demand growth, and (5) innovativeness of the utility and its management.[43]

The concept of market-based trading has many other potential applications for other pollutants. In 1995, Southern California established a so-called smog exchange where companies could trade pollution credits on the open market.[44] Such a system has also been proposed as a way to deal with automobile emissions.[45] However, the basic arguments that such trading is a more efficient solution and that industry participants will embrace it as more consistent with their norms and goals than are traditional command and control regulations are not sufficient. Barriers to implementation of a trading system need to be carefully identified and solutions to overcome these developed in order for a such a system to work appropriately.

Regulatory Negotiation and Consensus-Based Rule Making The use of consensus-based decision processes with respect to regulations is a strategy to reduce the usual adversarial relationship between business and government. Regulatory negotiation is the most formal of these approaches and is a process where the EPA and representatives of all the major groups affected by a particular regulation try to reach agreement on regulatory requirements. With respect to business participation, this process usually involves technical people instead of lobbyists who work on the legislative level. The process is said to improve the quality of rules by making them more cost-effective, and it increases public acceptance and minimizes litigation. Even when full agreement cannot be reached, regulatory negotiation can help identify issues and options, educate interested parties, and narrow areas of dispute.

The EPA has also experimented with other less formal methods to consult with affected parties, promote useful information exchange, and find common ground on controversial issues related to regulation. These methods range from continuous policy dialogue to ad hoc discussion forums to public meetings and focus groups. After a number of years of successful experimentation with consensus-based approaches, the EPA now routinely evaluates the appropriateness of using consensus-based rule making every time it issues or revises a regulation. As of June 1, 1995, the EPA was supposed to examine all regulations currently under development and identify candidates for regulatory negotiation and other forms of consensus-based decision making.[46]

Takings Legislation Such legislation deals with impacts of regulation on property owners. The idea behind this legislation is based on the Constitution, which says that government shall not take property from its citizens without fair compensation. When regulations affect the ability of a landowner to develop property and thus affect the value of those landholdings, this is in effect a taking, say proponents of such legislation, and the government should compensate the property holder the same as when land is taken for the building of a new highway. Thus if regulations pertaining to preservation of wetlands, for example, prevent a landowner from developing the property, the landowner should be compensated accordingly. Some proposals would require such compensation when any federal regulation lowers the value of property by 10 percent or more. Traditionally, courts have allowed compensation only in cases where government wiped out all, or nearly all, of a property's value.[47]

Critics of such legislation argue that government will not be able to afford this compensation, so implementation of regulations related to wetlands, for example, will be halted. Such legislation also does nothing about other kinds of takings such as those of corporations that affect the health and safety of citizens as well as property. And it says nothing about givings as when the government subsidizes logging, mining, and charges below market rates for grazing land in the western part of the country. It is also argued that such legislation will most likely become a welfare program for wealthy landowners and developers, who will file bogus claims relative to development of worthless property and claim compensation from the government.[48]

Environmental Justice The movement related to environmental justice raises questions about current methods of plant siting and placement of waste disposal facilities. These questions have to do with the proportion of the poor, particularly blacks, living near plants that emit dangerous pollutants and near waste disposal facilities.

Those concerned about environmental justice in this regard have two major concerns: (1) that a few individuals are forced to bear the external costs of industrial processes from which the public at large receives benefits, and (2) that a disproportionate percentage of these individuals are minority or low-income citizens.[49]

Sitings are currently done by both government and industry utilizing a DAD (decide, announce, and defend) strategy. Decisions to site risky facilities are often a *fait accompli* by the time communities learn about them. Intervenors are often then viewed as obstructionists rather than as representative of a reasonable point of view.[50] Proponents of environmental justice argue that compensating benefits ought to be provided to those who live near locally undesirable facilities. These benefits could take the form of (1) direct payments to affected landowners, (2) host fees paid into a community's general revenue fund that may be used to finance a variety of public projects or to lower property taxes, (3) grants for improving local health-care delivery systems and education, or (4) the provision of parks and other recreational amenities.[51]

THE PLURALISTIC PROCESS

The government, however, is not the only institution in our society that wields influence with respect to environmental policy. Our society is often characterized as pluralistic, meaning that it is relatively open as far as public participation is concerned. A pluralist society is composed of a number of groups, all of which, to varying degrees, wield influence in the policy-making process. These organizations can quite properly be called interest groups because they form around shared interests. People organize such groups and join or support them because they share common attitudes and values on a particular problem or issue and believe they can advance their interests better by organizing themselves into a group rather than pursuing their interests individually. These groups compete for access to formal institutions of decision making such as government and compete for the attention of key policy makers in the hope of producing outcomes that favor their interests.

Such interest groups are thus conveyors of certain kinds of demands that are fed into the public policy process. They fill a gap in the formal political process by representing interests that are beyond the capacities of individuals acting alone or representatives chosen by the people. At times they perform a watchdog function by sounding an alarm whenever policies of more formal institutions threaten the interests of their members. They generate ideas that may become formal policies of these institutions and help to place issues on the public policy agenda. The importance of these interest groups or associations in American life was recognized by Alexis de Tocqueville many years ago in his famous book on American democracy.

> Americans of all ages, all conditions, and all dispositions, constantly form associations. They have not only commercial and manufacturing companies, in which all take part, but associations of a thousand other kinds—religious, moral, serious, futile, extensive or restricted, enormous or diminutive. The Americans make associations to give entertainments, to found establishments for education, to build inns, to construct churches, to diffuse books, to send missionaries to the antipodes, and in this manner they found hospitals, prisons, and schools.[52]

The way a problem gets identified in a pluralistic system is for people who are concerned about the problem to organize themselves or join an existing organization to pursue their particular interests in the problem. If the problem is of widespread concern, and the group or groups dealing with it can attract enough financial and other kinds of support, the problem may eventually become public as people become aware of it and show varying degrees of support. Eventually government or other institutions may pick up on the problem and translate the issues being raised into formal legislation or other policy actions. These interest groups then continue to exercise influence in helping to design public policies to deal with these problems.

Thus in the pluralist model, problems are identified and policies designed in a sort of bottom-up fashion—concern about a problem can begin anywhere at the grassroots level in society and eventually grow into a major public issue that demands attention. Public policy reflects the interests of these groups, to some extent, and as groups gain and lose influence, public policy is altered to reflect the changing patterns of group influence. The public interest takes shape through the pulling and tugging that goes on between special interests. Public policy is the result of the relative influence of groups in the policy-making process, and results from a struggle of these groups to win public and institutional support. As one theorist claims, "What may be called public policy is the equilibrium reached in this [group] struggle at any given moment, and it represents a balance that the contending factions or groups constantly strive to weigh in their favor."[53]

In theory, a pluralistic system is an open system. Anyone with a strong enough interest in a problem can pursue this interest as far as it will take him or her. Membership in a particular social or income class or of a particular race does not shut one out from participating in the public policy process. Power is diffused in a pluralistic system and dominant power centers are hard to develop in such a competitive arrangement. The existence of many interest groups also provides more opportunities for leadership, making it possible for more people with leadership ability to exercise these talents. There are many opportunities for people to become political entrepreneurs who perform an organizing function by bringing people together with similar interests, a process analogous to that of economic entrepreneurs.[54]

But interest groups themselves, particularly as they become large, tend to be dominated by their own leadership. This leadership usually formulates policy for the group as a whole, and the public stance of an interest group often represents the views of a ruling elite within the interest group itself rather than all of the rank and file membership. Many interest groups may provide few, if any, opportunities for members to express their views on issues facing the group. Interest groups in many cases also draw most of their membership from better educated, middle- or upper-class segments of society. Many minorities and particularly the poorer elements of society are not adequately represented. Their problems are likely to be ignored, and some groups cannot advance their interests even in a pluralistic system unless championed by other people who are more likely to participate in public policy making.

Improved public policy decisions should also result from such a structure, since more people, particularly those closest to the problem, have an input in decision making. Yet a pluralistic system is a system of conflict because interest groups compete for attention and influence in the public policy process, and such competing interests do not necessarily result in the best public policy decisions. Conflict can get out of control and result in social fragmentation, making a policy decision for society difficult to reach. This is particularly true when interest groups are unwilling to compromise, in which case reaching a public policy decision for society as a whole may be impossible. Furthermore, some interests, as stated above, are not adequately represented.

A pluralistic system does seem to allow for more interests to be represented than alternative models of society, as more people should have a chance to promote their particular values and interests and have a chance to govern society. This is a mixed blessing, however, as the more pluralistic a society becomes, the more diverse will be the interests represented, and the less clear will be the direction in which society is moving. The lack of central direction for society, which an elite provides, can be a disadvantage of pluralism as society is pulled to and fro by the competition of many different interests with varying degrees of power and influence. Thus a society may find it increasingly difficult to formulate possible solutions to complex policy questions.

Some observers have characterized our society as one of interest group pluralism, whereby the federal government is subject to the pressures of special interest groups. Because of the changes in the seniority system in Congress and the proliferation of subcommittees, Congress has become a collection of independent power centers. The interest groups can thus take their case directly to individual congressmembers and establish close working ties with the subcommittee(s) in their areas of interest. The result is the infamous "iron triangle" composed of the interest groups, the congressional subcommittee, and the relevant federal agency, which becomes the focus of public policy making. This kind of process encourages government to act on individual measures without attention to their collective consequences. Policy is not made for the nation as a whole, but for narrow autonomous sectors defined by the special interests. While these groups may claim to be acting in the public interest, such claims are suspect.

> . . . the problem with the so-called public interest groups is not their venality but their belief that they alone represent the public interest. The confidence these groups have had in pursuing their numerous and sometimes far-reaching missions is not always warranted, especially when their activities—and their demands—are scrutinized in the context of the full effects of the government regulations which they so often instigate or endorse with tremendous zeal.[55]

Another problem with interest group pluralism is the removal of public policy making from public scrutiny. Decisions are made behind closed doors, effectively removed from popular control. As stated by Everett Carl Ladd, "The public cannot hope to monitor the policy outcomes that result from the individual actions of 535 Senators and Representatives operating through a maze of iron triangles."[56] The solution to this fractionalism, according to some observers, is a revitalized party system where the claims of interest groups can be adjusted to mesh with a coherent program that represents more of a national interest. The proliferation of interest groups makes necessary strengthened parties that can cope with the multiple organized pressures of interest group pluralism.

ENVIRONMENTAL GROUPS

Several environmental groups have been influential in the public policy process to place environmental issues on the agenda and support legislation and regulation to deal with environmental problems. Many of the ideas that these groups have generated have found their way into laws and thousands of regulations that affect business. They employ lawyers, economists, ecologists, and systems analysts to press for additional laws and regulations to promote their interests. And they challenge official interpretations of environmental impacts regarding government projects and plans that affect the environment.

These groups that are dedicated to protection and preservation of the environment often seem to wield an influence far beyond what their numbers would suggest. They have pressed for more and more environmental legislation and have successfully expanded the scope of environmental regulations. Through strong lobbying and litigation efforts, they managed to dramatically expand the role of government in virtually all aspects of environmental decision making. Many of them form coalitions that can be very effective in preventing environmental laws from being repealed or watered down by a hostile Congress or administration and try to mitigate the effects of staff and budgetary cutbacks at the agency level. A brief discussion of several of these groups will help to understand their interests and how they operate in the policy-making process.

Audubon Society: The Audubon Society was founded in 1905 as an organization that fights to preserve thousands of acres of wildlife habitat, works to protect endangered species, and teaches generations of children about the wonders of nature through a series of ecology camps. In the mid-1980s, the organization tried to diversify into new issues such as pollution and birth control in order to broaden its appeal. This change has created something of an image problem for the organization as it has strayed from its original mission of concern for wildlife.

Environmental Defense Fund: Best known for its analysis and sponsorship of novel solutions to environmental problems, such as market-based incentives for pollution control. In recent years, EDF has placed less emphasis on lawsuits and regulation and has stressed market-oriented solutions to environmental problems. Worked with McDonald's in getting the hamburger chain to abandon its "clamshell" sandwich containers. In 1992, EDF and General Motors announced a technical and policy dialogue, the first result of which was a proposed cash-for-clunkers program.

The Friends of the Earth: Founded in 1971, the Friends of the Earth is a tax-funded environmentalist organization that is similar in outlook and activities to the Sierra Club. The group is a major participant in the no-growth movement and supports government ownership of land to ensure that wealth is wisely passed from generation to generation. It has been lobbying for a national population policy and eventual population stabilization.

Greenpeace USA: An activist group that often takes direct action against organizations that inflict damage on the environment and wildlife. Often works in conjunction with its larger umbrella organization, Greenpeace International, which operates in 30 countries, has 1,000 employees, and 5 million contributors. Responding to criticism that some of its actions are irrelevant to the policy-making process, Greenpeace has recently engaged in more low-profile work such as lobbying and producing scientific research. It has also created alliances with citizens, consumers, and business and industry in order to produce better results with respect to environmental protection.

The National Wildlife Federation: This organization is one of the largest environmental groups having about 1.3 million members in 1994 with a staff of around 650 people. Many of its achievements include legislation that prohibits individuals from owning and using private property, by banning various areas from private ownership and development. Other efforts have to do with blocking the use of coal, nuclear power, natural gas, and hydroelectric power. Works to build bridges to business organizations through the Corporate Conservation Council that issues awards for outstanding environmental performance and engages in other activities related to the business sector.

Natural Resources Defense Council: The principal strategy of this group is to block economic development in the courts by suing firms for failure to pay adequate attention to the hundreds of environmental laws and regulations affecting their operations. The organization also sues the federal government over improper preparation of an environmental impact statement or for violations of the Endangered Species Act as well as for alleged violation of other laws and regulations. Involved in the controversy over the pesticide Alar by releasing a report raising concerns about the pesticide that received national attention.

The Nature Conservancy: Rather than lobby or engage in other political action, this group raises money to buy land or accepts gifts of land to hold in its natural state. The organization tries to protect places that desperately need protection and preservation by caring for the land it comes to own and keeping it from being developed. In this manner the organization protects wetlands, remote desert areas, pristine lakes, and other areas of concern. Through gifts, trades, and purchases of ecologically sensitive habitat, the organization has amassed the largest private sanctuary system in the world, managing over 2 million acres around the country.

The Sierra Club: This group lobbies Congress on dozens of different issues related to the environment ranging from nuclear energy to wetlands preservation. It opposes the licensing, construction, and operation of new nuclear fission plants, and favors the reduction of society's dependence on nuclear power. The club also has expanded its agenda to include endangered animals, hazardous waste disposal, and landfill-permitting cases. The organization is also a major player in various free trade debates. The Sierra Club is one of the most influential environmental groups in the country.

Wilderness Society: This organization is dedicated to preserving wilderness and wildlife, and protecting the country's prime forests, parks, rivers, and deserts. Recently closed some of its field offices where much of the work of checking up on the Forest Service and other government agencies was accomplished. Concerned about logging, grazing, and mining reform, and lobbies the federal government to promote its interests in protecting wilderness areas. Has also branched out into other areas such as global warming.

World Wildlife Fund: Originally chartered by international aristocrats to keep big game from being hunted to extinction, it is presently concerned with protecting large animals all over the world from poaching and development. Lobbies for trade sanctions against countries that promote imports of animal parts such as elephant tusks for ivory carvers. Has recently touted a human-oriented approach that tries to make preservation economically palatable to local farmers and hunters.

In recent years, many of these groups have strayed from their core competencies by diversifying the range of issues they address in hopes of appealing to a wider audience. As more environmental organizations began to address the same issues, competition for members and funds intensified, leading to a slowing and in some cases a reversal of growth. Diversification has also led to more levels of bureaucracy and a less than through understanding of relevant issues, all of which has diminished the popularity of these groups as well as their influence on the policy-making process. Much like their corporate counterparts, many of these organizations could benefit from downsizing and concentrating their efforts on core competencies.[57]

CONFLICT RESOLUTION

Because the public policy process is open to public participation, and because of the existence of powerful environmental groups mentioned above, there is often a great deal of conflict over the policies and programs of the government, as well as the shape of legislation and regulation that affects business. Major sources of conflict involve competing resource demands, differences in values regarding the relative worth of resources, and uncertainties regarding the costs, benefits, and risks involved in proposed actions. Because of the potential for conflict on environmental issues, techniques for resolution of conflict become of great importance in order to make efficient and effective decisions that will benefit the public at large and preserve our environmental resources. The choice of a decision-making technique can have a major effect on whether the decision achieves its objectives and is accepted by the publics that have a stake in the outcome.

There are several categories of conflict resolution as shown in Exhibit 4.2.[58] The notion of conflict anticipation has to do with the early identification of potential sources of conflict so that these problems may be studied and mitigated if possible before positions become hardened and an adversarial situation develops. The scoping

EXHIBIT 4.2 Types of Environmental Conflict Resolution

Type	*Definition*	*Examples*
Conflict anticipation	A third party identifies potential disputes before opposing positions are fully identified	Scoping or screening process in impact assessment identifies likely problems and affected groups
Joint problem solving	Ongoing group meetings discuss and clarify issues and resolve differences; agreements reached are informal	Structured workshops; adaptive environmental assessment; environmental planning citizens' advisory committees
Mediation	Formal negotiations between empowered representatives of constituencies; mediator facilitates but does not impose settlement	Technical meetings to seek settlements; facilitator uses a variety of negotiating and mediating techniques
Policy dialogues	Meetings to discuss and resolve differences between conflicting policy-making agencies; results become advisory to official policy-making bodies	Interagency advisory committees; ad hoc meetings between members of different governmental agencies
Binding arbitration	Formal arguments presented by opposing parties; arbitrator imposes settlement that parties have previously agreed to abide by	Labor-management contract arbitration; court arbitration hearings

Source: Walter E. Westman, *Ecology, Impact Assessment, and Environmental Planning* (New York: Wiley, 1985), p. 120.

process in environmental impact assessment is an example of this kind of conflict resolution. Identifying potential conflicts before they get out of hand gives the disputing parties an opportunity to work out compromises and solutions at early stages of a project proposal. Even if disputes are not resolved at this stage, at least differences can be explored and put on the table, so to speak.

The technique of joint problem solving involves the making of an informal agreement among the contending parties, which can then be considered in a more formal sense for possible adoption by decision makers. This process typically starts early and continues throughout the full term that is necessary for a decision to be made. Thus ongoing group meetings are often held throughout the decision-making process to clarify issues and resolve differences informally. In this way the various positions can be aired without formal commitment, and the parties to the dispute can develop an acquaintance with each other and perhaps even come to develop some degree of trust, which would serve them well at later stages when formal decisions have to be made.

Environmental mediation is a formal process of negotiation among officially recognized representatives of affected constituencies. There has to be a shared willingness among the parties to a dispute to attempt negotiation or else the technique may not resolve differences. The mediator may be asked to clarify areas of agreement and disagreement and suggest possible solutions to the conflict and ways to implement these solutions. However, the final agreement must be made by a separate decision-making body in order for the decision to be binding on the parties involved. The mediator facilitates but cannot impose a settlement on the disputing parties. Mediation is often used in labor disputes and may also be of use in environmental conflicts.

Policy dialogues involve informal forums for discussion where differences regarding governmental policies may be resolved and where advice may be provided to government agencies. The discussants could be representatives for the different agencies that are involved in the policy making on an issue or outside experts who have been asked to submit a report to the policy-making body. Interagency advisory committees are an example of this type of conflict resolution. These committees are often formed when agencies differ over proposed actions and have no other way to resolve the dispute.

Finally, binding arbitration involves an agreement among the opposing parties to abide by the decision of the arbitrator. Thus the arbitrator imposes a decision on the parties to the dispute. This process gives the disputing parties a chance to have their side of the story heard, and the arbitrator makes the decision after he or she has heard all the contending positions. Choice of the arbitrator is crucial in this technique, which may be why it is not used very often to settle environmental disputes. It may be difficult to find objective arbitrators who don't already have some fairly strong positions on many environmental issues.

Questions for Discussion

1. Why are markets themselves not able to respond effectively to environmental issues? Is public policy a more useful concept to understand how the commons is managed in our society? Why or why not?
2. Which definition of public policy makes the most sense to you personally? What essential elements are there to your definition? How does public policy differ from business policy?

3. What values can be expressed through the public policy process that cannot be expressed through markets? Is there a difference between the way people act as consumers and the way they act as citizens?
4. Comment on the statement that "ideas or convictions that can be supported by reasons in the political process are different from wants and interests satisfied in markets." Do you agree or disagree with this statement?
5. What is the function of a political process? Why do people participate in this process? What is the task of a political system? What skills and abilities are necessary to function effectively in this process?
6. What are public goods and services? How do these differ from private goods and services? Why can't public goods and services be provided through market exchange? Give examples to support your answer.
7. How are value conflicts resolved in the public policy process? How does this process differ from the way value conflicts are resolved by the market? Which process is more efficient?
8. Describe the administrative process. How does administrative law differ from statutory law? Does business have to concern itself with both levels of lawmaking? Why or why not?
9. Why are administrative agencies created? What functions do they perform? What laws or statutes govern the behavior of these agencies? What checks and balances exist as far as agency power is concerned?
10. How are regulatory activities carried out? What types of activities are generally used with respect to environmental regulation? How do these activities work and where can business executives and citizens have an input to the process?
11. What is a pluralistic system? How does this kind of system work with regard to public policy? What are interest groups and what functions do they perform?
12. What are the advantages and disadvantages of a pluralistic system? On the whole, do you believe better public policy comes from this kind of system? What is interest group pluralism? Does this term adequately characterize our society?
13. What is conflict resolution? What are the different categories of conflict resolution presented in the chapter? Which seem most effective from your point of view? Which are most relevant to environmental policy making?
14. Looking at the public policy process as a whole, would you say that the commons is managed as well as could be expected through this process? What are the major problems with regard to the public policy process? How could the system be improved?

Endnotes

1. Mark Sagoff, *The Economy of the Earth* (Cambridge: Cambridge University Press, 1988), p. 70.
2. James E. Anderson, David W. Brady, Charles Bullock III, *Public Policy and Politics in America* (North Scituate, MA: Duxbury Press, 1978), pp. 4–5.
3. Ibid., p. 5.
4. Theodore J. Lowi, *Incomplete Conquest: Governing America,* 2nd ed. (New York: Holt, Rinehart and Winston, 1981), p. 423.
5. Thomas R. Dye, *Understanding Public Policy,* 3rd ed. (Englewood Cliffs, NJ: Prentice Hall, 1978), p. 3.
6. Lee E. Preston and James E. Post, *Private Management and Public Policy: The Principle of Public Responsibility* (Englewood Cliffs, NJ: Prentice Hall, 1975), p. 11.
7. Ibid.
8. Ibid.
9. Anderson, Brady, Bullock, *Public Policy,* p. 6.

10. B. Guy Peters, *American Public Policy: Promise and Performance,* 2nd ed. (Chatham, NJ: Chatham House, 1986), p. vii.
11. Ibid.
12. Dye, *Understanding Public Policy,* p. 23.
13. John Rawls, *A Theory of Justice* (Cambridge, MA: Harvard University Press, 1971), p. 266.
14. Gerald Sirkin, *The Visible Hand: The Fundamentals of Economic Planning* (New York: McGraw-Hill, 1968), p. 45.
15. James Buchanan, *The Demand and Supply of Public Goods* (Chicago: Rand McNally, 1968), p. 8.
16. See Michael J. Mandel, "How Much Is a Sea Otter Worth?" *Business Week,* August 21, 1989, pp. 59, 62.
17. William R. Greer, "Pondering the Value of a Human Life," *New York Times,* August 16, 1984, p. 16.
18. Alasdair MacIntyre, "Utilitarianism and Cost-Benefit Analysis: An Essay on the Relevance of Moral Philosophy to 'Bureaucratic Theory,'" Donald Scherer and Thomas Attig, eds. *Ethics and the Environment* (Englewood Cliffs, NJ: Prentice Hall, 1983), pp. 145–46.
19. See Douglas G. Hartle, *Public Policy Decision Making and Regulation* (Montreal, Canada: The Institute for Research on Public Policy, 1979), pp. 213–218.
20. There is a school of thought called public choice theory that looks at government decision makers as rational, self-interested people who are just like the rest of us, and view issues from their own perspective and act in light of the personal incentives. While voters, politicians, and bureaucrats may desire to reflect the "public interest" and often advocate it in support of their decisions, this desire is only one incentive among many with which they are faced and is likely to be outweighed by more powerful incentives related to self-interest of one sort or another. See Steven Kelman, "Public Choice and Public Spirit," *The Public Interest,* No. 87 (Spring 1987), pp. 80–94, for an interesting critique of the public choice school of thought.
21. Aaron Wildavsky, *Speaking Truth to Power: The Art and Craft of Policy Analysis* (Boston: Little, Brown, 1979), pp. 253–254.
22. Dye, *Understanding Public Policy,* p. 20.
23. John D. Blackburn, Elliot I. Klayman, and Martin H. Malin, *The Legal Environment of Business: Public Law and Regulation* (Homewood, IL: Irwin, 1982), p. 65.
24. Robert E. Healy, ed., *Federal Regulatory Directory* 1979–80 (Washington, D.C.: Congressional Quarterly, Inc., 1979), p. 3.
25. Blackburn, Klayman, and Malin, *Legal Environment,* pp. 67–68.
26. Ibid., pp. 70–71.
27. Healy, ed., *Regulatory Directory,* p. 25.
28. Ibid., p. 31.
29. Blackburn, Klayman, and Malin, *Legal Environment,* p. 77.
30. Charles T. Goodsell, *The Case for Bureaucracy: A Public Administration Polemic,* 2nd ed. (Chatham, NJ: Chatham House, 1985), p. 126.
31. Ibid., p. 133.
32. Office of the Federal Register, *The United States Government Manual 1995/1996* (Washington, D.C.: U.S. Government Printing Office, 1995), p. 524.
33. Timothy Noah, "Clinton Establishes White House Office on Environment," *Wall Street Journal,* February 9, 1993, p. B6.
34. *The United States Government Manual 1995/1996,* pp. 139–140.
35. Ibid., pp. 324–325.
36. Ibid., p. 326.
37. Ibid., pp. 328–329.
38. Ibid., p. 331.
39. John Carey, "Are Regs Bleeding the Economy?" *Business Week,* July 17, 1995, pp. 75–76.

40. *Reinventing Environmental Regulation: National Performance Review,* March 16, 1995, p. 3.
41. Science Advisory Board, *Reducing Risk: Setting Priorities and Strategies for Environmental Protection* (Washington, D.C.: Environmental Protection Agency, 1990), p. 3.
42. United States Environmental Protection Agency, *Preserving Our Future Today* (Washington, D.C.: EPA, 1991), p. 18.
43. Douglas J. Lober, "Implementing a Market-Based Environmental Policy: Utility Behavior in the Sulfur Dioxide Allowance Trading Program," Paper Presented at the Resources for the Future Seminar Series, Washington, D.C., December 7, 1994.
44. Benjamin A. Holden, "Dirt in Hollywood? Californians Have Pollution-Rights Market Ready for It," *Wall Street Journal,* April 12, 1995, p. B2.
45. See Robert W. Hahn, "Let Markets Drive Down Auto Emissions," *Wall Street Journal,* October 17, 1994, p. A14.
46. *Reinventing Environmental Regulation,* p. 25.
47. Christopher Georges, "Wider Property-Owner Compensation May Prove a Costly Clause in the Contract with America," *Wall Street Journal,* December 30, 1994, p. A10.
48. David Frum, "The GOP's Takings Sell-Out," *Wall Street Journal,* March 16, 1995, p. A20. See also Florence Williams, "The Compensation Game," *Wilderness,* Fall 1993, pp. 29–33.
49. Thomas Lambert and Christopher Boerner, *Environmental Inequity: Economic Causes, Economic Solutions* (St. Louis, MO: Washington University Center for the Study of American Business, 1995), p. 18.
50. K. S. Shrader-Frechette, "Lay Risk Evaluation and the Reform of Risk Management," in *Future Risks and Risk Management,* Berndt Brehmer and Nils-Eric Sahlin, eds. (Dordrecht, Netherlands: Kluwer Academic Publishers, 1994), p. 215.
51. *Environmental Inequity,* p. 17.
52. Alexis de Tocqueville, *Democracy in America* (New York: Schocken, 1961), Vol. II, p. 128.
53. Earl Latham, *The Group Basis of Politics* (New York: Octagon Books, 1965), p. 36 as quoted in Anderson, Brady, and Bullock, *Public Policy,* p. 416.
54. Andrew S. McFarland, "Public Interest Lobbies Versus Minority Faction," *Interest Group Politics,* Allan J. Cigler and Burdett A. Loomis, eds. (Washington, D.C.: Congressional Quarterly, 1983), p. 327.
55. Murray L. Weidenbaum, *The Future of Business Regulation* (New York: AMACOM, 1979), p. 146.
56. Everett Carl Ladd, "How to Tame the Special-Interest Groups," *Fortune,* 102, no. 8 (October 20, 1980), 72.
57. See Christopher Boerner and Jennifer Chilton Kallery, *Restructuring Environmental Big Business* (St. Louis, MO: Washington University Center for the Study of American Business, 1995).
58. Walter E. Westman, *Ecology, Impact Assessment, and Environmental Planning* (New York: Wiley, 1985), pp. 120–123.

Suggested Reading

Bacow, Lawrence, and Michael Wheeler. *Environmental Dispute Resolution.* New York: Wiley, 1993.

Boyle, A., and Norton Rose, eds. *Environmental Regulation and Economic Growth.* New York: Oxford University Press, 1995.

Bromley, Daniel W. *Environment and Economy: Property Rights and Public Policy.* New York: Blackwell, 1991.

Bryant, Bunyon, ed. *Environmental Justice: Policies and Solutions.* Washington, D.C.: Island Press, 1995.

Buchanan, James. *The Demand and Supply of Public Goods.* Chicago: Rand McNally, 1968.

Caldwell, Lynton K., ed. *Environment as a Focus for Public Policy.* College Station, TX: Texas A&M University Press, 1995.

Canter, Larry W. *Environmental Impact Assessment,* 2nd ed. New York: McGraw-Hill, 1992.

Cherchile, Richard A., et al., eds. *Environmental Decision Making: A Multidisciplinary Perspective.* New York: Van Nos Reinhold, 1991.

Crowfoot, James E., and Julia M. Wondolleck. *Environmental Disputes: Community Involvement in Conflict Resolution.* Washington, D.C.: Island Press, 1990.

Dietz, F. J. *Environmental Policy in a Market Economy.* New York: St. Martins, 1991.

Diwan, Paras. *Environmental Protection: Problems, Policy, Administration, Law.* New York: St. Martins, 1990.

Draan, J., and R. J. Veld, eds. *Environmental Protection: Public or Private Choice.* Netherlands: Kluwer Academic Press, 1991.

Dye, Thomas R. *Understanding Public Policy,* 6th ed. Englewood Cliffs, NJ: Prentice Hall, 1987.

Felder, David W. *Environmental Conflicts.* New York: Felder Books, 1995.

Hartle, Douglas G. *Public Policy Decision Making and Regulation.* Montreal: The Institute for Research on Public Policy, 1979.

O'Leary, Rosemary. *Environmental Change: Federal Courts and the EPA.* Philadelphia: Temple University Press, 1993.

Olson, Mancur. *The Logic of Collective Action.* Cambridge, MA: Harvard University Press, 1977.

Pertikin, Jonathan S., ed. *Environmental Justice.* New York: Greenhaven, 1993.

Peters, B. Guy. *American Public Policy: Promise and Performance,* 2nd ed. Chatham, NJ: Chatham House, 1986.

Portney, Kent E. *Approaching Public Policy Analysis.* Englewood Cliffs, NJ: Prentice Hall, 1986.

Preston, Lee E., and James E. Post. *Private Management and Public Policy.* Englewood Cliffs, NJ: Prentice Hall, 1975.

Reeve, Roger N., ed. *Environmental Analysis.* New York: Wiley, 1994.

Sirkin, Gerald. *The Visible Hand: The Fundamentals of Economic Planning.* New York: McGraw-Hill, 1968.

Teubner, Gunther, et al., eds. *Environmental Law and Ecological Responsibility.* New York: Wiley, 1994.

Tietenberg, Tom. *Environmental Economics and Policy.* New York: Harper College, 1994.

Weiskel, Timothy C., and Richard A Gray. *Environmental Decline and Public Policy: Pattern, Trend, and Prospect.* New York: Pierian, 1992.

Zeff, Robin E. *Environmental Action Groups.* New York: Chelsea House, 1993.

CHAPTER 5

Global Environmental Problems

The environmental problems that were of concern in the 1970s were global in the sense that every industrial society had some of the same problems. Air pollution existed in every country that had factories and automobiles, and water pollution was a problem in societies that had manufacturing companies and municipalities with large quantities of waste to dispose of in lakes and rivers. The disposal of solid and hazardous waste began to pose serious problems for many countries as the 1970s drew to a close. But these problems were dealt with largely on a national basis, often in cooperation with state and local authorities. The United States, for example, passed laws and regulations related to air and water pollution that were largely implemented by the states. The problem of waste disposal was also dealt with on a federal level through laws and regulations related to solid and hazardous waste disposal. Every country that became concerned about these types of pollution passed some kind of laws or regulations to deal with the problem.

Theoretically, one could escape most of these problems by moving to a part of the country that was still in something of a natural state where the air was not polluted and the water was clean enough to drink without treatment. In the 1980s, however, problems appeared that were truly global in nature in that they affected people all over the world and required international cooperation to deal with effectively. No one can escape the effects of such problems as global warming, because there is nowhere to hide from such problems. One cannot go to Walden Pond to escape this type of problem, because the whole earth may be warming, not just certain parts of it, and Walden Pond will also be affected.

GLOBAL WARMING

Two problems that are truly global in nature that became concerns in the 1980s are global warming and ozone depletion. The phenomenon that goes by the name of global warming, sometimes also called the greenhouse effect, is believed to be caused by changes in the earth's atmosphere as a result of industrial processes. Although there is some controversy about the causes of global warming and, indeed, whether global warming is actually taking place, many scientists believe that there is some linkage between global warming and changes in the composition of the atmosphere. These changes in the atmosphere do not stem from modification in the major constituents of the atmosphere, but from increases in the levels of several of the atmosphere's minor constituents or trace gases. These trace gases include carbon dioxide, nitrous oxides, methane, and several compounds of chloroflorocarbons.[1] Exhibit 5.1 provides some information related to the buildup of these trace gases.

These trace gases have increased as a result of increased industrial activity such as the combustion of fossil fuels for energy, industrial and agricultural practices, burn-

EXHIBIT 5.1 Greenhouse Gases at a Glance

CARBON DIOXIDE (CO_2)

❑ Accounts for about 50% of the greenhouse effect.
❑ Is generated from burning fossil fuels, especially coal, and from deforestation.

CHLOROFLUOROCARBONS (CFCs)

❑ Account for about 15% of the greenhouse effect.
❑ Leak into upper atmosphere from refrigerators, air conditioners, styrofoam packaging; remain for 75 to 100 years.
❑ Thousands of times more heat-absorbing than CO_2
❑ Also responsible for nibbling away at the ozone layer, which protects the earth from too much ultraviolet radiation.
❑ In September 1987, 50 countries agreed to cut CFC production in half.

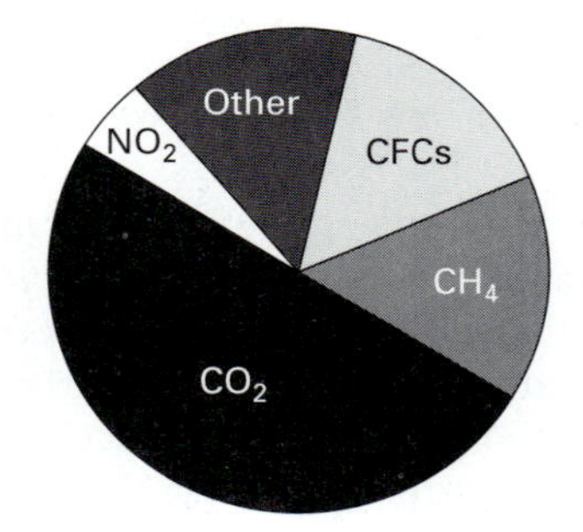

METHANE (CH_4)

❑ Accounts for 15–20% of the greenhouse effect.
❑ Is a by-product of burning wood, and is released from natural sources such as cattle, wetlands, rice paddies and termite mounds.

NITROUS OXIDES (NO_2)

❑ Accounts for about 5% of the greenhouse effect.
❑ Is formed when chemical fertilizers break down and when coal is burned.

OTHER

❑ Ozone (O_3), halons (CFC-like synthetic compounds), water vapor and other airborne particles.

Source: Sierra Club, *Global Warming,* May 1989, p. 2.

ing of vegetation, and deforestation. These activities are not only changing the chemistry of the atmosphere but also may be driving the earth toward a climatic warming of unprecedented magnitude. Unwelcome surprises are a possibility as human activities continue to affect an atmosphere whose inner mechanisms and interactions with living organisms and nonliving materials are incompletely understood.[2] The magnitude of the climatic change, its distribution, the speed with which it will occur, and the changes it will bring about are all known only with large degrees of uncertainty.

The theory about global warming holds that these trace gases form a shield around the earth that prevents some of the infrared waves from escaping into the atmosphere. These gases are relatively transparent to sunshine, which heats the earth, but trap heat by more efficiently absorbing the longer wavelength infrared radiation released by the earth.[3] These gases thus cause a warming effect by acting much like a greenhouse in keeping the heat of the sun in rather than letting it escape. As trace gases increase, particularly carbon dioxide, their heat-trapping ability also increases, further warming the earth. While there is some controversy over many of the particulars as well as whether a linkage actually exists between fluctuations in trace gases and global warming, the process has been described as an ongoing geophysical experiment.[4]

Warnings about this problem appeared as far back as 1890, when Swedish chemist Svante Arrhenius began to fear that the massive burning of coal during the industrial revolution, which pumped unprecedented amounts of CO_2 into the atmosphere, might cause problems. At that time he made a prediction that a doubling of atmospheric CO_2 would eventually lead to a 9 degree Fahrenheit warming of the earth, and suggested that glacial periods might have been caused by a diminished level of carbon dioxide. While his contemporaries scoffed at such an idea, he was right on track. During his time, CO_2 concentration was about 280 to 290 parts per million, but in recent years the count has risen to 340 parts per million. [5] Scientists have documented a 25 percent increase in carbon dioxide over the interglacial level in the past 100 years, and some scientists expect the present level to double by the year 2050 (Figure 5.1) Atmospheric methane has doubled during this same time period.[6]

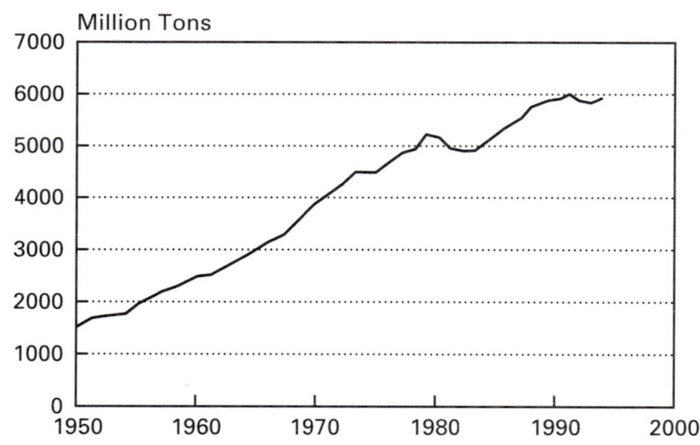

Source: From *Vital Signs 1995: The Trends That Are Shaping Our Future,* by Lester R. Brown et al., eds. Copyright © 1995 by Worldwatch Institute. Reprinted by permission of W. W. Norton & Co., Inc.

FIGURE 5.1 World Carbon Emissions from Fossil Fuel Burning, 1950–1994

The major cause of global warming thus seems to be the burning of fossil fuels, particularly coal, which releases carbon dioxide into the atmosphere. Deforestation also adds carbon dioxide to the atmosphere, as trees and other vegetation absorb CO_2 as they grow, and release an equal amount when they are burned or decay naturally. Further concentrations of carbon dioxide depend on the consumption of fossil fuels and the rate of deforestation. Growth in the use of fossil fuels will reflect population growth and the rate at which alternative energy sources and conservation measures are adopted. The rate of deforestation will depend on government policies with respect to continued exploitation of these resources around the globe.

Assuming that fossil fuel consumption will continue to increase at about its present pace, carbon dioxide emissions are expected to increase between 0.5 and 2.0 percent a year for the next several decades. Other trace gases could contribute as much to global warming as carbon dioxide because they are much better able to absorb infrared radiation even though they are emitted in much smaller quantities. But, predicting future emissions of these other gases is even more difficult than it is for carbon dioxide.[7] Thus there is some uncertainty regarding the levels of trace gases to be expected in the future.

Carbon emissions because of the burning of fossil fuel rose slightly in 1994 to 5.925 billion tons, which continued a six-year period in which such emissions have remained essentially constant. Deforestation added an additional 1.1 to 3.6 tons of carbon to the atmosphere. Each ton of carbon emitted into the air results in 3.7 tons of carbon dioxide. Rising carbon emissions at one time were assumed to rise along with economic growth. Estimated at 93 million tons in 1860, emissions rose to 525 million tons by 1900, and to 1.62 billion tons by 1950, which is still less than a third of today's level.[8] Since 1950, carbon dioxide emissions themselves have tripled. However, growth rates in emissions slowed from 4.6 percent annually in the 1950s and 1960s, to 2.5 percent in the 1970s and 1.2 percent in the 1980s. The main reasons for this decline were sluggish economic expansion in developing countries and higher oil prices, which encouraged greater energy efficiency.[9]

The Third World currently burns fossil fuels at far lower rates than in the industrial world, but if and when these countries overcome their economic problems, the potential for growth in fossil fuel use is enormous. Thus carbon emissions are expected to rise in the next few years despite the recent plateau. For example, emissions in 1993 grew 7 percent in China, 17 percent in India, and 28 percent in South Korea. Because of China's size, however, its 43 million-ton increase actually outweighed those of the other two countries. Rapid increases of coal use in China have made it the world's second largest emitter of carbon behind the United States, and at current growth rates, it could pass the United States in a few years.[10] Many developing countries now add far more carbon to the atmosphere through deforestation than through fossil fuel consumption. Brazil, for example, contributes some 336 billion tons of carbon each year through deforestation, over six times as much as through burning fossil fuels.[11]

Some of this increase will be absorbed by natural processes. Carbon dioxide in the atmosphere is continually being absorbed by green plants and by chemical and biological processes in the oceans. Because carbon dioxide is a raw material of photosynthesis, an increased concentration might speed the uptake by plants, which would counter some of the buildup. Increased uptake by the oceans may also slow the buildup to some extent. But the increased buildup of trace gases may also cause positive feed-

backs that would add to the atmospheric burden. Rapid climate change could disrupt forests and other ecosystems and reduce their ability to absorb carbon dioxide from the atmosphere. Global warming could also lead to release of the vast amount of carbon held in the soil as dead organic matter.[12]

Scientists estimate that to stabilize the amount of carbon dioxide in the air and avert the risk of climate change, global emissions will have to fall at least 60 percent. Such a decrease would require rich and poor nations alike to adopt new policies that reduce dependence on oil and coal as primary energy sources. However, the geopolitics of the carbon dioxide problem indicates that one group with the fewest people emits the most carbon at present, and the other countries with the most people appear bent on matching the first group's prolificacy as quickly as possible. It thus seems incumbent upon the rich minority of countries to demonstrate how modern societies can reduce carbon emission through the more efficient use of energy and the use of renewable sources.[13]

Predictions

What effect will the continued buildup of these trace gases have on the earth's climate? To answer this question scientists rely on mathematical climate models. History offers no clear answer to this question nor can climate be reproduced in a laboratory experiment. These models consist of expressions for the interacting components of the ocean-atmosphere system and equations representing the basic physical laws governing their behavior. To determine the effect of trace gas buildup, scientists simply specify the projected amount of greenhouse gases and compare the model results with a controlled simulation of the existing climate, based on present atmospheric composition.[14]

Making a model of anything involves making assumptions about how the world works, and it is these assumptions that are critical. The more factors that are included in the model, the more assumptions have to be made, and the more complex the model becomes. In predicting climate change, scientists have to construct a model of the atmosphere. These models go under the general name of "General Circulation Models" (GCMs) and may contain as many as 100,000 computer instructions. Because of these complexities, skeptics believe that climate models have not yet been validated, cannot yet be trusted, and therefore cannot serve as guides for public policy decisions.[15]

There are several shortcomings to this modeling technique. They are said to be inaccurate when it comes to dealing with atmospheric turbulence, precipitation, and cloud formation. Clouds, for example, have a net cooling effect reflecting sunlight back to outer space. Present models reproduce only average cloudiness, but climate change may actually cause incremental change in cloud characteristics altering the nature and amount of feedback processes. Another problem is with the oceans, which may act as a thermal sponge slowing any initial increase in global temperature. Because of the complexities involved, the dynamics of oceans are simplified, treated at course resolution, or left out entirely.[16]

The main effect of an ocean on climate is the redistribution of heat on the earth's surface. Such heat can flow horizontally in large ocean currents, or up and down vertically, moving between the surface and the depths of the ocean. In most models, these effects have not been included, or have been included in such a way that they were not allowed to change in response to changes in the atmosphere. The same is true with respect to cloud formation. Many scientists have argued that if the earth warms, more water vapor will evaporate to form more clouds, which in turn will reflect more sunlight away from earth, so that the greenhouse effect will be reduced or completely nullified.[17]

Given all these problems, many scientists believe that since the models are well enough validated and other evidence of greenhouse-gas effects on climate is strong enough, that the increases in average surface temperature predicted by the models for the next 50 years is valid within a rough factor of two, meaning that it is a better than even bet that the changes will take place. Most of the models are in rough agreement that a doubling of carbon dioxide or an equivalent increase in other trace gases would warm the earth's average surface temperature by between 3.0 and 5.5 degrees Celsius. Such a change would be unprecedented in human history.[18]

Actual Warming

About half a degree of real warming has taken place in the last 100 years but the 1980s appear to be the warmest decade on record, with 1988, 1987, and 1981 being the warmest years, in that order. There apparently was a rapid warming of the atmosphere before World War II, a slight cooling trend through the mid-1970s, and a second period of warming since then[19] (Figure 5.2). The new decade continued this trend, as scientists announced that 1990 was the earth's warmest year in the nearly two centuries that records have been kept. This meant that seven of the ten warmest years on record occurred since 1980. Temperature in the 1980s was about one third of a degree warmer than the globe's average temperature of the previous 30 years. This was a sharp increase when compared with long-term averages. And 1990 was another one third of a degree warmer than 1980.[20]

The eruption of Mount Pinatubo in the Philippines injected millions of tons of dust into the upper atmosphere. This dust spread around the globe and blocked enough sunlight to depress temperatures by about half a degree Celsius for several years in the early 1990s, but by early 1994, these effects began to wear off and temperatures began

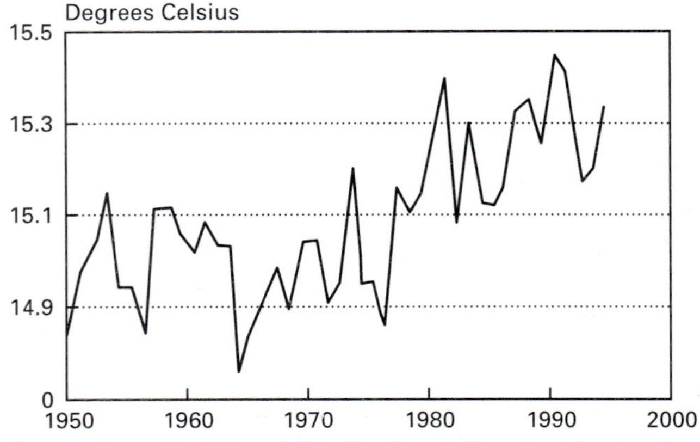

Source: From *Vital Signs 1995: The Trends That Are Shaping Our Future* by Lester R. Brown, Nichols Lenssen and Hal Kane, eds. Copyright © 1995. by Worldwatch Institute. Reprinted by permission of W. W. Norton & Company, Inc.

FIGURE 5.2 Global Average Temperature, 1950–1994

to return to their pre-eruption highs.[21] Preliminary figures for 1995 indicate that this year will be the warmest globally since records began to be kept beginning in 1856, and taking the effects of the eruption of Mount Pinatubo into account, that the years 1991–1995 were warmer than any similar five-year period, including the two half-decades of the 1980s, which was the warmest decade to date.[22]

There are two sets of long-term data by which surface temperatures are tracked. One comes from the British Meteorological Office and the University of East Anglia, which are based on land and sea measurements around the world. The second set of data is maintained by the NASA Goddard Institute for Space Studies in New York, based on a somewhat different combination of surface temperature observations around the world. Critics of this manner of collecting temperature data argue that temperature rises might be due to increasing urbanization where the temperatures are recorded, as city temperatures are higher than the surrounding countryside because of the concentration of people and their activities.[23]

In order to get a more accurate reading of actual climate change, scientists are experimenting with undersea sound waves to measure warming. At the beginning of 1991, scientists switched on a loudspeaker in the ocean near Heard Island, which is 2,550 miles southwest of Perth, Australia. Since the speed of sound varies with temperature, measuring the pace of this ocean music for 10 to 20 years may tell us whether the oceans are warming. Small variations in sea temperature are one factor in global weather, and ocean current changes can have drastic effects.[24]

Effects

What would be the effects of such climate change? Should warming continue, coastal areas would undoubtedly face a rise in sea levels. A global temperature increase of a few degrees Celsius over the next 50 or 100 years would raise sea levels by between 0.2 and 1.5 meters as a result of the thermal expansion of the oceans, the melting of mountain glaciers, and the possible retreat of the Greenland ice sheet's southern margins. Many cities near low-lying coastal areas would be flooded, and people would have to either erect sea walls or move to another location. Delta areas along the Atlantic and Gulf coasts of the United States would be flooded, and worldwide, more than 100 million people could be displaced.[25]

Such a change would mean hot, dry summers for many parts of the world. Hot spells like the one that killed hundreds in the upper Midwest in 1995 would become more frequent and more severe. There could be a decline in agricultural productivity in the Middle West and Great Plains as the corn and grain belts moved north by several hundred kilometers. Temperate zones would move farther north, leaving areas that were primary agricultural zones, such as the midsection of the United States, useless as far as growing crops is concerned. Such shifts would cause great hardships for some people and benefits for others.[26]

A rising sea level could flood dry lands in the Southeast, and increase the usage of water for irrigation in California's Central Valley, as spring runoff from the Sierras decreases. Costs to protect developed shorelines could reach $111 billion through the year 2100 if sea level rises one meter. Despite these efforts, an area the size of Massachusetts could be lost to water. And more summer heat will increase demand for electricity, which could require spending $325 billion to build new power plants. Animals and plant life will also be affected. While climate zones may move hundreds of miles to the north, animals and plants may not be able to migrate as quickly.[27]

Scientists have also linked global warming to bursts of sudden rainfall that have increasingly replaced consistent patterns of annual rainfall. Incidents of light and moderate rainfall have declined sharply in recent years in the United States, for example, while the number of extreme or heavy downpours have increased substantially, according to data analyzed by the U.S. National Oceanic and Atmospheric Administration. Analyzing rainfall and temperature records going back to 1910 from countries all over the world, these scientists have found that as average global temperatures increase so do one-day bursts of heavy rain. These heavy downpours can be very damaging, destroying crops and triggering violent floods.[28]

The west Antarctic ice shelf is also thought to be melting. Observations conducted in early 1995 showed that parts of the Larsen ice shelf had broken up into rubble and it was torn by a 40-mile crack in one place. In January an iceberg 23 miles wide and 48 miles long, almost as big as the state of Rhode Island, had broken off from the shelf and was slowly floating out to sea. The disintegration of the Larsen ice shelf was thought to be the result of a change in the temperature of Antarctica, as over the past 50 years, the average temperature on the Antarctica peninsula has risen 2.5 degrees Celsius, a greater increase than anywhere else in the world. While not everyone agreed, some scientists thought this increase was another indication of global warming.[29]

The insurance industry has begun to take global warming seriously because of big losses from violent storms that have hit various parts of the United States in recent years. Storms, floods, and droughts are hitting populated areas with greater frequency and severity than was predicted in the actuarial tables. Natural disasters during the 1980s were 94 percent more frequent than in the 1970s, which conforms with patterns predicted for global warming. Insurance companies are beginning to take projections of global warming into account by raising insurance premiums and canceling some policies, particularly for residents in coastal areas where changes in weather patterns can have major economic consequences.[30]

The Debate

James Hansen, an atmospheric scientist who heads NASA's Goddard Institute, made national headlines when he testified before Congress in 1988 that the greenhouse effect had already begun and was no longer just a theory. During the first five months of 1988, he testified, average worldwide temperatures were the highest in all the 130 years of record keeping. He also stated that he was 99 percent certain that the highest temperatures were not just a natural phenomenon but were the direct result of a buildup of carbon dioxide and other trace gases from manmade sources. His findings were based on monthly readings at 2,000 meterological stations around the world.[31]

Most world climate models indicate that the world is already committed to a warming of several degrees over the next few decades, which represents a warming unprecedented in human history. There is something of a delayed reaction to the buildup of greenhouse gases, in that it takes 20 years for these gases to have an effect. So even if we stopped burning fossil fuels tomorrow, we would still be in for some degree of warming in future years. Thus any strategies developed to deal with the problem will have to provide plans to adapt to a warmer globe and prevent greater warming in the future.[32]

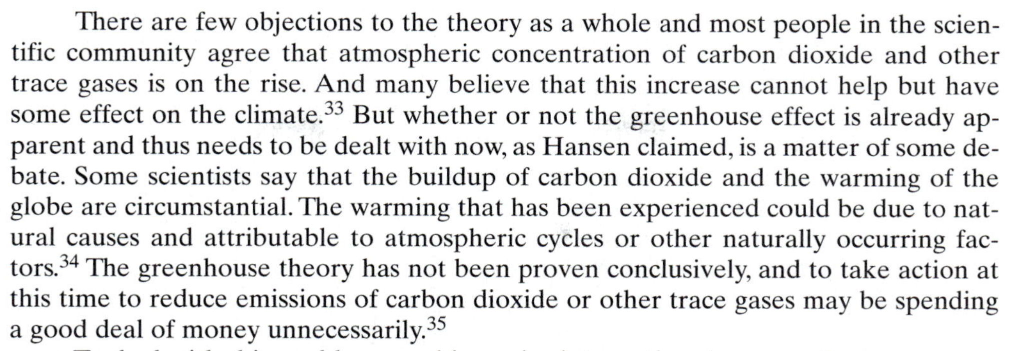

There are few objections to the theory as a whole and most people in the scientific community agree that atmospheric concentration of carbon dioxide and other trace gases is on the rise. And many believe that this increase cannot help but have some effect on the climate.[33] But whether or not the greenhouse effect is already apparent and thus needs to be dealt with now, as Hansen claimed, is a matter of some debate. Some scientists say that the buildup of carbon dioxide and the warming of the globe are circumstantial. The warming that has been experienced could be due to natural causes and attributable to atmospheric cycles or other naturally occurring factors.[34] The greenhouse theory has not been proven conclusively, and to take action at this time to reduce emissions of carbon dioxide or other trace gases may be spending a good deal of money unnecessarily.[35]

To deal with this problem would require international cooperation, as one nation could not solve the problem by itself. All nations of the world would have to agree to take steps to limit the release of carbon dioxide and other trace gases to have a major effect on global warming. While many policymakers would like to take a "wait and see" attitude and continue to study the problem, waiting for more direct evidence of the greenhouse effect could involve a commitment to greater climatic change than if action were taken now to slow the buildup of trace gases. In other words, waiting to solve the problem if the theory proves to be true, would cost more money in the future than taking some steps now to cut back on carbon dioxide and other emissions.

This phenomenon of global warming thus presents policy makers with a dilemma that puts them between a rock and a hard place. Waiting for conclusive and more direct evidence of global warming that would either confirm or deny the results of computer modeling is not a cost-free proposition. By then the world will be committed to greater climatic change than it would be if action were taken now to slow the buildup of greenhouse gases. But proposals for immediate action to slow the buildup of trace gases entail large investments as insurance against future events whose details are far from certain. Much money could be spent on such strategies that may not prove to have been necessary.[36]

Under the auspices of the United Nations, representatives from more than 130 countries met to start work on an international treaty aimed at slowing the buildup of greenhouse gases. This treaty was introduced at the 1992 Rio Earth Summit, where under the UN Framework Convention on Climate Change, the industrialized countries committed themselves to reducing emissions of greenhouse gases such as carbon dioxide to 1990 levels by the end of the decade. While the treaty required no specific timetables for reducing emissions and made no commitments to achieving specific levels of emissions, it still marks the first international effort to control greenhouse gases. A stronger treaty was not politically possible because of U.S. opposition.[37]

When President Clinton took office, the United States signed Agenda 21, an 800-page blueprint to combat global warming on an international level as well as other environmental evils. This action committed the United States to reduce greenhouse gases to 1990 levels from the increases projected for the end of the century. Meeting this commitment was not expected to be easy, as the United States produces nearly 25 percent of global carbon emissions. Continued growth depends on the use of cheap and seemingly unlimited supplies of carbon-based fuels.[38] At the end of 1993, the Clinton administration announced a voluntary seven-year energy efficiency program to reach this goal of reducing greenhouse-gas emissions. The plan contained 50 initiatives to limit future air pollution.[39] With the election of a Republican-controlled Congress in 1994, however, the attainment of this goal became more politically questionable.

In late 1995, a scientific report from the UN-sponsored Intergovernmental Panel on Climate Change (IPCC), which earlier had recommended a policy of energy conservation and reforestation that made economic sense even without climate change, issued a report stating that the global mean surface temperature of the earth had risen between 0.3 and 0.6 degrees Celsius over the past century, and that some of this change was caused by human activities. This was the first time the panel said the warming trend is unlikely to be entirely due to natural causes and that a pattern of climatic response to human activities is identifiable in the climatological record. The report also stated that a further rise of 1.0 to 3.5 degrees Celsius may be expected by 2100 with continued fossil fuel burning. This report combined the efforts of three working groups who had been studying climate change predictions, environmental impact, and public policy, and was expected to increase pressure on countries around the world to reduce greenhouse-gas emissions.[40]

Strategies

Coping effectively with global warming will force advanced societies to reverse trends that have dominated the industrial age ever since it began. The challenge cannot be met, according to some scholars, without a strong commitment on the part of both individual consumers and governments to promote greater energy efficiency and renewable energy sources, to institute a carbon tax on fossil fuels, to reverse tropical deforestation, and to advance a more rapid elimination of CFC production.[41] The EPA estimates that to stabilize atmospheric concentration of CO_2 at current levels, carbon emissions must be cut by 50 to 80 percent, taking them back to the level of the 1950s. Even a 20 percent cut would require a dramatic change in energy planning and land-use policies around the world.[42]

Any realistic strategy, according to Christopher Flavin, must start with the fact that one fourth of the world's population accounts for nearly 70 percent of the fossil fuel–based carbon emissions. The planet will never be able to support a population of 8 billion people generating carbon emissions at the rate of Western Europe today. Such emissions result in a concentration of CO_2 three times preindustrial levels, which is well above doomsday scenarios. But limiting carbon emissions in developing countries remains an extraordinarily difficult issue. No one has begun to wrestle with the critical equity issues this strategy entails.[43] The United States contributes about 25 percent of the world's CO_2 emissions with only 6 percent of the world's population. Thus the United States is in a good position to take global leadership in combatting the greenhouse effect with an ambitious energy efficiency campaign and the export of new energy-efficient technology.[44]

The atmospheric buildup of carbon dioxide results from burning gasoline in our cars and coal and oil in our power plants. These activities are responsible for about half of the greenhouse effect, according to some sources. Burning coal is said to produce twice as much carbon dioxide per unit of heat as natural gas, and a third more than the burning of oil. It is said that no global warming solution will be successful unless carbon dioxide emissions from coal combustion are controlled. The best solution is to limit the burning of coal through greater energy efficiency and the use of natural gas and renewable energy resources like solar and wind power.[45]

One scientist recommends that tie-in strategies be adopted, taking actions that will yield benefits even if climactic change does not materialize as predicted. Such a tie-in strategy is the pursuit of energy efficiency. More efficient use of fossil fuels will slow the buildup of carbon dioxide in addition to curbing acid rain, reducing urban air pollution, and lessening our dependence on foreign sources of supply. These benefits would accrue to our society even if the effects of carbon dioxide on global warming would prove to be overstated. The same tie-in strategies would hold true for the development of

Generic Strategies

WAIT AND SEE

This strategy is basically the one the U.S. has adopted. There is time for further research, and indeed, further research is needed before policies are adopted. This strategy has the advantage of causing little economic dislocation at present and not taking actions that would later prove to have been unnecessary. Much money could be spent needlessly and jobs lost that could be avoided if more study were done. However, if the most pessimistic predictions about global warming prove to be true, the warming may be more severe if we wait and it may be more costly to deal with it in the future.

ASSUME THE WORST AND ACT ACCORDINGLY

Such a strategy would probably involve a significant reduction in the use of fossil fuels, perhaps as much as 50 percent, a halt to deforestations, and the reforesting of millions of square miles of the earth's surface. This approach has the advantage of dealing with the worst case scenario immediately, and suffering whatever changes are necessary. But this strategy could cause enormous economic hardship, which may turn out to have been unnecessary if less pessimistic predictions turn out to be right.

NO REGRETS

This approach involves taking actions that make sense whether the warming predictions turn out to be true or false, in other words, this strategy is the same as the tie-in strategy mentioned earlier. Almost every effort to reduce air pollution and acid precipitation will also help reduce trace gas emissions. Further improvements in energy efficiency can make a big difference. Reduction of our dependence on foreign oil makes sense even without consideration of the greenhouse effect. This strategy has the advantage of avoiding unnecessary costs connected with the worst case strategy. But if the latter proves to be true, the "no regrets" response will be inadequate.

Source: James Trefil, "Modeling Earth's Future Climate Requires Both Science and Guesswork," *Smithsonian,* 21, no. 9 (December 1990), 36.

alternative energy sources, the revision of water laws, the search for drought-resistant crop strains, and other such strategies. These steps would offer widespread benefits even in the absence of any climactic change.[46]

Another recommendation is for the implementation of a carbon tax on fossil fuels, which would allow market economies to consider the global environmental damage of fossil fuel combustion. Coal would be taxed the highest, oil next, and natural gas would follow. In countries that do not wish to raise total taxes, the carbon tax could be offset by a reduction in other taxes. Picking the correct tax would be a complicated undertaking, but the environmental costs of using a particular fuel could be assessed and then internalized though taxes. Such taxes would have to be agreed on internationally. Tax revenues could be used to develop permanent and stable funding for improving energy efficiency and developing renewable sources.[47]

The first step in a world-wide strategy, according to one expert, is the establishment of ambitious but practical goals for the reduction of carbon emissions, particularly in those countries that currently use fossil fuels most heavily. Industrial countries will have to take the lead and reduce CO_2 emissions 20 to 35 percent over the next 10 years. If emissions are kept at today's levels, Third World increases would raise global emissions 20 to 30 percent by the end of the decade, and 50 to 70 percent in 20 years.[48] Overall, carbon emissions would have to be reduced to a maximum of 2 billion tons per year in order to stabilize atmospheric concentrations of greenhouse gases by the middle of the twenty-first century. Production of CFCs must be ended, and global carbon emissions must be cut by 10 to 20 percent over the next decade.[49]

The costs of some of these plans, however, are astronomical. According to economists Alan Manne of Stanford University and Richard Richels of the Electric Power Research Institute, costs to the United States alone would run from $800 billion to $3.6 trillion in cumulative costs over the next century. Annual costs could run as high as 5 percent of the gross national product. By way of contrast, the nation is currently spending $85 billion a year, or about 2 percent of GNP, on all other environmental protection measures.[50] Another economist estimated that the cost of stabilizing these emissions would be about $50 billion a year for the United States and $150 billion annually for the rest of the world.[51] Even the modest voluntary plan to promote energy efficiency announced by the Clinton administration was estimated to cost taxpayers $1.9 billion by the year 2000 with expectations that the private sector would spend more than $60 billion on environmental technologies.[52]

What Can Be Done

Several things can be done to deal with global warming, most of which would be of benefit to societies whether or not global warming proves to be of long-lasting concern. Thus they are examples of techniques or technologies that could be implemented under a tie-in strategic approach to the overall problem of global warming. Most of these technologies are already available or are well along as far as research and development are concerned. Thus they can be implemented rather quickly if the proper incentives are provided.[53]

Cleaning Up Coal Coal is used to produce over half the world's power, but pound for pound, coal can be up to 100 times as dirty, depending on the pollutant, as oil and natural gas. But technology does exist to reduce that pollution. Fluidized-bed combustion, for example, sharply reduces oxides of nitrogen and sulfur dioxide emitted from coal burning. Massive fans keep powdered coal suspended in midair so it burns cleaner with less energy loss. Plants being built in Stockholm using a variant of this technique are estimated to achieve an efficiency of 85 percent by capturing waste heat and emitting one-tenth the sulfur dioxide of plants in the United States and one-sixth the oxides of nitrogen.

The integrated gasification-combined cycle technology combines two concepts with regard to the burning of coal: (1) Coal is turned from a solid into a gas, removing much of the sulfur in the process; and (2) the gas is used to run two turbines, one powered by the hot combustion gases and the other by steam. Finally, add-on cleaners can be used to clean up coal burning. Scrubbers are systems that can remove up to 95 percent of the sulfur from stack gases, and have been in widespread use throughout the world for nearly two decades. Technology for removing nitrogen pollution, such as selective catalytic reduction, is of more recent origin.

Natural Gas Natural gas contains only half the carbon of coal and none of the sulfur and less nitrogen. Simple conversion to natural gas will thus dramatically reduce unwanted trace gas buildup. The use of natural gas also allows utilities to use new energy-efficient super turbines and plants. These turbines work like a jet engine bolted to the ground, and already turn 47 percent of their fuel into electricity, compared to 38 percent efficiency for the best coal-fired plants. Several such super turbines are already in operation, under construction, or on order, and more than 200 slightly less efficient versions are already running.

Conservation Conservation measures can be applied either where electricity is produced or where it is used. Cogeneration is a demonstrated method for raising the efficiency of power plants. This technique involves putting heat that would otherwise be wasted to some use in warming homes and offices or even running manufacturing processes. Cogeneration is already common in Sweden and other parts of Europe where efficiency can rise to as high as 85 percent. Conservation measures by consumers can involve turning thermostats down and using more efficient light bulbs. According to some estimates, conservation measures by consumers alone could save one third to one half of the electricity produced in the United States. This kind of conservation would involve no reduction in living standards (see the box that follows).

Energy Efficiency More efficient use of energy has the immediate potential to cut fossil fuel use at least 3 percent annually in industrial countries. In developing nations, efficiency can be used to limit fossil fuel growth while economic development continues. Efficiency improvements worldwide between 1990 and 2010 could make a 3 billion ton difference in the amount of carbon being released into the atmosphere each year. Such a strategy implies an annual rate of energy efficiency improvement of 3 percent. No other approach, according to one source, offers as large an opportunity for limiting carbon emissions in the next two decades.[54]

Fuel Cells Conventional power plants make steam to spin turbine blades that generate electricity, while fuel cells depend on chemistry rather than mechanics. The chemicals used vary, but the principle is the same as the fuel reacts chemically to decompose

Your Contribution to Global Warming

Industry isn't the only culprit; every light bulb you burn can send carbon dioxide into the atmosphere

This rough guide to the emissions you may be causing is based on power requirements of common appliances, the average energy produced by eastern and western coal, and the assumption that all the electricity comes from coal.

How much coal does it take to turn on your light bulb? And how much of the carbon dioxide in our atmosphere can we blame on your light bulb? Too bad there are no punch lines; this is no joke. More than half the electricity consumed in this country is powered by coal. Coal burning is our main source of carbon dioxide, the primary greenhouse gas out of the quartet that also includes carbon monoxide, nitrogen oxides, and chlorofluorocarbons.

Scientists estimate there is 25 percent more carbon dioxide in our atmosphere now than there was two centuries ago. They expect the present level to double by the year 2050. The subject, however, is not completely bleak. Our very ability to calculate how much coal we burn helps us understand how to burn less and reduce harmful emissions.

To figure out how your electricity use contributes to the greenhouse effect that many scientists believe is already warming our planet, you might first want to understand how your energy use is calculated. Your electric company bills you

ELECTRICAL CONNECTIONS TO POLLUTION

Electrical Appliances		*Pounds of Carbon Dioxide Added to Atmosphere**
Color Television	per hour	.64
Steam Iron	per hour	.85
Vacuum Cleaner	per hour	1.70
Air Conditioner, room	per hour	4.00
Toaster Oven	per hour	12.80
Ceiling Fan	per day	4.00
Refrigerator, frostless	per day	12.80
Waterbed Heater	per day	24.00
with thermostat	per day	12.80
Clothes Dryer	per load	10.00
Dishwasher	per load	2.60
Toaster	per use	.12
Microwave Oven	per 5-min use	.25
Coffeemaker	per brew	.50

*At room temperature and sea level, every pound of carbon dioxide occupies 8.75 cubic feet, about half the size of a refrigerator.

for kilowatts, or 1,000-watt units of electric power. If you leave a 100-watt light bulb on for 10 hours, you use one kilowatt hour (KWH) of electricity.

Let's suppose you use a 100-watt bulb as an outside night light for eight hours a night, 365 days a year. This bulb consumes 292 KWH a year. The energy probably comes from coal, especially since your light is on at night; during the day, oil and gas supplement coal at peaks of energy demand. The other two sources of electric power, hydro and nuclear, supply about a quarter of our electricity. But for the purposes of this exercise, let's assume that your light's power is from coal.

Each KWH used in your home requires the burning of an average of 1.28 pounds of coal (eastern coal has a slightly higher energy content than western coal). Since your light uses 292 KWH a year, it requires the burning of about 375 pounds of coal. And since carbon dioxide is made up of one carbon atom and two oxygen atoms—the coal is as much as 84 percent carbon—the total emission of carbon dioxide is about 1.8 times the weight of the coal, say 675 pounds for your light. Every pound of carbon dioxide occupies 8.75 cubic feet at room temperature and sea level. That means you are responsible for emitting about 6,000 cubic feet of carbon dioxide every year. And that's just one light. If you use 1,000 KWH a month, you add 28,062 pounds of carbon dioxide—or 245,540 cubic feet (one football field 5 feet deep)—to the atmosphere each year.

What can you do to help? Use other fuels? The burning of oil or natural gas also releases carbon dioxide and other pollutants, although neither releases as much carbon dioxide as coal. Nuclear power plants don't release carbon dioxide, but they are far more costly, release more waste heat, and pose serious environmental risks. Even if nuclear energy were perfectly safe, it takes about ten years to build a nuclear power plant. We can't wait that long to solve the greenhouse problem.

Here's how you can help. Replace that 100-watt night-light bulb with a tiny screw-in fluorescent tube that requires only 13 watts of power and uses 13 percent of the energy required by the old one—about 38 KWH a year instead of 292 KWH. Outdoors at night, you won't notice the slight difference in the amount of light. You'll reduce the amount of coal burned and the resulting carbon dioxide emissions by 87 percent—and save as much as $20 a year.

One light bulb may not sound like much, but if all the 100 million or so households in the United States did the same, we would prevent the annual release of 29.7 million tons of carbon dioxide. We could reduce our need for power by 8,500 megawatts—the equivalent of 17 coal-fired power plants of 500 megawatts each!

You can save even more energy and money with improved insulation and efficient major appliances. Buy the most energy-efficient refrigerators, freezers and air conditioners you can find. An air conditioner with an energy efficiency ratio (EER) of 12 will use only two-thirds as much electricity as one with an EER of 8.

Remember that one light bulb may not make much difference, but when we multiply by 100 million, perhaps we can have a last laugh after all.—George Burnwell

Source: National Wildlife, 28, no. 2 (February–March 1990), 53.

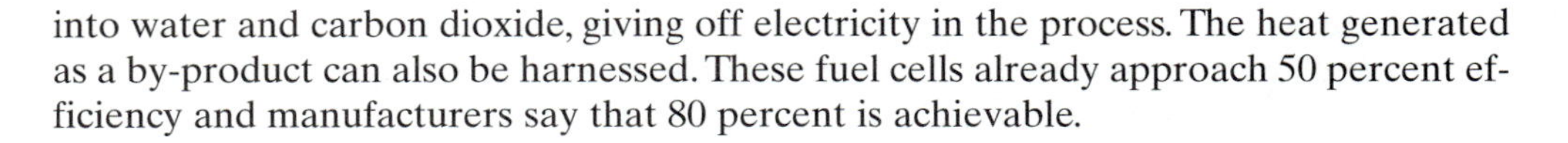

into water and carbon dioxide, giving off electricity in the process. The heat generated as a by-product can also be harnessed. These fuel cells already approach 50 percent efficiency and manufacturers say that 80 percent is achievable.

Renewable Sources Renewable energy technology such as wind power, geothermal, solar thermal, photovoltaic, and various biomass technologies, are likely to advance most rapidly in the years ahead.[55] These sources have several advantages over fossil fuels and nuclear power. They are inexhaustible, produce little or no pollution or hazardous waste, and pose few risks to public safety. They are also entirely a domestic resource and thus would be virtually immune to foreign disruptions like the recent conflict in the Gulf produced. They also provide a hedge against energy-price inflation caused by the depletion of fossil fuel reserves. And large-scale renewable-energy technologies can be installed relatively quickly in six months to two years.

Building Cleaner Cars According to some sources, the potential for reducing carbon dioxide from cars and trucks is at least as great as from power plants. In the United States, where new cars are the least fuel efficient in the world, an increase in the average fuel efficiency of just one mile per gallon would reduce carbon dioxide emissions by about 40 billion pounds per year. This reduction is equivalent to the closing of six coal-fired plants. The current new-car auto fleet averages about 26 miles per gallon, but there have been attempts to raise this average significantly. The technology for improvements in car mileage is here, as already in 1984, a Volvo design team produced a prototype four-passenger car that achieved 83 miles per gallon on the highway while designed to meet U.S. crash and pollution standards.

Reversing Deforestation Tropical deforestation is estimated to be the source of 20 to 25 percent of the gases that are warming the atmosphere. Policies to stop destruction of these forests would make a major contribution to slowing down the greenhouse effect. Even where slash-and-burn techniques have to be continued, carbon dioxide emissions can be reduced by between 90 and 98 percent by using different strategies. Villagers can be encouraged to reclaim land that had been cleared earlier and then abandoned. Younger forests contain less carbon than older more primary forests, thus torching second-growth trees can reduce pollution by about 90 percent. Second, farmers can be persuaded to use simple techniques to keep their land continually productive in a system of sustainable cropping. If existing land could be made more productive, it would not have to be abandoned to burn virgin forest elsewhere, cutting pollution another 90 percent.

Countries in temperate regions can also help restore the earth's carbon balance by planting trees. The challenges to a global reforestation plan include finding sufficient and appropriate land, raising needed financing, and mobilizing social institutions to accomplish reforestation. Large areas of marginal crop and grazing lands could be converted to trees, which stabilizes soils at the same time it increases the rate of carbon fixing. Tree planting can improve an urban environment by moderating summer heat and improving aesthetics. The real challenge facing these programs is maintaining the trees once they are in the ground.[56]

Phasing Out Chlorofluorocarbons These compounds were responsible for an added or estimated 25 percent of added greenhouse effect during the 1980s. In the United

States, CFCs constituted 40 percent of the country's greenhouse emissions.[57] They are being phased out, however, and thus their effect in producing global warming will be gradually reduced. The timetable originally agreed upon at Montreal was revised to speed up the phaseout of CFCs both to reduce global warming and to protect the ozone layer. Companies that produced CFCs have adopted a strategy of phasing them out as soon as possible and have found some substitutes that are being implemented.

Until modern times, the earth's climate has been largely self-correcting and stable, letting in just the right amount and type of solar energy and providing just the right balance of temperature and moisture to sustain living things. Alternating cycles of warming and cooling, and lesser or greater concentrations of trace gases have forced some species into extinction. But these same changes have helped other species to evolve. Just as we have begun to discover the climatic rhythms that have gone on for hundreds of millions of years, we may also have begun to change them irrevocably. And the crucial question is how many more unexpected changes in climate are waiting to be discovered.[58] Thus, in some sense, the world is being subjected to a great experiment as long as trace gases continue to increase.

OZONE DEPLETION

Another global environmental problem that became of concern in the 1980s is depletion of the ozone layer in the stratosphere. This layer absorbs most of the ultraviolet radiation that comes from the sun, and depletion of this layer allows higher levels of ultraviolet radiation to reach the surface of the planet. Too much ultraviolet radiation can damage plant and animal cells, cause skin cancer and eye cataracts in humans, reduce crop yields, deplete marine fisheries, cause damage to materials of various kinds, and kill many smaller and more sensitive organisms. Each 1 percent drop in ozone is projected to result in 4 to 6 percent more cases of skin cancer. Increased exposure to radiation also depresses the human immune system, lowering the body's resistance to attacking organisms.[59]

Terrestrial and aquatic ecosystems are also affected by ultraviolet radiation. Screenings of more than 200 different plant species, most of them crops, found that 70 percent were sensitive to ultraviolet radiation. Increased exposure may decrease photosynthesis, water use efficiency, yield, and leaf area. Aquatic ecosystems are also threatened, as phytoplankton would decrease their productivity 35 percent with a 25 percent reduction in the ozone layer. Destruction of upper-level ozone could augment the damage done by lower levels of the same substance, which is a major air pollutant. As more ultraviolet radiation hits the earth's surface, the photochemical process that creates smog will accelerate. Overall, the risks to aquatic and terrestrial ecosystems and to human health because of an increase in ultraviolet radiation are enormous[60] (Exhibit 5.2).

Ozone (O_3) is a form of oxygen that rarely occurs naturally in the lower atmosphere. It is created when ordinary oxygen molecules (O_2) are bombarded with ultraviolet rays in the stratosphere. This radiation breaks the oxygen molecules apart and some of the free oxygen atoms recombine with O_2 to form ozone. This configuration gives oxygen a property it ordinarily does not have; that is, the ability to absorb ultraviolet rays. Thus the ozone layer is able to protect oxygen at lower levels from being

EXHIBIT 5.2 What If the Shield Erodes?

The sun bombards the earth with photons, light rays of varying wavelengths. The shorter the wavelength, the more damage the rays can inflict. Among the most dangerous are wavelengths of ultraviolet light (UV) measuring from 200 to about 315 nanometers (a nanometer is a billionth of a meter).

The ozone layer blocks all of the most damaging UV light, the band from 200 to 290 nanometers known as UV-C. Without that protection, life on earth could not survive. Fortunately, only a drastic reduction of the ozone shield would allow any UV-C to filter through. THe ozone also screens out much of the UV between 290 and 315 nanometers, the band called UV-B. But enough gets through to cause, even now, millions of cases of skin cancer every year.

It's the UV-B that has scientists worried. For every 1 percent drop in ozone, there's a 2 percent increase in UV-B intensity at the earth's surface. Most plants and animals have evolved under relatively constant UV-B. Any prolonged increase could set in motion far-reaching—and unpredictable—ecological changes.

When UV-B radiation strikes a living thing, it is absorbed by the outer layers of cells. Microscopic plants and animals lack such protection. Single-celled plants at the bottom of the marine food chain are the base on which much of the world's population ultimately depends for protein. Those plants suffer a drastic drop in photosynthesis when exposed to UV-B at levels found under the Antarctic ozone hole.

Nor is the effect limited to single-celled plants. When exposed to increased UV-B in experiments, many crop plants react negatively. Peas, beans, squash, cabbage, and soybeans show reduced nutrient content, slower growth, and lower yields, as well as impaired photosynthesis. Plants vary in their response to UV-B, though. Some weeds, especially, seem to love it.

In humans, exposed body parts usually have some protection from UV-B. Pigmentation in the skin's surface layer stops UV rays before they reach lower layers. In the eye, the cornea and lens block UV from reaching the retina, the light-sensitive tissue that UV would damage.

But inborn defenses against UV go only so far. Some sun-induced damage still occurs.

Skin cancer. Exposure to UV-B is linked to all forms of skin cancer—basal-cell carcinoma, squamous-cell carcinoma, and malignant melanoma. The first two are quite common, and usually curable. But melanoma, which represents only 3 percent of all skin cancers, accounts for two-thirds of skin-cancer deaths. The EPA projects more than 60 million additional cases of skin cancer and about one million additional deaths among Americans alive today or born by the year 2075 if CFC usage continues to grow at its present rate.

Eye damage. Under extremely bright conditions, such as sunlight on snow, prolonged UV exposure can cause painful snowblindness, a temporary inflammation of the cornea. A rise in UV radiation may increase the frequency of such effects. Even more worrisome is the potential effect on the lens. Growing evidence suggests that chronic, lifetime exposure to UV contributes to certain types of cataracts—opaque regions in the lens that interfere with vision. The EPA projects some 17 million additional cases in the future caused by CFC damage to the ozone layer.

Some scientists believe that increased UV-B exposure might affect certain immune-system processes, but any danger is still speculative. No doubt exists, however, that UV causes sunburn and premature aging of the skin. A rise in UV will probably increase the incidence and seriousness of those conditions.

Source: "Can We Repair the Sky?" *Consumer Reports,* May, 1989, p. 323.

broken up and keeps most of the harmful rays of the sun from penetrating to the earth's surface.[61] These vital functions are performed without most people even being aware that such an ozone shield exists.

The Culprit

The culprit in ozone depletion was identified in 1974 by Mario Molina and Sherwood F. Rowland, two chemists at the University of California at Irvine, who theorized that chlorofluorocarbons (CFCs) eventually drifted up to the stratosphere to react chemically with ozone molecules in a destructive fashion. While many chemicals that are released into the atmosphere decay in weeks or months, CFCs are so chemically inert that they often can stay intact for a century. Normal disposal mechanisms had little or no effect on CFCs, as rain, for example, would not wash them out of the atmosphere since they were not soluble in water. These characteristics gave them ample time to rise through the atmosphere to reach higher altitudes and do their damage to the ozone layer.

Because of the nature of the chemical reactions, CFCs release chlorine atoms when finally broken down by ultraviolet radiation from the sun, and these chlorine atoms act as a catalyst in a series of reactions that convert ozone into oxygen (Figure 5.3). Because the chlorine acts as a catalyst rather than as a reagent, a single molecule of chlorine can destroy thousands of ozone molecules before it eventually gets washed out of the atmosphere. Another family of compounds called halons, which contain bromines, were discovered to be a hundred times more efficient that the chlorine compounds at ozone destruction.[62]

CFCs often take six to eight years to reach the upper layer of the atmosphere. The latest ozone measurements reflect only the response of the ozone layer to gases released through the mid- to late 1980s. The gases now rising through the lower atmosphere will take six to eight years to reach the stratosphere to do their damage. An

EXHIBIT 5.3 Ozone Destroyers

CFCs (or chlorofluorocarbons) are a whole family of chemicals that contain chlorine and fluorine. Some are more damaging to the ozone layer than others. Here are the most commonly used kinds:

CFC 11, CFC 12	The most common CFCs and the most destructive to the ozone layer. Also the most stable, with atmospheric lifetimes of 75 and 110 years, respectively.
CFC 22	Used in air conditioners and refrigerators. Poses less danger to the ozone layer than CFCs 11 & 12 because it breaks down more rapidly.
CFC 113	A solvent for metals and circuit boards. Its use has increased dramatically in the last decade.
Halons	Technically not CFCs, because they contain bromine. But represent a new and growing threat because bromine destroys ozone even faster than chlorine. Used in fire extinguishers.
"Safe" CFCs	Materials that contain hydrogen, making them break down faster. They include CFCs 123, 124, 133a and 502. A related safe chemical, which has no chlorine or bromine, is FC134a.

Source: Douglas Starr, "How to Protect the Ozone Layer," *National Wildlife,* December–January 1988, p. 27.

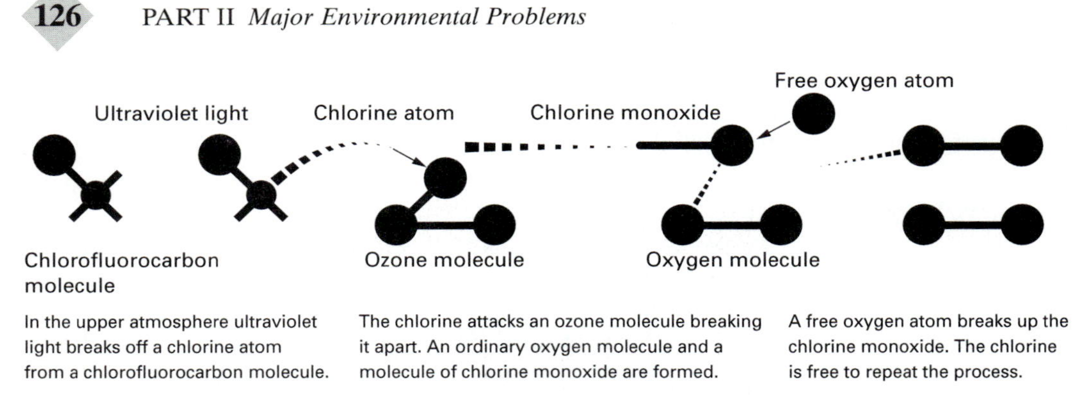

Source: United States Environmental Protection Agency, *Environmental Progress and Challenges: EPA's Update* (Washington, D.C.: U.S. Government Printing Office, 1988), p. 38.

FIGURE 5.3 How Ozone Is Destroyed

additional 2 million tons of substances containing chlorine and bromine were estimated to be trapped in insulation foams and appliances and firefighting equipment. Chlorine concentrations were expected to triple by 2075, according to some experts, while bromine concentrations were expected to grow faster, exhibiting a tenfold increase from current levels.[63]

When first discovered, CFCs proved to be remarkable compounds. Since they were inert, they did not react with other chemicals with which they were mixed. They were also neither toxic nor flammable at ground level. Chemists at General Motors are given credit for first discovering these compounds. The refrigeration industry needed a new refrigerant to survive in the 1920s, and in 1928, chemists at GM focused their attention on the strong carbon-fluorine bond of certain carbon-based compounds know as fluorocarbons. These fluorocarbons are chemically similar to hydrocarbon molecules, except that one or more hydrogen atoms are replaced by chlorine, fluorine, or bromine atoms. CFCs are fully chlorinated fluorocarbons, which means that they have no hydrogen atoms and are made of only chlorine, fluorine, and carbon.

After this initial discovery, the number of CFC compounds grew quickly into the dozens, and were used as a universal coolant, refrigerating 75 percent of the food consumed in the United States, as a blowing agent in rigid insulation forms, as an aerosal propellent, as a solvent to remove glue, grease, and soldering residues from microchips and other electronic products, and as a component of foam packaging containers (Table 5.1). Between 1958 and 1983, the average production of some forms of CFC compounds grew 13 percent a year, and could continue to grow more or less indefinitely.[64]

When Molina and Roland developed their theory, empirical validation was unavailable because of the difficulties of measuring actual levels of stratospheric ozone. But almost as soon as news of their work hit the popular press, consumers began switching to nonaerosol packaging for common household products such as deodorants. In 1978, the United States banned the use of CFCs as an aerosol propellent, but most of the rest of the world continued to use them for this purpose. At that time, the United States consumed about half of all CFCs manufactured worldwide, and aerosol uses accounted for about half of this consumption. Manufacturers of personal care products

TABLE 5.1 Global CFC Use by Category 1985

Aerosols	25%
Rigid Foam Insulation	19
Solvents	19
Air Conditioning	12
Refrigerants	8
Flexible Foam	7
Other	10
Total	100%

Source: From *State of the World 1989: A Worldwatch Institute Report on Progress Toward a Sustainable Society* by Lester R. Brown, et al., eds. Copyright © 1989 by Worldwatch Institute. Reprinted by permission of W. W. Norton & Company, Inc.

and of aerosol containers used in industry switched to other propellents, which included carbon dioxide and simple aliphatic hydrocarbons like propane and butane.[65]

Discovery of the Ozone Hole and Its Effects

Government and industry built computer models to simulate the chemical and physical processes that determine ozone levels in the stratosphere, some of which led scientists to believe that Molina and Rowland had overstated the problem. Thus no international action was taken to limit CFC usage until the discovery of the ozone hole over Antarctica. This phenomenon was first discovered in 1983 when the British Antarctic Survey discovered that concentrations of ozone in the stratosphere were dropping over Antarctica at a dramatic rate each austral spring to be replenished again by the end of the fall season. This discovery led to a $10 million scientific mission carried out by the United States under the combined sponsorship of NASA, the National Oceanic and Atmospheric Administration, and the Chemical Manufacturers Association to find out more about this phenomenon.[66]

By the spring of 1987, the average ozone concentration over the South Pole was discovered to be down 50 percent, and in isolated spots it had actually disappeared (Figure 5.4). The report on this discovery also indicated that the ozone layer around the entire globe was eroding much faster than any model had predicted. Ozone depletion was said to be occurring far more rapidly and in a different pattern than had been forecast. While the role of CFCs in ozone depletion had been hotly contested after the theory was formulated in 1974, within a matter of weeks the report's conclusions were widely accepted and the need for immediate policy decisions became apparent to many of the world's leaders.[67]

In the spring of 1985, representatives of 21 nations met under the auspices of the United Nations Environment Program to consider worldwide restrictions of CFC usage. The Soviet Union, Japan, and the European Economic Community argued for a freeze on existing CFC levels only, but the United States, Canada, and Scandinavian countries pressed for a virtual phaseout of CFCs by the year 2000. The phaseout proposal, a personal initiative of the EPA administrator, was opposed by the Reagan White

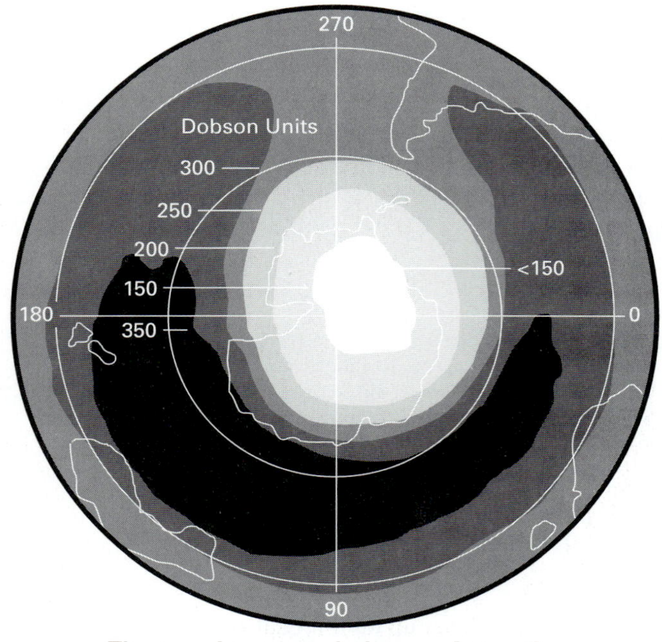

The growing ozone hole over Antarctica

Source: United States Environmental Protection Agency, *Preserving Our Future Today* (Washington, D.C.: EPA, 1991), p. 27.

FIGURE 5.4 Satellite Image of Antarctica

House. Interior Secretary Donald Hodel suggested that everyone could stay indoors or wear hats, sunglasses, and sunscreen for protection against increased ultraviolet radiation. The ridicule that greeted this proposal spurred a Senate resolution urging a two-stage phaseout of CFCs and halons. As a compromise, U.S. negotiators proposed a 50 percent reduction in CFC production and a freeze on halons. This plan formed the basis for the Montreal treaty.[68]

On September 16, 1987, after years of debate and heated negotiation, the Montreal Protocol on Substances that Deplete the Ozone Layer was signed by 24 countries. By mid-November of 1988, that total had increased to 35 countries. The agreement included a freeze on CFC production at 1986 levels to be reached by 1989, a 20 percent decrease in production by 1993, and another 30 percent cut by 1998. Halon production was also subject to a freeze based on 1986 levels starting in 1992. In order to obtain this many signatures, the treaty included extended deadlines for some countries, allowances to accommodate industry restructuring, and loose definitions of products that could legitimately be traded internationally. Developing countries were given a ten-year grace period past the industrial country deadline during which CFC production could be increased to meet "basic domestic needs."[69]

Six months after the protocol was signed, the U.S. Senate approved the protocol by a vote of 83 to zero. After this action, President Reagan promptly signed the ratification agreement, making the United States the second nation to ratify the agreement. The treaty entered into force on January 1, 1989. The treaty did try to distribute economic burdens fairly and was sensitive to special situations. It established periodic scientific, economic, and technical assessments so that specific provisions could be adapted to evolving conditions. There were even provisions for emergency meetings of signatories in case of unexpected and fast-breaking developments.[70]

The cumulative effect of the loopholes meant that even with widespread participation, the protocol's goals of halving CFC use by 1998 would not be attained. The agreement would most likely not arrest ozone depletion, merely slow its acceleration unless it was strengthened.[71] Since it takes CFCs several years to drift up to the ozone layer, there were still many in the pipeline, so to speak, that had yet to do their damage. But this treaty was an unprecedented effort to deal with a global problem and provided a model for agreements to deal with other global problems such as the greenhouse effect.[72]

In June 1990, representatives from 75 counties met in London to sign an accord that strengthened provisions of the treaty. Projections showed that if industrialized countries phased out CFCs as scheduled, but the less developed countries did not go along, these countries' use of CFCs would soar from 15 percent to 50 percent by the end of the century. That increase would leave chlorine levels slightly above the current level even with reductions by developed countries. Thus the pact called for eliminating CFC usage worldwide in a decade, and it set up an international fund of $200 billion to help less developed countries join the campaign. This aid would help subsidize purchase of CFC substitutes by less developed countries, and build new plants to produce refrigerators and other products that use CFC substitutes.[73]

In 1989, the U.S. Congress enacted an excise tax on ozone-depleting chemicals (ODCs) with far-reaching effects. Congress hoped the tax would discourage the use of ozone-depleting chemicals and encourage an expedited search for safe substitutes. The tax applied to any substance, which at the time of sale or use by a manufacturer, producer, or importer, was listed as an ODC and which is manufactured or produced in the United States or entered into the United States for consumption, use, or warehousing. The tax was initially set at $1.37 per pound, increasing to $1.67 per pound in 1992, and $2.65 per pound in 1993. Some ODCs were considered less depleting than others and were subject to a scaled-down rate. The tax was effective January 1, 1990, and affected manufacturers, wholesalers and retailers, and businesses that retain stocks of chemicals.[74]

The impact these measures would have on industry was expected to be severe. The estimated annual world production of the chemicals was worth about $2.2 billion and the industries that used them had annual sales of many additional billions of dollars. In the United States alone, CFCs represented a $28 billion industry that employed about 715,000 people in 5,000 companies.[75] Manufacturers in the United States sold about $750 million of the compounds annually to about 5,000 customers in the refrigeration, air-conditioning, automotive, plastic foam, and electronics industries. These industries, in turn, produced $27 billion each year in goods and services directly dependent on CFCs. Some $135 billion of installed equipment and products required the availability of CFC compounds for maintenance and repair. Thus measures to phase out production of CFCs jeopardized the life of existing equipment and forced industry to spend a good bit of money on engineering new products.[76]

CFCs were pervasive in the United States, contained in 100 million refrigerators, 90 million cars and trucks, 40,000 supermarket display cases, and 100,000 commercial building air-conditioners. Du Pont estimated that banning CFCs would render useless or require altering capital equipment valued at $135 billion in the United States alone. Most substitutes at the time were either toxic to humans or corroded the metal machinery in which they were used, or were simply inefficient coolants. Others were potentially explosive. Companies would have to invest millions of dollars in new plant construction to make substitute products so they wanted to be sure the substitutes were not toxic and did not fail to serve the purpose.[77]

Subsequent Developments

In February 1992, then-President Bush announced a speedup in the phaseout of ozone-destroying chemicals by U.S. manufacturers. Responding to a report from NASA stating that chlorine monoxide reached record levels over Canada, the United States, and Europe, the President said that the United States would phase out production of ozone-destroying chemicals, mainly CFCs, by the end of 1995, which was five years earlier than agreed upon under the international treaty. Chlorine monoxide reacts in the presence of sunlight to cause a thinning of the ozone layer. By measuring the level of this chemical, the level of ozone destruction can be predicted.[78]

Responding to a European study that showed ozone depletion much greater than earlier believed, the European Community took steps in 1992 to ban production and consumption of ozone-destroying chemicals by 1995, and reduce their use by 85 percent at the end of the following year. These bans were also five years ahead of the Montreal Protocol and two years earlier than previous bans issued by the Community. Some European nations decided to move even more quickly, as Germany and the Netherlands were considering comprehensive bans by 1993.[79] Other industrialized countries agreed to follow suit at an international conference in Copenhagen in late 1992 where an agreement was reached to speed up phasing out CFCs by 1996, and limits were proposed on the use of a class of substitute chemicals called hydrochorofluorocarbons (HCFCs) that are less destructive of the ozone shield than CFCs but are still harmful.[80]

Later that same year, satellite measurements by NASA showed that the ozone hole over Antarctica reached record levels, covering an area almost three times larger than the land area of the United States, some 8.9 million square miles.[81] In 1993, the ozone layer showed a sudden and steep reduction, falling 2 to 3 percent below the lowest levels observed in earlier years.[82] Then in the winter of 1993–1994, ozone over the United States recovered from the record low levels of the previous winter rising to slightly above normal. Some scientists thought that chemicals from the eruption of Mount Pinatubo in the Philippines helped cause the unusual decline the year before, and as that material settled out of the air, ozone levels recovered.[83]

Developments such as the latter caused something of a backlash regarding phasing out CFCs to protect the ozone layer. The backlash seemed to center on two points: (1) that CFCs do not destroy the ozone layer, and (2) that even if ozone is thinning, it poses no threat to human health. Regarding the former point, a Nobel Prize–winning chemist at Texas A&M University signed a petition calling for repeal of the international treaty phasing out CFCs even though he hadn't looked into the issue scientifically.[84] Other scientists suggested that the ozone hole is a natural and transitory phenomenon related to sea temperatures, volcanic eruptions, tropical wind patterns, and sunspot cycles.[85]

This backlash picked up steam in 1994 with the election of a Republican Congress, which picked up on these suggestions. Some wanted to move the deadline for a total ban on production of CFCs back to the end of the decade. Others wanted to repeal the ban entirely.[86] Many who thought the battle over CFCs was over, had to regroup to counter the backlash with new evidence of their own to support the ban on CFCs and reaffirm the seriousness of the situation. NASA released new satellite information, for example, that proved to some scientists that manmade chemicals caused the ozone hole over the South Pole and the global thinning of the protective ozone layer. The amount of hydrogen fluoride measured by the satellite corresponded directly to the amount of CFCs in the atmosphere, which meant that the thinning of the ozone layer cannot be caused by chemicals from volcanoes or other natural sources, according to the project scientist.[87]

Regarding the latter point, a UN panel stated that ozone levels could drop by 3 percent during the next decade, leading to a 10 percent rise in skin cancer. Depletion of the ozone layer during the summer means more ultraviolet radiation will reach the surface of the earth when people are most likely to be basking in the sunshine.[88] Meanwhile, the number of skin cancer cases in the United States continued to increase, as more new cases of skin cancer were diagnosed every year than all other types of cancer combined. As of 1994, there were between 900,000 and 1.2 billion new cases of nonmelenoma skin cancer a year, and some doctors said there was no evidence that the epidemic of skin cancer in the United States had peaked.[89]

Strategies

Long-term reductions of CFCs involve alternative product disposal methods, use of substitute chemicals, and development of new process technologies. Use of CFC propellents in aerosols, which is still a large source of CFC emissions worldwide, can be banned. Rapid evaporation of cleaning solvents can also be eliminated. Since for this use CFCs are not incorporated into the final product, emissions from this use can be reduced immediately. Capturing CFC emissions from flexible foam manufacturing can also be accomplished fairly quickly, but it requires an investment in new ventilation systems. Another area that offers significant savings at low cost is improved design, operating, and maintenance standards for refrigeration and air-conditioning equipment. Refrigerants that are drained during systems' recharging can be recovered. But to recover CFCs from junked automobiles and other appliances years after they are produced requires either a collection system or a bounty scheme to encourage reclamation by salvagers.[90]

Almost 8 million refrigerators and freezers are thrown away in the United States every year, and without refrigerant recovery, about 4 million pounds of ozone-depleting chemicals escape from these appliances every year. In 1990, Congress amended the Clean Air Act to include laws that prohibit the release of ozone-depleting refrigerants into the atmosphere during the service, maintenance, or disposal of air-conditioning and refrigeration equipment. As of July 1, 1992, refrigerants must be recovered from the appliance before it is disposed of to avoid potential penalties of up to $25,000 per day. Recovery equipment must be used that draws the refrigerant into a holding tank where it is purified and sold for reuse, and only then can the appliance be discarded without harmful effects.[91]

Over the longer term, however, phasing out the use and emission of CFCs will require the development of chemical substitutes that do not harm the ozone layer (Exhibit 5.4). The challenge for business and government is to find alternatives that perform the same function at a reasonable cost, that do not require major equipment modifications, that are nontoxic to workers and consumers, and that are environmentally benign. One of the major delays associated with new chemical compounds is the need for extensive toxicity testing that can run from five to seven years. In some cases, it may be possible that a new product design can eliminate or at least reduce the need for CFCs or substitute chemicals while providing additional benefits. Manufacturers must ask if the functions performed by CFCs are really necessary.[92]

Chemical companies such as Du Pont, the world's largest producer of CFCs, and Imperial Chemical Industries (ICI) in Britain are now in a race to find and develop substitutes that can be mass produced at affordable costs to industry.[93] Approximately 14 major chemical companies worldwide are working on studies of possible substitutes. The key to substitutes or alternatives to CFCs may be in the use of hydrogen. If one or more hydrogen atoms are substituted for a chlorine or fluorine atom in a CFC molecule, the desirable thermodynamic properties may be retained. The new compound that

EXHIBIT 5.4 Ozone Uses and Substitutes

Uses	*Chemical*	*What Industry Can Do*	*What You Can Do*
Blowing bubbles into flexible foam for furniture, bedding, carpet padding, dashboards	CFC 11	Substitute methylene chloride (a suspected carcinogen) or CFC 123 as blowing agent.	Buy spring mattresses and non-foam furniture.
Blowing bubbles into rigid foams for egg cartons, coffee cups, home insulation	CFC 11 CFC 12	Use pentane (which is flammable) for blowing or improve the recovery of CFC vapors.	Buy food in cardboard packaging. Use fiberglass or cellulose insulation.
Solvent for dry-cleaning leather goods, cleaning printed circuit boards, degreasing metal	CFC 113	Substitute water for cleaning circuit boards and use other solvents for dry-cleaning, or recycle CFCs.	Shop for washable fabrics.
Refrigerant in automobile air conditioners, the greatest single consumer source of CFCs	CFC 12	Substitute CFC 22 (which requires higher pressures) or FC134a (not yet commercially available).	Replace air-conditioner hoses every three years. Ask mechanics to drain coolant into bottles, rather than letting it evaporate.
Used in fire extinguishers for high-tech equipment	Halon	No good substitutes, but industry can recycle unused extinguishers.	Buy other types of extinguishers for home applications.

Source: Douglas Starr, "How to Protect the Ozone Layer," *National Wildlife,* December–January 1988, p. 28.

is formed in this manner is less stable that CFCs, and most of it will break down in the lower atmosphere before reaching the stratospheric ozone layer. But if too many hydrogen atoms are introduced, the compound may become flammable.

Five such hydrochlorofluorocarbons (HCFCs) have been developed and are used widely. Compared with CFC compounds, HCFCs have far less effect on the ozone layer, something in the range of 2 to 5 percent of the ozone-depletion potential on a weight basis. Since chlorine is the element that causes the major problem with CFCs, the ideal substitute would be comprised of only carbon, hydrogen, and fluorine. These compounds called hydrofluorocarbons (HCFs) have no potential to deplete the ozone layer. To date, at least three HCF compounds have been developed in limited quantities. They are being developed by Du Pont as a replacement for CFC usage in automobile air-conditioning systems. At the present time, there are toxicity as well as lubrication problems associated with the compound, but if these problems can be solved, commercial production is scheduled for the immediate future.[94]

In mid-1990, Du Pont announced plans to design four world-scale plants to produce its line of HCFs, which would become operational somewhere around the middle of the decade. These plants will be capable of producing more than 140 million pounds of HCFs annually and could supply most of the world's needs for the compounds through the end of the century. The company said that overall costs of commercializing its line of CFC alternatives could total more than $1 billion in research, development, and capital projects, including $240 million invested through 1990 and projected costs over the next ten years.[95]

Refrigerator makers were planning to make their product without using CFCs, a challenge to most engineers that will not be easy to solve. In a refrigerator that weighs 250 pounds and is priced at $700, CFCs weigh less than three pounds and cost no more than $5 to install initially. These CFCs served two important functions: (1) They circulated between the appliance's compressor and evaporator as a refrigerant, and (2) they made up part of the rigid foam insulation in the walls of the cabinet. Many of the substitutes reduced energy efficiency at a time when the industry was being asked to reduce the product's energy consumption. Recent gains in energy efficiency were largely a result of increased use of CFCs in the product. Other substitutes involved extensive retooling of compressors and potential liability from flammability.[96]

In 1989, the Natural Resources Defense Council, a few utilities, and the EPA got together to explore ways to reduce energy consumption, knowing that refrigerator makers would soon be revamping product lines to comply with EPA-mandated improvements in energy efficiency and the impending ban on CFC usage. Instead of thinking about new laws, they developed a contest called the Super Efficient Refrigerator Program (SERP) funded entirely by 24 utilities. The prize was $30 million in the form of a cash rebate to the manufacturer per refrigerator sold and bragging rights for the manufacturer that could produce a refrigerator that was both environmentally friendly and energy efficient. The new refrigerator had to eliminate CFCs entirely and boost energy efficiency by at least 25 percent over 1993 federal standards.[97]

Early in 1990, Carrier Corporation announced a recovery and recycling system for refrigerants that are responsible for 10 percent of CFC emissions in the country. Carrier manufactures cooling and heating systems. The plan called for the company to invest $750 million to help phase out the use of CFCs in air-conditioning refrigerants and to recover and recycle existing CFC refrigerants. Most of the CFCs coming from this source are released into the air during routine maintenance of large cooling units. The new equipment, which looks like a home hot-water tank turned on its side and mounted on wheels, is intended to be used by air-conditioning maintenance crews to drain, contain, and recycle refrigerants.[98]

In January 1991, Du Pont announced a family of refrigerator and air-conditioner coolants as substitutes for CFCs that were to be used by General Motors in their 1994 models. The substitutes go under the trade name "Suva" and required an investment of $240 million in ten facilities that were either in operation or under construction. The substitute was expected to sell for around $10 a pound in comparison to the CFC compound previously used, which sold for $2.40 a pound including a $1.37 special tax instituted by Congress. To accommodate the new coolant, GM will have to develop a larger evaporator and condenser and redesign the compressor on its air conditioners. It was estimated that GM will have to spend up to $1 billion over the next two to three years to convert to the new coolant.[99]

As a result of these efforts, CFC production declined for the sixth consecutive year in 1994, bringing total production levels down by 77 percent from the 1988 peak. A panel of scientists from around the globe concluded that these efforts meant that concentrations of chlorine and bromine had reached a peak in the lower atmosphere. Given the time lag before these chemicals reach the stratospheric ozone layer, depletion of this layer would continue to get worse until the end of the century. But if countries comply with their treaty obligations, the atmosphere should then begin to heal, and by about 2050, chlorine levels should return to where they were when the Antarctic ozone hole first appeared in the late 1970s.[100]

One impediment to these projections is that a black market for CFCs has developed. This black market topped 22,000 tons in 1994 according to some estimates, and is expected to continue at high levels. Some experts estimate that half the CFCs sold in 1995 may be contraband. Importers bring tons of CFCs from Europe into the United States, where most wind up with auto air-conditioner rechargers. Rather than using recycled CFCs as the law requires, some shops use the cheaper black market CFCs for recharging and save money by avoiding the federal excise tax of $5.35 per pound.[101]

Another problem is that other chlorine-based chemicals can continue to cause trouble if not controlled. They are methyl chloroform, used as a degreasing agent for metals and electronics-equipment parts, and carbon tetrachloride, used in the production of CFCs as well as in pesticides and certain kinds of dyes. Over the next century, methyl chloroform will account for 35 percent of the buildup of chlorine in the atmosphere unless it is controlled.[102] Other gases that may contribute to the problem are hydrogen chloride and hydrogen fluoride, which are emitted from volcanic eruptions. Volcanoes eject an average of about 11 million tons of hydrogen chloride and 6 million tons of hydrogen flouride into the atmosphere annually.[103]

Questions for Discussion

1. What are global problems? What makes these problems different from the more traditional environmental problems that have been around for some time? How can global problems be solved?
2. Describe the greenhouse effect. What are trace gases? Have these gases increased over the past several years? What predictions are made about future levels of these gases?

3. What are general climate models? What do they generally show with regard to global warming? What problems do these models have that raise some doubts about the validity of their predictions?
4. Given the evidence, has global warming actually taken place over the past decade? If so, is this warming attributable to the greenhouse effect? Why or why not? What other events could cause global warming?
5. What would be the effects of global climate change on the scale predicted by the models? What areas of the world would be most affected? What areas might benefit from such changes? Would you like to see these changes happen? Why or why not?
6. What are the risks involved in waiting any longer to take steps that would reduce trace gas emissions? What are the risks involved in cutting back on carbon dioxide emissions at this time, given the current state of knowledge? Are more studies necessary? What kind of studies would be helpful to policy makers?
7. What is a tie-in strategy? Does this approach to global warming make sense? What specific measures would you advocate to implement this strategy? What other strategies would you recommend? What should business do about the problem at this time?
8. Prioritize the specific measures to deal with global warming described in the chapter. Which of these measures are most important to begin immediately and which can wait until further information is discovered? Which would be most costly? Which would impact business organization most severely?
9. What functions does the ozone layer perform? What will happen if this layer continues to deteriorate? What are some of the cost estimates regarding the damage that would be done by further depletion?
10. What are CFCs? What uses have they had in our society? Why were they such a useful substance? Was there any way business could have tested for the ozone-depletion effects these substances are believed to possess?
11. Describe the story relative to the discovery of the effect CFCs have on the ozone layer. What role did theory and empirical evidence play in this story? Are there any lessons to be learned here that could apply to global warming? Do you believe the full story has been discovered?
12. Describe the events that led to the Montreal Protocol. What impacts was the treaty predicted to have on industry? Why was it revised soon after it was passed and ratified? How did Du Pont Company react to the treaty? Was Du Pont a responsible corporate citizen?

Endnotes

1. Another class of trace gases called fully fluorinated compounds (FFCs) has recently been identified. They are said to trap more heat than other trace gases and last forever by any human time scale. Companies are taking steps to limit and control their usage so they do not become a major problem. See Elizabeth Cook, "Curbing 'Immortal' Greenhouse Gases," *Tomorrow,* V, no. 4 (December 1995), 64.
2. Thomas E. Graedel and Paul J. Crutzen, "The Changing Atmosphere," *Scientific American,* 261, no. 3 (September 1989), 66.
3. Stephen H. Schneider, "The Changing Climate," *Scientific American,* 261, no. 3 (September 1989), 70.
4. Ibid.
5. Michael D. Lemonick, "The Heat Is On," *Time,* October 19, 1987, p. 63.
6. Schneider, "The Changing Climate," p. 72.
7. Ibid., p. 73.
8. Lester R. Brown et al., *Vital Signs 1995* (New York: W. W. Norton, 1995), pp. 66–67.
9. Christopher Flavin, "Carbon Rate Takes a Breather," *World Watch,* May–June 1992, pp. 33–34.

10. Lester R. Brown et al., *Vital Signs 1994* (New York: W. W. Norton, 1994), p. 68. See also Nicholas Lenssen, "All the Coal in China," *World Watch,* March–April 1993, pp. 22–29.
11. Christopher Flavin, "Slowing Global Warming," *State of the World 1990* (Washington, D.C.: Worldwatch Institute, 1990), p. 20.
12. Schneider, "The Changing Climate," p. 73.
13. Brown et al., *Vital Signs, 1995,* p. xx.
14. Schneider, "The Changing Climate," p. 74–75.
15. James Trefil, "Modeling Earth's Future Climate Requires Both Science and Guesswork," *Smithsonian,* 21, no. 9 (December 1990), 33. See also Carolyn Lockhead, "Global Warming Forecasts May Be Built on Hot Air," *Insight,* April 16, 1990, pp. 14–18.
16. Schneider, "The Changing Climate," p. 75.
17. Trefil, "Modeling Earth's Future Climate," p. 34. Scientists have recently improved climate models by plugging in new information about the role clouds play in absorbing sunlight, how much carbon dioxide is absolved by forests and oceans, the cooling effect of pollution and volcanic eruptions, and historical climate. These models, while still speculative, predict lower potential warmings, about 2 degrees Fahrenheit to 4 degrees Fahrenheit further into the future. See Emily T. Smith, "Global Warming: The Debate Heats Up," *Business Week,* February 27, 1995, pp. 119–120.
18. Schneider, "The Changing Climate," p. 75.
19. Ibid., p. 72.
20. "1990 Was Hot Year for Earth," *Times-Picayune,* January 10, 1991, p. A10. See also "Hot Times," *Time,* January 21, 1991, p. 65.
21. Brown et al., *Vital Signs,* 1995, pp. 64–65.
22. William K. Stevens, "Global Temperature Rise Boosts Warming Theory," *Times-Picayune,* January 4, 1996, p. A4.
23. Daniel Haney, "Study Bucks Theory of Global Warming," *Times-Picayune,* February 18, 1990, p. B8.
24. John Carey, "Is the World Heating Up? Well, Just Listen," *Business Week,* February 4, 1991, pp. 82–83. The effort to measure ocean temperatures in this manner ran into difficulty over the issues of animal rights and noise pollution. Fears surfaced that the noise these devices would produce could deafen marine creatures and drown out mating calls, for example. See Tracie Cone, "Sound-Wave Experiment Sparks Whale of a Fight," *Times-Picayune,* May 1, 1994, p. A20.
25. Schneider, "The Changing Climate," p. 77.
26. Ibid.
27. Paulette Thomas, "EPA Predicts Global Impact from Warming," *Wall Street Journal,* October 21, 1988, p. B4.
28. Amal Kumar Haj, "Weather Scientists See Rainfall Shift to Downpours," *Wall Street Journal,* September 21, 1995, p. B7.
29. David Bjerklie, "One Big, Bad Iceberg," *Time,* March 20, 1995, p. 65; Sharon Begley, "Ice Cubes for Penguins," *Newsweek,* April 3, 1995, p. 56.
30. Eugene Linden, "Burned by Warming," *Time,* March 14, 1994, p. 79.
31. David Brand, "Is the Earth Warming Up?" *Time,* July 4, 1988, p. 18. See also Bob Davis and David Wessel, "NASA Aide Says White House Made Him Dilute Testimony on Greenhouse Effect," *Wall Street Journal,* May 9, 1989, p. B4.
32. Sierra Club, "Global Warming," May 1989, p. 3.
33. Bill McKibben, *The End of Nature* (New York: Random House, 1989), p. 29.
34. See Eugene Linden, "Big Chill for the Greenhouse," *Time,* October 31, 1988, p. 90. Other complications appeared when a scientist argued that there is another cycle of seasons or another year hidden in the weather records. This year was labeled the anomalistic year and is determined by the distance of the earth from the sun, which varies during the year be-

cause of the earth's orbit and apparently has some effect on seasonal temperatures in various localities around the earth. Around 1940, this temperature cycle began showing a sharp change that went well beyond what might be caused by the annual changes in distances. See Jerry E. Bishop, "Long-Ignored Cycle in Climate Suggests Worse Greenhouse Effect Than Thought," *Wall Street Journal,* April 11, 1995, p. B5.

35. See Carolyn Lockhead, "The Alarming Price Tag on Greenhouse Legislation," *Insight,* April 16, 1990, pp. 10–13.
36. Schneider, "The Changing Climate," p. 78.
37. Rose Gutfeld, "Earth Summitry: How Bush Achieved Global Warming Pact with Modest Goals," *Wall Street Journal,* May 27, 1992, p. A1.
38. Philip Elmer-Dewitt, "Not Just Hot Air," *Time,* May 3, 1993, p. 60; Emily T. Smith, "One Year After Rio: The U.S. Waves a Green Flag," *Business Week,* June 14, 1993, pp. 90–92.
39. Robert S. Boyd, "Middle-of-Road Plan to Tackle Global Warming," *Times-Picayune,* October 20, 1993, p. A12.
40. Charles L. Harper, Jr., "Time to Phase Out Fossil Fuels?" *Wall Street Journal,* December 26, 1995, p. A6. See also Eduardo Lachica, "Asia Faces Increasing Pressure to Act as Global Warming Threatens Its Coasts," *Wall Street Journal,* August 23, 1994, p. B7.
41. Flavin, "Slowing Global Warming," p. 18.
42. Ibid., p. 20.
43. Ibid., p. 22.
44. Sierra Club, "Global Warming," May 1989, p. 2.
45. Ibid.
46. Schneider, "The Changing Climate," p. 78.
47. Flavin, "Slowing Global Warming," p. 28. See also Martin Feldstein, "The Case for a World Carbon Tax," *Wall Street Journal,* June 4, 1992, p. A10.
48. Ibid., pp. 35–36.
49. Ibid., p. 37.
50. Lockhead, "The Alarming Price Tag on Greenhouse Legislation," p. 10.
51. Bob Davis, "Bid to Slow Global Warming Could Cost U.S. $200 Billion a Year, Bush Aide Says," *Wall Street Journal,* April 16, 1990, p. B5.
52. Boyd, "Plan to Tackle Global Warming," p. A12.
53. Curtis A. Moore, "Fixing the Atmosphere," *International Wildlife,* May–June 1989, pp. 19–23.
54. Flavin, "Slowing Global Warming," pp. 22–23.
55. Ibid., p. 25.
56. Ibid., p. 31.
57. Ibid., p. 32.
58. Lemonick, "The Heat Is On," p. 67.
59. Cynthia Pollock Shea, "Protecting the Ozone Layer," *State of the World 1989* (New York: W. W. Norton, 1989), p. 82.
60. Ibid., pp. 83–85.
61. Lemonick, "The Heat Is On," p. 61.
62. McKibben, *The End of Nature,* pp. 39–40. See also Amal Kumar Naj, "Bromines May Be Harming Ozone Layer as Much as Fluorocarbons, Report Says," *Wall Street Journal,* August 30, 1988, p. 37.
63. Shea, "Protecting the Ozone Layer," p. 87.
64. McKibben, *The End of Nature,* p. 39.
65. Forest Reinhardt, "Du Pont Freon Products Division (A)," in Managing Environmental Issues: A Casebook, Rogene A. Buchholz, Alfred A. Marcus, and Jame E. Post, eds. (Upper Saddle River, NJ: Prentice Hall, 1992), p. 267.
66. Lemonick, "The Heat Is On," p. 59.
67. Shea, "Protecting the Ozone Layer," p. 81.

68. "Can We Repair the Sky?" *Consumer Reports,* May 1989, p. 324.
69. Shea, "Protecting the Ozone Layer," pp. 93–94.
70. Richard Benedick, "Diplomacy and the Ozone Crisis," *The GAO Journal,* Summer 1989, p. 37.
71. Shea, "Protecting the Ozone Layer," p. 94.
72. Laurie Hays, "Du Pont Plans to Complete Phase-Out Of Chlorofluorocarbons by Year 2000," *Wall Street Journal,* July 29, 1988, p. 16.
73. Vicky Cahan, "Fixing the Hole Where the Rays Come In," *Business Week,* July 2, 1990, p. 58.
74. Grant Thornton, *National Tax Alert,* February 23, 1990, pp. 1–3.
75. Laurie Hays, "CFC Curb to Save Ozone Will Be Costly," *Wall Street Journal,* March 28, 1988, p. 5.
76. Elliott D. Lee, "Limits Worry Chlorofluorocarbon Firms," *Wall Street Journal,* September 15, 1987, p. 6.
77. Amal Kumar Naj, "As CFC Phase-Out Looms, Doubts on Substitutes Arise," *Wall Street Journal,* March 6, 1989, p. B4.
78. Terrance Hunt, "Bush Speeds Up Chemical Plan to Save Ozone," *Times-Picayune,* February 12, 1992, p. A3.
79. Mark Fritz, "Europeans Rush to Ban Ozone-Eating Chemicals," *Times-Picayune,* March 5, 1992, p. A21.
80. Barbara Rosewicz, "Industrialized Nations to Hasten End to Use of Ozone-Damaging Chemicals," *Wall Street Journal,* November 27, 1992, p. B4.
81. Paul Recer, "Polar Ozone Hole Grows to 3 Times Land Area of U.S.," September 30, 1992, p. A3.
82. Amal Kumar Naj, "Ozone Problem in Atmosphere Becomes Worse," *Wall Street Journal,* April 23, 1993, p. B7.
83. Randolph E. Schmid, "Protective Ozone Layer Recovers from Record Low," *Times-Picayune,* August 27, 1994, p. A5.
84. Sharon Begley, "Is the Ozone Hole in Our Heads?" *Newsweek,* October 11, 1993, p. 71.
85. Paul Craig Roberts, "What's Flying Out the Ozone Hole? Billions of Dollars," *Business Week,* June 13, 1994, p. 22.
86. John Carey, "Shy Business Doesn't Back the GOP Backlash on the Ozone," *Business Week,* July 24, 1995, p. 47.
87. Paul Recer, "Ozone Hole Caused by CFCs, NASA Says," *Times-Picayune,* December 20, 1994, p. D14.
88. "Ozone Loss: Cancer Risk Could Accelerate," *Times-Picayune,* October 23, 1991, p. A11.
89. "Number of Skin Cancer Cases Soars," *Times-Picayune,* May 3, 1994, p. A4.
90. Shea, "Protecting the Ozone Layer, pp. 89–90.
91. United States Environmental Protection Agency, "Disposing of Appliances with Refrigerants: What You Should Know," Office of Air and Radiation, May 1993.
92. Shea, "Protecting the Ozone Layer, pp. 91–92.
93. See Joseph Weber, "Quick, Save the Ozone," *Business Week,* May 17, 1993, pp. 78–79.
94. Concerns were raised in 1995 that some CFC replacements when released into the atmosphere where they broke down produced a by-product that was carried back to the earth by rain and was found to inhibit plant growth in certain concentrations. See Amal Kumar Naj, "Replacements for CFCs Cause Controversy," *Wall Street Journal,* August 22, 1995, p. B13.
95. "Du Pont Co. Plans 4 Hydrofluorocarbon Alternatives Plants," *Wall Street Journal,* June 22, 1990, p. A2.
96. Richard Koenig, "Refrigerator Makers Plan for Future Without CFCs," *Wall Street Journal,* December 15, 1989, p. B1.
97. James B. Treece, "The Great Refrigerator Race," *Business Week,* July 5, 1993, pp. 78–81.

98. Charles W. Stevens, "Carrier Unveils Recycling Unit for Refrigerants," *Wall Street Journal,* January 9, 1990, p. B3.
99. Amal Kumar Naj, "Du Pont Unveils Coolants to Substitute for CFCs; GM Plans Use in 1994 Cars," *Wall Street Journal,* January 22, 1991, p. B4.
100. Brown et al., *Vital Signs* 1995, pp. 62–63.
101. Sharon Begley, "Holes in the Ozone Treaty," *Newsweek,* September 25, 1995, p. 70; Rae Tyson, "Nearly Banned Coolants Are Hot Illegal Imports," *USA Today,* December 21, 1995, p. 5A.
102. Vicky Cahan, "Just When the Ozone War Looked Winnable . . . ," *Business Week,* June 12, 1989, p. 56.
103. "A Natural Culprit in Ozone Depletion," *Insight,* September 5, 1988, p. 57.

Suggested Reading

Abrahamson, Dean Edwin, ed. *The Challenge of Global Warming.* Washington, D.C.: Island Press, 1989.

Barker, Terry, et al., eds. *Global Warming and Energy Demand.* New York: Routledge, 1994.

Benarde, Melvin A. *Global Warning: Global Warming.* New York: Wiley, 1992.

Benedick, Richard E. *Ozone Diplomacy: New Directions in Safeguarding the Planet.* Cambridge: Harvard University Press, 1991.

Berhard, Harold W., Jr. *Global Warming Unchecked: Signs to Watch For.* Cambridge: Cambridge University Press, 1993.

Cogan, Douglas G. *Stones in a Glass House: CFCs and Ozone Depletion.* Washington, D.C.: Investor Responsibility Research Center, 1988.

Dornbusch, Rudiger, and James M. Poterba. *Global Warming: Economic Policy Responses.* Cambridge: MIT Press, 1991.

Fisher, David E. *Fire and Ice: The Greenhouse Effect, Ozone Depletion, and Nuclear Winter.* New York: Harper & Row, 1990.

Flavin, Christopher. "Slowing Global Warming." *State of the World 1990.* Washington, D.C.: Worldwatch Institute, 1990.

Litfin, Karen. *Ozone Discourse: Science and Politics in Global Environmental Cooperation.* New York: Columbia University Press, 1994.

McKibben, Bill. *The End of Nature.* New York: Random House, 1989.

Mintzer, Irving M., et al. *Protecting the Ozone Shield: Strategies for Phasing Out CFCs During the 1990s.* Washington, D.C.: World Resources Institute, 1989.

Mitchell. George J. *World on Fire: Saving an Endangered Earth.* New York: Scribner, 1991.

National Academy of Sciences. *Global Environmental Change.* Washington, D.C.: National Academy Press, 1989.

Oppenheimer, Michael, and Robert H. Boyle. *Dead Heat: The Race Against the Greenhouse Effect.* New York: Basic Books, 1990.

Peters, Robert L., and Thomas E. Lovejoy. *Global Warming and Biological Diversity.* New Haven: Yale University Press, 1992.

Roan, Sharon L. *Ozone Crisis.* New York: Wiley, 1989.

Schneider, Stephen H. *Global Warming: Are We Entering the Greenhouse Century?* San Francisco: Sierra Club Books, 1989.

Shea, Cynthia Pollack. *Protecting Life on Earth: Steps to Save the Ozone Layer.* Washington, D.C.: Worldwatch Institute, 1988.

Silver, Cheryl Simon, and Ruth S. Defries. *One Earth, One Future: Our Changing Global Environment.* Washington, D.C.: National Academy Press, 1990.

United Nations Environmental Program. *The Greenhouse Gases.* Washington, D.C.: UNEP, 1988.

United States Environmental Protection Agency, *The Potential Effects of Global Climate Change on the United States.* Washington, D.C.: EPA, 1988.

CHAPTER 6 Air Pollution

The air we breathe is just like any other aspect of the environment as far as pollution is concerned. It has a certain dilutive capacity; that is, air will absorb a certain amount of pollution without the quality of the air being seriously affected. But modern industrial processes have used the air to dump such large amounts of certain pollutants that its dilutive capacity has been exceeded and the quality of the air has been seriously affected. The air in certain places can become unhealthy to breathe and can cause serious illnesses if breathed continuously. Thus air pollution is a serious problem in many places of the world and can no longer be ignored if human health is of consideration.

About 99 percent of the air that we breathe is gaseous nitrogen and oxygen. But as air moves across the earth's surface, it picks up trace amounts of various chemicals produced by natural events and human activities, minute droplets of various liquids, and tiny particles of various solid materials. These pollutants mix vertically and horizontally and often react chemically with each other or with natural components of the atmosphere. Continuous or repeated exposure to some of these pollutants can damage lung tissue, plants, buildings, metals, and other materials. The movement of the air and atmospheric turbulence help dilute many of these pollutants, but some of the more long-lived pollutants can be transported great distances before they return to the earth's surface.[1]

These pollutants can be classified as primary or secondary pollutants. The definition of a primary pollutant is one that enters the air as a result of natural events or human activities such as industrial processes. Substances such as carbon monoxide, carbon dioxide, sulfur dioxide, nitrogen oxide, most hydrocarbons, and most suspended particulate matter are examples of primary pollutants. Secondary air pollutants are those formed in the air itself through a chemical reaction between a primary pollutant and one or more components of the air. Typical secondary pollutants are ozone, most nitrates and sulfates, and liquid droplets of chemicals such as sulfuric acid.[2]

Natural sources of air pollutants include forest fires, dispersal of pollen, volcanic eruptions, sea spray, bacterial decomposition of products of organic matter, and natural radioactivity. With the exceptions of large volcanic eruptions and buildup of radioactive radon gas inside buildings, most atmospheric emissions from widely scattered natural sources are diluted and dispersed throughout the world and rarely reach concentrations high enough to cause serious damage to the environment or to human health.

Major Air Pollution Problems

There have been several times in recent history when major problems with air pollution killed hundreds of people. In 1952, an air pollution incident in London killed 4,000 people, and further disasters in 1956, 1957, and 1962 killed a total of about 2,500 people. The first major air pollution disaster in the U.S. occurred in 1948 in the town of Donora in Pennsylvania's Monongahela Valley when fog laden with sulfur dioxide vapor and suspended particulate matter from nearby steel mills hung over the town for five days. About 6,000 of the town's 14,000 inhabitants fell ill and 20 were killed as a result of the pollution. In 1963, high concentrations of air pollutants accumulated in the air over New York City, killing about 300 people and injuring thousands.

Source: G. Tyler Miller, Jr., *Living in the Environment* (Belmont, CA: Wadsworth, 1990), p. 487. © Wadsworth Publishing Co.

The most serious threat comes from pollutants released into the air as a result of human activities. Much of the outdoor pollution in the United States comes from six types of pollutants including ozone, carbon monoxide, nitrogen dioxide, particulate matter, sulfur dioxide, and lead. The nature and health effects of these pollutants are varied (see Exhibit 6.1).

Years of exposure to these pollutants can overload or deteriorate the natural defenses of the human body and cause or contribute to a number of respiratory diseases such as lung cancer, chronic bronchitis, and emphysema. Elderly people, infants, pregnant women, and persons with heart disease, asthma, or other respiratory diseases are especially vulnerable to air pollution. The World Health Organization (WHO) estimates that nearly 1 billion urban dwellers, or almost one of every five people on earth, are being exposed to health hazards from air pollutants.[3] Despite improvements in air quality in the United States since 1970, the EPA estimated that in 1994, 62 million people still lived in counties where air quality levels exceeded the national air quality standards for at least one of the six principal pollutants.[4]

Some of the air pollutants also cause direct damage to leaves of plants and trees and cause crop damage that can run into the billions of dollars. Chronic exposure to air pollutants interferes with photosynthesis and plant growth, reduces nutrient uptake, and causes leaves or needles to turn yellow and or brown, and in some cases, to drop off altogether. Some kinds of air pollution can also leach vital plant nutrients from the soil and kill essential soil microorganisms. Trees can also be made more vulnerable to drought, frost, insects, fungi, mosses, and diseases by some kinds of pollutants. The effects of such chronic exposure of trees to multiple air pollutants may not be visible for several decades, but then suddenly large numbers begin dying because of nutrient depletion and increased susceptibility to other environmental elements.[5]

Air pollutants also damage various materials to the tune of millions of dollars. The fallout of soot and grit on buildings and clothing requires costly sandblasting and cleaning to restore to a decent condition. Irreplaceable marble statues, historic buildings, and stained-glass windows throughout the world have been pitted and discolored

EXHIBIT 6.1 Major Air Pollutants and Their Health Effects

Pollutant	*Sources*	*Effects*
Ozone. A colorless gas that is the major constituent of photochemical smog at the earth's surface. In the upper atmosphere (stratosphere), however, ozone is beneficial, protecting us from the sun's harmful rays.	Ozone is formed in the lower atmosphere as a result of chemical reactions between oxygen, volatile organic compounds, and nitrogen oxides in the presence of sunlight, especially during hot weather. sources of such harmful pollutants include vehicles, factories, landfills, industrial solvents, and numerous small sources such as gas stations, farm and lawn equipment, etc.	Ozone causes significant health and environmental problems at the earth's surface, where we live. It can irritate the respiratory tract, produce impaired lung function such as inability to take a deep breath and cause throat irritation, chest pain, cough, lung inflammation, and possible susceptibility to lung infection. Smog components may aggravate existing respiratory conditions like asthma. It can also reduce yield of agricultural crops and injure forests and other vegetation. Ozone is the most injurious pollutant to plant life.
Carbon Monoxide. Odorless and colorless gas emitted in the exhaust of motor vehicles and other kinds of engines where there is incomplete fossil fuel combustion.	Automobiles, buses, trucks, small engines, and some industrial processes. High concentrations can be found in confined spaces like parking garages, poorly ventilated tunnels, or along roadsides during periods of heavy traffic.	Reduces the ability of blood to deliver oxygen to vital tissues, affecting primarily the cardiovascular and nervous systems. Lower concentrations have been shown to adversely affect individuals with heart disease (e.g., angina) and to decrease maximal exercise performance in young, healthy men. Higher concentrations can cause symptoms such as dizziness, headaches, and fatigue.
Nitrogen Dioxide. Light brown gas at lower concentrations; in higher concentrations becomes an important component of unpleasant-looking brown, urban haze.	Result of burning fuels in utilities, industrial boilers, cars, and trucks.	One of the major pollutants that causes smog and acid rain. Can harm humans and vegetation when concentrations are sufficiently high. In children, may cause increased respiratory illness such as chest colds and coughing with phlegm. For asthmatics, can cause increased breathing difficulty.
Particulate Matter. Solid matter or liquid droplets from smoke, dust, fly ash, and condensing vapors that can be suspended in the air for long periods of time.	Industrial processes, smelters, automobiles, burning industrial fuels, woodsmoke, dust from paved and unpaved roads, construction, and agricultural ground breaking.	These microscopic particles can affect breathing and respiratory symptoms, causing increased respiratory disease and lung damage and possibly premature death. Children, the elderly, and people suffering from heart or lung disease (like asthma) are especially at risk. Also damages paint, soils clothing, and reduces visibility.

(continued)

EXHIBIT 6.1 *(Continued)*

Pollutant	*Sources*	*Effects*
Sulfur Dioxide. Colorless gas, odorless at low concentrations but pungent at very high concentrations.	Emitted largely from industrial, institutional, utility, and apartment-house furnaces and boilers, as well as petroleum refineries, smelters, paper mills, and chemical plants.	One of the major pollutants that causes smog. Can also, at high concentrations, affect human health, especially among asthmatics (who are particularly sensitive to respiratory tract problems and breathing difficulties that SO_2 can induce). Can also harm vegetation and metals. The pollutants it produces can impair visibility and acidify lakes and streams.
Lead. Lead and lead components can adversely affect human health through either ingestion of lead-contaminated soil, dust, paint, etc., or direct inhalation. This is particularly a risk for young children, whose normal hand-to-mouth activities can result in greater ingestion of lead-contaminated soils and dusts.	Transportation sources using lead in their fuels, coal combustion, smelters, car battery plants, and combustion of garbage containing lead products.	Elevated lead levels can adversely affect mental development and performance, kidney function, and blood chemistry. Young children are particularly at risk due to their greater chance of ingesting lead and the increased sensitivity of young tissues and organs to lead.
Toxic Air Pollutants. Includes pollutants such as arsenic, asbestos, and benzene.	Chemical plants, industrial processes, motor vehicle emissions and fuels, and building materials.	Known or suspected to cause cancer, respiratory effects, birth defects, and reproductive and other serious health effects. Some can cause death or serious injury if accidentally released in large amounts.
Stratospheric Ozone Depleters. Chemicals such as chlorofluorocarbons (CFCs), halons, carbon tetrachloride, and methyl chloroform that are used in refrigerants and other industrial processes. These chemicals last a long time in the air, rising to the upper atmosphere where they destroy the protective ozone layer that screens out harmful ultraviolet (UV) radiation before it reaches the earth's surface.	Industrial household refrigeration, cooling and cleaning processes, car and home air conditioners, some fire extinguishers, and plastic foam products.	Increased exposure to UV radiation could potentially cause an increase in skin cancer, increased cataract cases, suppression of the human immune response system, and environmental damage.
Greenhouse Gases. Gases that build up in the atmosphere that may induce global climate change—or the "greenhouse effect." They include carbon dioxide, methane, and nitrous oxide.	The main man-made source of carbon dioxide emissions is fossil fuel combustion for energy-use and transportation. Methane comes from landfills, cud-chewing livestock, coal mines, and rice paddies. Nitrous oxide results from industrial processes, such as nylon fabrication.	The extent of the effects of climate change on human health and the environment is still uncertain, but could include increased global temperature, increased severity and frequency of storms and other "weather extremes," melting of the polar ice cap, and sea-level rise.

Source: United States Environmental Protection Agency, *What You Can Do to Reduce Air Pollution* (Washington, D.C.: EPA, 1992), pp. 4–5.

by air pollutants. Unless painted and maintained properly, iron and steel used in railroad tracks and to support bridges can become corroded and seriously weakened by air pollutants reacting with the metal. And various pollutants also damage leather, rubber, paper, paint, and certain kinds of fabrics.[6]

The total amount of pollution in the air over the United States at any given time adds up to hundreds of millions of tons. Table 6.1 shows the total emissions over several decades of the six most pervasive pollutants. The amount of these pollutants spewed into the air during many of these years was over 200 million metric tons (a metric ton is about 2,200 pounds), nearly a ton for every man, woman, and child in the country. Table 6.2 shows the source of these pollutants for 1993, the latest year for which figures are available. A glance at this table shows clearly that certain kinds of sources predominate for different types of pollution. Highway vehicles in 1993, for example, accounted for 62 percent of carbon monoxide emissions. Stationary sources of fuel combustion such as electric utilities accounted for 88 percent of sulfur dioxide emissions.

TABLE 6.1 National Air Pollutant Emissions, 1940–1993

[**In thousands of tons.** PM-10 = Particulate matter of less than ten microns. Methodologies to estimate data for 1900 to 1984 period and 1985 to present emissions differ. Beginning with 1985, the estimates are based on a modified National Acid Precipitation Assessment Program inventory]

Year	*PM-10*	*PM-10 Fugitive Dust*[1]	*Sulfur Dioxide*	*Nitrogen Dioxide*	*Volatile Organic Compounds*	*Carbon Monoxide*	*Lead*
1940	15,956	(NA)	19,954	7,568	17,118	90,865	(NA)
1950	17,133	(NA)	22,384	10,403	20,856	98,785	(NA)
1960	15,558	(NA)	22,245	14,581	24,322	103,777	(NA)
1970	12,838	(NA)	31,096	20,625	30,646	128,079	219,471
1980	6,928	(NA)	25,813	23,281	25,893	115,625	74,956
1983	5,849	(NA)	22,471	22,364	24,607	115,334	49,232
1984	6,126	(NA)	23,396	23,172	25,572	114,262	42,217
1985	3,676	44,701	23,148	22,853	25,417	112,072	20,124
1986	3,679	49,940	22,361	22,409	24,826	108,070	7,296
1987	3,630	42,131	22,085	22,386	24,338	105,117	6,840
1988	3,697	59,975	22,535	23,221	24,961	106,100	6,464
1989	3,661	53,323	22,653	23,250	23,731	100,806	6,099
1990, prel.	4,229	44,929	22,261	23,192	24,276	103,753	5,635
1991, prel.	3,902	49,127	22,149	22,977	23,508	99,898	5,020
1992, prel.	3,676	44,953	21,592	22,991	23,020	96,368	4,741
1993, prel.	3,688	41,801	21,888	23,402	23,312	97,208	4,885

NA Not available.

[1]Sources such as agricultural tilling, construction, mining and quarrying, paved roads, unpaved roads, and wind erosion.

Source: U.S. Bureau of the Census, *Statistical Abstract of the United States 1995,* 115th ed. (Washington, D.C.: U.S. Government Printing Office, 1995), p. 233.

TABLE 6.2 Air Pollution Emissions by Pollutant and Source, 1993

[In thousands of tons. See headnote, table 374]

Source	*Particulates*[1]	*Sulfur Dioxide*	*Nitrogen Oxide*	*Volatile Organic Compounds*	*Carbon Monoxide*	*Lead*
Total	**45,489**	**21,888**	**23,402**	**23,312**	**97,208**	**4,885**
Fuel combustion, stationary sources	1,212	19,266	11,690	648	5,433	497
Electric utilities	270	15,836	7,782	36	322	62
Industrial	219	2,830	3,176	271	667	18
Other fuel combustion	723	600	732	341	4,444	417
Residential	674	178	(NA)	310	4,310	9
Industrial processes	553	1,852	905	3,091	5,219	2,281
Chemical and allied product manufacturing	75	450	414	1,811	1,998	109
Metals processing	141	580	82	74	2,091	2,118
Petroleum and related industries	26	409	95	720	398	(NA)
Other	311	413	314	486	732	54
Solvent utilization	2	1	3	6,249	2	(NA)
Storage and transport	55	5	3	1,861	56	(NA)
Waste disposal and recycling	248	37	84	2,271	1,732	518
Highway vehicles	197	438	7,437	6,094	59,989	1,383
Light-duty gas vehicles and motorcycle	(NA)	(NA)	3,685	3,854	39,452	1,033
Light-duty trucks	(NA)	(NA)	1,387	1,612	14,879	(NA)
Heavy-duty gas vehicles	(NA)	(NA)	304	314	4,292	(NA)
Diesels	(NA)	(NA)	2,061	315	1,366	(NA)
Off highway[2]	395	278	2,986	2,207	15,272	206
Miscellaneous[3]	42,828	11	296	893	9,506	(NA)

NA Not available.

[1]Represents both PM-10 and PM-10 fugitive dust; see table 374.

[2]Includes emissions from farm tractors and other farm machinery, construction equipment, industrial machinery, recreational marine vessels, and small general utility engines such as lawn mowers.

[3]Includes emissions such as from forest fires and various agricultural activities, fugitive dust from paved and unpaved roads, and other construction and mining activities, and natural sources.

Source: Source of tables 374 and 375: U.S. Environmental Protection Agency, *National Air Pollutant Emission Trends, 1900–1993.* U.S. Bureau of the Census, *Statistical Abstract of the United States 1995,* 115th ed. (Washington, D.C.: U.S. Government Printing Office, 1995), p. 233.

Public policy measures designed to reduce air pollution date from the Air Pollution Act of 1955, which authorized the Public Health Service to undertake air pollution studies through a system of grants. This act created the first federally funded air pollution research activity. The Clean Air Act of 1963, which replaced the 1955 act, was aimed at the control and prevention of air pollution. It permitted legal steps to end specific instances of air pollution and authorized grants to state and local governments to initiate control programs. The 1965 amendments to the Clean Air Act (called the National Emissions Standards Act) gave the federal government authority to curb motor vehicle emissions and set standards, which were first applied to 1968 model vehicles.

The Air Quality Act of 1967 required the states to establish air quality regions with standards for air pollution control and implementation plans for their accomplishment. The Clean Air Act Amendments of 1970 provided the legal basis for a new system of national air quality standards to be set by the federal government and called for a rollback of auto pollution levels. In the Clean Air Act Amendments of 1977, new deadlines were set for the attainment of air quality standards. The Clean Air Act of 1990 focused on reduction of smog in cities across the country, toxic air emissions, and instituted a market-based approach for reducing sulfur dioxide emissions.

Thus the regulatory approach to pollution control is many-faceted, and management of the commons we call the atmosphere is a very complicated matter. The management system that has evolved over the years to control pollution and improve the quality of the air that we breathe is complex and deals with various levels of control. The Clean Air Act Amendments of 1970 and 1977 laid out several areas where different kinds of management methods are used to accomplish the goals of pollution control legislation with respect to air quality. These areas form a framework for understanding how the commons is being managed in this country.

- Air quality management (ambient air quality)
- Limitations on industrial growth in areas where the air quality is better than the national standards, termed Prevention of Significant Deterioration (PSD)
- Restrictions on industrial growth and expansion in areas where national air quality standards have not been met (nonattainment areas)
- Limits on emissions from stationary sources (factories and power plants)
- Limitations on toxic emissions from industrial sources (primarily chemical plants)
- Limits on emissions from mobile sources (primarily cars and trucks)

AMBIENT AIR QUALITY MANAGEMENT

To control ambient air quality, which refers to the air that surrounds us, the EPA sets primary and secondary standards for six principal pollutants knows as "criteria" pollutants. Five of these pollutants were initially identified by the EPA in 1971 as being the most pervasive of artificial pollutants and in need of immediate reduction and control. In 1978, lead was added to the list of harmful pollutants. These six pollutants are thus sulfur dioxide, particulates, carbon monoxide, ozone, nitrogen dioxide, and lead. It should be noted that ozone is not directly emitted into the air, but is formed by sunlight acting on emissions of nitrogen oxides and volatile organic compounds (VOC).

The primary standards concern the minimum level of air quality necessary to keep people from becoming ill and are aimed at protecting human health. These primary standards are intended to provide an "adequate margin of safety," which has been defined to include a "representative sample" of so-called sensitive populations such as the elderly and asthmatics. These standards are to be set without regard to cost or availability of control technology. The secondary standards are aimed at the promotion of public welfare and the prevention of damage to animals, plant life, and property generally. These standards are based on scientific and medical studies that have been made of the pollutant's effects. Table 6.3 shows the current standards that are in effect. As can be seen, the primary and secondary standards for some of the six pollutants are the same.

Because air pollution problems vary from place to place throughout the country, a regional concept was adopted for air pollution control through the establishment of 247 air quality control regions. These air quality control regions are useful units for management and control as each region has individual problems and individual characteristics of pollution control. An air quality control region is defined by the EPA "as an area with definite pollution problems, common pollution sources, and characteristic weather."[7] The states were given responsibility for drawing up plans called state implementation plans (SIPs) to attain the standards for the air quality control regions within their boundaries. A state implementation plan is a collection of the regulations a state develops to clean up polluted areas. Individual states may have stronger pollution controls, but none can have weaker pollution controls than those set for the whole country.

The primary standards were initially to be attained by mid-1975 as required by the 1970 Clean Air Act. When that deadline came, however, only 69 of the 247 air quality control regions were in compliance with all the antipollution standards then in existence. Sixty regions failed the standards for particulates, 42 for sulfur dioxide, 74 for ozone, 54 for carbon monoxide, and 13 for nitrogen dioxide. Some interesting headlines appeared in the newspapers when these goals were not attained, among them, "AIR TO BE ILLEGAL—BUT BREATHE ANYWAY."

Obviously some adjustments had to be made. These were finally worked out in the 1977 amendments. Under these amendments, the primary standards were to be attained as expeditiously as possible but not later than December 31, 1982, with exten-

TABLE 6.3 National Quality Standards for Ambient Air

Pollutant	*Averaging Time*	*Primary Standards (Health)*	*Secondary Standards (Welfare, Materials)*
Particulates	annual	75 $\mu g/m^3$	60 $\mu g/m^3$
	24 hour	260 $\mu g/m^3$	150 $\mu g/m^3$
Sulfur dioxide	annual	80 $\mu g/m^3$ (0.03 ppm)	
	24 hour	365 $\mu g/m^3$ (0.14 ppm)	
	3 hour	—	1,300 $\mu g/m^3$ (0.5 ppm)
Carbon monoxide	8 hour	10 mg/m^3 (9 ppm)	same as primary
	1 hour	40 mg/m^3 (35 ppm)	
Nitrogen dioxide	annual	100 $\mu g/m^3$ (0.05 ppm)	same as primary
Ozone	1 hour	240 $\mu g/m^3$ (0.12 ppm)	same as primary
Lead	3 month	1.5 $\mu g/m^3$ (0.006 ppm)	

[1]In micrograms or milligrams per cubic meter—$\mu g/m^3$ and mg/m^3—and in parts per million—ppm.

Source: United States Environmental Protection Agency, *Cleaning the Air: EPA's Program for Air Pollution Control* (Washington, D.C.: EPA June 1979), p. 10.

sions until 1987 for two pollutants most closely related to transportation systems (carbon monoxide and ozone) if the state required an annual automobile inspection of emission controls. Each state was also required to draw up specific plans for bringing each nonattainment region up to standard and for maintaining the purity of air in regions that already met the standards.

In order to prevent the states from failing to meet the deadlines again, the EPA was authorized to impose sanctions on states failing to meet the standards. The law allowed the EPA to withhold millions of federal dollars for highway construction, new sewage treatment plants, and clean-air planning grants as well as to ban construction or modification of most factories, power plants, and other major sources of air pollution. This "action-forcing" strategy has been used to force some states and cities into compliance.[8] New deadlines were set in the Clean Air Act of 1990 and the EPA was given new enforcement powers. The EPA can now fine violators much like a police officer issuing traffic tickets, rather than having to go to court for even minor violations.[9]

Ambient air quality is measured by using a pollutant standards index (PSI), which provides the EPA with a uniform system of measuring pollution levels for the major regulated air pollutants. Once these levels are determined, the PSI figures are reported in all metropolitan areas of the country where the population exceeds 200,000 people. These index figures enable the public to determine whether air pollution levels in a particular location are good, moderate, unhealthful, or worse. The PSI converts the measured pollutant concentration in a community's air to a number on a scale of 0 to 500, with the most important number being 100 since that number corresponds to the established standard. A PSI level in excess of 100 means that the pollutant is in the unhealthful range on a given day, and a PSI level at or below 100 means that a pollutant reading is in the satisfactory range. The following intervals and terms describe the PSI air quality levels.[10]

From 0 to 50	Good
From 50 to 100	Moderate
From 100 to 200	Unhealthful
From 200 to 300	Very Unhealthful
Above 300	Hazardous

These intervals relate to the potential health effects of the daily concentrations of each pollutant (Exhibit 6.2). Each value has built into it a margin of safety that protects highly susceptible members of the public. The EPA determines the index number of each of the "criteria" pollutants and then reports the highest of the figures for each metropolitan area and identifies which pollutant corresponds to the figure that is reported. On days when two or more pollutants exceed the standard of 100, the pollutant with the highest index level is reported, but information on any other pollutant above 100 may also be reported.[11]

The PSI places maximum emphasis on acute health effects occurring over very short time periods of 24 hours or less rather than on chronic effects occurring over months or years. By notifying the public when a PSI value exceeds 100, citizens of that area are given an adequate opportunity to react and take whatever steps they can to avoid exposure. However, the PSI does not take into account the possible adverse effects

EXHIBIT 6.2 General Health Effects and Cautionary Statements

Index Value	*PSI Descriptor*	*General Health Effects*	*Cautionary Statements*
Up to 50	Good	None for the general population.	None required.
50 to 100	Moderate	Few or none for the general population	None required.
100 to 200	Unhealthful	Mild aggravation of symptoms among susceptible people, with irritation symptoms in the healthy population.	Persons with existing heart or respiratory ailments should reduce physical exertion and outdoor activity. General population should reduce vigorous outdoor activity.
200 to 300	Very Unhealthful	Significant aggravation of symptoms and decreased exercise tolerance in persons with heart or lung disease; widespread symptoms in the healthy population.	Elderly and persons with existing heart or lung disease should stay indoors and reduce physical activity. General population should avoid vigorous outdoor activity.
Over 300	Hazardous	Early onset of certain diseases in addition to significant aggravation of symptoms and decreased exercise tolerance in healthy persons. At PSI levels above 400, premature death of ill and elderly persons may result. Healthy people experience adverse symptoms that affect normal activity.	Elderly and persons with existing diseases should stay indoors and avoid physical exertion. At PSI levels above 400, general population should avoid outdoor activity. All people should remain indoors, keeping windows and doors closed, and minimize physical exertion.

Source: United States Environmental Protection Agency, *Measuring Air Quality: The Pollutant Standards Index* (Washington, D.C.: EPA, 1994), pp. 4–5.

associated with combinations of pollutants, called synergism. Nor does it specifically take into account the damage air pollutants can do to animals, vegetation, and certain materials like building surfaces and statues.[12]

PREVENTION OF SIGNIFICANT DETERIORATION

The 1977 amendments strengthened efforts to maintain air quality in regions where the air was already cleaner than the standards allowed. Before these amendments, many areas of the country could be polluted up to the standards, something many environmentalists found unacceptable. Three kinds of regions were defined with respect to these areas. A Class I region includes all national parks and wilderness areas and may include further areas named by the states. In these regions, no additional sulfur or particulate sources are permitted. Class II areas encompass every other PSD region in the nation. In these areas, some industrial development is permitted up to a specified level. Class III areas can have about twice as much pollution from new sources, sometimes even up to the minimum federal standards. Any potential new pollution sources in these regions must obtain a permit before operating and meet a number of other conditions, such as using the best available control methods (BACT).

NONATTAINMENT AREAS

Nonattainment areas for each criterion air pollutant are identified by the EPA and state governors and then classified into five categories, ranging from marginal to extreme depending on the severity of the pollution. Cleanup requirements and deadlines to reach the standards are then tailored to these classifications. An offset policy has also been adopted for these nonattainment areas. New industrial development is permitted as long as offsetting reductions are made from existing sources for the pollutants to be emitted by the new facilities. These existing sources must reduce their emissions more than enough to compensate for new sources of pollution so that the area keeps moving toward attainment. These offsets can be accomplished within a plant facility or at another plant owned by the same or some other company in the area. Trading offsets among companies is also allowed, and is one of the market approaches adopted in the 1990 Clean Air Act.[13]

STATIONARY SOURCES

Stationary sources of air pollution, as distinguished from ambient air quality, are also controlled. Typical stationary sources are power plant and factory smokestacks, industrial vents for gases and dust, coke ovens, incinerators, burning dumps, and large furnaces. The state plans required by the 1977 amendments must inventory these sources and determine how they should be reduced to bring the regions into conformance with ambient air quality standards. Approximately 35 states adopted statewide permit programs to control stationary sources of air pollution. The Clean Air Act of 1990 instituted a national permit system for these larger sources of air pollution, that includes information on which pollutants are being released, how much may be released, and what kinds of steps the source's owner or operator is taking to reduce pollution.[14]

The law currently makes the EPA set emission limits for certain designated pollutants for selected categories of industrial plants and for those that are substantially modified. Control technique guidelines (CTG) focus on retrofitting technology for existing plants. In addition, the EPA sets standards for all major stationary new sources. These limits are called "new source performance standards" (NSPS) and are specific to each industry. These standards set the maximum amount of each kind of pollutant that can be emitted from a new plant's stacks for each unit of the plant's production.

Although all air pollutants are regarded as hazardous to some degree, some are considered so dangerous to human health that they are limited individually through the setting of hazardous emission standards, called National Emissions Standards for Hazardous Air Pollutants (NESHAP). Presently such limitations apply to any discharge of asbestos, beryllium, mercury, benzene, arsenic, vinyl chloride, radionuclides, and coke oven emissions. These substances have strict limits as to the amount that can be emitted into the atmosphere from stationary sources. Other substances are also being considered for this category of control, including carbon tetrachloride, chromium, chloroform, and other suspected carcinogens.

In 1989, the EPA announced new rules to protect the public from cancer-causing vapors of benzene. These new regulations forced a broad range of companies, including coke producers, petroleum and chemical plants, and about 200,000 gasoline service stations, to cut emissions of benzene that pose cancer risks to workers and people living near the facilities. The new rules were expected to cost these various industries more than $1 billion. Most of this cost was expected to fall on half of all gasoline stations in the United States and on 15,000 bulk gasoline plants and 1,500 gasoline storage terminals. These regulations were considered to be important because they set a new policy for regulating all sorts of toxic air emissions. The new policy set out to limit chemical emissions so that the greatest number of people possible face no more than a one-in-a-million risk of contracting cancer.[15]

In 1982, the EPA adopted a new policy toward stationary sources of pollution called the bubble concept, which gives business more flexibility in meeting EPA standards. Under this policy, the agency assumes that an area that might include several plants is covered by an imaginary bubble. Companies within the bubble are allowed to expand their industrial operations so long as total emissions within the bubble don't increase. This approach gives plant engineers an incentive to find or develop the most inexpensive methods of limiting plant-wide emissions of a particular pollutant to a level required in their permit to operate. Companies that reduce or limit their pollution will be allowed to sell or trade credits to other companies that are already violating clean air standards and yet want to expand operations.[16]

In 1986, the EPA expanded its bubble policy by announcing that it would approve bubbles in areas that had not met federal air pollution standards. Permission would be granted if pollution sources cut their emissions more than 20 percent below the allowable pollution level. Such bubbles must also be consistent with state air quality goals. Approval for new bubbles had ground to a halt over the previous 18 months as the EPA was in the process of revising the policy. When the policy was announced, about 125 bubble applications were pending in 29 states.[17]

TOXIC EMISSIONS

After the Bhopal, India, incident, there was increased pressure on the EPA to regulate toxic gas emissions, particularly from large plants. This incident brought long overdue attention to the health threat posed by airborne toxic chemicals from industrial sources. Such chemicals, which can cause cancer and birth and genetic defects, had often escaped regulation. In 1987, the EPA concluded that toxic air emissions may cause as many as 2,000 cancer deaths a year. Toxic air emissions, like acid depositions, can be carried great distances before falling to the ground, and are likely to rise rapidly in developing countries as new polluting factories are built.[18]

After several months of discussion, the EPA announced a plan to set up a new system to notify states about specific chemicals that may present substantial risk locally but little risk nationally. Each state would be required to monitor the major sources of these chemicals and publish the results, but would be free to regulate these emissions as it saw fit, meaning that regulations of a few hazardous pollutants could vary widely from state to state.[19] The EPA also had authority to list and regulate hazardous chemicals, but through 1990, the agency had listed and regulated only seven chemicals.

The 1990 Clean Air Act includes a list of 189 toxic pollutants selected by Congress on the basis of their potential health and/or environmental hazards that the EPA must regulate. New chemicals can be added to the list as necessary. With respect to these 189 chemicals, the EPA must identify categories of sources that release them and decide whether they are major (large) or area (small) sources. These categories could be gasoline service stations, coal-burning power plants, or chemical plants, for example. Once these categories of sources are listed, the EPA must issue regulations for the major sources first and then issue regulations to reduce pollution from small sources.[20]

Wherever possible, companies will be given flexibility to choose how they meet the regulations, and will have to use the maximum available control technology (MACT) to reduce pollutant releases. If a company reduces its releases of a hazardous air pollutant by about 90 percent before the EPA regulates the chemical, it will get extra time to finish cleaning up the remaining 10 percent. The EPA will also study whether and how to reduce hazardous air pollutants from small neighborhood polluters. Finally, the alternative fuels program for mobile sources described below, should significantly reduce toxic emissions from vehicles.[21]

Another aspect of controlling toxic air emissions was SARA Title III, known as the Emergency Planning and Community Right-to-Know Act, passed in 1986 as a response to the Bhopal accident. This title did not prescribe specific action that had to be taken by companies, but rather required them to reveal an unprecedented amount of information about the types and quantities of chemicals they produce, use, and routinely or accidentally release into the environment.[22] The first report of toxic emissions was released in 1989, and showed that emissions in eight states exceeded 100 million pounds. The total included 235 million pounds of carcinogens such as benzene and formaldehyde, and 527 million pounds of such neurotoxins as toluene and trichlorethylene. The EPA estimated that such air toxins cause more than 2,000 cases of cancer annually.[23]

The 1993 report showed Louisiana to be the worst polluter with 451 million pounds of releases to air, water, and land including underground injection (Figure 6.1). Excluding the latter, the state still had 293 million pounds of releases. The state, however, hoped to shed its top ranking in 1994, a position it had held for the previous four years, when its total releases were cut by two thirds to 152.7 million pounds. Credit for these improvements was given to companies located in the state who reportedly had spent billions of dollars implementing program improvements and changes. Dramatic reductions in releases to water and underground injection were achieved.[24]

This requirement enables government and the public to scrutinize company activities more closely, and has forced companies to communicate with the local communities in a way that they were not required to previously. As a result, companies are voluntarily taking actions to protect public health and safety from toxic emissions. Companies are driven by concern about their reputations and potential liability to better manage their activities that could result in serious environmental problems. Companies have also been motivated to provide more information about their products to consumers, enabling them to take more precautions in using the products.[25]

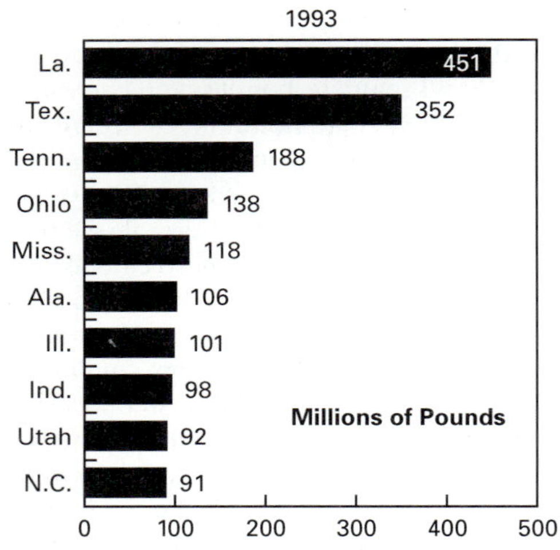

Source: United States Environmental Protection Agency, Office of Pollution Prevention and Toxics, *1993 Toxic Release Inventory* (Washington, D.C.: U.S. Government Printing Office, 1993), p. ES-6.

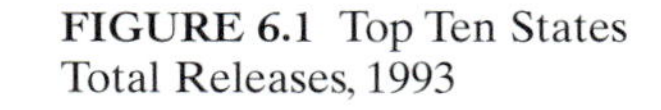
FIGURE 6.1 Top Ten States Total Releases, 1993

MOBILE SOURCES

For mobile sources of air pollution, the EPA has traditionally set standards for automobile, truck, and aircraft emissions. These motor vehicle emission standards are directed at controlling nitrogen oxides, carbon monoxide, particulate matter, and lead. Although volatile organic compounds (VOCs) are not regulated as a criteria pollutant, emissions of these compounds are also controlled. This approach was based on the belief that control of these pollutants can best be accomplished by using control devices such as catalytic converters and electronic carburetors on autos and other vehicles.

This approach, however, has not worked well enough to bring some of the pollutants, particularly smog, under control. An EPA report indicated that motor vehicles are responsible for up to half of the smog-forming volatile organic compounds (VOCs) and nitrogen oxides released into the air. They also release more than 50 percent of hazardous air pollutants and up to 90 percent of the carbon monoxide found in urban air. The problem is that more people are driving more cars more miles on more trips each year. Many people live some distance from their work and so have to commute by car because public transportation is either not available or not acceptable, and most people still drive to work alone. Also, as lead was phased out of gasoline, refiners changed formulas to make up for octane loss, which made gasoline more likely to release smog-forming VOC vapors in the air.[26]

In early 1990, the American Lung Association released a study stating that air pollution from motor vehicles is responsible for $40 billion to $50 billion in annual health-care costs, and as many as 120,000 unnecessary or premature deaths. Annual health-care costs actually ranged from as little as $4 billion to as much as $93 billion, but a $40 billion to $50 billion range appeared to be the most realistic estimate. Release of the study was said to represent an attempt to identify the costs associated with leaving the nation's clean air standards unchanged, and make a strong case for more stringent vehicle controls on emissions.[27]

Some 96 areas missed the deadline for meeting health standards for ozone, a main ingredient of smog. The Clean Air Act of 1990 addresses this problem by requiring that all but the worst nine areas not meeting the ozone standard comply by November 1999, all but Los Angeles, Baltimore, and New York by November 2005, Baltimore and the NYC area by November 2007, and Los Angeles by November 2010. Areas that are moderately polluted or worse must cut smog 15 percent within six years. After that, areas that are seriously polluted or worse must make 9 percent improvements every three years until they meet the standards (Exhibit 6.3).

Beginning in 1995, all gasoline sold in the nine cities in the United States with the highest ozone concentrations must be cleaner burning, reformulated gasoline that cuts emissions of hydrocarbons and toxic pollutants by 15 percent compared to gasoline sold in 1990. By the year 2000, the reductions must equal 20 percent. Other cities can "opt-in" to the reformulated gasoline program. Starting with 1998 models, a certain percentage of new vehicles purchased by owners of fleets of ten or more vehicles in the two dozen cities with the highest ozone and carbon monoxide concentrations must be capable of using clean fuels that will run considerably cleaner than today's autos. By the year 2001, even cleaner models must be produced.

Tougher tailpipe standards are phased in starting with 1994 models to cut nitrogen oxides by 60 percent and hydrocarbons by 30 percent. Even deeper cuts are required for 2003 models if the EPA finds they are cost-effective and needed. These standards have to be maintained for 10 years or 100,000 miles. Warranties on pollution control equipment must last eight years or 80,000 miles for catalytic converters, electronic emissions control units, and onboard emissions diagnostic equipment and two years or 24,000 miles for other pollution gear beginning with model year 1995. Special nozzles are required on gasoline pumps in almost 60 smoggy areas. Also, more effective fume-catching canisters may be phased in on all new cars, starting in the mid-1990s. Devices are also required on cars to alert drivers to problems with pollution control equipment.

By the model year 1996, car makers must begin producing at least 150,000 clean-fuel vehicles annually under a California pilot program requiring phase-in of even tougher emission limits. By the model year 1999, 500,000 such vehicles must be produced. These tougher standards can be met by any combination of vehicle technology or cleaner fuels. By 2001, these standards will become even stricter. The state itself issued a requirement that in 1998, for 2 percent of cars each automaker sells there must be "zero emission" vehicles, which will rise to 10 percent of unit sales or about 200,000 vehicles annually by the year 2003. Other states can "opt in" to the California program through incentives, not sales or production mandates.

In October 1991, 12 northeastern states and the District of Columbia promised to adopt tougher new rules related to smog reduction, similar to those of California that manufacturers claimed would add $550 to the cost of a new vehicle.[28] These states eventually petitioned the EPA for permission to apply a California-style program to the region. Automakers vigorously opposed that request, and the EPA tried to work out a compromise between them and northeastern environmental regulators who wanted to address the area's air quality problems.[29] But under threat of a suit by environmental and health groups as well as the attorney general of one of the states, the

EXHIBIT 6.3 Ozone Nonattainment Areas

EXTREME

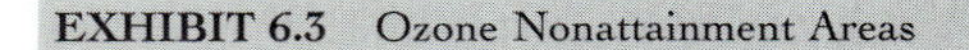

Ozone standard must be met by 2010

Los Angeles-South Coast Air Basin, CA

SEVERE

Ozone standard must be met by 2007

Chicago-Gary-Lake County, IL-IN

Houston-Galveston-Brazoria, TX

Milwaukee-Racine, WI

New York-N New Jer-Long Is. NY-NJ-CT

Southeast Desert Modified-Air Quality Maintenance Area, CA

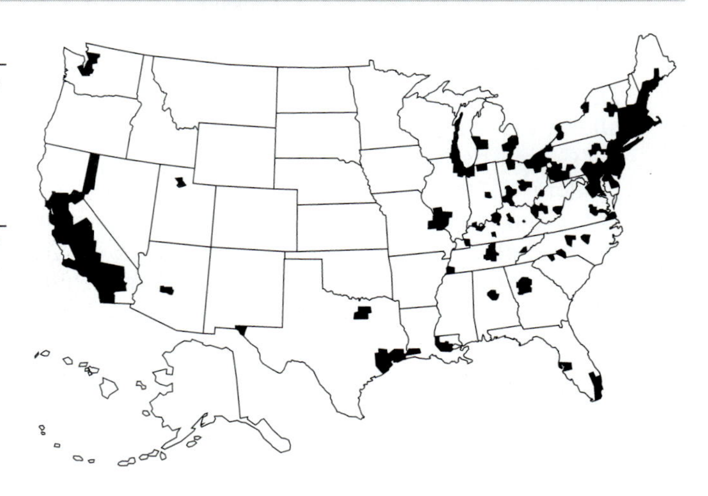

SEVERE

Ozone standard must be met by 2005

Baltimore, MD

San Diego, CA

Philadelphia-Wilmington-Trenton, PA-NJ-DE-MD

Ventura Co, CA

SERIOUS

Ozone standard must be met by 1999

Atlanta, GA

Baton Rouge, LA

Beaumont-Port Arthur, TX

Boston-Lawrence-Worcester (E.MS), MA-NH

El Paso, TX

Greater Connecticut

Portsmouth-Dover-Rochester, NH

Providence (All RI), RI

Sacramento Metro, CA

San Joaquin Valley, CA

Springfield (Western MA), MA

Washington, C-MD-VA

MODERATE

Ozone standard must be met by November 1996

Atlantic City, NJ

Charleston, WV

Charlotte-Gastonia, NC

Cincinnati-Hamilton, OH-KY

Cleveland-Akron-Lorain, OH

Dallas-Fort Worth, TX

Dayton-Springfield, OH

Detroit-Ann Arbor, MI

Grand Rapids, MI

Greensboro-Winston Salem-High Point, NC

Huntington-Ashland, WV-KY

Kewaunee Co, WI

Knox & Lincoln Cos, ME

Lewiston-Auburn, ME

Louisville, KY-IN

Manitowoc Co, WI

Miami-Fort Lauderdale-W. Palm Beach, FL

Monterey Bay, CA

Muskegon, MI

Nashville, TN

Parkersburg, WV

Phoenix, AZ

Pittsburgh-Beaver Valley, PA

Portland, ME

Raleigh-Durham, NC

Reading, PA

Richmond, VA

Salt Lake City, UT

San Francisco-Bay Area, CA

Santa Barbara-Santa Maria-Lompoc, CA

Sheboygan, WI

St. Louis, MO-IL

Toledo, OH

EXHIBIT 6.3 *(Continued)*

MARGINAL

Ozone standard must be met by 1993

Albany-Schenectady-Troy, NY
Allentown-Bethlehem-Easton, PA-NJ
Altoona, PA
Birmingham, AL
Buffalo-Niagara Falls, NY
Canton, OH
Columbus, OH
Door Co, WI
Edmonson Co, KY
Erie, PA
Essex Co (Whiteface Mtn), NY
Evansville, IN
Greenbrier Co, WV
Hancock & Waldo Cos, ME
Harrisburg-Lebanon-Carlisle, PA
Indianapolis, IN
Jefferson Co, NY
Jersey Co, IL
Johnstown, PA
Kent and Queen Anne's Cos, MD
Knoxville, TN
Lake Charles, LA
Lancaster, PA
Lexington-Fayette, KY
Manchester, NH
Memphis, TN
Norfolk-Virginia Beach-Newport News, VA
Owensboro, KY
Paducah, KY
Portland-Vancouver, OR-WA
Reno, NV
Scranton-Wilkes-Barre, PA
Seattle-Tacoma, WA
Smyth Co, VA (White Top Mtn)
South Bend-Elkhart, IN
Sussex Co, DE
Tampa-St. Petersburg-Cleanwater, FL
Walworth Co, WI
York, PA
Youngstown-Warren-Sharon, OH-PA

Source: United States Environmental Protection Agency, *The Plain English Guide to the Clean Air Act* (Washington, D.C.: EPA, 1993), p. 25.

EPA approved the program. Automakers were then faced with producing vehicles for this region that meet the California guidelines for emissions. And two of these states, Massachusetts and New York, issued mandates requiring the sale of a small number of "zero emission" electric vehicles beginning in 1998.[30]

Automakers have fought mandates for electric vehicles, doubting the feasibility of mass producing them and questioning whether sufficient demand exists. The Big Three mounted a campaign to persuade regulators in California and the Northeast to back off from the 1988 deadlines for mass production of electric vehicles, throwing their support behind the historic $1 billion partnership they agreed to with the Clinton administration to develop "clean cars" for the next century.[31] They issued an ultimatum to the Northeast saying its clean car program will have to be scrapped if sales mandates for electric cars continued in force, and pushed a plan to develop a so-called 49 state model that is cleaner than current vehicles and could be sold in every state but California.[32]

With respect to cleaner burning gas, the oil and gas industry originally estimated that cleaner or reformulated gasoline could require up to $30 billion in spending for new fuel tanks and retooled refineries. The development of alternative fuels is another problem (see box). These fuels, such as ethanol and methanol, improve combustion and reduce emissions of carbon monoxide and unburned hydrocarbons. But the alcohol in these fuels absorbs water when transported by pipeline, so it must be stored in separate tanks and blended with gasoline at terminals around the country. Ethanol, which comes from corn, could cost the oil companies $1 billion for each point of market share it captures.[33]

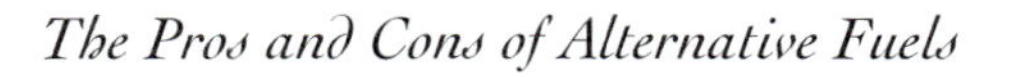

Ethanol: Made from corn and other carbohydrate sources, has a higher octane rating than gasoline. Its higher oxygen content permits fuel to burn more efficiently, reducing carbon monoxide, benzene, and other toxins. It would reduce imports and benefit the farm economy. But, at least for now, its higher cost requires subsidies. It has a lower energy content than gasoline, is hard to ignite in cold weather and is difficult to transport through pipelines.

Methanol: Made from natural gas and coal, reduces carbon monoxide and other pollutants, has more octane and mixes easily with gasoline. Its drawbacks include emissions of toxins like formaldehyde, its higher cost and lower energy content. Much of it is imported, so it would increase energy dependence.

Compressed Natural Gas: The least expensive alternative. It substantially reduces hydrocarbon and carbon monoxide emissions and requires no refining. There is an abundant domestic supply. But natural gas is less powerful. It has to be injected by hose in a slow process and requires large on-board compression tanks.

Electricity: Touted as the cleanest and quietest power source. Until recently it was considered a long way from commercial development. General Motors recently said it will mass-produce a battery-powered car, but they didn't say how soon. GM says the car, using an 850-pound battery that would charge in two hours, will accelerate from zero to 60 mph in eight seconds, and run 125 miles before recharging.

Source: Copyright 1990, *USA Today*. Reprinted with permission.

However, in 1991, the oil industry signed an agreement committing it to invent and begin selling a cleaner-burning gasoline by January 1995 in the nation's nine smoggiest cities, which accounted for 25 percent of gasoline sold in the country. The EPA estimated that gasoline would cost motorists four to five cents more a gallon and the oil industry $3 billion to $5 billion to retrofit refineries and storage systems.[34] As the deadline drew closer, it was estimated that gasoline prices could go up as much as nine cents a gallon when all the additional costs of producing cleaner gas were taken into account.[35]

Besides the nine cities that were required to use the new fuel, some 40 other regions had promised to consider selling reformulated gasoline. But as the program started, drivers complained about the higher prices and engine damage, and several of these regions pulled out of the program. The EPA allowed these regions to pull out without suffering sanctions, which infuriated the oil industry. The industry was faced with great difficulty in estimating the demand for the new gasoline not knowing how many regions would eventually participate. It thus cut back on production, not wanting to get stuck holding expensive inventories of reformulated gas, driving prices up even further. Refiners alone spent $5 billion upgrading existing facilities and building new ones to supply those regions required to use the new gasoline.[36]

There are many questions about these efforts to reduce emissions from mobile sources. A controversial proposal for auto emission control was made by a University of Denver chemist after several years of research. "Modern cars are so clean that tightening standards or switching fuels is a total waste of money," he insisted. His research showed that only a tiny fraction of the cars on the road, perhaps as small as 8 percent, account for more than half the pollution. Instead of mandating lower emissions for new cars, he believed remote sensors that he had developed and used for his research should be used to spot gross polluters. Owners of these cars could then be sent notices to have their cars fixed. The cars causing the most pollution are those with broken or disabled pollution controls or those in need of a tune-up badly. Supporters say that his approach could reduce pollution in some cities by 30 percent or more, and be cheaper than many inspection programs.[37]

Some studies question whether ozone poses as much a threat to health as is usually assumed, and argue that chronic effects of prolonged exposures to elevated ozone have not been demonstrated after two and a half decades of research.[38] Others argue for a market-based approach to drive down auto emissions by assigning credits to car companies based on the emissions expected from the sale of cleaner automobiles.[39] Meanwhile, research on electric vehicles continues as well as on other aspects of pollution control such as better control devices. And, in April 1995, General Motors announced the largest alternative fuel vehicle production undertaken in the country. Beginning in the 1997 model year, all four-cylinder light-duty Chevrolet and GMC S-series pickup trucks will be produced as ethanol flexible fuel vehicles, meaning motorists can use gasoline or ethanol fuels as they desire.[40]

ACID RAIN

Acid rain is believed to be largely a manmade problem, directly traceable to the burning of fossil fuels in power plants, factories, and smelting operations, and, to a somewhat lesser extent, the burning of gasoline in automobiles. The burning of these fuels releases sulfur dioxide, nitrogen oxides, and traces of such toxic metals as mercury and cadmium into the atmosphere to mix with water vapor. Acid rain then results from chemical reactions that follow to produce dilute solutions of nitric and sulfuric acids. These solutions, or acidic "depositions," come down to ground level in the form of hail, snow, fog, rain, or even in dry particles (Figure 6.2). These acid depositions are formally defined as having a pH level under 5.6 (a neutral solution has a level of pH 7) (Figure 6.3).[41]

Acid rain is said to cause many serious environmental problems. When it enters a body of water, acid rain carries a deadly burden of toxic metals that can stunt or kill aquatic life. In the Adirondack Mountains of New York, more than 150 lakes that had previously supported trout life were thought to be fishless due to the high acid content of the lakes.[42] A report released by the EPA in 1995 stated that many of these lakes might never recover from acid rain pollution. Projections indicated that nearly half of the 700 ponds and lakes involved in the study could be so affected by manmade pollution that they will be virtually lifeless by 2040. The report also stated that its scientists didn't know enough about acid rain to set meaningful standards.[43]

As the buffering effect of the acid-neutralizing minerals in the water diminishes, these lakes appear to die suddenly and turn clear and bluish. Surface waters that have a low acid-buffering capacity are unable to neutralize the acid effectively. Snowmelt in northern areas can quickly kill a lake as all acids accumulated in the snow are released at once. Because of the freezing point depression phenomenon

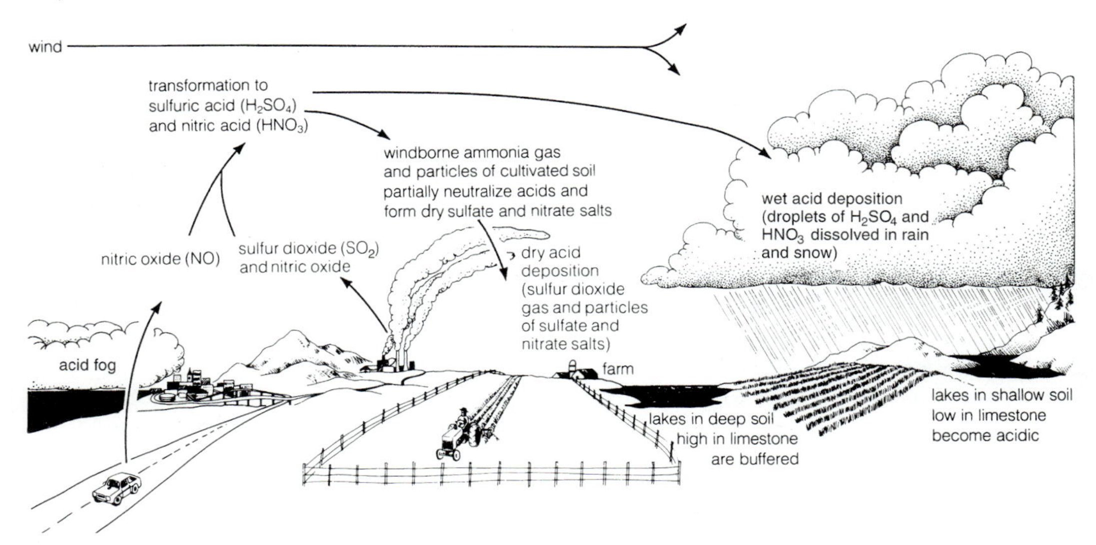

Source: G. Tyler Miller, Jr. *Living in the Environment,* 6th ed. (Belmont, CA: Wadsworth, 1990), p. 494. © Wadsworth Publishing Co.

FIGURE 6.2 Acid Rain

combined with the recrystallization of snow after it falls, the most acidic snow crystals will melt first, thereby releasing 50 to 80 percent of the acids in the first 30 percent of snowmelt.[44]

When acid rain is absorbed into the soil, it can rob plants of nutrients because it breaks down minerals containing calcium, potassium, and aluminum. This aluminum may eventually reach lakes through water tables and streams and further contribute to the suffocation of fish.[45] Acid rain is suspected of spiriting away mineral nutrients from the soil on which forests thrive. Areas with acid-neutralizing compounds in the soil can

FIGURE 6.3 How Acid Is Acid Rain?

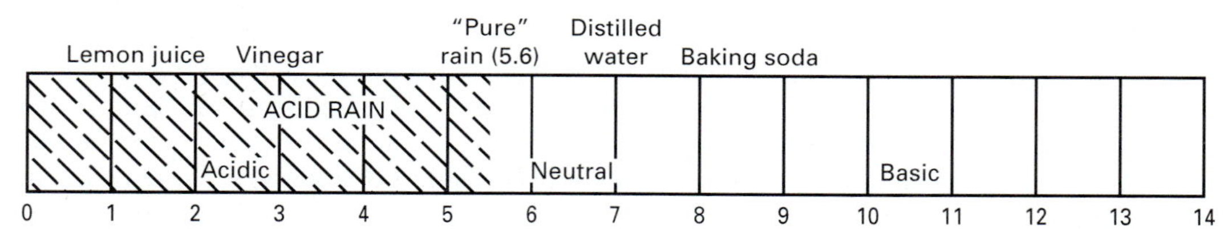

The pH scale ranges from 0 to 14. A value of 7.0 is neutral. Readings below 7.0 are acidic; readings above 7.0 are alkaline. The more pH decreases below 7.0 the more acidity increases.

Because the pH scale is logarithmic, there is a tenfold difference between one number and the next one to it. Therefore, a drop in pH from 6.0 to 5.0 represents a tenfold increase in acidity; while a drop from 6.0 to 4.0 represents a hundredfold increase.

All rain is slightly acidic. Only rain with a pH below 5.6 is considered "acid rain."

Source: United States Environmental Protection Agency, *Meeting the Environmental Challenge* (Washington, D.C.: U.S. Government Printing Office, 1990), p. 9.

Acid Rain

"Acid rain" is the term loosely used to refer to all forms of acid deposition which can occur in the forms of rain, snow, fog, dust or gas. Man-made emissions of sulfur dioxide (SO_2) and nitrogen oxides (NOx) are the principal causes. These pollutants are trans[f]ormed into acids in the atmosphere where they may travel hundreds of miles before falling in some form of acid rain. Acid rain has been measured with a pH of less than 2.0—more acidic than lemon juice. The political implications of acid rain are an important issue, as the pollutants causing acid rain may originate within the political boundary, yet the effects of these pollutants realized within another.

EPA research in the 1980s has increased scientific understanding of the effects of acid rain, including the sterlization of lakes and streams, detrimental reproductive effects on fish and amphibians, possible forest dieback and deterioration of man-made structures such as buildings and sculptures. These effects have been most obvious in the eastern U.S. and Canada, and in much of both western and eastern Europe. The Clean Air Act of 1970 helped to curb the growth of SO_2 and NOx emissions in the U.S., and the 1990 Clean Air Act Amendments will bring significant additional reductions.

Source: United States Environmental Protection Agency, *Meeting the Environmental Challenge* (Washington, D.C.: U.S. Government Printing Office, 1990), p. 9.

experience years of acid rain without serious problems. But the thin soils of the mountainous and glaciated Northeast have very little buffering capacity, which makes them vulnerable to damage from acid rain. Acid rain also has corrosive assault on buildings and water systems that costs millions of dollars annually. It may also pose a substantial threat to human health, principally by contaminating public drinking water.[46]

Over 80 percent of sulfur dioxide emissions in the United States originate in the 31 states east of or bordering the Mississippi River, and more than half the acid rain falling on the eastern United States originates from the heavy concentration of coal-and-oil burning power and industrial plants in seven central and upper midwestern states. Prevailing winds transport these emissions hundreds of miles to the Northeast across state and national boundaries (Figure 6.4). The acidity of the precipitation falling over much of this region has a pH of 4.0 to 4.2, which is 30 to 40 times greater than the acidity of the normal precipitation that fell on this region in previous decades. Acid rain costs the United States at least $6 billion a year according to the National Academy of Sciences, and costs will rise sharply if strong action is not taken. The cost of reducing acid rain could run from $1.2 billion to $20 billion, depending on the extent of cleanup and the technology employed.[47]

The phenomenon of acid rain has caused a problem in the relations between the United States and Canada. Environmental officials in Canada have projected the loss of 48,000 lakes by the end of the century if nothing is done to reduce the emissions that produce acid rain. Some 2,000 to 4,000 lakes in Ontario have become so acidified that they can no longer support trout and bass, and some 1,300 more in Quebec are said to

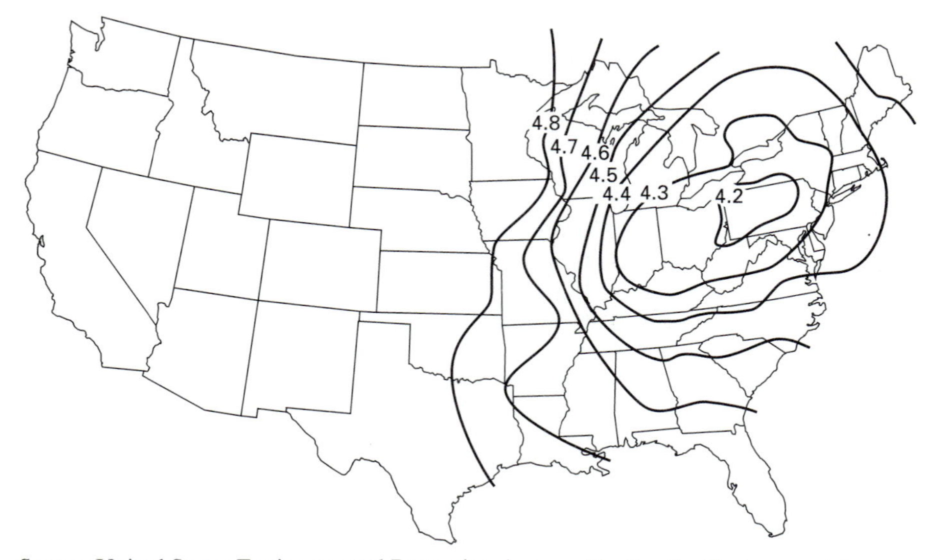

Source: United States Environmental Protection Agency, *Meeting the Environmental Challenge* (Washington, D.C.: U.S. Government Printing Office, 1990), p. 9.

FIGURE 6.4 Areas Where Precipitation in the East Is Below pH5

be on the brink of destruction. Canadians are also worried that acid rain will harm the forestry and related industries that provide jobs for one in every ten Canadians and earn $14 billion a year. The Canadian government contends that about 70 percent of the acid rain that it receives comes from the United States, primarily from heavily industrialized areas in the Midwest.[48]

Before the 1970 Clean Air Act, sulfur dioxide and nitrogen oxide emissions in the United States were increasing dramatically (Figure 6.5). Between 1940 and 1970, annual sulfur dioxide emissions had increased by more than 55 percent and nitrogen oxide emissions had almost tripled. By 1986, however, annual sulfur dioxide emissions had declined by 21 percent and nitrogen oxide emissions had increased only 7 percent, even though the economy and the combustion of fossil fuels had grown substantially over the same time period. More reductions, however, needed to be accomplished to solve the problem.[49]

In 1980, a National Acid Precipitation Assessment Program (NAPAP) was initiated to study the problem. Part of this program included a National Surface Water Survey, which found the four subregions with the highest percentages of acidic lakes were the Adirondacks of New York, where 10 percent of the lakes were found to be acidic; the Upper Peninsula of Michigan, where 10 percent of the lakes were also found to be acidic; the Okefenokee Swamp in Florida, which is naturally acidic; and the lakes in the Florida Panhandle, where the acidity is unknown. The 1988 Stream Survey, which was also part of the program, determined that approximately 2.7 percent of the total stream reaches sampled in the mid-Atlantic and Southeast are acidic. The major cause of sulfates in streams was found to be atmospheric deposition.[50]

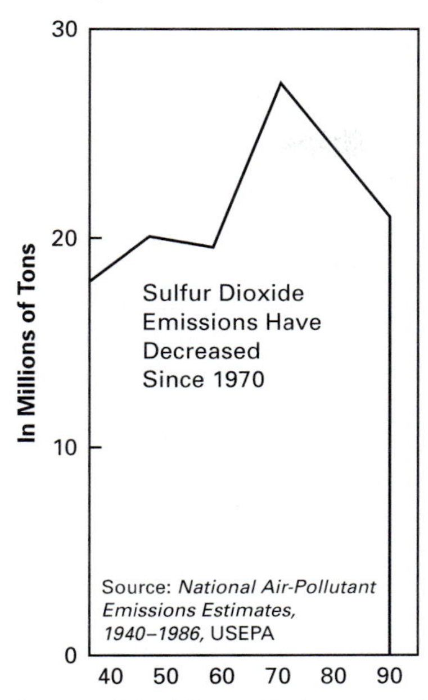

Source: United States Environmental Protection Agency, *Environmental Progress and Challenges: EPA's Update* (Washington, D.C.: U.S. Government Printing Office, 1988), p. 28.

FIGURE 6.5 Sulfur Dioxide Emissions, 1940–1986

The study also found that air pollution is a factor in the decline of both managed and natural forests. But there seemed to be no direct impacts to seedlings by acidic precipitation or gaseous sulfur dioxide and nitrogen oxides at ambient levels. Ozone was suspected to be the leading pollutant that stressed regional forests and reduced growth. The NAPAP assessment also indicated that there are no measurable consistent effects on crop yield from the direct effects of simulated acidic rain at ambient levels of acidity. Finally, the report concluded that there were too many uncertainties involved in assessing the effect of acid rain on materials or human health to be able to draw any firm conclusions.[51]

In general, the director of the project stated that "no apparent trend in the acidity of rainfall has been detected" in testimony before Congress. "Because of complex atmospheric reactions, percentage reductions in emissions may not result in similar percentage reduction in depositions," he added. These statements challenged the claims made in a 1983 National Academy of Science report, which held that the relationship between emissions and depositions was proportional.[52] This latter report was widely used as the basis for proposals to cut sulfur dioxide emissions, and the NAPAP, even though it concluded that acid rain was not the environmental catastrophe widely portrayed a decade earlier, did not calm fears about the effects of acid rain and the need to further cut sulfur dioxide and nitrogen oxide emissions.[53]

Thus the new Clean Air Act passed in 1990 contains provisions for large reductions in emissions of sulfur dioxide and nitrogen oxides to reduce acid rain to manageable levels. By the year 2000, SO_2 emissions are to be reduced nationwide by 10 million tons below 1980 levels, a 40 percent decrease. Emissions of nitrogen oxide will also be reduced by 2 million tons below levels that would occur in the year 2000 without new controls. This reduction represents about a 10 percent reduction from 1980 levels. These reductions will be achieved by instituting a variety of reforms aimed at limiting emissions after 1995 from electric power plants and other sources.

In the first phase, the 110 dirtiest power plants in 21 states in the Midwest, Appalachian, southeastern, and northeastern regions of the country must cut sulfur dioxide emissions by 1995 for a total cut nationwide of 5 million tons. Two-year extensions can be given to plants that commit to buy scrubbing devices that allow continued use of high-sulfur coal. In the second phase, more than 200 additional power plants must make sulfur dioxide cuts by 2000, for a total nationwide cut of 10 million tons. This deadline can be extended until 2004 for plants that use new clean coal technology. A nationwide cap on utility sulfur dioxide emissions is imposed after the year 2000. Utilities must also cut nitrogen oxide emissions by 2 million tons a year, or about 25 percent, beginning in 1995.

These reductions are obtained through a program of emission allowances where each utility can "trade and bank" its allowable emissions, something of a market-based approach to pollution control, which is hoped will achieve regional and national emission targets in the most cost-effective manner. Power plants covered by the program are issued allowances that are each worth one ton of sulfur dioxide released from smokestacks. To obtain reductions in sulfur dioxide pollution, these allowances are set below the current level of sulfur dioxide releases. Plants can release only as much sulfur dioxide as they have allowances. If a plant expects to release more sulfur dioxide than it has allowances, it has to buy allowances from plants that have reduced their releases below their number of allowances and therefore have allowances to sell or trade. These allowances can be bought and sold nationwide, and there are stiff penalties for plants that release more pollutants than their allowances cover.[54]

This program was expected to cost $4 billion annually by the year 2000 and increase electricity rates about 1.5 percent nationwide. But these added costs were said to be more than offset by the environmental benefits from acid rain control.[55] The first trade under the program took place in 1992 when Wisconsin Power & Light agreed to sell pollution credits to the Tennessee Valley Authority and the Duquesne Light Company of Pittsburgh.[56] The Chicago Board of Trade began offering futures contracts on sulfur dioxide credits in 1993 and had hoped to become the pollution clearinghouse to the nation, but a vibrant off-exchange pollution rights market also developed.[57]

By the end of 1995, compliance with the provisions of the new Clean Air Act had become a profitable experience for utilities mainly because of falling prices for low-sulfur coal. More than twice the number of utilities that were expected to participate in the initial phase of the program had signed up and were stockpiling allowances so they could defer more costly requirements of the law scheduled to go into effect in the year 2000. Burning low-sulfur coal, which became less expensive than high-sulfur coal, saved them $153 million in 1995, and allowed them to emit only about 60 percent of sulfur dioxide emissions required under the new standards. The stockpiled allowances will be spent in later years to keep older plants on line without investing in costly pollution control equipment or retiring them as the standards tighten.[58]

INDOOR AIR POLLUTION

According to the EPA, there is a growing body of scientific evidence indicating that the air within homes and other buildings can be more seriously polluted than outdoor air even in the largest and most industrialized cities that have serious problems with outdoor air pollution. Indoor air pollutants include radon, asbestos, tobacco smoke, formaldehyde, airborne pesticide residues, chloroform, perchlorethylene (associated with dry cleaning), paradichlorbenzene (from mothballs and air fresheners), and a broad array of airborne pathogens (Figure 6.6). The EPA has developed general information and specific guidelines designed to raise public consciousness of indoor air pollution and strategies to reduce and prevent such pollution. These guidelines include documents offering guidance on construction of new homes and rehabilitation of existing housing.[59]

FIGURE 6.6 Air Pollution in the Home

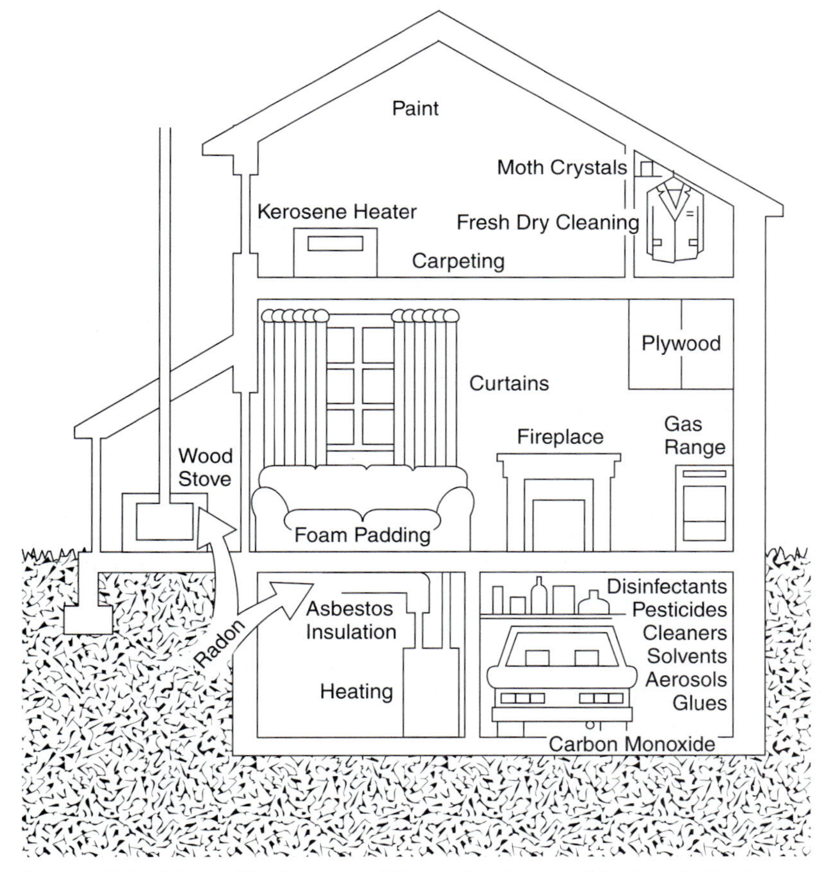

Source: United States Environmental Protection Agency, *Meeting the Environmental Challenge* (Washington, D.C.: U.S. Government Printing Office, 1990), p. 11.

The indoor air pollution problem is affected by the need to save energy by building tightly constructed homes that do not let heat and cold escape. This tightness means indoor air pollutants are trapped and cannot escape to the outside. Some experts believe our current indoor air pollution problems are partly the result of energy conservation measures implemented after the Arab oil embargo of 1973 when fuel prices skyrocketed. Homeowners and building managers sought ways to conserve energy resources. Ventilation standards were lowered in offices, for example, as five cubic feet of fresh air per minute per person became the recommended standard. The previous standard had been ten cubic feet per minute per person. Windows that opened were replaced with mechanical ventilation systems. Homeowners caulked and weather-stripped around windows and doors and used new insulation products such as foams containing formaldehyde. Tightening up buildings reduced fuel bills but also limited the natural ventilation that diluted stale and contaminated indoor air inside.[60]

Since most people spend 90 percent of their time indoors, many may be exposed to unhealthy concentrations of pollutants (Exhibit 6.4). People most susceptible to pollution—the aged, the ill, and children—spend nearly all of their time indoors. The degree of risk depends on how well buildings are ventilated and the type, mixture, and amount of pollutants in the building. Long-term effects of exposure range from impairment of the nervous system to cancer.[61] There is limited legislation to protect the general population from these non-job contacts with toxic substances.

The EPA has explored strategies to address high-risk indoor air problems, which may include not only regulations but also other approaches such as public education, technical assistance, and training programs.[62] Also needed are effective, easily operated, and commercially available devices to monitor personal exposure to indoor air pollution. The EPA has taken action on asbestos, volatile organic compounds (VOCs) in drinking water, and certain pesticides. Schools are required to inspect for asbestos, to prepare management plans where there is a problem, and to take action when they find friable (easily crumbled) asbestos. The EPA has also acted to control exposure to a number of pesticides believed to contribute to indoor air pollution.[63]

Exposure to indoor radon is also believed to be a serious environmental health problem, perhaps second only to smoking as a cause of lung cancer. Radon is a radioactive, colorless, ordorless, naturally occurring gas that results from the radioactive decay of radium-226 found in many types of rocks and soils, which seeps through the soil and collects in homes by entering cracks in the foundation. When inhaled, radon can adhere to particles and then lodge deep in the lungs, increasing the risk of cancer. Radon problems have been identified in every state, and millions of homes throughout the country, according to the EPA, have elevated radon levels (Figure 6.7). Already in 1988, the EPA and the Surgeon General recommended that all Americans (other than those living in apartment buildings above the second floor) test their home for radon.[64]

Regarding office buildings, something called the "sick building syndrome" began to emerge as a health hazard (Exhibit 6.5). Complaints about headache, nausea, sore throat, or fatigue were commonplace among workers at the office, which usually cleared up upon their leaving the building. Many buildings were found to have inadequate ventilation that allowed indoor contaminants from smoking and vapors from photocopying machines, cleaning liquids, and solvents to accumulate. The Environmental Protection Agency estimated that the economic cost of indoor air pollution totaled tens

EXHIBIT 6.4 Indoor Air Pollution: The Worst Offenders

Pollutant	*Danger*	*Source*	*Control*
Formaldehyde	Eye, nose, and throat irritation, headaches, nausea. Possible cause of nasopharyngeal cancer.	Thousands of products and furnishings, including furniture, carpet, plywood, particleboard and urea-formaldehyde foam insulation.	Increase home ventilation, and seal sources such as particleboard. Test and seal walls containing urea-formaldehyde foam.
Combustion Products	Headaches, dizziness, respiratory ailments. High concentrations of carbon monoxide can be fatal.	Gas stoves, furnaces, other fuel-burning appliances. Tobacco smoke.	Make sure appliances are properly maintained and vented to the outdoors. Install an exhaust fan above gas stove.
Household Products	Wide range of health effects, including respiratory irritation, damage to central nervous system, liver, or kidneys, possibly cancer.	Aerosol sprays, cleaning products, paints, solvents, glues, pesticides.	Follow the label and use with plenty of ventilation. Dispose of properly, preferably through a local household toxics collection program. Substitute nontoxic products.
Tobacco Smoke	"Passive smokers" may face greater risk of lung cancer, other cancers, heart disease. Children of smokers show increased prevalence of respiratory ailments.	Smokers.	Nonsmokers should avoid tobacco smoke. Smokers should isolate themselves, increase home ventilation and air filtration. Better yet, they can kick the habit.
Radon	Second leading cause of lung cancer in the United States.	Soil and rocks beneath home, some groundwater, some building materials.	Test your home, then seek professional guidance if necessary. An air-to-air heat exchanger can reduce radon concentrations. Sealing basement floors and openings around pipes and air ducts can prevent radon from entering homes.

Source: Mike Lipske, "How Safe Is the Air Inside Your Home?" *National Wildlife,* 25, no. 3 (April–May, 1987), 36–37.

of billions of dollars annually in lost productivity, direct medical care, lost earnings, and employee sick days.[65]

In 1994, the Occupational Safety and Health Administration proposed the first regulations governing indoor air quality in offices. The goal of the regulations was to protect the estimated 70 million people who work in nonindustrial buildings from secondhand smoke and the so-called sick building syndrome. In its investigations, OSHA found that 52 percent of indoor air problems were caused by inadequate ventilation; thus its regulations focused on ventilation systems and required that they be regularly inspected and cleaned, and circulate air according to existing codes. These regulations, OSHA contended, would eliminate 80 percent of several indoor air pollution problems including preventing 69,000 severe headaches and 105,000 respiratory problems each year.[66]

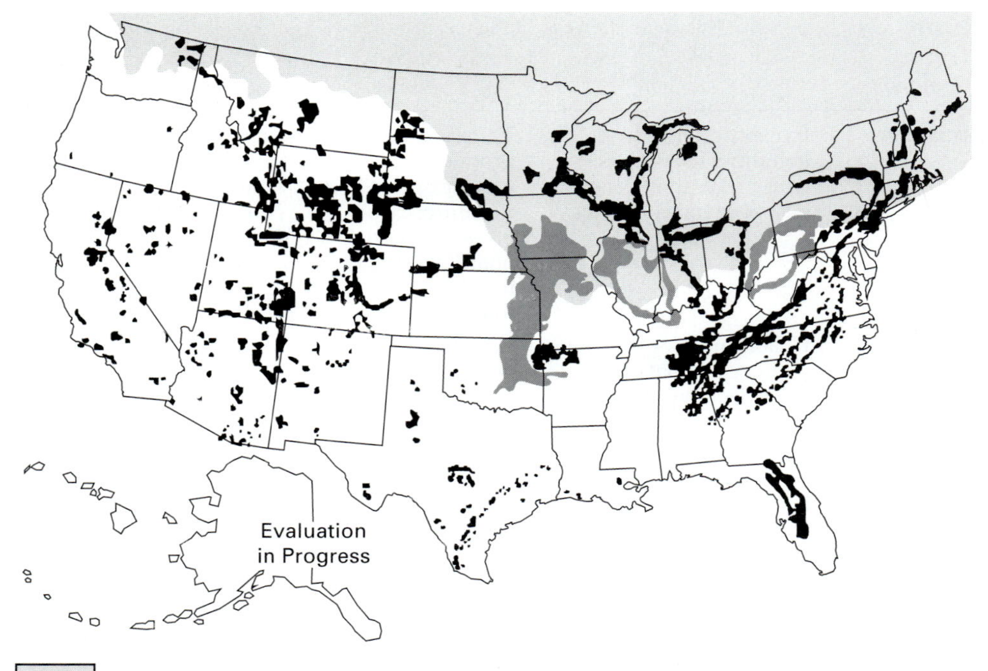

Extent of continental glaciation

Geologic areas with known or expected indoor radon levels: granitic rocks, black shales, phosphatic rocks, near surface distribution of NURE potential uranium sources.

Areas with scattered occurrences of uraniun bearing coals and shale

Source: United States Environmental Protection Agency, *Environmental Progress and Challenges: EPA's Update* (Washington, D.C.: U.S. Government Printing Office, 1988), p. 35.

FIGURE 6.7 Areas with Potentially High Radon Levels

PROGRESS IN THE PRESENT

Major improvements in air quality have been attained over the past several decades (Figure 6.8). Atmospheric levels of sulfur dioxide, carbon monoxide, total suspended particulates, and lead have all been reduced. Between 1970 and 1994, carbon monoxide emissions dropped 23 percent, volatile organic compounds 24 percent, particulates (PM-10) were down 78 percent, sulfur dioxide emissions dropped 32 percent, and lead emissions dropped a dramatic 98 percent because of phasing out leaded gasoline. According to the EPA, emissions of nitrogen oxides increased slightly, some 14 percent since 1970 because of increased processing or manufacturing by industry and increased amounts of fuels burned by electric utility plants.[67]

Over the last ten-year period (1985–1994), carbon monoxide emissions decreased 15 percent, volatile organic compounds 10 percent, particulates 12 percent, sulfur dioxide 9 percent, and lead 75 percent. Again, nitrogen oxides increased 3 percent for reasons stated earlier. Since 1970, overall emissions of the six pollutants decreased 24 percent,

EXHIBIT 6.5 What Are the Symptoms of a Sick Building?

The symptoms of "sick building" syndrome can mimic those of many diseases, from flu and colds to more serious lung disorders. Among the many signs of this malady are:

- Headaches
- Sinus problems
- Upper respiratory distress
- Eye irritation
- Runny nose
- Cough
- Dizziness
- Shortness of breath
- Nausea
- Tightness in the chest

One of the ways to help judge whether an illness might be related to a building or a home is to ask yourself the following questions:

- Do the symptoms go away when you leave your office? If so, there's a chance you are being exposed to some irritant in the work place.
- Are the symptoms worse as the week progresses, get better on the weekend and then become worse again Monday? Such a cycle could indicate a sick building-associated illness. One way to tell is by keeping a diary of your symptoms, recommends Dr. Rebecca Bascum, director of the University of Maryland's environmental research facility. "A lot of people get headaches and feel bad," Bascum says. "They notice their headaches more at work when they have to produce, but on looking at a diary realize that they have headaches other times and that it is not a building-related problem."
- Do any co-workers also suffer?
- Have you had any water problems in your office that might have left carpeting or a ceiling tile damp?
- Is your office new? New carpeting and furniture may emit formaldehyde or plasticizers for a few months.
- Has your office been painted recently, remodeled, cleaned extensively or exterminated? Any one of these can leave high levels of irritants that might be troubling you.
- Do you work with smokers?

If you are concerned about your home, ask yourself the same questions, only keep in mind a different time frame. Do your symptoms get better when you are away from home? Are they worse on weekends and better during weekdays? If so, you may have a problem in your house. But remember that what people sometimes attribute to being part of the sick building syndrome can also be traced to other problems. If you work at a computer terminal for eight hours, chances are your eyes will be tired at the end of the day, explains Dr. Michael Hodgson of the University of Pittsburgh, and it could have absolutely nothing to with indoor air pollution.—*Sally Squires*

Source: Mike Lipske, "How Safe is the Air Inside Your Home?" *National Wildlife,* 25, no. 3 (April–May, 1987), 36.

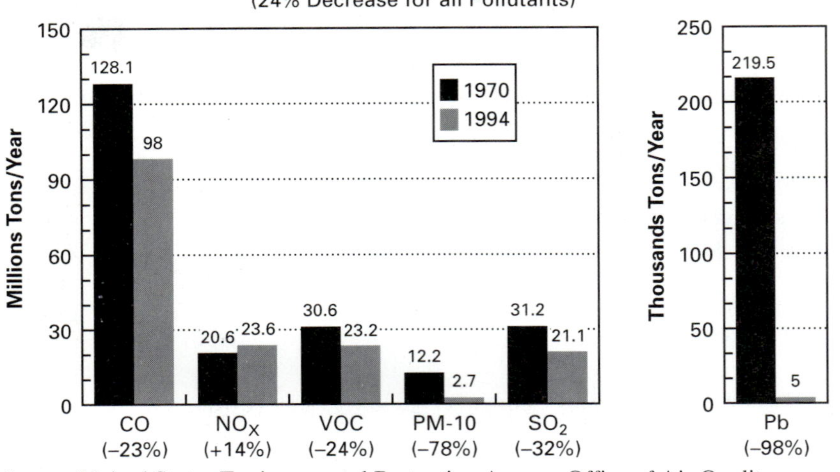

Source: United States Environmental Protection Agency, Office of Air Quality Planning and Standards, *Air Quality Trends* (Washington, D.C.: EPA, 1995), p. 4.

FIGURE 6.8 Comparison of 1970 and 1994 Emissions

while the United States population increased 27 percent, vehicle miles traveled increased 111 percent, and gross domestic product increased 90 percent. Thus the improvements in emissions and air quality occurred simultaneously with significant increases in economic growth and population. Nonetheless, in 1994 approximately 62 million people in the country lived in counties with air quality levels above the primary national air quality standards, with 50 million people living in areas that were above the ozone standard.[68]

Ozone has proven to be particularly difficult to control, and as indicated earlier, the 1990 Clean Air Act targets reduction of ozone levels in cities across the country. Studies released in 1986 suggested that ozone is harmful to human health at half the level previously thought, which would put a third of the United States population at risk.[69] Another study released in 1989 indicated that at the high levels existing in some cities, ozone could cause permanent damage, affecting lung tissue over many years leading to a precipitous and irreversible loss of breathing power. Children's lungs were found to be particularly susceptible to permanent damage at levels below the federal standard.[70] Other studies showed that children raised in ozone-polluted areas have unusually small lungs and adults lose up to 75 percent of their lung capacity. Autopsies of Los Angeles accident victims ranging in age from 14 to 25 showed that one in four had severe lung lesions of the sort caused by ozone, which gave evidence of "destructive, irreversible disease in young people."[71]

There has been considerable progress, according to the EPA, in controlling particulate matter, but smaller particles still require more rigorous controls. Approximately 13 million people live in areas where concentrations of smaller particles exceed the standards. In July 1987, the EPA revised its standards relative to particulate matter to monitor only those particles that pose a risk to health because they are small enough to penetrate the most sensitive regions of the respiratory tract. It replaced the old total suspended particulate (TSP) standard with a PM-10 standard, which includes particles with a diameter of 10 micrometers or less.[72] These smaller particles are said to be more hazardous than large ones because they lodge deeper in the lungs, causing chronic irritation that can trigger asthma attacks, aggravate other lung diseases, and interfere with the blood's ability to release carbon dioxide and take in oxygen.[73]

In late 1996, the EPA proposed more stringent regulations for emissions of these smaller particles. It would replace the 1987 standards with limits on particles smaller than 2.5 micrometers. The proposal would allow only a maximum annual average airborne particulate concentration of 15 micrograms per cubic meter and a daily maximum of 50 micrograms per cubic meter. The agency claimed that by 2007, this new standard would cause 4,000 to 17,000 fewer deaths and 63,000 fewer cases of chronic bronchitis annually. Taking into account these and other health and welfare benefits, the EPA calculated their annual economic value at between $58 billion to $119 billion by 2007, compared with an estimated cost to business of $6.3 billion annually to achieve partial compliance. The EPA also proposed to tighten the allowable amount of ground-level ozone from the current standard of 0.12 parts per million to 0.08 parts per million. This new standard would also be measured over an eight-hour period, rather than the current one-hour procedure.[74]

There was a good deal of controversy over these proposals, as 250 communities expected to find themselves in violation of the new standards for ozone and fine-particulate pollution. A coalition of governors and big-city mayors along with industry groups lined up to block these standards, and many questions were raised about the agency's estimates of benefits and costs, as well as the necessity for such regulations.[75] After weeks of intensive debate within the Clinton administration itself, the President finally approved the proposals in the summer of 1997, claiming that the hard edges of the original plan had been softened. A provision was added that blurs the deadlines for meeting the new standards and allows some industries to continue to pollute if others compensate by doing extra cleanup. States will have at least seven to ten years to rid the air of most smog-producing ozone and fine particulate matter. This decision still faced a congressional challenge, however, which can send any major federal regulation back to the administration and demand that it be reconsidered. Groups like the National Association of Manufacturers announced that it would continue to lobby Congress seeking a showdown over the proposal.[76]

The drop in lead levels over the past decade, which has been dramatic, is mainly the result of mandated use of unleaded gasoline and reductions in the amount of lead permitted in leaded gasoline. In the early 1970s, according to EPA figures, over 200 billion grams of lead were used in gasoline each year, but in 1989 less than one billion grams were used. This amounts to a reduction of over 99 percent. By 1992, about 95 percent of all gasoline sold in the country was lead-free, and over the next several years, the remainder will be phased out entirely. Lead emissions from stationary sources such as smelters and battery plants are now the major sources of lead in the air. The EPA has thus focused its efforts on those few sources, mostly smelters, that are not yet in compliance.[77]

Regarding toxic emissions, the EPA's Toxic Release Inventory has prompted actions by industries and communities to address the problem. The 1993 Toxic Release Inventory showed a decrease of 600 million pounds of toxic releases from 1989 levels of 33 percent, and a reduction of 110 million pounds or 8 percent from 1992 levels. These downward trends are expected to continue as the EPA attempts further reductions of the 189 toxic pollutants listed in the Clean Act Air of 1990. Sources of these regulated

The Dramatic Reduction of Lead in the Air

The dramatic reduction of lead in the air we breathe is one of the EPA's most important success stories. Lead has long been used in gasoline to increase octane levels to avoid engine knocking. Lead is a heavy metal that can cause serious physical and mental impairment. Children are particularly vulnerable to effects of high lead levels. Two efforts begun 15 years ago are responsible for a 95-percent reduction in the use of lead in gasoline.

Recognizing the health risks posed by lead, EPA in the early 1970s required the lead content of all gasoline to be reduced over time. The lead content of leaded gasoline was reduced in 1985 from an average of 1.0 gram/gallon to 0.5 gram/gallon, and still further in 1986 to 0.1 gram/gallon.

In addition to phasing down of lead in gasoline, EPA's overall automotive emission control program required the use of unleaded gasoline in many cars beginning in 1975. Currently, about 70 percent of the gas sold is unleaded.

These two efforts, combined with reductions in lead emissions from stationary sources such as battery plants and non-ferrous smelters, have substantially reduced lead levels. This success has been one of the greatest contributions EPA has made to the nation's health.

Source: United States Environmental Protection Agency, *Environmental Progress and Challenges: EPA's Update* (Washington, D.C.: U.S. Government Printing Office, 1988), p. 16.

pollutants will be required to achieve emission reductions comparable to similar facilities that have the best available controls. By 2005, the air toxins program is projected to reduce emissions of toxic air pollutants by at least 1 billion pounds.[78]

Efforts are also being made to control new sources of air pollution that were overlooked in previous programs. In 1994, the EPA proposed a first-ever regulation to limit air emissions from gasoline and diesel engines used on pleasure boats. Marine engines were said to account for 30 percent of all hydrocarbon emissions and 16 percent of all nitrogen oxide emissions from nonroad engines nationwide.[79] Also in 1994, the EPA proposed the first nationwide emission standards for lawn mowers, garden tractors, and other gas-powered machinery such as leaf blowers and weed cutters, which were estimated to produce 5 percent of air pollution in the country overall and a good deal more in cities. A dirty, inefficient gas mower was said to emit the same amount of hydrocarbons in one hour as would driving a new car 340 miles.[80]

STRATEGIES FOR THE FUTURE

Most of the approaches taken to date to reduce or control air pollution are considered by some experts to be nothing more than technological Band-Aids rather than efforts to address the roots of the problem, said to be inappropriate energy, transportation, and industrial systems. The most widespread technological innovation, for example, has been the introduction of electrostatic precipitators and baghouse filters for the control of particulate emissions from power plants. The predominant technique used to reduce sulfur dioxide has been to put flue-gas desulfurization technology (scrubbers) on coal-burning power plants. These scrubbers can remove as much as 95 percent of a given plant's sulfur dioxide emissions. For the control of nitrogen oxide emissions from power plants, a variety of approaches have been used with mixed results. Clean coal technologies that lower emissions of both sulfur dioxide and nitrogen oxides during combustion are under investigation.[81]

These technologies provide necessary immediate reductions to air pollution, but they are not the ultimate solution, according to some experts. They create environmental problems of their own, in many cases, and do little, if anything, to reduce carbon dioxide emissions, which may contribute to global warming. They are best viewed as a bridge to the day when energy-efficient societies are the norm instead of the exception, and when renewable sources such as solar, wind, and water power provide the bulk of the world's energy.[82] The ultimate solution to the air pollution problem involves a major change in industrial societies and the industrial systems in these societies.

Efficient use of energy is an essential strategy for reducing emissions from power plants, which are a major source of many pollutants. Equally important, according to some experts, are the savings resulting from avoided power plant construction, which in some cases can more than offset the cost of emission controls at existing plants. It is essential that governments put economic incentives for energy reform into place as part of their air quality strategies. Industrial societies may also have to change their transportation systems. Reducing urban air pollution may require a major shift away from automobiles as the cornerstone of transportation systems. This may involve more experiments with alternative forms of transportation. In the meantime, societies must encourage the manufacture and purchase of automobiles that are both low in emissions and higher in fuel economy.[83]

Efforts to control toxic emissions will be most successful if they focus on waste minimization rather than simply on control of emissions after they have been created. Perhaps the most effective incentive for waste reduction is strict regulation regarding the disposal of these wastes into land, air, and water. This practice will force up the price of disposal, making it cost-effective for industries to reduce waste generation. Public access to information about what chemicals a plant is emitting, can be instrumental in spurring a response to this problem.[84]

Few policy makers, however, are considering implementation of comprehensive strategies that are necessary to reduce pollution for future generations. According to one expert, air pollution is an eminently solvable problem, but simply tinkering with the present system will not be adequate. The only strategy that will work is a comprehensive approach to the problem that focuses on pollution prevention rather than pollution control. As society is faced with ever-mounting costs to human health and the environment, the question is not how society can afford to control air pollution; the appropriate question to ask is how we can afford not to deal with the problem.[85]

Questions for Discussion

1. Distinguish between primary and secondary pollutants. What are natural pollutants? What substances are considered to be the most harmful as far as outdoor pollution is concerned?
2. What effects do these pollutants have on human health? How many people are exposed to health hazards because of these pollutants? What other damage is caused by these pollutants?

3. Describe the various levels of regulation with respect to air pollution. Are all these levels of control necessary? Why or why not? What would you recommend as an alternative system? How is ambient air quality controlled? How are the regulations with respect to ambient air implemented?
4. Why are toxic emissions of concern? What does the new law (SARA) require with respect to these emissions? What long-term effect has these requirements had on companies? Is there a lesson to be learned here with respect to pollution control?
5. Study the new Clean Air Act in detail. What are its major provisions? What impact do you predict this bill will eventually have on industry? How will if affect your pocketbook? Will it affect your lifestyle? If so, in what ways?
6. What is acid rain? Is it as much of a problem as was initially thought? What technologies exist to deal with the problem? How effective are these technologies? How is the new Clean Air Act going to deal with acid rain? What kind of system is set up in the legislation and how well do you think it will work?
7. What are common sources of indoor pollution? How serious a problem is this kind of air pollution? What can be done about the problem? How serious a problem is radon gas? What new evidence exists with respect to this problem?
8. What progress has been made with regard to outdoor air pollution? What problems remain to be solved? Will the new Clean Air Act adequately address these problems in your opinion? What else needs to be done? Are drastic changes needed in the system and in our lifestyle for real improvements to be realized?

Endnotes

1. G. Tyler Miller, Jr., *Living in the Environment* (Belmont, CA: Wadsworth, 1990), pp. 484–485.
2. Ibid., p. 485.
3. Ibid., pp. 497–498.
4. United States Environmental Protection Agency, Office of Air Quality Planning and Standards, *Air Quality Trends* (Washington, D.C.: EPA, 1995), p. 1.
5. Miller, *Living in the Environment,* p. 498.
6. Ibid., p. 501.
7. *Cleaning the Air: EPA's Program for Air Pollution Control* (Washington, D.C.: EPA, 1979), p. 9.
8. See "EPA Proposes End to Highway Funds for Chicago Area," *Wall Street Journal,* May 2, 1984, p. 6; and "Detroit Faces Cutoff of Highway Funding Over EPA Standards," *Wall Street Journal,* June 12, 1984, p. 3. See also Robert E. Taylor, "U.S. Is Likely to Impose Growth Curbs in Areas Not Meeting Ozone Standard," *Wall Street Journal,* February 20, 1987, p. 6.
9. United States Environmental Protection Agency, Office of Air and Radiation, *The Plain English Guide to the Clean Air Act* (Washington, D.C.: EPA, 1993), p. 4.
10. United States Environmental Protection Agency, Office of Air Quality Planning and Standards, *Measuring Air Quality: The Pollutant Standards Index* (Washington, D.C.: EPA, 1994), pp. 1–2.
11. Ibid., pp. 2–3.
12. Ibid., pp. 6–7.
13. EPA, *Plain English Guide,* pp. 6–7.
14. Ibid., p. 3.
15. Barbara Rosewicz, "EPA Announces Steps to Protect Public from Cancer-Causing Vapors of Benzene," *Wall Street Journal,* September 1, 1989, p. A3.
16. "EPA Is Set to Allow Factory Trade-Offs for Air Pollution," *Wall Street Journal,* April 2, 1982, p. 5.

17. Robert E. Taylor, "EPA Is Expanding Its 'Bubble' Policy for Air Pollution," *Wall Street Journal,* November 16, 1986, p. 18.
18. Hilary F. French, "Clearing the Air," *State of the World 1990* (Washington, D.C.: Worldwatch Institute, 1990), p. 103–104.
19. Robert E. Taylor, "EPA Is Planning to Leave Regulation of Most Toxic-Gas Emissions to States," *Wall Street Journal,* June 4, 1985, p. 5.
20. EPA, *Plain English Guide,* p. 8.
21. Ibid., pp. 8–9
22. "SARA Title III Brings Chemical Risk Issues to the Forefront of Corporate Management," *CEM Report,* Vol. 9 (Summer 1990), p. 1.
23. Sharon Begley, "Is Breathing Hazardous to Your Health?" *Newsweek,* April 3, 1989, p. 25.
24. Mark Schleifstein, "La. Toxic Emissions Fall by Two Thirds in 1994," *Times-Picayune,* December 5, 1995, p. A1.
25. "SARA Title III," p. 7. See also Mary Beth Regan, "An Embarrassment of Clean Air," *Business Week,* May 31, 1993, p. 34.
26. EPA, *Plain English Guide,* p. 9.
27. "Study: Auto Pollution Drives Up Health Costs," *Times-Picayune,* January 20, 1990.
28. David Woodruff and Thane Peterson, "Here Come the Greenmobiles," *Business Week,* November 11, 1991, pp. 46–48.
29. Oscar Suris, "EPA Is Negotiating with Auto Makers over Northeast Air-Quality Proposal," *Wall Street Journal,* July 25, 1994, p. A3.
30. Oscar Suris, "EPA Backs Auto-Emissions Plan Sought by Northeast Despite Industry Protests," *Wall Street Journal,* December 20, 1994, p. B6.
31. Oscar Suris, "Big Three Discuss a Joint Crusade Against California's Electric-Car Rule," *Wall Street Journal,* October 25, 1993, p. A4. See also Oscar Suris, "Detroit Steps Up Push for Delay on Electric Cars," *Wall Street Journal,* February 7, 1994, p. A4.
32. Oscar Suris, "Big Three Fight Sales Mandates for Electric Cars," *Wall Street Journal,* July 3, 1995, p. A2. Chrysler broke ranks with the industry by agreeing to offer electric minivans for sale in California by 1998. See Oscar Suris, "Chrysler to Sell Electric Minivans in '98 in California, Giving in to Regulators," *Wall Street Journal,* April 6, 1994, p. A7.
33. Mark Ivey, "Fuel Wars: Big Oil Is Running Scared," *Business Week,* June 4, 1990, p. 132. See also Caleb Solomon, "Shell Pumps Cleaner Gas in Dirtiest Cities in U.S.," The *Wall Street Journal,* April 12, 1990, p. B1.
34. Barbara Rosewicz, "Oil Industry Signs Accord for Gasoline That Burns Cleaner to Be at Pumps by '95," *Wall Street Journal,* August 19, 1991, p. C12.
35. Allanna Sullivan, "Gasoline Prices Appear Poised for Big Rise," *Wall Street Journal,* August 22, 1994, p. C1.
36. Allanna Sullivan, "Confusion Over EPA Fuel Plan Leads to Rise in Gasoline Prices," *Wall Street Journal,* August 22, 1995, p. B1. See also Mary Beth Regan, "May Old Clean-Air Laws Be Forgot . . . ," *Business Week,* December 26, 1994, p. 64.
37. John Carey, "If Don Stedman Is Right, The Clean Air Act Is All Wrong," *Business Week,* October 1, 1990, p. 40.
38. Kenneth Chilton and Christopher Boerner, *Smog in America: The High Cost of Hysteria* (St Louis, MO: Washington University Center for the Study of American Business, 1996).
39. Robert W. Hahn, "Let Markets Drive Down Auto Emissions," *Wall Street Journal,* October 17, 1994, p. A14.
40. General Motors, Environmental Report, 1995, p. 25. See also David Woodruff, "Shocker at GM: People Like the Impact," *Business Week,* January 23, 1995, p. 47.
41. "Storm Over a Deadly Downpour," *Time,* December 6, 1982, pp. 84–86.
42. Anne LaBastille, "Acid Rain, How Great a Menace?" *National Geographic,* 160, no. 5 (November 1981), 653.

43. "Acid Rain Damage in Adirondacks May Endure," *Times-Picayune,* November 4, 1995, p. A12.
44. Betty Hielman, "Acid Precipitation," *Environmental Science and Technology,* 15, no. 10 (October, 1981), 1123.
45. Ibid., p. 1122.
46. "Storm Over a Deadly Downpour," pp. 84–86.
47. Miller, *Living in the Environment,* pp. 496–497.
48. "Storm Over a Deadly Downpour," pp. 84–86.
49. United States Environmental Protection Agency, *Environmental Progress and Challenges: EPA's Update* (Washington, D.C.: U.S. Government Printing Office, 1988), p. 28.
50. Ibid., p. 29.
51. Ibid., pp. 29–30.
52. S. Fred Singer, "The Answers on Acid Rain Fall on Deaf Ears," *Wall Street Journal,* March 6, 1990, p. A20.
53. See Edward S. Rubin, *Global Warming Research: Learning from NAPAP's Mistakes* (St Louis, MO: Washington University Center for the Study of American Business, 1992).
54. EPA, *Plain English Guide,* p. 15.
55. "Acid Rain Fight May Raise Rates," *Times-Picayune,* October 30, 1991, p. A7.
56. "Pollution Swap," *Time,* May 25, 1992, p. 22.
57. Jeffrey Taylor, "CBOT Plan for Pollution-Rights Market Is Encountering Plenty of Competition," *Wall Street Journal,* August 24, 1993, p. C1.
58. Jeff Bailey, "Utilities Overcomply with Clean Air Act, Are Stockpiling Pollution Allowances," *Wall Street Journal,* November 15, 1995, p. A8.
59. United States Environmental Protection Agency, *Meeting the Environmental Challenge* (Washington, D.C.: U.S. Government Printing Office, 1990), p. 11.
60. Mike Lipske, "How Safe Is the Air Inside Your Home?" *National Wildlife,* 25, no. 3 (April–May 1987), 37–39.
61. EPA, *Environmental Progress and Challenges,* p. 32. See also William Wanago, "Environmental Hazards at Home: What Is Your Risk?" *Health & Fitness,* Winter 1994, pp. 18–20.
62. EPA, *Meeting the Environmental Challenge,* p. 11.
63. EPA, *Environmental Progress and Challenges,* p. 34.
64. EPA, *Meeting the Environmental Challenge,* p. 10.
65. Amy Docker Marcus, "In Some Workplaces, Ill Winds Blow," *Wall Street Journal,* October 9, 1989, p. B1. See also David Holzman, "Elusive Culprits in Workplace Ills," *Insight,* June 26, 1989, pp. 44–45.
66. Mitchell Pacelle, "Plan to Clear the Office Air Spurs a Battle," *Wall Street Journal,* December 6, 1994, p. B1.
67. EPA, *Air Quality Trends,* pp. 3–4.
68. Ibid.
69. Robert E. Taylor, "New Studies Indicate Ozone Is Harmful at Half Level Previously Thought Safe," *Wall Street Journal,* April 24, 1986, p. 4.
70. David Stipp, "Breathing Ozone at Cities' Current Levels May Injure Lungs, Research Indicates," *Wall Street Journal,* September 18, 1989, p. B4. See also Andy Pasztor, "Children's Health Is Called at Risk Due to Pollution," *Wall Street Journal,* October 28, 1993, p. B7.
71. Curtis A. Moore, "Poisons in the Air," *International Wildlife,* September–October, 1995, p. 42.
72. EPA, *Air Quality Trends,* p. 9.
73. See Michael Castleman, "Tiny Particles, Big Problems," *Sierra,* November–December, 1995, pp. 26–27.
74. Thomas D. Hopkins, "Proof? Who Need Proof? We're the EPA!" *Wall Street Journal,* May 21, 1997, p. A14; John J. Fialka, "Group Gears Up to Block EPA Proposals on National Air-Quality Standards," *Wall Street Journal,* November 29, 1996, p. A8.

75. Ibid. See also Stephen Huebner and Kenneth Chilton, *EPA's Case for New Ozone and Particulate Standards: Would Americans Get Their Money's Worth?* (St Louis, MO: Washington University Center for the Study of American Business, 1997).
76. Heather Dewar, "Clinton Imposes New Limits on Pollution, Alarms Industry," *Times-Picayune,* June 26, 1997, p. A1; John J. Fialka, "Clinton Backs Rules Covering Air Pollution," *Wall Street Journal,* June 26, 1997, p. A4.
77. EPA. *Air Quality Trends,* p. 6.
78. Ibid., p. 12. In 1996, the EPA proposed to extend the reporting requirements for toxic releases to seven previously exempt categories of industrial activity, and expand by one third the total number of businesses annually required to report their toxic releases to the public. See Timothy Noah, "EPA Proposal Expands List of Industries Subject to Disclosing Toxic Releases," *Wall Street Journal,* June 27, 1996, p. B8.
79. Timothy Noah, "EPA Proposes to Curb Air Emissions from Boats, Gasoline, Diesel Engines," *Wall Street Journal,* November 1, 1994, p. B7.
80. Victoria Balfour, J. Howard Green, and Jay Peterzell, "The Backyard Besieged," *Time,* July 4, 1994, p. 62.
81. French, "Clearing the Air," pp. 110–111.
82. Ibid., p. 111.
83. Ibid., p. 114.
84. Ibid. See also Amal Kumar Naj, "Some Companies Cut Pollution by Altering Production Methods," *Wall Street Journal,* December 24, 1990, p. A1.
85. French, "Clearing the Air," p. 118.

Suggested Reading

Barker, Jerry, and David T. Tinsey. *Air Pollution Effects on Biodiversity.* New York: Chapman and Hall, 1992.

Colbeck, I., and A. R. MacKenzie. *Air Pollution by Photochemical Oxidents.* New York: Elsevier, 1994.

Consumer Product Safety Commission. *The Inside Story: A Guide to Indoor Air Quality.* Washington, D.C.: CPSC, 1988.

DeNevers, Noel. *Air Pollution Control Engineering.* New York: McGraw-Hill, 1995.

Elliott, Thomas C., and Robert G. Schwieger, eds. *The Acid Rain Sourcebook.* New York: McGraw-Hill, 1984.

French, Hilary F. "Clearing the Air." *State of the World 1990.* Washington, D.C.: Worldwatch Institute, 1990.

Gay, Kathlyn. *Air Pollution.* New York: Watts Publishing, 1991.

MacKenzie, James J., and Mohamed T. El-Ashry. *Air Pollution's Toll on Forests and Crops.* New Haven: Yale University Press, 1993.

MacKenzie, James J. *Breathing Easier: Taking Action on Climate Change, Air Pollution, and Energy Efficiency.* Washington, D.C.: World Resources Institute, 1989.

Majumder, Shyamal K., et al., eds. *Air Pollution: Environmental Issues and Health Effects.* Philadelphia: Penn Science, 1991.

Mellanby, K., ed. *Air Pollution: Acid Rain and the Environment.* New York: Elsevier, 1989.

Regens, James L., and Robert W. Rycroft. *The Acid Rain Controversy.* Pittsburgh, PA: University of Pittsburgh Press, 1988.

Ross, R. *Air Pollution and Industry.* New York: Van Nos Reinhold, 1995.

Rousseau, David, et al. *Your Home, Your Health, and Well Being.* Vancouver, B.C.: Enwright, Hartley, and Marks, 1988.

Snodgrass, M. E., ed. *Environmental Awareness: Air Pollution.* New York: Bancroft-Sage, 1991.

United States Environmental Protection Agency. *Environmental Progress and Challenges: EPA's Update.* Washington, D.C.: U.S. Government Printing Office, 1988.

United States Environmental Protection Agency. *Meeting the Environmental Challenge.* Washington, D.C.: U.S. Government Printing Office, 1990.

Wark, Kenneth, and Cecil P. Warner. *Air Pollution: Its Origin and Control,* 2nd ed. New York: Harper College, 1990.

Welburn, Alan. *Air Pollution and Climate Change: The Biological Impact.* New York: Halsted Press, 1994.

Zennati, Paolo et al., ed. *Air Pollution.* New York: Elsevier, 1993.

CHAPTER 7

Water Pollution

Water is essential to all life and makes up 50 to 97 percent of the weight of all plants and animals and about 70 percent of the human body. Water is also a vital resource for agriculture, manufacturing, transportation, and many other human activities. Despite its importance, water is said to be one of the most poorly managed resources in the world. The human race wastes water indiscriminately, pollutes it with various contaminants, and charges too little for making it available for irrigation and other purposes. Consequently, even greater waste and pollution of this resource are encouraged.[1]

Water covers about 71 percent of the earth's surface, making it our most abundant resource. About 97 percent of this amount is salt water, most of which is contained in the oceans of the world. Only 3 percent of all the water in the world is freshwater, and only a small amount of this percentage is usable. The rest is highly polluted, lies too far under the earth's surface to be extracted at an affordable cost, or is locked up in glaciers, polar ice caps, atmosphere, and soil in various places of the world. The freshwater that is usable still amounts to an average of 8.4 million liters (2.2 million gallons) for each person on earth. This supply of freshwater is continually collected, purified, and distributed in natural cycles, which works as long as humans don't use water faster than it can be replenished or overload it with waste material.[2]

The freshwater that we have available for use comes from two sources, groundwater and surface-water runoff. Precipitation that does not soak into the ground or return to the atmosphere by evaporation is called surface water and becomes runoff that flows from the earth's surface into streams, rivers, lakes, wetlands, and reservoirs. This surface water can be withdrawn from these sources and used for human activities, but only part of the total annual runoff is available for such purposes. Some of it flows in rivers to the oceans of the world, and some must be left in lakes and streams for natural purposes.[3]

Some precipitation, under the influence of gravity, slowly soaks deeper into the earth where it fills pores and fractures in spongelike, or permeable, layers of sand, gravel, and porous rock such as sandstone. These porous, water-bearing layers of underground rock are called aquifers, and the water they contain is called groundwater. Aquifers are recharged or replenished naturally by precipitation, but this recharging process is usually quite slow compared to the more rapid replenishment of surface-water supplies. If the withdrawal rate of an aquifer exceeds its recharge rate, the water in the aquifer is no longer a renewable resource.[4]

Water usage is measured in two ways, water withdrawal and water consumption. Water withdrawal occurs when water is taken from a groundwater or surface-water source and transported to a place where it is used in some fashion. Water consumption occurs when the water that has been withdrawn is not available for reuse in the area from which it was withdrawn. About three fourths of the water withdrawn worldwide each year is used for irrigation. The rest is used in industrial processes, in cooling electric power plants, and in homes and businesses. This use varies widely from country to country. In the United States, about three fourths of the freshwater withdrawn each year comes from rivers, lakes, and reservoirs, while the rest comes from groundwater aquifers. Almost 80 percent of this water is used for cooling electric power plants and for irrigation.[5]

Water withdrawal in the United States has more than doubled since 1950, because of increases in population, urbanization, and economic activity. About one fourth of this water is consumed. The remaining three fourths returns to replenish surface-water or groundwater supplies.[6] Worldwide, global water use has more than tripled since 1950, and what is removed from rivers, lakes, and groundwater amounts to 30 percent of the world's stable renewable supply. Demand for water has been growing faster than population because of improved living standards. Per capita use of water is nearly 50 percent higher than it was in 1950 and continues to climb in most of the world.[7]

Between 1985 and 2020, worldwide withdrawal of water for irrigation is projected to double, primarily because of increasing population pressures in less developed nations. Withdrawal for industrial processing and cooling electric power plants is projected to increase 20 times because of increasing industrialization in these countries, and withdrawal for public use in homes and businesses is projected to increase fivefold.[8] A report released in late 1993, projected that by 2025, one out of three people will be living in countries with inadequate freshwater supplies. And over the next three decades, between 46 and 52 countries will be either "water stressed" or "water-scarce."[9]

CONTROLLING WATER POLLUTION

Freshwater can become so contaminated by human activities that it is no longer useful for some purposes and can be harmful to living organisms. Some of the fertilizers and pesticides that are applied to croplands run off into nearby surface waters or leach into aquifers far below the surface. Poor land-use policies accelerate the natural erosion of the soil that pollutes surface waters with sediment. Some of the sludge and other wastes we produce on land are dumped into the world's waters where it causes serious pollution problems. There are many different types and effects of water pollution as shown in the list below.[10]

The world's rivers receive enormous amounts of natural sediment runoff, industrial discharges, human sewage, and surface runoff from urban and agricultural uses of land resources. Because rivers are moving bodies of water, many of them can recover rapidly from some forms of pollution, such as excess heat and degradable oxygen-demanding wastes, as long as they are not overloaded. Slowly degradable and nondegradable pollutants, however, are not eliminated by these natural purification processes. The ability of a river to recover depends on its volume, flow rate, tempera-

Major Types and Effects of Water Pollutants

Disease-Causing Agents: Bacteria, viruses, protozoa, and parasitic worms that enter water from domestic sewage and animal wastes and cause diseases.

Oxygen-Demanding Wastes: Organic wastes, which when degraded by oxygen-consuming bacteria, can deplete water of dissolved oxygen.

Water-Soluble Inorganic Chemicals: Acids, salts, and compounds of toxic metals such as lead and mercury. High levels of such dissolved solids can make water unfit to drink, harm fish and other aquatic life, depress crop yields, and accelerate corrosion of equipment that uses water.

Inorganic Plant Nutrients: Water-soluble nitrate and phosphate compounds that can cause excessive growth of algae and other aquatic plants, which then die and decay, depleting water of dissolved oxygen and killing fish. Excessive levels of nitrates in drinking water can reduce the oxygen carrying capacity of the blood and kill unborn children and infants.

Organic Chemicals: Oil, gasoline, plastics, pesticides, cleaning solvents, detergents, and many other water-soluble and insoluble chemicals that threaten human health and harm fish and other aquatic life. Some of the more than 700 synthetic organic chemicals found in trace amounts in surface and underground drinking-water supplies in the United States can cause kidney disorders, birth defects, and various types of cancer in laboratory test animals.

Sediment or Suspended Matter: Insoluble particles of soil, silt, or other solid inorganic and organic materials that become suspended in water and that in terms of total mass are the largest source of water pollution. In most rivers, sediment loads have risen sharply because of accelerated erosion from cropland, rangeland, forestland, and construction and mining sites. Suspended particulate matter clouds the water, reduces the ability of some organisms to find food, reduces photosynthesis by aquatic plants, disrupts aquatic food webs, and carries pesticides, bacteria, toxic metals, and other harmful substances. Bottom sediment destroys feeding and spawning grounds of fish and clogs and fills lakes, reservoirs, river and stream channels, and harbors.

Radioactive Substances: Radioisotopes that are water soluble or capable of being biologically amplified in food chains and webs. Ionizing radiation from such isotopes can cause DNA mutations, leading to birth defects, cancer, and genetic damage.

Heat: Excessive inputs of heated water used to cool electric power plants. The resulting increases in water temperatures lower dissolved oxygen content and make aquatic organisms more vulnerable to disease, parasites, and toxic chemicals.

Source: Adapted from G. Tyler Miller, *Living in the Environment* (Belmont CA: Wadsworth, 1990), pp. 518–20. © Wadsworth Publishing Co.

ture, acidic level, and the volume of incoming degradable waste material. Rivers that move slowly can easily be overloaded with oxygen-demanding wastes.[11]

Lakes and reservoirs, on the other hand, act as natural traps, collecting nutrients, suspended solids, and toxic chemicals in bottom sediments. They can take from one to one hundred years to flush themselves compared to days and weeks for most rivers. This fact makes lakes more vulnerable to contamination with plant nutrients, oil, pesticides, and toxic substances that can destroy life at the bottom of the lake and poison fish in the lake. Eutrophication of lakes is a natural process, but the addition of phosphates and nitrates as a result of human activities can produce in a few decades the same degree of plant nutrient enrichment that may take thousands to millions of years by natural processes.[12]

Such cultural eutrophication is a major problem for shallow lakes and reservoirs. When large masses of floating algae die, dissolved oxygen in the surface layer of water is depleted as they fall to the bottom and are decomposed by aerobic bacteria. Then important game and commercial fish such as lake trout and smallmouth bass die of oxygen starvation, leaving the lake populated by carp and other less desirable species that need less oxygen. If excess nutrients continue to flow into the lake, the bottom water becomes foul and almost devoid of animals, as anaerobic bacteria take over and produce their smelly decomposition products.[13]

Controlling water pollution involves efforts in two areas of concern: (1) reducing the pollution of free-flowing surface waters and protecting their uses, and (2) maintaining the quality of drinking water. In the early 1970s, the impact of conventional pollutants on surface waters was recognized and programs were developed for their control. Later, the dangers posed by toxic pollutants on the nation's waters were identified and steps were taken to eliminate their discharge. The need to protect drinking water became apparent in the mid-1970s as over 50 percent of the nation's drinking water was threatened by contamination from various sources such as underground storage tanks, fertilizers, pesticides, hazardous waste sites, and other sources.[14]

Thus a series of laws and regulations has been developed over the past several decades to protect our water resources. However, some of the laws relating to water pollution go back to the early years of the century or before. Laws to control water pollution actually began with the Rivers and Harbors Act of 1899, which prohibited discharge of pollutants or refuse into or on the banks of navigable waters without a permit. The next public policy measure on water pollution was the Oil Pollution Act of 1924, which prohibited the discharge of refuse and oil into or upon coastal or navigable waters of the United States. These laws are still enforced in situations where their application is appropriate.

Modern efforts to control water pollution began with the Water Pollution Control Act of 1948, which declared that water pollution was a local problem and required the U.S. Public Health Service to provide information to the states that would help them coordinate research activities. The Water Pollution Control Act of 1956 contained enforcement provisions by providing for a federal abatement suit at the request of a state pollution control agency. The Water Pollution Control Act Amendments of 1961 broadened federal jurisdiction and shortened the process of enforcement by stating that where health was being endangered, the federal government did not have to receive the consent of all the states involved.

The Water Quality Act of 1965 provided for the setting of water quality standards that were state and federally enforceable. These became the basis for interstate water quality standards. This act also created the Water Pollution Control Administration within the Department of Health, Education and Welfare. The Clean Water Restoration Act of 1966 imposed a fine of $100 per day on a polluter who failed to submit a required report. Finally, the Water Quality Improvement Act of 1970 prohibited discharge of harmful quantities of oil into or upon the navigable waters of the United States or their shores. It applies to offshore and onshore facilities and vessels. The act also provided for regulation of sewage disposal from vessels.

SURFACE WATER

Pollution of surface water occurs when the quantity of wastes entering a body of water overwhelms its capacity to assimilate the pollutants these wastes contain. Thus the natural cleansing ability of oxygen contained in the water is compromised and the water can no longer break down organic pollutants. Excessive nutrients from agricultural activities and municipal sewage also cause eutrophications, which is a state of ecological imbalance where algae growth is favored at the expense of other forms of aquatic life. Large algae formations at the surface of the water deplete available oxygen and prevent sunlight from reaching submerged vegetation. Photosynthesis is seriously hampered, which reduces both support for aquatic life and the assimilative capacity of the water.

The major sources of surface-water pollution are (1) organic wastes from urban sewage, farms, and industries; (2) sediments from agriculture, construction, and logging; (3) biological nutrients, such as phosphates in detergents and nitrogen in fertilizers; (4) toxic substances from industry and synthetic chemicals such as those found in pesticides, plastics, and detergents; (5) acid and mineral drainage from open-pit and deep-shaft mining; and (6) runoff containing harmful chemicals and sediment drained from streets and parking lots[15] (Figure 7.1).

There are both point and nonpoint sources of water pollution. Point sources are places where polluting substances enter the water from a discernible, confined, and discrete conveyance such as a sewer pipe, culvert, tunnel, or other channel or conduit.

FIGURE 7.1 Water Pollutants and Their Sources

	Common Pollutant Categories							
	BOD	***Bacteria***	***Nutrients***	***Ammonia***	***Turbidity***	***TDS***	***Acids***	***Toxins***
Point Sources								
Municipal Sewage Treatment Plants	•	•	•	•				•
Industrial Facilities	•							•
Combined Sewer Overflows	•	•	•	•	•	•		•
Nonpoint Sources								
Agricultural Runoff	•	•	•		•	•		•
Urban Runoff	•	•	•		•	•		•
Construction Runoff			•		•			•
Mining Runoff					•	•	•	•
Septic Systems	•	•	•					•
Landfills/Spills	•							•
Silviculture Runoff	•		•		•			•

Abbreviations: Biological Oxygen Demand, BOD; Total Dissolved Solids, TDS.

Source: United States Environmental Protection Agency, *Environmental Progress and Challenges: EPA's Update* (Washington D.C.: U.S. Government Printing Office, 1988), p. 70.

Point sources are those that come from industrial facilities and municipal sewage systems. Pollutants can also wash off, run off, or seep from broad areas of land. These are called nonpoint source pollutants because they cannot be located with much precision. Degradation of water from nonpoint sources is caused by the cumulative effect of all the pollutants that originate from large land areas within a single watershed. Common pollutants of the latter type are sediment eroded from soil exposed during construction of buildings and pesticides and fertilizers washed off cropland by rainwater (Figure 7.2).

The wastewater from municipalities consists primarily of water from toilets and so-called "gray water" from sinks, showers, and other uses. The wastewater that runs through city sewers may be contaminated by organic materials, nutrients, sediment, bacteria, and viruses. Toxic substances used in the home such as paint and household cleaners and pesticides may also find their way into sewers. In some towns and cities, industrial facilities are hooked into the municipal discharge system and dump some of their wastes into this system. Finally, storm water sometimes enters the municipal system through street sewers, particularly in those old cities that do not have a dual system, and may carry with it residues, toxic chemicals, and sediments, and in worst cases untreated sewage.[16]

Industrial sources such as the manufacturing of steel and chemicals, produce billions of gallons of wastewater every day. Some of these pollutants are similar to those in municipal sewage, but often are more concentrated. Other pollutants from industrial sources are more exotic and include a great variety of heavy metals and synthetic organic substances. These pollutants may present serious hazards to human health and the environment, particularly when large quantities are discharged. The quantity of water that is discharged by industrial sources varies depending on the type of manufacturing process employed.[17]

FIGURE 7.2 Point and Nonpoint Sources of Water Pollution

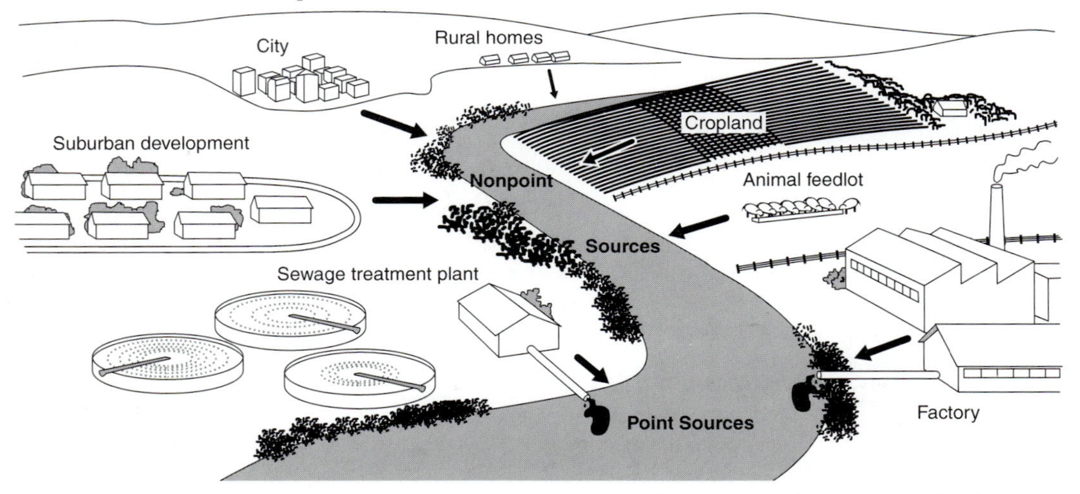

Source: G. Tyler Miller, *Living in the Environment* (Belmont CA: Wadsworth, 1990), p. 521. © Wadsworth Publishing Co.

The current system of water pollution control for surface water was established by the Federal Water Pollution Control Act Amendments of 1972, which mandated a sweeping federal–state campaign to prevent, reduce, and eliminate water pollution. This law proclaimed two general goals for the United States: (1) to achieve wherever possible by July 1, 1983, water that is clean enough for swimming and other recreational uses, and clean enough for the protection and propagation of fish, shellfish, and wildlife; and (2) by 1985, to have no discharges of pollutants into the nation's waters.[18] The goal of this act is to restore and maintain the chemical, physical, and biological integrity of the nation's waters. With this mandate, regulations and programs were developed to reduce pollutants entering all surface water, including lakes, rivers, estuaries, oceans, and wetlands.

Point Sources

The act established a National Pollutant Discharge Elimination System (NPDES), which required permits for all point sources of pollution, providing the first major direct enforcement procedure against polluters. Under the system, it is illegal for any industry or municipality to discharge any pollutant into the nation's waters without a permit from EPA or from a state that has an EPA-approved permit program. When issued, the permit regulates what may be discharged (see box) and the amount of each identified pollutant allowed from a facility. The discharger must monitor its wastes and report on discharges, and comply with all applicable national effluent limits and with state and local requirements that may be imposed. If a plant cannot comply immediately, the permit contains a compliance schedule of firm dates by which the pollutants will be reduced or eliminated.

By 1988, 39 states were issuing permits under the NPDES structure with the EPA itself issuing permits in the remaining states and on Indian reservations. While about 48,400 industrial and 15,300 municipal facilities have NPDES permits, the EPA estimated that about 10 percent of the major facilities are in significant noncompliance with their permit conditions. These facilities are subject to federal and state enforcement action, which can range all the way from an informal telephone call to formal judicial proceedings with possible financial penalties.[19]

This act was amended by the Clean Water Act of 1977, which made over 50 changes in the 1972 law. The most important from a business point of view was a change in the classification system of industrial pollutants and the establishment of new deadlines. This change resulted in a much greater emphasis on the control of toxic pollutants. Toxic substances such as heavy metals and synthetic chemicals have rapidly contaminated the nation's waters. One major source of these substances is industrial discharges; therefore, more attention to toxins was given in the amendments. These new categories and their deadlines as amended in 1987 are described below.

> *Conventional Pollutants:* These include BOD (biological oxygen demand), suspended solids, fecal coliforms, pH (acidity), and other pollutants so designated by the EPA. Industry is to have installed the "best conventional" technology (BCT) as expeditiously as practicable but in no case later than March 31, 1989, to control these pollutants.
>
> *Toxic Pollutants:* The 1977 amendments specify an "initial list" of toxic substances to which the EPA may add or subtract. Industry is to have installed the "best available" technology

National Pollutant Discharge Elimination System

What Is an NPDES Permit? Under the Clean Water Act, the discharge of pollutants into the waters of the United States is prohibited unless a permit is issued by EPA or a state under the National Pollutant Discharge Elimination System (NPDES). These permits must be renewed at least once every five years.

What Do NPDES Permits Contain? An NPDES permit contains effluent limitations and monitoring and reporting requirements. Effluent limitations are restrictions on the amount of specific pollutants that a facility can discharge into a stream, river, or harbor. Monitoring and reporting requirements are specific instructions on how sampling of the effluent should be done to check whether the effluent limitations are being met. Instructions may include required sampling frequency (i.e., daily, weekly, or monthly) and the type of monitoring required. The permittee may be required to monitor the effluent on a daily, weekly, or monthly basis. The monitoring results are then regularly reported to the EPA and state authorities. When a discharger fails to comply with the effluent limitations or monitoring and reporting requirements, EPA or the state may take enforcement action.

How Are These Effluent Limitations Developed? Congress recognized that it would be an overwhelming task for EPA to establish effluent limitations for each individual industrial and municipal discharger. Therefore, Congress authorized the Agency to develop uniform effluent limitations for each category of point sources such as steel mills, paper mills, and pesticide manufacturers. The Agency develops these effluent limitations on the basis of many factors, most notably efficient treatment technologies. Once EPA proposes an effluent limit and public comments are received, EPA or the states issue all point sources within that industry category NPDES permits using the technology-based limits. Sewage treatment plants also are provided with effluent limitations based on technology performance.

What Are Water Quality–Based Limits? Limitations that are more stringent than those based on technology are sometimes necessary to ensure that state developed water quality standards are met. For example, several different facilities may be discharging into one stream, creating pollutant levels harmful to fish. In this case, the facilities on that stream must meet more stringent treatment requirements, known as water quality–based limitations. These limits are developed by determining the amounts of pollutants that the stream can safely absorb and calculating permit limits such that these amounts are not exceeded.

Source: United States Environmental Protection Agency, *Environmental Progress and Challenges: EPA's Update* (Washington, D.C.: U.S. Government Printing Office, 1988), p. 50.

(BAT) not later than three years after a substance is placed on the toxic pollutant list and in no case later than March 31, 1989, to control toxic substances.

Nonconventional Pollutants: This category includes "all other" pollutants; that is, those not classified by the EPA as either conventional or toxic. The treatment required is the "best available" technology (BAT) as expeditiously as possible or within three years of the date the EPA established effluent limitations, but no later than March 31, 1989. A modification of these requirements is available under certain circumstances.

Rather than regulate surface-water pollution on a substance by substance basis, as was done in air pollution, or establishing effluent limitations for each individual industrial and municipal discharger, Congress developed technology-based standards that apply to the broad categories of pollutants as mentioned. While there is some disagreement as to exactly what these standards mean, in general "best conventional technology" means the average level of technology that an industry has installed to control that category of pollutants, while "best available technology" means the most sophisticated technology currently available, regardless of its cost and whether or not it is recognized as an industry standard.

In early 1987, Congress approved further amendments to the Clean Water Act by passing a $20 billion bill over the President's veto. The bill authorized $9.6 billion in grants and $8.4 billion in revolving construction projects for wastewater treatment plants; as much as $2 billion to clean up specific lakes, rivers, and estuaries; $400 million in grants to help states plan ways to reduce the toxic runoff from farmland and city streets; and funds to eliminate "hot spots" of toxic chemicals in waterways. The bill was backed by a coalition of construction companies, municipalities, and environmentalists.[20]

Nonpoint Sources

Nonpoint sources of pollution are regulated under Section 208 of the Clean Water Act. These nonpoint sources of pollution are a much more difficult problem to control. They generally cannot be collected and treated in some fashion, but can only be reduced by greater care in the management of water and land resources. One reason the United States did not meet the goals of the Federal Water Pollution Control Act was because a technology still does not exist to control nonpoint sources of pollution. These sources pour as much as 79 percent of all nitrates and 92 percent of all suspended solids into surface waters. Some major nonpoint sources of pollution are the following:

- Urban storm water: water running off buildings and streets, carrying with it oil, grease, trash, salts, lead, and other pollutants.
- Agricultural runoff: rain washing fertilizers, pesticides, and topsoil into water.
- Construction runoff: earth washed into streams, rivers, and lakes from erosion.
- Acid mine drainage: water seeping through mined areas.
- Forestry runoff: water washing sediments from areas where the earth has been disturbed by logging and timber operations.[21]

Section 208 requires that states and localities establish programs to control nonpoint source pollution. In contrast to point sources of pollution where uniform national standards for controlling them have been developed, state and local governments have been assigned the major burden and responsibility for developing nonpoint source pollution controls. The reason for this approach is that soil conditions and types, climate, and topography (which are primary determinants of nonpoint source pollution) vary throughout the country. State and local authorities are required to develop a process for identifying significant nonpoint sources of pollution and to set forth control procedures, including, where appropriate, land-use regulations.

The 1987 amendments require states to identify specific sources of nonpoint pollutants and develop Best Management Programs (BMPs) that control and reduce nonpoint pollution. Storm water also received more attention as municipalities with populations of 100,000 or more and over 100,000 private-sector sources are required to obtain NPDES permits for storm-water runoff. In 1990, the EPA issued a storm-water report describing how 173 cities and 47 counties can obtain permits for discharging storm water into municipal sewage systems. This effort has proven controversial, however, as implementing an NPDES permit for a community of 250,000 costs some $500,000 on average. The permit approach also limits the flexibility of the EPA, state agencies, and municipalities to solve storm-water runoff problems through alternative strategies that may prove to be more cost-effective.[22]

Progress and Problems

The efforts made over the past few years to improve the quality of surface water have been encouraging. The EPA reports that in 1972, 36 percent of the nation's rivers that were assessed by the states met their water quality standards, but by 1988, that figure had increased to 70 percent. These rivers supported such beneficial uses as fishing and swimming. Almost 96 percent of the lakes that were assessed were meeting both fishable and swimmable goals. Between 1977 and 1988, the number of people served by adequate sewage treatment plants, which means secondary treatment or better, increased 84 percent, from 75 million to 138 million. Secondary treatment means 85 percent removal of conventional pollutants such as oxygen-demanding materials and suspended solids.[23]

However, the EPA reports that poorly treated sewage continues to cause pollution problems in many areas. Combined sewers are still found in 1,100 communities around the nation. Combined sewer overflow occurs during rainstorms when sewer systems connected to storm-water drains overflow and empty into waterways. The cost of addressing this problem has been estimated at between $100 billion and $200 billion nationwide. In 1994, the EPA issued a new plan to deal with this problem that the EPA claimed would cost cities only $41 billion because it would give localities more flexibility. Cities must immediately maximize sewage flow into treatment plants and prohibit combined sewer overflows during dry weather. But they were also given more leeway in performing secondary treatment of sewage overflow.[24]

New pollutants are gaining attention, such as minute amounts of toxic chemicals that are harder to identify and control. In 1992, the EPA imposed water-cleanliness standards on twelve states and two territories that failed to meet federal rules with respect to toxic pollutants. These states and territories had to follow federal standards for emission levels of specific toxic chemicals when issuing permits to businesses and municipalities that discharge waste into the nation's waters. These federal standards restrict levels of arsenic, asbestos, and benzene based on their harm to both human and aquatic life. Ordinarily, states set their own levels with respect to toxic substances.[25]

Nonpoint sources of pollution continue to be a problem. Toxins and other pollutants often come from many small sources that are widely dispersed and very difficult to control, such as urban runoff and drainage of pesticides, fertilizers, and animal wastes from farmland. These sources now appear responsible for most of the remaining damage to the nation's rivers, streams, and lakes. While industrial and municipal point-source discharges have decreased significantly over the past two decades, conventional water pollutants from nonpoint sources have been steadily increasing. In some cases, these nonpoint sources of pollution have more than offset point-source improvements.[26]

Agricultural runoff is the largest unregulated source of water pollution, affecting between 50 and 70 percent of all surface water and groundwater in the country that have been evaluated for water quality. From 1990 to 1992, the EPA awarded $140 million in grants to all 57 states and territories to help them focus on key nonpoint source problems within their watersheds. About half of this money was granted to address agricultural concerns.[27] The EPA also issued regulations in 1992 that affected thousands of businesses from auto junkyards to chemical plants. These facilities were required to file plans with the EPA to control storm-water discharges. Every business covered by the permit process was required to submit a description of its pollution sources and control measures.[28] In later years, the EPA wanted even tighter controls on industrial and agricultural runoff because it claimed that more than a third of the nation's rivers and streams were unfit for fishing and swimming.[29]

The Great Lakes contain one fifth of the world's freshwater, making concern for this important natural resource understandable. Because many large cities such as Detroit and Chicago are located on the shores of these lakes, they receive high quantities of water discharged from industrial and municipal sources. States bordering the lakes evaluated 87 percent of the Great Lake's shoreline miles in 1988, and found that only 8 percent were fully supporting their designated uses, and 4 percent of these were threatened. However, contaminant levels in fish and wildlife had declined as a result of pollution controls.[30] In 1993, the EPA released a proposal under court order for further cleaning up pollution of these lakes. The guidelines were the first attempt to make standards for chemical discharges the same in all eight states bordering the Great Lakes. The proposal was estimated to cost factories, municipal wastewater plants, and other polluters between $80 million and $505 million.[31]

DRINKING WATER

About one out of every two Americans and 95 percent of those living in rural areas depend on groundwater supplies for drinking water. About 75 percent of cities in the United States depend on groundwater for their supply of drinking water. In 1982, the EPA found that 45 percent of the large public water systems served by groundwater were contaminated with synthetic organic chemicals that posed potential health threats. In 1984, at least 8,000 water wells throughout the nation were considered to have unusable or degraded water.[32] Nearly one third of Americans believe their drinking water is either contaminated or are very concerned it may become contaminated in the future, according to a 1993 Roper survey.[33]

Also in 1993, over 400,000 residents in Milwaukee became ill from drinking water contaminated by the microbe cryptosporidium and some 100 deaths were attributed to that contamination. Later that same year, New York City and Washington, D.C., advised their residents to temporarily boil their drinking water to protect against the risk of contamination. In 1994, some 30 million Americans were served by drinking water systems that violated one or more public standards. And states identified over 1,000 community water systems serving approximately 13 million people that needed to install filters to protect their water supplies against microbial threats.[34]

Some bacteria and most suspended solid pollutants are removed as contaminated surface water seeps through the soil into underground aquifers. But this natural process of purification can also become overloaded by large volumes of wastes, and the effectiveness of this process varies with the type of soil involved. No soil is effective in filtering out viruses and some synthetic organic chemicals. Once such contaminants reach groundwater supplies, they are usually not effectively diluted and dispersed because the movement in most aquifers is slow and is not turbulent. Degradable organic wastes are not broken down as readily as in rapidly flowing surface waters because groundwater has little dissolved oxygen and a fairly small population of anaerobic, decomposing bacteria. Thus it can take hundreds or thousands of years for contaminated groundwater to cleanse itself of degradable waste material. Slowly, degradable and nondegradable waste material can permanently contaminate aquifers.[35]

Two major sources of groundwater contamination are leaks of hazardous organic chemicals from underground storage tanks and seepage of hazardous organic chemicals and toxic heavy metal compounds from landfills, abandoned hazardous waste dumps, and industrial waste storage lagoons located above or near aquifers (Figure 7.3). Another source is accidental leaks from wells used to inject almost 60 percent of the country's hazardous wastes deep underground. Laws regulating injection of these wastes are weak and poorly enforced, making this a particularly threatening source of groundwater contamination. Wastes can escape during this process and permanently contaminate aquifers; thus some environmentalists and others believe this disposal process should be banned.[36]

FIGURE 7.3 Sources of Groundwater Contamination

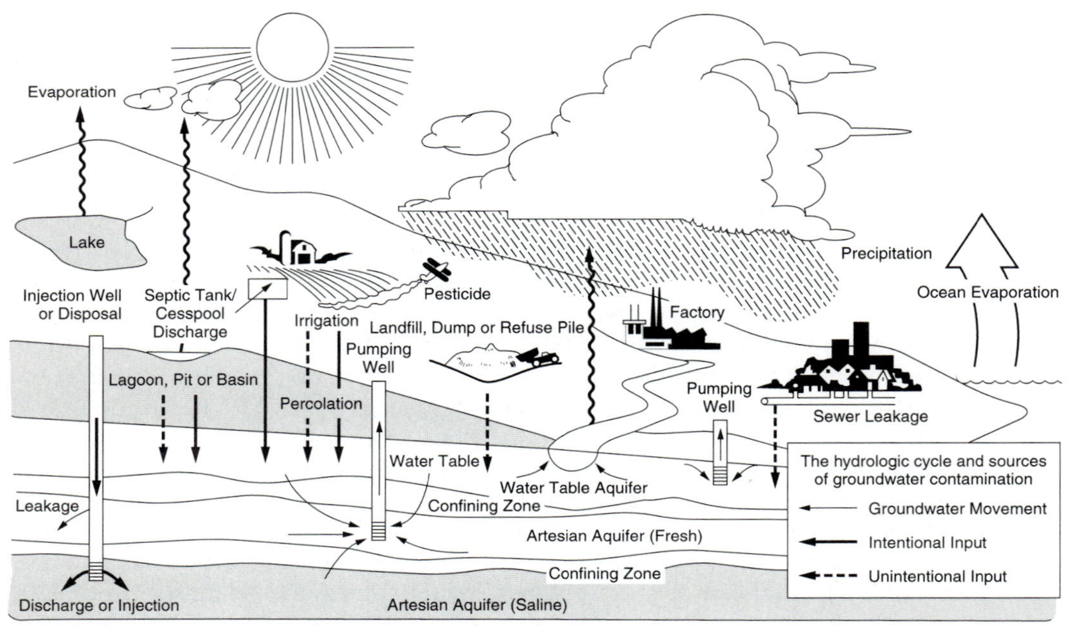

Source: United States Environmental Protection Agency, *Environmental Progress and Challenges: EPA's Update* (Washington, D.C.: U.S. Government Printing Office, 1988), p. 53.

Drinking Water Regulations

The quality of drinking water is regulated by the Safe Drinking Water Act of 1974 as amended in 1977 and 1986, which gives the EPA authority to set national standards to protect drinking water. These standards represent the maximum contaminant levels (MCL) allowable and consist of numerical criteria for specified contaminants. The states bear primary responsibility for enforcing drinking water standards, assisted in part with federal funds.[37] The EPA also issues rules to protect underground sources of drinking water (aquifers) from contamination by underground injection of wastes and other materials.

The EPA took over the setting of drinking water standards from the U.S. Public Health Service in 1975, and in a 10-year period, set limits for only 25 contaminants. Critics complained that hundreds of synthetic chemicals that were widely used had not been included in drinking water standards. In response to this criticism and growing evidence of drinking water contamination, the EPA in late 1985, proposed standards to restrict the levels of eight additional chemicals found in public water systems, and 39 other compounds and 4 microbes found in drinking water. These standards were based on health effects of the chemicals, cost considerations, and feasibility of maintaining the limits.[38]

Revisions to the Safe Drinking Water Act passed in 1986 provided money to protect aquifers that are the sole source of drinking water for an area and authorized the EPA to review any federally funded projects that may threaten or affect their quality. The revisions also required states to develop plans to safeguard other public water supplies, and gave the EPA 3 years to set standards for 83 contaminants, including 26 for which the agency had already set enforceable levels. The EPA was also required to issue regulations within 18 months requiring public water systems to test for contaminants not yet regulated. Public systems would have to test their water at least once every 5 years thereafter for such contaminants.[39]

Continuing Problems

In early 1990, the EPA did issue new regulations for 25 additional chemicals, which would cover an additional 2.5 million metric tons of waste or wastewater generated by 17,000 businesses each year.[40] But the EPA came under fire in the summer of 1990 and was accused of doing little to protect some 30 million Americans from exposure to potentially contaminated drinking water. The National Wildlife Federation (NWF) sued the EPA for relaxing rules that required states to comply with the Safe Drinking Water Act. This lawsuit was supported by a General Accounting Office report, which concluded that the water safety system is plagued by unreliable statistics on contamination, serious underreporting of violations, and falsification of data.[41]

In 1995, environmental groups strongly recommended tougher standards to keep carcinogens out of drinking water, arguing that millions of people drink water that contains harmful levels of arsenic, radon, and chlorine. Their recommendations were supposedly based on a survey of more than 100 water systems, which showed significant levels of these cancer-causing agents. Arsenic and radon generally come from natural sources. The presence of some carcinogens in water has been known for years, the groups claimed, but only recently has there been an attempt to more closely examine the extent of contamination and possible health effects.[42]

In response to these concerns, the EPA established a Wellhead Protection Program designed to prevent pollution before it fouls groundwater and community water wells. The program encourages community involvement in identifying primary risks to local water supplies and in developing local programs for preventing pollution. Between 1989 and 1991, 17 states had developed federally approved wellhead protection programs.[43] In 1995, this concept was expanded to include surface water sources of drinking water, and called source water protection. The purpose of this program was to find potential contamination sources and determine the best way to manage them.[44]

In 1987, the EPA began to pay more attention to underground storage tanks that were said to be leaking motor fuels and chemical solvents into groundwater. Most of these tanks were made of bare steel without any corrosion protection, and were nearing the end of their useful lives. About 400,000 of these underground storage tanks were thought to be leaking. Rules were issued that required existing tanks to be monitored and repaired or replaced if leaking. The rules banned installation of new bare steel tanks that lacked corrosion protection, and required double-walled tanks for hazardous chemicals.[45]

Changes to these regulations in 1988 required phasing in of new tanks within five years instead of ten with the oldest tanks forced to comply within the first year. The new tanks were also to be installed with monitoring devices.[46] In 1992, the EPA estimated that about 1.7 million underground storage tanks contained petroleum and chemicals, and that more than 100,000 leaks from these tanks had been reported. The agency also reported that 17 states had developed or proposed programs to provide assistance to help owners and operators of these underground storage tanks.[47]

Lead poisoning also became of concern in 1987, as an EPA study estimated that some 42 million Americans may drink water that exceeded the standard the EPA was proposing. Unlike other chemical contaminations of the water supply, the introduction of lead usually occurs between the treatment facility and the tap where water is drawn. The major culprit is lead solder in water coolers and the use of lead in plumbing solder.[48] The 1986 amendments banned the use of lead solders, flux, and pipes in the installation or repair of public water systems and drinking water plumbing connected to these systems. States are responsible for enforcing this ban, and the EPA can withhold grant money if it determines the state is not enforcing the requirements.

In 1991, the EPA enacted new standards lowering the maximum average lead content of drinking water from 50 to 15 parts per billion. These standards were expected to cost water users $500 million to $800 million annually.[49] In 1992, drinking water in 130 cities was found to be higher than this new standard. In 10 of these cities, lead was above 70 parts per billion, with Charleston, South Carolina, posting the worst level of 211 parts per billion.[50] In 1992, the EPA stated that more than 10 percent of the population gets its drinking water from systems containing unsafe levels of lead. This statement was based on a study of 6,400 water systems around the country, and found that 819 of them serving 30 million people exceed the legally permissible lead level.[51]

There seems to be no end to problems that threaten drinking water supplies.[52] In 1990, the Science Advisory Board to the EPA cited drinking water contamination as one of the highest ranking environmental risks in the country. The board identified four trends for drinking water that will have significant implications for future protection efforts: (1) increased population growth resulting in diminishing supplies of good quality water sources, (2) public demand for increasingly cleaner drinking water, (3) a changing

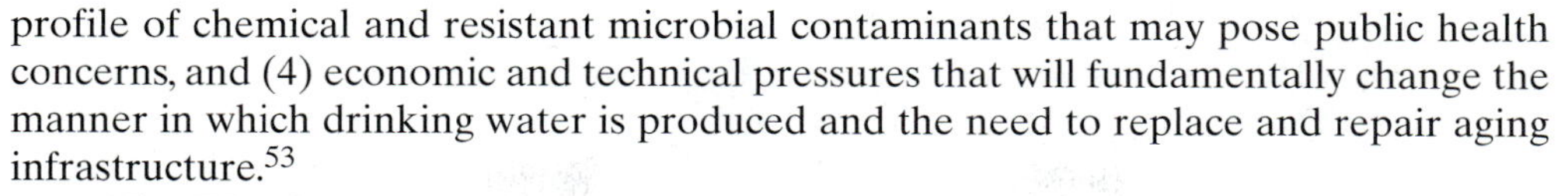

profile of chemical and resistant microbial contaminants that may pose public health concerns, and (4) economic and technical pressures that will fundamentally change the manner in which drinking water is produced and the need to replace and repair aging infrastructure.[53]

The EPA has adopted an agenda for action to address these concerns that proposes changes in the nation's system of drinking water protection. These changes include (1) engaging consumers in drinking water protection, (2) better targeting of priority health risks, (3) stronger community preventive approaches, (4) streamlining and focusing implementation of existing safety standards, and (5) investing in the nation's drinking water infrastructure. As a first step, the EPA began holding public meetings to obtain people's views and suggestions regarding these changes it is initially proposing and other possible actions it could take in the future.[54]

OCEANS AND COASTAL WATERS

The deterioration of oceans and coastal waters was highlighted in 1988 and 1989 when beaches were closed that were littered with medical waste and contaminated with fecal coloform bacteria. Shellfish beds approved as safe for consumption have decreased steadily as 17 of 22 coastal states lost shellfish acreage to pollution from 1986 to 1990, amounting to almost 40 percent of all shellfish beds in the country. As a result, there has been a rapid decline in many marine species, which in 1990 were at their lowest levels in history resulting in millions of dollars of lost revenues. Many of the nation's estuarine waters have elevated levels of toxic substances, and eutrophication is increasing the number of dead zones where fish cannot survive. Coastal fisheries, wildlife, and waterfowl populations have declined while population and industrial growth along coastal areas have increased dramatically over the past decades. In 1990, almost half of all Americans lived within 50 miles of the coast, and continued growth of coastal populations (Figure 7.4) means that more and more of the U.S. population will live within 50 miles of the coast in future years.[55]

The nation's near coastal waters encompass inland waters from the coast itself to the head of the tide, defined as the farthest point inland at which the influence of tides on water level is detected. These waters include bays, estuaries, and coastal wetlands, and the coastal ocean out to the point at which it is no longer affected by land and water uses in the coastal drainage basin. When considered as a whole, these ecosystems support a wide range of ecological, economic, recreational, and aesthetic uses that depend upon good water quality.[56] But these near coastal areas bear the brunt of massive inputs of wastes into the ocean, which can overwhelm the natural dilution and degradation processes of these areas and destroy them as sources of food and recreational pleasure.

These coastal waters and accompanying wetlands are home to many ecologically and commercially valuable species of fish, shellfish, birds, and other wildlife. Some 85 percent of nation's commercially harvested fish are dependent on near coastal waters at some point in their life cycle. Billions of dollars a year are generated as income from commercial and recreational fishing, tourism and travel, urban waterfront and private real estate development, and recreational boating, marinas, harbors, all involving coastal waters. Millions of people use the bays, beaches, and coastal ocean for swimming, boating,

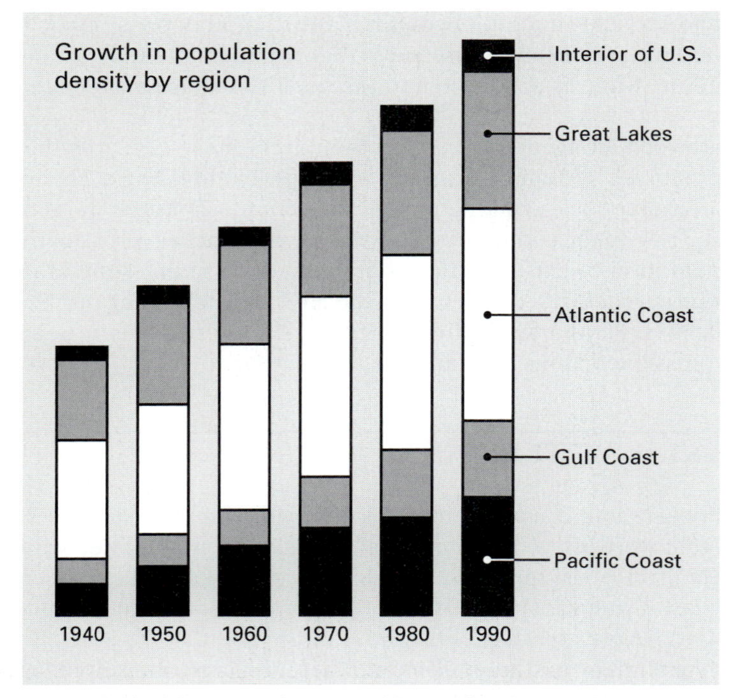

Source: United States Environmental Protection Agency, *Securing Our Legacy* (Washington, D.C.: EPA, 1992), p. 30.

FIGURE 7.4 America Moves to the Coasts

fishing, hiking, and for open space. These areas are thus very valuable to the nation as a whole.[57]

These coastal environments are susceptible to contamination because they act as sinks for large quantities of pollution discharged from municipal sewage treatment plants, industrial facilities, and hazardous waste disposal sites nearby. Nonpoint source runoff from agricultural lands, suburban development, city streets, and combined sewer and storm-water overflows pose an even more significant problem than point sources of pollution. Activities such as dredging, draining and filling, dam construction, and the building of shorefront houses may further degrade these environments. Debris on beaches from sewer and storm drain overflows can cause public safety and aesthetic concerns and can result in major economic losses for coastal communities during the tourist season. Growing population pressures will continue to subject these coastal ecosystems to further stress. Thus coastal waters are subject to a number of serious environmental problems including toxic contamination, eutrophication, pathogen contamination, habitat loss and alteration, and changes in living resources[58] (Exhibit 7.1).

As just one example of what can happen to seafood from chemical pollutants (Exhibit 7.2), there is a windswept inlet about halfway between Galveston and Corpus Christi, Texas, called Lavaca Bay, that used to be a fisherman's paradise. But in 1989, a five-square-mile section of the bay became so polluted with mercury that the finfish and crabs caught there were unfit for human consumption. The Texas Department of

EXHIBIT 7.1 Primary Causes and Effects of Marine Pollution

Type	*Primary Source/Cause*	*Effect*
Nutrients	Runoff approximately half sewage, half from upland forestry, farming, other land uses; also nitrogen oxides from power plants, cars, and so on.	Feed algal blooms in coastal waters. Decomposing algae depletes water of oxygen, killing other marine life. Can spur toxic algal blooms (red tides) that release toxicants into the water that can kill fish and poison people.
Sediments	Runoff from mining, forestry, farming, other land uses; coastal mining and dredging.	Cloud water. Impede photosynthesis below surface waters. Clog gills of fish. Smother and bury coastal ecosystems. Carry toxicants and excess nutrients.
Pathogens	Sewage; livestock.	Contaminate coastal swimming areas and seafood, spreading cholera, typhoid, and other diseases.
Persistent Toxicants (PCBS, DDT, heavy metals, and so on)	Industrial discharge; wastewater from cities; pesticides from farms, forests, home use, and so on; seepage from landfills.	Poison or cause disease in coastal marine life. Contaminate seafood. Fat-soluble toxicants that bioaccumulate in predators.
Oil	46 percent runoff from cars, heavy machinery, industry, other land-based sources; 32 percent, oil tanker operations and other shipping; 13 percent, accidents at sea; also offshore oil drilling and natural seepage.	Low-level contamination can kill larvae and cause disease in marine life. Oil slicks kill marine life, especially in coastal habitats. Tar balls from coagulated oil litter beaches and coastal habitat.
Introduced Species	Several thousand species in transit every day in ballast water; also from canals linking bodies of water and fishery enhancement projects.	Outcompete native species and reduce marine biological diversity. Introduce new marine diseases. Associated with increased incidence of red tides and other algal blooms.
Plastics	Fishing nets; cargo and cruise ships; beach litter; wastes from plastics industry and landfills.	Discarded fishing gear continue to catch fish. Other plastic debris entangles marine life or is mistaken for food. Litters beaches and coasts. May persist for 200-400 years.

Source: From *State of the World 1994: A Worldwatch Institute Report on Progress Toward a Sustainable Society* by Lester R. Brown, et al., eds. Copyright © 1994 by Worldwatch Institute. Reprinted by permission of W. W. Norton & Company, Inc.

Health posted a warning declaring the area off limits to fishing. The source of the mercury was a huge alumina plant on the bay operated by the Aluminum Company of America, which legally discharged mercury into the water from 1966 until 1970, an act that was thought harmless at the time but was later found to have long-lasting impacts. When ingested in sufficient amounts, mercury can damage the human nervous system and kidneys.[59]

EXHIBIT 7.2 Contaminants Found in Seafood

Contaminant	*Source*	*Possible Health Effects*[1]
Methyl mercury	Industrial discharges	Impairment of immune system; kidney damage; central nervous system damage; reproductive system damage; hearing and vision loss
Lead	Industrial discharges; naturally occurring	Impairment of immune system nervous system damage; liver disease; reproductive damage; possible link to cancer
DDT[2]	Agricultural and urban runoff	Liver cancer; central nervous system damage; reproductive damage; tumor promotion; impairment of immune system
Chlordane[3]	Agricultural and urban runoff	Headaches; dizziness; excitability; confusion; lack of coordination; seizures; convulsions
Hepatitis A virus	Sewage	Acute inflammation of the liver; extreme fatigue; jaundice; nausea; vomiting and diarrhea
Vibrio bacteria	Naturally occurring	Mild to severe diarrhea; severe blood poisoning; skin contact with water can cause mild to severe infection of an open wound

[1]Many of the effects are based on animal studies. All depend on a variety of factors, including the form of the substance, the dose, the duration and frequency of exposure, the sensitivity of the individual, preexisting health conditions, age and genetic makeup. Many effects "probably would not occur upon the occasional ingestion of the toxicant in quantities commonly associated with food products," according to the Texas Department of Health.

[2]DDT was banned for use in the United States in 1972, but lingers in the environment and may still be used in Mexico and other countries.

[3]Banned in 1988.

Source: Texas Department of Health as reported in Jim Morris, "A Breeding Ground for Disease," *Dallas Times Herald,* July 9, 1989, p. A17.

The shores of the United States are also being inundated by marine debris. In one day in September 1987, volunteers collected 307 tons of litter, two thirds of which was plastic, on the sands of the Texas Gulf Coast. This litter included 31,733 bags, 30,295 bottles, and 15,631 six-pack yokes.[60] In 1991, the 12 most frequently collected marine debris items were cigarette butts, plastic pieces, foamed plastic pieces, paper pieces, plastic food bags and wrappers, glass pieces, plastic caps and lids, plastic straws, glass beverage bottles, metal beverage cans, foamed plastic cups, and plastic cups and utensils (Table 7.1). This debris comes from beachgoers, storm-water sewers and combined sewer overflow, ships and other vessels, industrial facilities, and offshore oil and gas platforms.[61]

This debris, particularly plastic material, poses a serious problem for wildlife. Nonbiodegradable debris can cause injury and death of fish, marine mammals, and birds. As many as 2 million seabirds and 100,000 marine mammals may die every year after eating or becoming entangled in the plastic debris. Sea turtles choke on plastic bags they apparently mistake for jellyfish. Sea lions become ensnared when they playfully poke their noses into plastic nets and rings and, unable to open their jaws in some cases, starve to death. Brown pelicans become enmeshed in fishing line and sometimes hang themselves. In Florida, people have seen them hanging from tree branches.[62]

TABLE 7.1 The Dirty Dozen

Rank	*Debris Item*	*Total Number Reported*	*Percent of Total Debris Collected*
1.	cigarette butts	940,430	18.08
2.	plastic pieces (fragments of larger objects)	344,268	6.62
3.	foamed plastic pieces	289,802	5.57
4.	paper pieces	225,297	4.33
5.	plastic food bags/wrappers	220,945	4.25
6.	glass pieces	219,468	4.22
7.	plastic caps/lids	212,852	4.09
8.	plastic straws	191,401	3.68
9.	glass beverage bottles	189,855	3.65
10.	metal beverage cans	189,447	3.64
11.	foamed plastic cups	125,008	2.40
12.	plastic cups/utensils	118,597	2.28
	TOTAL	**3,267,371**	**62.83**

Source: Center for Marine Conservation, *1991 International Coastal Cleanup Results* (Washington D.C.: Center for Marine Conservation, 1991), p. 16.

Suffocating and sometimes poisonous blooms of algae, which are called red and brown tides, regularly blot the nation's coastal bays and gulfs. These blooms are cropping up in new places and formerly nontoxic algae are turning toxic, suggesting to some scientists that they are becoming a "major planetary trend." They usually leave a trail of dying fish and contaminated mollusks and crustaceans behind, which can transfer these poisons to people by concentrating toxins in their internal organs.[63] Patches of water known as dead zones, where the oxygen in the water has almost been totally depleted, are proliferating. As many as 1 million fluke and flounder were killed in the summer of 1988 when they became trapped in anoxic water in the Raritan Bay in New Jersey. During the same summer, another dead zone said to be 300 miles long and 10 miles wide was adrift in the Gulf of Mexico.[64]

Regarding pollution of the oceans, dumping of dredged material, sewage sludge, and industrial wastes are major sources of pollution. Sediments dredged from urban harbors are often highly contaminated with heavy metals and toxic chemicals like PCBs and petroleum hydrocarbons. When dumped in the ocean, these contaminants can be absorbed by marine organisms.[65] The oceans also receive agricultural and urban runoff, atmospheric fallout, garbage and untreated sewage from ships, accidental oil spills from tankers and offshore oil drilling platforms, and intentional discharges of oil by tankers when they empty or clean their bilges.[66]

The dumping of sewage sludge into the ocean continued to increase until 1990, when the EPA negotiated consent agreements with local jurisdictions to phase out ocean dumping of sludge (Figure 7.5). Evidence continued to mount that millions of tons of this foul-smelling waste reached the ocean floor and contaminated animal life there. Previously, it had been assumed that sewage sludge dumped in the ocean would disperse and pose no environmental hazard.[67] Two years after implementation of this ban,

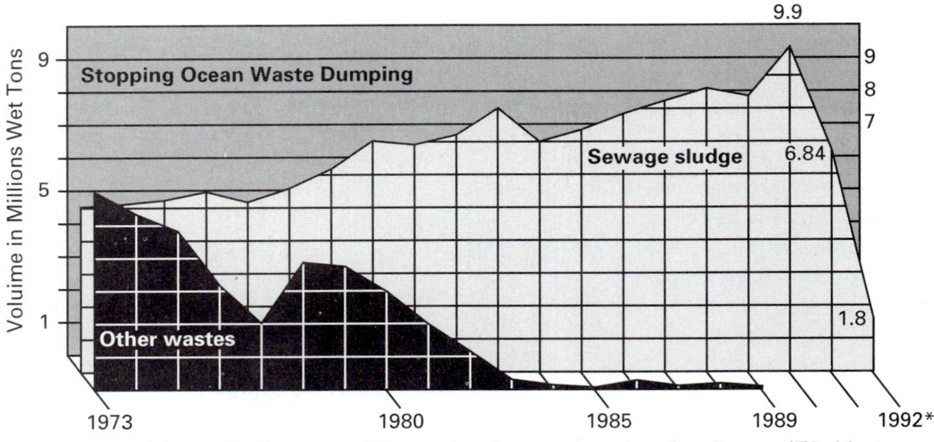

Source: United States Environmental Protection Agency, *Securing Our Legacy* (Washington, D.C.: EPA, 1992), p. 29.

FIGURE 7.5 Stopping Ocean Waste Dumping

however, an international investment group prodded Congress to allow tests of a new ocean disposal technology that supposedly isolate waste material on the deep sea floor to prevent it from migrating because of ocean currents.[68]

Crude oil as it comes from the ground and refined petroleum can be accidentally or deliberately released into the ocean from a number of sources. Tanker accidents and blowouts at offshore drilling rigs are the best-known sources because they receive a great deal of publicity when they occur. But about half of the oil reaching the oceans is said to come from the land as a result of runoff and dumping of waste oil by cities and industrial facilities. Tanker accidents account for only 10 to 15 percent of oil spilled into the oceans each year, but such spills can have severe ecological and economic impacts on the region where they occur. There has been a decrease since 1973 in the average annual number of major oil tanker accidents due to improved safety measures, better navigational equipment, and better training.[69] All of these factors were sadly lacking, however, during the *Valdez* oil spill in 1989, which was the largest such spill the United States had experienced in its waters.

Such accidents helped to keep the focus on oil tankers and the dangers of transporting oil in this manner. In response to the *Valdez* oil spill, Congress passed the Oil Pollution Act of 1990, which affects oil transportation, storage, and handling operations within all navigable waters under United States supervision. The act requires that by 2015, all tankers will be equipped with double hulls to reduce the possibility of major oil spills. The law also increases the liability of owners and operators of oil-handling facilities and vessels, and requires screening of certain crew members for alcohol and drug-related offenses. Oil spill response planning was improved, and greater responsibility was given to the federal government to direct oil spill cleanups.[70]

Meanwhile, the oil industry funded its own rapid-deployment force to clean up spills wherever they occurred in U.S. waters. Twenty firms agreed to create a $35 million Marine Spill Response Corporation (MSRC) to reduce the harmful effects of spills and the financial costs to shippers. The organization operates out of five regional response centers with 22 preposition sites where equipment and sometimes vessels and personnel will be located. In 1992, MSRC launched the first of 16 new vessels that will form the backbone of its national network of response to major oil spills. The vessels are designed to hold 4,000 barrels of recovered oil temporarily and provide a command and control center for spill-fighting operations.[71]

The effects of oil on ocean ecosystems depend on the type of oil, the amount released, the distance of the release from coastal areas, the time of year, weather conditions, the average water temperature, and ocean and tidal currents. After an oil spill, hydrocarbons such as benzene and toluene affect shellfish and nonmigratory fish, sometimes killing them in great numbers. Other chemicals remain on the surface and form floating tarlike blobs that adhere to marine birds, sea otters, seals, sand, rocks, and almost all objects it encounters. This oily coating destroys the animals' natural insulation and buoyancy, and most drown or die of exposure. Heavy oil components that sink to the ocean floor or wash into estuaries can have the greatest long-term ecosystem impact. These components can kill bottom-dwelling organisms such as crabs, oysters, mussels, and clams or make them unfit for human consumption. Oil slicks that wash onto beaches can have serious economic effects on coastal residents who lose income from fishing and tourist activities.[72]

In order to deal with pollution of the oceans and near coastal areas, the EPA established an Office of Marine and Estuarine Protection in 1984 to administer all the agency's ocean and coastal programs. Some of the most important achievements of this office include (1) continuation of the Great Lakes restoration program and start-up of programs in the Chesapeake Bay and 17 estuaries that are part of the National Estuary Program; (2) progress toward a ban on ocean dumping of sewage, sludge, and industrial waste; (3) the creation of a Coastal and Marine Policy to promote coordination of coastal programs conducted by different federal and state agencies; and (4) the implementation of a "near coastal waters" plan for managing environmental problems in waters that are not being addressed by the ongoing bay and estuary programs.[73]

Over the next several years, the EPA intends to work with state and local governments to increase the acres of shellfish beds open to harvest, reduce fishery bans and advisories due to contamination, decrease beach closures, and eliminate ocean dumping of sewage and industrial wastes. The EPA also wants to strengthen nonpoint source management programs in all coastal counties and tighten controls on point source discharges of toxics, nutrients, and other pollutants to restore the quality of coastal waters. Storm-water discharge permits will be required for large cities in all coastal counties, and help will be provided to smaller municipalities with storm-water problems. State and local governments are being encouraged to manage coastal development so that it proceeds in an environmentally sound direction.[74]

Regarding oceans, the EPA is asking all types of offshore activities, such as oil and gas operations, to help protect marine waters and surrounding ecosystems from degradation. It is also taking enforcement actions to eliminate any illegal ocean disposal of waste material. Finally, the EPA and other federal agencies are working with international bodies such as the United Nations to assess the health of the world's oceans and develop an integrated approach to preventing further degradation of the oceans. Such an approach is needed if we are to halt pollution of the oceans and ensure survival of the human race. As Jacques-Yves Cousteau commented, "The very survival of the human species depends upon the maintenance of an ocean clean and alive, spreading all around the world. The ocean is our planet's life belt."[75]

RESPONSES TO WATER POLLUTION PROBLEMS

The potential solutions that exist with regard to the provision of safe drinking water and improving the quality of surface water are not cheap or, in most cases, easy to implement. But it seems that the best solutions focus on inputs that cause contamination rather than on trying to clean up contaminated water. Such an approach is more ecologically and, in most cases, economically sound than allowing water to become contaminated in the first place. Most people would agree that the most economical and surest way of ensuring safe drinking water, for example, is to make sure that existing wells and surface-water sources do not become contaminated. Several things can be done within this general approach.

1. Reduce or eliminate the use of pesticides and fertilizers in agriculture. According to a report published by the Office of Technology Assessment, 260,000 tons of active ingredients in pesticides and 42 million tons of fertilizer are spread annually over the equivalent of 280 million acres across the country.[76] Contamination usually occurs through conventional application on farmland and an increasingly common method of irrigation called chemigation, in which water is mixed with pesticides. In addition, fertilizers are also contaminating the groundwater in many communities that rely on agriculture to support the economy. Contamination in states such as Iowa are particularly worrisome to federal and state officials because they usually rely on groundwater for over three quarters of the drinking water supply.[77]

Alternative means of pest control are available. One study done by the University of Nebraska has shown that crop yield can actually be increased without the use of pesticides.[78] This method involves a constant monitoring of insect and bird seasonal variations, planting techniques, and alternate field cycles of usage. It is not yet known if this method can be used on a large scale, but the results look promising. Alternatives to the use of fertilizer also exist that promise crop yields comparable to yields using fertilizer. If pesticide and fertilizer usage could be reduced, this action would have a major impact on the pollution problem regarding surface and drinking water.

2. Increase the amount of banned substances that can be disposed of in our nation's waterways. Of the 65,000 chemicals in existence today, 40,000 are suspected to be cancer causing or carcinogenic. Yet only a handful have been banned from disposal in our nation's waterways. While the EPA plans on adding approximately 15 to 25 new chemicals to the banned list each year, this number pales in comparison with the thousand or so new chemicals introduced into the environment each year. The testing procedure must be speeded up so that more chemicals can be banned if necessary, and industry must take more precautionary steps to make sure the new chemicals they generate are environmentally safe and will not cause a water pollution problem.

3. Industrial pretreatment for discharge and reuse should be required. Pretreatment is a fancy word for cleaning up the discharge from industrial plants before it is discharged

into municipal sewage systems, a practice that exists in many localities. Without pretreatment, many adverse effects can occur, including the corrosion of collection systems or of the sewage treatment system due to corrosives in the wastewater, exposure of workers to toxic substances and hazardous fumes, limited or more expensive sludge disposal options, or the passing through of hazardous toxins to the water supply.

Such pretreatment programs can have beneficial effects to companies as well as to the environment. When a wastewater treatment facility treats water, it basically takes the unwanted organic material as well as some toxins out of the water and sends the treated water to customers or discharges it into lakes and rivers. What is left behind is a material called sludge, the organic and toxic waste taken from the water. This sludge has to be disposed of somehow, probably in a landfill or in an incinerator now that ocean dumping is prohibited. Both of these methods of disposal have problems, but if this sludge does not contain any toxic material, it can be disposed of much more easily.

4. Water recycling should be promoted throughout the country. Water supplies for cities and industries are typically taken from a river or aquifer, used in a factory or home, and then released as "wastewater" to the nearest water source. As the demand for water increases, this approach easily overtaxes existing water sources.[79] Not only must large quantities of freshwater be made available, but natural waterways are used to dilute the discharged wastes. This traditional way of thinking about and setting up industrial processes cannot continue as we are running out of water resources.

Industrial activities account for nearly one quarter of the world's water usage. In contrast to agriculture, only a small fraction of this water is actually consumed. Most of the water in industry is used for cooling, processing, and other activities that may heat and pollute the water, but do not use it up in any sense. This type of usage allows a factory to recycle and reuse its water supplies, perhaps many times over during the course of its usage period. Given incentives to do so, some companies have demonstrated that they can cut their water usage 40 to 90 percent with available technologies and practices.[80]

Recycling is thus an alternative that can conserve water supplies. Some companies, such as the 3M Corporation, have already successfully employed a waste-recycling system. In such a closed-loop system, waste isn't simply dumped into a nearby stream or lake, but instead, chemical wastewater is separated and the different chemical compounds are returned to their original form. In this way, the recovered chemicals can be used to create new products.[81]

An Armco steel mill in Kansas City, Missouri, which manufactures steel bars from recycled ferrous scrap, draws into the mill only nine cubic meters of water per ton of steel produced, compared with as much as 100 to 200 cubic meters per ton in many other steel plants. Besides cutting its total water needs by using recycled iron scrap rather than new metal, the Armco plant uses each liter of water 16 times before releasing it, after final treatment, to the river. There are other such success stories, such as a paper mill in Hadera, Israel, that requires only 12 cubic meters of water per ton of paper, whereas many of the world's paper mills use 7 to 10 times this amount.[82]

Even though several stories such as these can be told, the recycling of water in manufacturing plants has not been attempted on a large-scale basis. But this may change. Many industrial pollution control processes already recycle water by design. And because wastewater must be treated to meet environmental standards, recycling partially treated water within a plant may become more economical than paying the high costs associated with treatment of discharges to the level required. As pollution control standards are made more stringent, the costs of industrial recycling tend to go down and become more economically feasible.[83]

Several cities are also planning to recycle their water supplies, by reclaiming wastewater for drinking purposes. More and more cities are preparing to use supertreated wastewater to supplement the reservoirs and aquifers that supply household needs where it mixes with water from other sources. No city is currently suggesting that treated wastewater be pumped directly into water intake lines. There are a host of questions involving reusage of wastewater in this manner, including cost, reliability, and perhaps most important, public perception and acceptance. The latter is especially important given recent outbreaks of waterborne diseases in several cities. Traditional wastewater treatment involves removing solids with screening, settling, and skimming; biological treatment to remove organic matter; and disinfection, usually with chlorine. Advanced treatment involves passing the water through additional filters and chemical processes to remove trace organic wastes, disease carriers, and chemical contaminants.[84]

5. A related strategy is an attempt, particularly by cities and urban areas, to curb waste in the water system itself. As water systems deteriorate because of age or lack of maintenance, large amounts of water can be lost through broken pipes and other faults in the distribution network. Finding and fixing these leaks result in water savings and with a quick payback on its investment.[85] Just as energy planners have discovered it is often cheaper to promote energy saving than build more power plants, water planners realize that efficiency measures related to repairing the system may realize savings that makes new dams, reservoirs, wells, and treatment plants unnecessary. The city of Boston, for example, launched a leak repair and detection program in 1988 to save water from leaking out of its system. Later it was determined that this program reduced water losses in the greater Boston area by more than 80 percent, from more than 25 million gallons per day to fewer than 4 million gallons.[86]

6. Public conservation of water should also be encouraged for the same reasons. Probably one of the most ignored and forgotten methods of increasing the supply of fresh drinking water is the notion of conservation. Historically, conservation has been used only during short-term crises, such as drought-induced water shortages. But as cities grow and face severe physical and financial constraints to supply water to thirsty residents, conservation on a long-term basis makes more sense. Building new wastewater treatment facilities is an expensive proposition, as is the search for ever more distant water supplies and the transportation of those supplies where they are needed. Slowly the idea is spreading that managing demand, rather than continuously striving to meet it, is a surer path to water security.

The U.S. Congressional Budget Office estimates that of the nation's 756 large urban water systems (those that serve more than 50,000 people), 170 will need additional water supplies by the year 1992. Arid Las Vegas grew 62 percent in the 1980s, for example, and adds some 4,000 newcomers a month. The city has applied for annual rights to more than a billion cubic meters of water from nonrenewable aquifers hundreds of miles to the north.[87] The city of Los Angeles faces severe water shortages due to its tremendous growth over the past several decades, and the fact that it has to get its water from sources many miles away. The city is facing challenges from small communities that also rely on the same water sources. Drought conditions only amplify an already serious problem.[88]

Washington, D.C. was able to conserve water through better management techniques. The city put into place a conservation-oriented rate structure and a public awareness program, implemented a new efficient water-delivery network, and diverted some flood storage capacity to the water supply. These four programs made a $250 million investment unnecessary. Florida, which was hard hit over several winters with little rainfall, instituted fines and penalties to violators of the newly implemented state water resource laws, designed to curb waste and encourage recycling.[89]

The use of water-efficient fixtures can also save a great amount of water each day. The typical U.S. toilet, for example, turns about five gallons of high-quality water into wastewater each time it is flushed, a needless waste of water. There are currently a variety of fixtures on the market that can greatly reduce water usage in showers and other places. Substituting the most common water-saving varieties for a conventional model could reduce household water use by one fifth. Use of extremely low-water fixtures could cut existing levels of water usage by as much as 50 to 70 percent, according to some estimates.[90]

7. Some experts recommend the use of markets to allocate shrinking water supplies and encourage conservation. In many instances, it is said, that concerns about water do not stem from an absence of supplies, but from the absence of a proper market to insure a balance between consumer demand and those supplies. If government regulations and subsidies distorting water use were to be eliminated, water markets would allow excess supplies to be sold or leased to those who were willing to pay for more water. Water quality could also be improved if Congress were to introduce a system of tradable pollution discharge permits, to be required for all public and private pollution discharges. The federal government would establish the maximum level of discharges for each water basin or drainage area, and then allow the states to meet discharge goals through the most cost-effective methods.[91]

A market for water would work like any other market, as owners of water supplies would offer their water for sale and users would offer to buy it, with the two parties negotiating an agreeable price. Only by removing federal subsidies will the groups interested in water be forced to evaluate the true costs and benefits of projects. Besides achieving an equilibrium between existing supply and demand for water, markets would also (1) encourage conservation by requiring users to pay the full market price of supplies, which means prices would rise when supplies are short; (2) when facing high market costs for limited supplies, potential users, such as developers, would be forced to consider whether the cost of water made the development worthwhile; and (3) a market price would encourage users of artifically cheap water, such as farmers, to consider selling water to those who value it highly but cannot obtain supplies under the current system.[92]

8. What do people do if the water supply is already contaminated? Many wastewater treatment plants are ill equipped to remove toxic substances, as they were designed to remove mostly organic materials. What do people do if they suspect the water is unsafe to drink? Bottled water is one solution. In the United States, about 475 bottling plants in operation represent approximately 600 different brands. Americans spent about $2 billion on bottled water in 1989, making it the fastest growing segment of the entire American beverage industry.[93] But questions have to be raised about the safety of this

bottled water. Much of it comes from the same aquifers that produce tap water. No one can say with confidence that the gallon of spring water that we buy at the market isn't as contaminated as tap water.

For one thing, the Food and Drug Administration (FDA), not the EPA, regulates the bottled water industry. The FDA standards for bottled water are not as stringent as the EPA's standards for tap water, and the testing done by bottlers under FDA guidelines is not as frequent as testing done by the EPA for groundwater and surface water.[94] The point is that most water ultimately comes from the same source, and if one portion of this source is contaminated, the possibility exists that the entire supply for a region of the country is contaminated. Thus the use of bottled water may give consumers a false sense of security. To cope with these problems, the FDA tightened the rules on bottled water in 1995 to promote honesty and fair dealing in the marketplace.[95]

The point-of-use water purification market is one area that looks promising. This market involves the use of household water purification devices. There are three different technologies on the market today: (1) an activated carbon filtration system, in which contaminants attach themselves directly to the carbon's porous surface; (2) distillers in which the water is heated until it vaporizes and bacteria and suspended matter are removed; and (3) the reverse osmosis system in which water is forced through a semipermeable membrane that catches most dissolved components in the water. All three types of filtering techniques can be quite effective, but unfortunately, not all units will catch all different types of contaminants. The point-of-use market was expected to grow from its current level of $1.7 billion a year to over $3.8 billion by 1995, according to some reports.[96]

9. Regarding surface water, alternatives for cleaning up contaminated water are being developed. Scientists at the Sandia National Laboratories have devised a way to clean up polluted water with a sun-powered detoxification system. The unit is believed to be able to remove most organic materials, including most industrial solvents, pesticides, dioxins, PCBs, and other chemicals. The process is radically different from conventional wastewater treatment processes that remove only toxins from water. This process actually breaks the toxins down into smaller, safer molecules that can then be released into the water system. The unit takes ultraviolet light from the sun and concentrates it on the polluted water, which then separates into water, carbon dioxide, and some very dilute acids that can very easily be neutralized. The system cleans about 30 gallons of water a minute, and scientists believe that they can increase this by a factor of 2 to 3 within a few years.[97]

Another alternative to conventional sewage systems is already being used in some places. "Natural" sewage systems modeled after nature's own purifying powers are already treating the sewage of some communities. Cattails, reeds, and rushes in man-made marshes are extracting toxic chemicals and metals from highly polluted wastewater; canal lilies are cleaning up the discharge from backyard septic tanks; and in a New England greenhouse, snails, microorganisms, marsh plants, and fish are transforming raw sewage into freshwater. Thus nature's way may provide some alternatives that more communities can adopt.[98]

10. Since farming accounts for some 70 percent of global water use, irrigation is responsible for many environmental problems. Waterlogged and salted land, declining and contaminated aquifers, shrinking lakes and inland seas, and the destruction of aquatic

habitats are some of the problems caused by excessive irrigation. Degradation of irrigated land from poor water management is forcing some land to be retired completely. Contamination of land and water by salts and toxic chemicals is only one indication that some irrigation is unstainable. The falling water table is a signal that groundwater withdrawals are exceeding the rate of replenishment. Visible ecological damage from large-scale irrigation projects has spawned strong opposition to the construction of new dams and diversion projects. New concerns are centered on the loss of free-flowing rivers, the destruction of fisheries from stream flow depletion, and damage to the marine and wildlife habitat.[99]

Making irrigation more efficient is also a top priority in attaining more sustainable water usage. Possible savings range from 10 to 50 percent and constitute a large new source of supply. Reduction of irrigation needs by only 10 percent, for example, would free up enough water roughly to double domestic water usage worldwide. There is a wide variety of measures that can boost agriculture's water productivity, including new and improved irrigation technologies such as new sprinkler designs and low pressure sprinklers, better management practices by farmers and water managers, and changes in the institutions that govern the distribution and usage of irrigation water.[100]

In conclusion, one expert recommends that any water management project should lean toward increasing the efficiency of water consumption rather than toward increasing the supply of water, as increasing the supply is often more costly and merely postpones a crisis situation. Mining of groundwater in order to increase the supply should be avoided at all costs, it is recommended, unless it can be guaranteed that the aquifer from which the water is being taken will be replenished in a reasonable period of time. Prevention of pollution and restoration of bodies of water that are already polluted should take precedence over the development of purification technologies. Purifying technology is becoming more complex and costly as the number of pollutants in the water increases. End-of-pipe remedies for industrial water pollution should be replaced by recycling and reuse. If sound principles of this kind are not followed in a water management system, it is all too easy to predict what will eventually happen to our water supply.[101]

Questions for Discussion

1. Where does the freshwater that we use come from, and how much is available on a per capita basis? How can freshwater become contaminated? What are some of the main types of water pollutants and what are some of their effects? What factors affect the ability of a lake or river to recover from contamination by these pollutants?
2. What are the major sources of surface water pollution? Distinguish between point and nonpoint sources of water pollution. How are these kinds of sources controlled? Describe the National Pollutant Elimination Discharge System. How does it work and what is required of business organizations? What kinds of sources are regulated in this manner?
3. Give some examples of nonpoint sources of pollution. How are these sources controlled? Are these types of controls easy to apply? Why or why not? What incentives exist to comply with the regulations? What kind of a system might work better to control these sources of water pollution?
4. What are the major sources of groundwater contamination? How are these sources of pollution controlled? What kind of standards are in force to control these sources? Are you confident the system works well enough to protect your local drinking water supplies?

5. What kind of pressures exist with respect to our coastal waters? What pollution threats to coastal waters exist because of these pressures? What valuable economic benefits are provided by coastal waters? What ecological functions do they perform?
6. What is happening to our oceans? What are the most important sources of pollution as far as our oceans are concerned? How can such pollution be controlled? What threats do oil spills pose to oceans and how can they be prevented?
7. Evaluate the solutions to water pollution presented in the chapter. Which strike you as most practical? Which are likely to be most effective? Which will impact business most severely? What would you recommend to deal with our water pollution problem?
8. What does it mean to say that we should change our thinking from one of trying to expand existing sources of supply to meet demand to one of managing demand for water in light of existing and projected shortages? What policies and practices stem from this approach to demand management?

Endnotes

1. G. Tyler Miller, *Living in the Environment,* 6th ed. (Belmont CA: Wadsworth, 1990), p. 238.
2. Ibid.
3. Ibid., pp. 238–239.
4. Ibid., p. 240.
5. Ibid., p. 242.
6. Ibid.
7. Sandra Postel, "Facing a Future of Water Scarcity," *USA Today Magazine,* September 1993, p. 68.
8. Miller, *Living in the Environment,* p. 242.
9. Bob Engelman, "Thirst: World Needs More Water," *Times-Picayune,* November 8, 1993, p. A5.
10. Miller, *Living in the Environment,* p. 518.
11. Ibid., pp. 521–522.
12. Ibid., pp. 523–524.
13. Ibid., p. 525.
14. United States Environmental Protection Agency, *Environmental Progress and Challenges: EPA's Update* (Washington, D.C.: U.S. Government Printing Office, 1988), p. 45.
15. *Setting the Course: Clean Water* (Washington, D.C.: National Wildlife Federation, undated), p. 5.
16. United States Environmental Protection Agency, *Environmental Progress and Challenges: EPA's Update* (Washington, D.C.: U.S. Government Printing Office, 1988), p. 46.
17. Ibid.
18. *A Guide to the Clean Water Act Amendments* (Washington, D.C.: Environmental Protection Agency, 1978), pp. 1–2.
19. EPA, *Environmental Progress and Challenges,* p. 72.
20. Robert E. Taylor, "Senate Approves Clean Water Act, 86–14, Joining House in Overriding Reagan Veto," *Wall Street Journal,* February 5, 1987, p. 5.
21. *Clean Water and Agriculture* (Washington, D.C.: Environmental Protection Agency, 1977), pp. 2–3.
22. James Lis and Kenneth Chilton, *Clean Water—Murky Policy* (St. Louis, MO: Washington University Center for the Study of American Business, 1992), pp. 43–44.
23. Environmental Protection Agency, *Meeting the Environmental Challenge* (Washington, D.C.: U.S. Government Printing Office, 1990), p. 2.
24. "EPA Act to Speed Treatment of Sewage Overflowing in Rains," *Wall Street Journal,* April 12, 1994, p. A22.

25. "EPA Imposes Rules on Water Cleanliness on Certain States," *Wall Street Journal,* December 4, 1992, p. B6.
26. United States Environmental Protection Agency, *Environmental Investment: The Cost of a Clean Environment* (Washington, D.C.: EPA, 1990), pp. 10–17.
27. United States Environmental Protection Agency, Office of Communications, Education, and Public Affairs, *Securing Our Legacy* (Washington, D.C.: EPA, 1992), p. 24.
28. Bruce Alpert, "EPA's Storm-Water Rules Target Array of Businesses," *Times-Picayune,* September 4, 1992, p. C8.
29. Bruce Alpert, "EPA Wants More Limits on Industrial Runoff," *Times-Picayune,* February 3, 1994, p. A7.
30. Lis and Chilton, *Clean Water—Murky Policy,* p. 35.
31. Katherine Rizzo, "Great Lakes Cleanup Pact Set," *Times-Picayune,* April 1, 1993, p. A10.
32. Miller, *Living in the Environment,* p. 536.
33. United States Environmental Protection Agency, Office of Water, *Strengthening the Safety of Our Drinking Water* (Washington, D.C.: EPA, 1995), p. i.
34. Ibid., pp. i–ii.
35. Ibid., p. 537.
36. Ibid., pp. 537–538.
37. Andy Pasztor, "EPA Will Let States Retain Responsibility for Safety of Underground Water Supply," *Wall Street Journal,* December 30, 1983, p. 28.
38. Robert E. Taylor, "EPA's Plan to Regulate Contaminants of Water Isn't Seen Satisfying Congress," *Wall Street Journal,* October 14, 1985, p. 5.
39. United States Environmental Protection Agency, *Safe Drinking Water Act: 1986 Amendments* (Washington, D.C.: U.S. Government Printing Office, 1986), pp. 1–5. See also United States Environmental Protection Agency, Office of Water, *Is Your Drinking Water Safe?* (Washington, D.C.: EPA, 1991), pp. 2–3.
40. "EPA Expands Rules in Battle to Control Water Contamination," *Wall Street Journal,* March 7, 1990, p. A8.
41. "The Year of the Deal," *National Wildlife,* 29, no. 2 (February–March 1991), 36.
42. H. Joseph Hebert, "Millions Drink Harmful Tap Water, Groups Claim," *Times-Picayune,* October 27, 1995, p. A16.
43. EPA, *Securing Our Legacy,* p. 25.
44. United States Environmental Protection Agency, Office of Water, *Source Water Protection: Protecting Drinking Water Across the Nation* (Washington, D.C.: EPA, 1995). Unnumbered.
45. Robert E. Taylor, "EPA Plans to Require the Replacement of Many Storage Tanks Within 10 Years," *Wall Street Journal,* April 3, 1987, p. 4. See also "Costly Cleanups at the Gas Pump," *Business Week,* April 20, 1987, pp. 28–29.
46. Paulette Thomas, "EPA Issues Rules to Prevent Leaks in Storage Tanks," *Wall Street Journal,* September 14, 1988, p. 52.
47. Edgar Poe, "Monitoring Underground Oil/Chemical Storage Tanks," *Times-Picayune,* May 13, 1992, p. B7.
48. Barbara Rosewicz, "Electric Coolers May Add Unsafe Levels of Lead to Drinking Water, Study Finds," *Wall Street Journal,* December 10, 1987, p. 11. See also Barbara Rosewicz, "Water Coolers Focus of Inquiry on Lead Risk," *Wall Street Journal,* February 4, 1988, p. 21.
49. Bruce Alpert, "Tougher Lead Laws Spark Debate," *Times-Picayune,* September 29, 1991, p. B1; Barbara Rosewicz, "EPA Issues Rules to Reduce Lead Levels in Drinking Water of American Homes," *Wall Street Journal,* May 8, 1991, p. B6.
50. Rose Gutfeld, "Lead in Water of Many Cities Found Excessive," *Wall Street Journal,* October 21, 1991, p. B7.
51. Timothy Noah, "EPA Study Finds Unsafe Lead Levels in 819 Water Systems Across the U.S.," *Wall Street Journal,* May 12, 1993, p. A2.

52. United States Environmental Protection Agency, Office of Water, *Report to the Congress on Radon in Drinking Water* (Washington, D.C.: EPA, 1994), pp. i–v.
53. EPA, *Strengthening the Safety of Our Drinking Water,* p. 7.
54. Ibid., p. 10
55. EPA, *Securing Our Legacy,* pp. 29–30, 34.
56. EPA, *Environmental Progress and Challenges,* p. 65.
57. Ibid.
58. Ibid., pp. 65–68.
59. Jim Morris, "A Breeding Ground for Disease," *Dallas Times Herald,* July 9, 1989, p. A17.
60. Anastasia Toufexis, "The Dirty Seas," *Time,* August 1, 1988, p. 47.
61. Center for Marine Conservation, *1991 International Coastal Cleanup Results* (Washington, D.C.: Center for Marine Conservation, 1991), p. 16.
62. Toufexis, "The Dirty Seas," p. 47.
63. David Stipp, "Toxic Red Tides Seem to Be on the Rise, Increasing the Risks of Eating Shellfish," *Wall Street Journal,* November 22, 1991, p. B1.
64. Toufexis, "The Dirty Seas," p. 46.
65. EPA, *Environmental Progress and Challenges,* p. 68.
66. Miller, *Living in the Environment,* p. 529.
67. Melanie Burney, "Debate on Ocean Dumping Renewed," *Times-Picayune,* November 14, 1992, p. A12.
68. J. Scott Orr, "Ocean Dump Plan Making Waves," *Times-Picayune,* December 13, 1992, p. A18.
69. Miller, *Living in the Environment,* p. 533.
70. "The Genesis and Effects of the OPA," *BIC U.S.,* March–April 1991, p. 25.
71. "National Network for Major Spill Response Launches First Vessel," *BIC U.S.,* March–April 1991, p. 25.
72. Miller, *Living in the Environment,* pp. 534–535.
73. EPA, *Meeting the Environmental Challenge,* p. 4.
74. Ibid., p. 5.
75. Ibid.
76. "No More Pesticides?" *CBS News Magazine,* 60 Minutes, March 24, 1990, p. 14.
77. Ibid.
78. "University of Nebraska Study Shines New Light on Alternatives to Pesticides," *Agribusiness News,* June 1987, p. 27.
79. Sandra Postel, "Water for the Future: On Tap or Down the Drain," *The Futurist,* March–April, 1986, p. 18. See also Center for Marine Conservation, *1991 International Coastal Cleanup Results,* p. 34.
80. Postel, "Facing a Future of Water Scarcity," p. 70.
81. See Robert P. Bringer, "Pollution Prevention Plus," *Pollution Engineering,* XX, no. 10 (October 1988), 84–89.
82. Postel, "Water for the Future," p. 18.
83. Ibid.
84. Robert Tomsho, "Cities Reclaim Waste Water for Drinking," *Wall Street Journal,* August 8, 1994, p. B1.
85. Postel, "Facing a Future of Water Scarcity," p. 71.
86. Sandra Postel, "Plug the Leak: Save the City," *International Wildlife,* January–February 1993, pp. 38–41.
87. Postel, "Water for the Future," pp. 19–20; Lester R. Brown et al. *Vital Signs 1995* (New York: W. W. Norton, 1995), p. 123.
88. Richard Martin, "A Fight to Rescue a Dying Lake," *Insight,* October 17, 1988, pp. 20–21.
89. Postel, "Water for the Future," p. 21.

90. Ibid., p. 19.
91. Kent Jeffreys, "How Markets for Water Would Protect The Environment," *The Heritage Foundation State Backgrounder,* September 26, 1989, pp. 1–10.
92. Ibid. See also Robert D. Hof, "California's Next Cash Crop May Soon Be . . . Water?" *Business Week,* March 2, 1992, p. 76–78.
93. Gina Bellafante, "Bottled Water: Fads and Facts," *Garbage,* January–February 1990, pp. 46–50.
94. Ibid. See also "Tap Water May Be Safest, Officials Say," *Times-Picayune,* April 11, 1991, p. D1.
95. "FDA Tightening Rules on Bottled Water," *Times-Picayune,* November 8, 1995, p. C1.
96. U.S. Water News, March 1990, p. 75.
97. Robert Pool, "Sun-Powered Pollution Clean Up," *Science,* Vol. 245, March 1988, p. 23.
98. Janet Marinelli, "After the Flush, The Next Generation," *Garbage,* January–February 1990, pp. 24–35.
99. Sandra Postel, "Saving Water for Agriculture," *State of the World 1990* (Washington, D.C.: Worldwatch Institute, 1990), p. 47.
100. Postel, "Facing a Future of Water Scarcity," p. 69.
101. J. W. Maurits la Riviere, "Threats to the World's Water," *Scientific American,* 261, no. 3 (September 1989), 94.

Suggested Reading

Anderson, Terry L. *Water Crisis: Ending the Policy Drought.* Baltimore, MD: Johns Hopkins, 1993.

Baker, Brian. *Groundwater Protection from Pesticides.* New York: Garland Publishers, 1990.

Borgese, Elisabeth Mann. *The Future of the Oceans.* New York: Harvest House, 1986.

Cheremisinoff, Paul M. *Water Management and Supply.* Upper Saddle River, NJ: Prentice Hall, 1993.

Clarke, Robin. *Water: The International Crisis.* Cambridge: MIT Press, 1993.

Conservation Foundation. *Groundwater Pollution.* Washington, D.C.: Conservation Foundation, 1987.

Currie, J. C. and A. T. Pepper, eds. *Water and the Environment.* New York: Routledge, Chapman and Hall, 1993.

Dunne, Thomas, and Luna B. Leopold. *Water in Environmental Planning.* Boston: W. H. Freeman, 1995.

Feldman, David L. *Water Resource Management: In Search of an Environmental Ethic.* Baltimore: Johns Hopkins, 1994.

Gay, Kathlyn. *Water Pollution.* New York: Watts Publishing, 1991.

Getchen, David H. *Water Law in a Nutshell.* St. Paul, MN: West, 1991.

James, Jody, ed. *Environmental Awareness: Water Pollution.* New York: Bancroft-Sage, 1991.

Kato, Ichrio. *Water Management and Environmental Protection.* New York: Columbia University Press, 1995.

Linsley, Ray K. *Water Resources Engineering,* 4th ed. New York: McGraw-Hill, 1993.

National Academy of Sciences. *Oil in the Sea.* Washington, D.C.: National Academy Press, 1985.

National Academy of Sciences. *Drinking Water and Health.* Washington, D.C.: National Academy Press, 1986.

Office of Technology Assessment. *Wastes in Marine Environments.* Washington, D.C.: U.S. Government Printing Office, 1987.

Patrick, R. E. Ford, and J. Quarles, eds. *Groundwater Contamination in the United States.* Philadelphia: University of Pennsylvania Press, 1987.

Tarlock, A. Dan, et al. *Water Resources Management,* 4th ed. Washington, D.C.: Foundation Press, 1993.

Thanh, N. C. and Asit K. Biswas, eds. *Environmentally Sound Water Management.* New York: Oxford University Press, 1991.

Thomas, Patricia, ed. *Water Pollution Law and Liability.* Netherlands: Kluwer Academic Press, 1993.

United States Environmental Protection Agency. *Environmental Progress and Challenges: EPA's Update.* Washington, D.C.: U.S. Government Printing Office, 1988.

United States Environmental Protection Agency. *Meeting the Environmental Challenge.* Washington, D.C.: U.S. Government Printing Office, 1990.

United States Environmental Protection Agency. *Securing Our Legacy: An EPA Progress Report.* Washington, D.C.: U.S. Government Printing Office, 1992.

Wekesser, Carol, ed. *Water: Opposing Viewpoints.* New York: Greenhaven, 1994.

CHAPTER 8

Pesticides and Toxic Substances

The two topics that are the subject of this chapter are related in that they both involve the use of chemicals, some of which have proven dangerous to human health and the environment. Usage of pesticides and other chemical substances have increased over the past several decades as chemicals have proven useful for a multitude of purposes. But they also have toxic effects, in some cases, that need to be investigated and regulated if necessary to protect human beings and the environment from serious damage. Even though pesticides and toxic substances are subject to different sets of laws and regulations, they are similar enough to warrant treatment in the same chapter.

PESTICIDES

Pests destroy crops worth billions of dollars each year, and with a steadily expanding population and a decrease in available land, the world has used more and more pesticides to control these pests and maintain high crop yields. A pest is considered to be any unwanted organism that directly or indirectly interferes with human activity. Weeds are also a problem when they invade gardens, crop fields, and other areas and compete for soil nutrients and water. Since about 1945, gardens and fields of crops have been treated with a variety of chemicals called pesticides. These substances can kill organisms and weeds that are considered to be undesirable. The most widely used types of pesticides are the following.

- Herbicides: Used to kill weeds, which are unwanted plants that compete with crop plants for soil nutrients.
- Insecticides: Used to kill insects that consume crops and food and transmit diseases to humans and livestock.
- Fungicides: Used to kill fungi that damage crops.
- Rodenticides: Used to kill rodents, mostly rats and mice.[1]

Many people believe that there is no alternative to pesticide use, given existing technology, to raise crops on the scale that is required to feed an ever-growing human population. Besides helping in the production of greater quantities of food, pesticides also help reduce loss of food in storage and control disease carriers. Each year, about 3 billion pounds of pesticides are used for these and other purposes. Agriculture accounts for the greatest percentage; followed by industry, forestry, and government; and home gardening and lawn usage.[2] Pesticides are used for many purposes, as the following list illustrates, and not all pesticides are used in agriculture, as is commonly thought.

Yet in poisoning pests, human beings may also be poisoning themselves. Pesticides persist in the environment for long periods of time, and move up through the food chain from plankton or insects to animals and humans, making dietary exposure unavoidable in many situations. They also move downward through soil to contaminate groundwater used for drinking. Through these exposures, pesticides pose a threat to human health by causing cancer or birth defects or other health and environmental problems. In 1994, global sales of pesticides reached a record $25 billion. About 80 percent of pesticides are used in industrialized countries, but sales in these countries have leveled off, while use in developing countries is projected to grow rapidly.[3]

Common Usages of Pesticides

- Fiber crops—cotton and hemp, for example.
- Specialized field crops, such as tobacco.
- Crops grown for oil, such as castor bean and safflower.
- Ornamental shrubs and vines, like mistletoe.
- General soil treatments, such as manure and mulch.
- Household and domestic dwellings.
- Processed non-food products—textiles and paper, for example.
- Fur and wool-bearing animals such as mink and fox, laboratory and zoo animals, pet sprays, dips, collars, litter and bedding treatments.
- Dairy farm milk-handling equipment.
- Wood production treatments on railroad ties, lumber, boats, and bridges.
- Aquatic sites, including swimming pools, diving boards, fountains, and hot tubs.
- Uncultivated non-agricultural areas, such as airport landing fields, tennis courts, highway rights-of-way, oil tank farms, ammunition storage depots, petroleum tank farms, saw mills, and drive-in theaters.
- General indoor/outdoor treatments, in bird roosting areas, or mosquito control.
- Hospitals, including syringes, surgical instruments, pacemakers, rubber gloves, bandages, and bedpans.
- Barber shops and beauty shops.
- Mortuaries and funeral homes.
- Preservatives in paints, vinyl shower curtains, and disposable diapers.
- Articles used on the human body, such as human hair wigs, contact lenses, dentures, and insect repellents.
- Specialty uses, such as mothproofing and preserving specimens in museums.

Source: United States Environmental Protection Agency, *Environmental Progress and Challenges: EPA's Update* (Washington, D.C.: U.S. Government Printing Office, 1988), p. 128.

Pesticide use has increased as industrialized agricultural practices have become more widespread throughout the world. Diverse ecosystems, containing small populations of many species, are replaced with greatly simplified agricultural ecosystems that contain large populations of only one or two desired plant species. In such simplified ecosystems, some organisms that would be controlled naturally in more diverse systems can grow in number and achieve the status of pests that pose serious threats to crops that feed the world's expanding population. As a result, the human race has to spend a great deal of time and money controlling these pests through the use of pesticides.[4]

In the United States, about 700 biologically active ingredients and 1,200 inert ingredients are mixed to make some 50,000 individual pesticide products. About 77 percent of these products are applied to commercial cropland, 11 percent to government and industrial lands, another 11 percent is used by households, and 1 percent is used in forest lands. Herbicides account for 85 percent of all pesticide use the the United States and 88 percent of the the pesticides used for farmland. Insecticides make up about 10 percent of pesticide use in the United States, and fungicides another 5 percent. About 20 percent of the pesticides used each year in the United States are applied to lawns, gardens, parks, and golf courses. About 91 percent of all U.S. households use pesticides indoors, with the average homeowner applying about five times more pesticide per unit of land area than farmers.[5]

One of the first people to point out the dangers of pesticide use was Rachel Carson. In her book *Silent Spring,* information about the dark side of pesticide use was presented to the public.[6] Before this, pesticides were by and large seen as an unqualified benefit, but after her book, fear began to spread throughout society that pesticides were unmanageable poisons. Concerns increased as more studies were done that substantiated the dangers of unregulated pesticide usage. The World Health Organization (WHO), for example, estimated that at least 1 million people worldwide are poisoned by pesticides. Of these people, about three quarters suffer chronic health problems such as dermatitis, nervous disorders, and cancer, and between 4,000 and 19,000 of them die each year because of exposure.[7] At least half of those poisoned and 75 percent of those killed are farm workers in developing nations where warnings are not adequate and regulation is either lax or nonexistent.[8]

In 1993, a released report showed that 21 percent of Indonesian farmers studied exhibited three or more symptoms of pesticide poisoning during the spray period. An estimated 10,000 farmers died in China in 1993 from pesticide poisoning.[9] A National Cancer Institute study in 1992 showed that farmers have higher risks of several forms of cancer, possibly because of exposure to pesticides. Cancers that are higher in farmers were found to be on the rise in the rest of the population as well, and pesticides were one likely suspect, as their usage has spread to urban and suburban areas.[10]

Some 20,000 Americans, the majority of whom are children, become sick because of unsafe use or storage of pesticides in and around the home each year. Pesticides are the second most frequent cause of poisoning in young children after medicines. Accidents and unsafe practices in plants manufacturing pesticides can expose workers, their families, and sometimes entire communities to harmful levels of pesticides or chemicals used in their manufacture.[11] Scientists are concerned about possible effects of continuous, long-term exposure to very low levels of pesticides. These chronic effects, if they exist, will not show up for several decades after exposure, making early detection impossible.

It has been found that traces of almost 500 of the 700 active ingredients used in pesticides in the United States show up in the food most people eat every day. Pesticide residues are likely to be found on tomatoes, grapes, apples, lettuce, oranges, potatoes, beef, and dairy products. In 1987, the National Academy of Sciences reported that exposure to pesticides in food could cause up to 20,000 cases of cancer a year in the United States, in a worst-case scenario. The same year the EPA ranked pesticide residues in foods as the third most serious environmental problem in the country.[12] In 1993, emerging evidence suggested that a pesticide called endosulfan approved for use on fruits and vegetables by the EPA may be linked to breast cancer. While definite proof was lacking, the EPA considered the matter to be potentially significant and deserve further study.[13]

Concerns about pesticide residues erupted in the spring of 1989, when the National Resources Defense Council (NRDC) released a report stating that apples treated with the pesticide Alar were exposing children to dangerously high levels of daminiozide, a possible carcinogen. The NRDC claimed that daminiozide use may cause one case of cancer for every 4,200 preschoolers, 240 times the acceptable standard.[14] The story was shown on *60 Minutes* and Meryl Streep made several appearances on talk shows and Capitol Hill attacking pesticides. Soon after these incidents, apples were ordered removed from school cafeterias in New York City, Los Angeles, and Chicago. Other school systems followed suit, and signs were posted above produce bins all over the country advertising Alar-free apples. The state of Washington, where 50 percent of the nation's apples are grown, faced huge economic losses estimated at $100 million.[15] There was a good deal of controversy over this incident, and 11 apple growers eventually brought a class action suit against CBS and the NRDC charging them with tortiously interfering with their business expectations by publishing knowingly false statements about the safety of eating apples.[16]

Scientists also believe that some pesticides may be drifting around the world to distant places where they have never been used. They found potentially harmful compounds from insecticides in the Arctic and Antarctic parts of the world. Most of the insecticides found had been banned in the United States and Europe some time ago, but were still being used in developing countries. The data showed that the distillation effect was very real, according to these scientists, and that certain volatile insecticides will travel to the cooler regions of the Northern Hemisphere no matter where they are used in the world.[17]

Regulation

As a direct result of this growing concern about the dangers of unregulated pesticide usage, federal pesticide regulation was toughened and enforcement responsibility transferred from the Department of Agriculture, which promoted chemical pest control, to the Environmental Protection Agency (EPA). Congress also passed the Federal Insecticide, Fungicide, and Rodenticide Act (FIFRA) in 1972, which required that all commercial pesticides be approved for general or restricted usage. Since the passage of this law, the EPA has banned over 40 pesticides because of their potential hazards to human health. The use of DDT was banned in 1972, and several other pesticides such as aldrin, dieldrin, toxaphene, and ethylene dibromide have been suspended or banned since that time[18] (Exhibit 8.1). Over the past 20 years, the EPA has

EXHIBIT 8.1 Pesticides Taken Off the Market

Pesticides	*Use*	*Concerns*
Aldrin	Insecticide	Oncogenicity
Chlordane (agricultural uses; termiticide uses suspended or cancelled)	Insecticide/Termites, Ants	Oncogenicity; reduction in non-target and endangered species
Compound 1080 (Livestock collar retained, rodenticide use under review)	Coyote control; Rodenticide	Reductions in non-target and endangered species; no known antidote
Dibromochloropropane (DBCP)	Soil Fumigant—Fruits and vegetables	Oncogenicity; mutagenicity; reproductive effects
DDT and related Compounds	Insecticide	Ecological (eggshell thinning); carcinogenicity
Dieldrin	Insecticide	Oncogenicity
Dinoseb (in hearings)	Herbicide/Crop dessicant	Fetotoxicity; reproductive effects; acute toxicity
Endrin (Avicide use retained)	Insecticide/Avicide	Oncogenicity; teratogenicity; reductions in non-target and endangered species
Ethylene Dibromide (EDB) (Very minor uses and use on citrus for export retained)	Insecticide/Fumigant	Oncogenicity; mutagenicity; reproductive effects
Heptachlor (Agricultural uses; termiticide uses suspended or cancelled)	Insecticide	Oncogenicity; reductions in non-target and endangered species
Kepone	Insecticide	Oncogenicity
Lindane (Indoor smoke bomb cancelled; some uses restricted)	Insecticide/Vaporizer	Oncogenicity; teratogenicity; reproductive effects; acute toxicity; other chronic effects
Mercury	Microbial Uses	Cumulative toxicant causing brain damage
Mirex	Insecticide/Fire Ant Control	Non-target species; potential oncogenicity
Silvex	Herbicide/Forestry, rights-of-way, weed control	Oncogenicity; teratogenicity; fetotoxicity
Strychnine (Rodenticide use and livestock collar retained)	Mammalian predator control; rodenticide	Reductions in non-target and endangered species
2,4,5,-T	Herbicide/Forestry, rights-of-way, weed control	Oncogenicity; teratogenicity; fetotoxicity
Toxaphene (Livestock dip retained)	Insecticide—Cotton	Oncogenicity; reductions in non-target species; acute toxicity to aquatic organisms; chronic effects on wildlife

Oncogenicity—Causes tumors
Mutagenicity—Causes mutation
Carcinogenicity—Causes cancer
Teratogenicity—Causes major birth defects
Fetotoxicity—Causes toxicity to the unborn fetus

Source: United States Environmental Protection Agency, *Environmental Progress and Challenges: EPA's Update* (Washington, D.C.: U.S. Government Printing Office, 1988), p. 118.

canceled the registrations of 35 potentially hazardous pesticides and eliminated the use of 60 toxic inert ingredients in pesticide products.[19]

Pesticides are thus regulated by FIFRA, where the EPA was assigned the responsibility for protecting human health from any commercially available product used to kill germs, insects, rodents, and other animal pests, as well as weeds and fungi. Such products cannot be sold until they are first registered with the agency.[20] If test data show that a pesticide may be harmful to human health or the environment, the EPA can refuse to register it, restrict its use to certain applications, or require that only certified applicators apply the pesticide. Once they are registered, manufacturers of the product must use appropriate labels showing the approved uses of the pesticide. Over 50,000 pesticides have been registered since FIFRA was enacted.[21]

Before a pesticide is registered for use on food or feed crops, a "tolerance" or legally enforceable residue limit must be set by the agency. Both domestically produced and imported foods are monitored to make sure they comply with the established tolerances. This procedure was instituted to protect consumers from exposure to unsafe levels of pesticide residue on the food they purchase in the marketplace. While the EPA initially was able only to determine tolerances for the general population, its new system takes into account differences in susceptibilities within the general population such as differences in age and geographic location.[22]

There were also different standards for different types of food. For example, raw foods were evaluated by weighing the risks of pesticide exposure against the benefits of pesticide usage. Processed foods, on the other hand, were subject to a more stringent standard, which stated that food additives including pesticide residues must pose zero risk of cancer regardless of their benefit. Because of these differences, some existing chemicals in raw foods were prohibited in processed foods.[23] If confidence can be established in this kind of program, it is hoped the fear consumers have about the safety of some kinds of products may be reduced. Progress was made in the exposure of humans to pesticides that had been banned.

Amendments to the act in 1978 required the EPA to reregister the 35,000 pesticides previously registered and already on the market, because the long-term health and environmental effects of many of these pesticides were poorly understood. The EPA was required to do a benefit-cost analysis on these products, taking into account new information. If this analysis revealed that a particular product posed an unreasonable risk to human health or the environment when weighed against its benefits to agriculture and society, it had to be removed from the marketplace or restrictions had to be placed on its use as a pesticide.

The EPA can remove a pesticide from the market in three ways. If the EPA determines that a pesticide poses "unreasonable adverse effects," it can issue a notice of intent to cancel the pesticide, at which time the affected registrant may request a hearing before an administrative law judge. If the judge disagrees with the EPA, the administrator of the EPA can make the final decision to cancel without any obligation to indemnify holders of a canceled product. During this decision-making process, which can take several years, the pesticide can continue to be produced, sold, and used. Even after a final decision, remaining stocks of the pesticide can generally be used.[24]

The EPA can also suspend a pesticide's registration, but in order to do so, it must determine that the pesticide poses an "imminent hazard," meaning that the short-term risks of continued use outweigh any possible benefits. Registrants can dispute the suspension, but production of the pesticide would have to cease until final action is taken. Sales and use of the pesticide could continue during this period. If the final decision is to prohibit sale and use of the product, the agency must indemnify at market value prior to the suspension holders of any remaining stocks.[25]

Finally, the EPA can order an emergency suspension if it finds that the pesticide poses an imminent hazard to humans and the environment. This suspension order precedes any hearing that is taking place and prohibits the sale and use of the pesticide immediately. After taking a final action to suspend the pesticide because it poses this kind of hazard, the EPA must also indemnify holders of remaining stocks of the pesticide. This method of removing a pesticide from the market has been used on only a few occasions to remove pesticides such as 2,4,5,T/Silvex, EDB, and Dinoseb from the market.[26]

The 1972 law required the EPA to proceed on a product-by-product basis, even though many products have similar chemical properties. Further amendments to FIFRA in 1978 allowed the EPA to take a "generic" approach to registering pesticides. Using this approach, the agency is able to make one regulatory decision for an entire group of pesticides that have similar chemical ingredients rather than looking at them separately. It was estimated that using this method the agency would have to consider fewer than 600 active ingredients contained in the 45,000 different commercial products on the marketplace. Standards could be set for these 600 ingredients, and products would be registered according to whether they measure up to these standards.

Further amendments passed in 1988 were designed to speed up reregistration and remove potential threats to the budget of the EPA that could thwart its efforts to protect the public. These amendments require the EPA to complete reregistration by 1997, at least ten years earlier than would have been required without the amendments. This expedited reregistration requirement and the financial commitment to make it possible are the most important provisions of the amendments. The amendments also involve changes in the procedures for dealing with the recall, storage, and disposal of older pesticides when cancellation or the demands of reregistration force them off the market. Most of the financial burden of disposal was shifted to the pesticide industry, and the EPA was given additional authority to require holders of canceled pesticides to dispose of the products and specify the means of disposal.[27]

In 1989, the agency was attacked for its system of setting pesticide standards. Critics contended that the agency was slow to revise standards for older pesticides and took a far too limited view in evaluating the dangers of new pesticides. When it set pesticide tolerances, some experts claimed that the agency was grossly underestimating the risk to the public. Children, in particular, were said to be exposed to dangerously high cancer risks. One of the problems concerned the treatment of inert ingredients used to dissolve, dilute, or stabilize the active agents. These inerts are excluded from the EPA's risk calculations, and only those ingredients the manufacturers list as active are examined. Chemical makers claimed that these inerts were trade secrets and so shouldn't be reviewed, but certain experts believed that some of these inerts were often more toxic than the active agents.[28]

While the initial health concern regarding the regulation of pesticides was whether a chemical used in pesticides could cause cancer, pesticides are now tested for a variety of potential problems including reproductive, immunological, and neurological effects. All new and previously registered pesticides are also screened for their potential to contaminate groundwater. Groundwater may be more vulnerable to contamination in some areas because of geological or other factors. In 1986, a survey indicated that 30 states had found wells contaminated with one or more of 60 different pesticides. These concerns were one of the major reasons that ethylene dibromide (EDB) and dibromochloropropane (DBCP) were taken off the market.[29] In 1990, the EPA completed its first national survey of 127 pesticides and nitrates in drinking water. This information will be used to evaluate regulatory and state plans for protecting drinking water from pesticide pollution.[30]

Pesticides also harm fish and wildlife, which come into contact with pesticides by feeding on contaminated fields and water, or by preying on other contaminated organisms. Some chemicals such as DDT were found to persist in the environment and accumulate in the tissues of wildlife, causing the thinning of eggshells, which prevented the successful hatching of chicks. Since the 1960s, the use of a number of chemicals that were harmful to fish and wildlife have been canceled, and some species threatened with extinction have since shown signs of recovery (see box). These include the California brown pelican, the bald eagle, and the peregrine falcon.[31] Yet a report released in 1993 accused a new generation of so-called "soft" pesticides to replace those that were canceled, such as azinphos-methyl (AZM), of killing fish and wildlife in large numbers.[32]

The agency is currently evaluating some chemicals because of concerns about effects on fish and wildlife, rather than solely being concerned about whether a chemical could cause harm to humans. Manufacturers are required to submit information regarding the effects of pesticides on wildlife through laboratory studies and demonstration of the actual impacts of a pesticide in the environment. The EPA is also developing better methods for determining the effects of pesticides on entire ecosystems as well as on individual organisms. The agency has begun a five-year research program to improve computer models of ecosystems in different environments across the country.[33]

Under the EPA's pesticide program, more than a million private users, mostly farmers, and 250,000 commercial applicators have been trained and certified in the safe use of pesticides. This program trains people in the proper use, handling, storage, and disposal of pesticides, including provisions for protective clothing, warning about treated areas, and waiting periods after spraying. Only such certified users are allowed to use pesticides that have a "restricted use only" classification. In late 1984, the EPA adopted a tougher policy toward pesticides thought to be hazardous, by banning or restricting the use of eight high-volume pesticides, and proposing more stringent controls on about 25 other widely used pesticides that were suspected of posing health hazards for consumers and farm workers.[34]

Despite these efforts, the risks to farmworkers continued. It was estimated that in the course of a 30-day harvesting season, the typical farmworker might be exposed to 15 different compounds at various times, making the process of identifying which pesticide would be to blame for a particular health problem impossible. Little is known about the chronic effects of many pesticides, and many of the studies that have been conducted can be faulted for one reason or another, thus making proof of a link between a specific pesticide and cancer difficult to establish. The benefit of the doubt has always gone in favor of manufacturers and users of pesticides rather than the workers,

Endangered Species Return but Still Require Protection from Pesticides

By the early 1970s, the bald eagle had all but vanished from many areas of its natural range. At that time, only about 1,000 nesting pairs of eagle were found in the entire United States. Too few offspring were hatching successfully, and evidence pointed to the use of DDT and other persistent pesticides as the likely factor responsible for this condition. These pesticides and their by-products had affected calcium metabolism, making eggshells so thin that they broke under the weight of the nesting birds. Our national symbol, the bald eagle, became an endangered species.

Although still endangered, the bald eagle has shown remarkable signs of recovery since the ban on DDT in 1972. Today, there are almost 2,000 nesting pairs of bald eagles in the United States. The majority of birds seem to be producing normal eggs, even though problems with eggshells have been reported for some nests. The California brown pelican and peregrine falcon, also threatened with extinction by the persistent pesticides, have been recovering slowly as well.

Although the bald eagle and other birds have been recovering since the ban on DDT, there are still many thousands of species listed as endangered. These species usually have become endangered for a variety of reasons other than pesticides contamination; they nonetheless are particularly vulnerable to added stresses such as pesticides. Pesticides may kill wildlife directly, or may contaminate the food, water, and habitat of the wildlife.

Under the Endangered Species Act, EPA is required to consult with the Fish and Wildlife Service to ensure that pesticides do not jeopardize endangered species and their habitat. If the Fish and Wildlife Service determines that a pesticide is likely to be harmful to endangered species, it suggests alternatives to EPA for preventing damage to the species. It usually recommends not using the pesticide where endangered species would be exposed.

EPA is developing an approach to implementing these restrictions through the pesticide label. As proposed in the Federal Register, pesticide labels would list the counties that have limitations on use in endangered species habitats. Labels would refer users to county bulletins that contain maps indicating the portions of the county where pesticide use is limited. At first these limitations would apply to the four groups of pesticides that already have been evaluated: certain crop pesticides, pasture and rangeland pesticides, forestry pesticides, and mosquito larvicides. Information currently is being gathered for maps that would identify the location of potentially affected endangered species.

Because the presence of endangered species may vary within a state and even within a single county, states will have a particularly important role in determining where limitations are needed. Several states already have proceeded to develop recommendations for implementing the program. By working with agriculture and wildlife experts and government officials, we hope to ensure that pesticides pose minimal threats to endangered species.

Source: United States Environmental Protection Agency, *Environmental Progress and Challenges: EPA's Update* (Washington, D.C.: U.S. Government Printing Office, 1988), p. 136.

and pesticides have often been assumed safe until proven otherwise. To cope with this problem, the EPA issued new regulations to protect farmworkers in 1988, which some believed were still inadequate.[35]

In October 1984, the EPA took its first step toward regulating pesticides created by gene-splicing or other methods of biotechnology. The EPA issued rules requiring all companies or individuals planning to test the effectiveness of pesticides produced by genetic manipulation to notify the agency at least 90 days before the tests are conducted. The EPA can then deny experimental permits for tests that the agency determines may pose "substantial health concerns." These rules were the first phase of a comprehensive set of proposed policies and regulations that would spell out the EPA's authority to regulate the commercial testing and use of all such pesticides in the future.[36] A framework for coordinating federal regulation of biotechnology was established in 1986 that built on existing legislation and practices, but imposed additional levels of federal review for certain applications, particularly relating to new microorganisms.[37]

Under both the Toxic Substances Control Act and FIFRA, the EPA has instituted a process for carefully evaluating proposals for field testing of genetically altered products. Before permission for such an experiment is granted, information about the experiment is evaluated and shared with other agencies. If the experiment is approved, conditions are specified for conducting the experiment and results are required to be submitted to the EPA so that effects can be monitored and the decision reevaluated. Several tests of genetically engineered microorganisms outside the laboratory have been approved including tests on strawberries and potatoes using a microorganism modified to retard frost formation on plants. Many developments in biotechnology (see box on p. 221) promise great benefits to the public but also need to be regulated to prevent adverse effects to humans and the environment.[38]

The National Academy of Science issued a report in the late 1980s recommending that a "negligible risk" standard be adopted with respect to pesticide residues. This report challenged the zero-risk standard of the Delaney Clause and would replace it with a standard that allowed approval of pesticides in cases where they would produce tumors in fewer than one out of every million people exposed. The panel that produced the report stated that the new standard would eliminate 98 percent of the cancer risk from 28 pesticides that the EPA has linked to cancer. Current standards were said to eliminate only about half the estimated cancer risk from these compounds.[39]

The Clinton administration introduced legislation in 1994 to overhaul pesticide regulation based upon this "negligible risk" standard where the product could only be used in amounts where there is a "reasonable certainty of no harm." This legislation would replace a confusing set of different standards for raw and processed foods with one health-based standard for determining how much of a pesticide is safe, and would replace the Delaney Clause that prohibits even minute amounts of pesticide residues known to "induce cancer in man or animal." Under the new legislation, pesticides could be used that were found to cause cancer in laboratory animals, but the allowed levels would be so low so as to pose no more risk of cancer to humans.[40]

In the summer of 1996, such legislation finally was signed into law by the President. This legislation repealed the Delaney Clause and replaced it with a standard based upon the government's assurance that raw and processed foods will pose minimal cancer risk, defined as not more than one incidence of cancer per million people

Some Developments in Biotechnology

There are many potential uses of genetically altered microorganisms. Some of the most exciting developments in biotechnology are described below.

- ***Tracking the release of genetically altered bacteria***

A microorganism has been developed that can be tracked in the environment to provide information on its behavior. The microbe is formed by inserting two genes from a common bacterium into another microorganism. When these genes are present, the bacteria form blue colonies when exposed to a certain sugar, thereby allowing scientists to follow the survival of the genetically changed microbes both inside and outside the test area. The marking system can be used to mark other microorganisms and should help allay public concerns about potential consequences of releasing such microorganisms in the environment. EPA approved field tests of the bacteria on wheat and soybeans.

- ***Developing bacteria to protect plants against frost***

EPA approved field tests of a bacteria designed to protect strawberry and potato plants from mild frosts. The new bacteria are the same as those that normally colonize these plants, except that they lack a protein that promotes the formation of ice crystals. Scientists expect that these new bacteria can help plants resist frost if the bacteria are applied before the normal bacteria can establish themselves.

- ***Using bacteria to enhance alfalfa yield***

EPA also is examining the design of an experiment to test the effectiveness of genetically engineered bacteria to enhance the yield of alfalfa. This experiment will be conducted in Pepin County, Wisconsin and will use an altered form of the bacteria that occur naturally in the soil. These altered bacteria work together with the roots of legumes (such as alfalfa, soybeans, and peas) to convert nitrogen gas into a form that can be used by the plants.

- ***Using dead bacteria as a pesticide***

An innovative approach to pesticide development involves the insertion into another microbe of the genes that contain the code for a protein toxin. These altered microbes subsequently are grown in cultures to produce large quantities of the toxic protein. When these bacteria are killed, they can be administered as a pesticide. Small-scale field trials currently are underway to determine the effectiveness of these dead bacteria as pesticides.

- ***Using bacteria for toxic waste disposal***

Researchers currently are working on the development of bacteria that can metabolize specific compounds in toxic wastes, such as PCBs, dioxin, and oil spills. For example, an EPA scientist has developed a strain of bacteria that can metabolize several components of crude oil. This development may enable us to control oil spills using only one bacteria rather than several different types. Bacteria also are being developed to extract toxic metals from landfills, mines, and wastewater.

Source: United States Environmental Protection Agency, *Environmental Progress and Challenges: EPA's Update* (Washington, D.C.: U.S. Government Printing Office, 1988), p. 140.

exposed. The new bill also requires the EPA to publish information about the risks and benefits posed by chemicals in pesticides and recommendations to consumers for reducing dietary exposure to pesticides. This information will be distributed to large retail grocers for public display. While easing regulation of processed foods, the law tightened regulation of raw foods by setting allowable levels of pesticide residues in food to help guarantee that fruits, vegetables, and grain are free of dangerous pesticide residues. This provision was particularly focused on protecting the health of infants and children who are more susceptible to such residues.[41]

Pesticide use appears to have leveled off in recent years after steadily increasing in the 1960s and 1970s (Figure 8.1). To encourage this trend, the EPA is supporting the development of new integrated pest management practices, which will hopefully reduce the reliance on chemicals by using a variety of pest control methods. A new farm bill could also further reduce agricultural use of pesticides by promoting more environmentally sound crop rotation practices, increasing funds for sustainable agriculture, promoting research and education, and providing incentives for farmers to adopt more environmentally compatible farming methods.[42] There is supposedly a correlation between farm subsidies and increased chemical usage, and elimination of subsidies could result in a 35 percent reduction in chemical use per acre. Currently, farmers use more than 500 million pounds of pesticides, some 2.3 pounds per acre, and the United States spends $1.4 billion to control pesticide pollution and an additional $600 million to control agricultural runoff.[43]

Regarding use of pesticides overseas, the EPA claims to be making some efforts to prevent pesticide misuse and overuse in other countries. The United States is an important exporter of pesticides and, in some cases, has sold pesticides overseas that have been banned or restricted in this country. But, since the United States is also a major importer of food commodities, oftentimes pesticide use overseas returns to the United States as residues on the food that is imported.[44] Thus the United States has an interest in ensuring that pesticides are used responsibly throughout the world. To achieve this level of protection, the EPA has developed goals for international pesticide activities and wants to harmonize U.S. and international pesticide standards. The agency has

FIGURE 8.1 U.S. Pesticide Usage

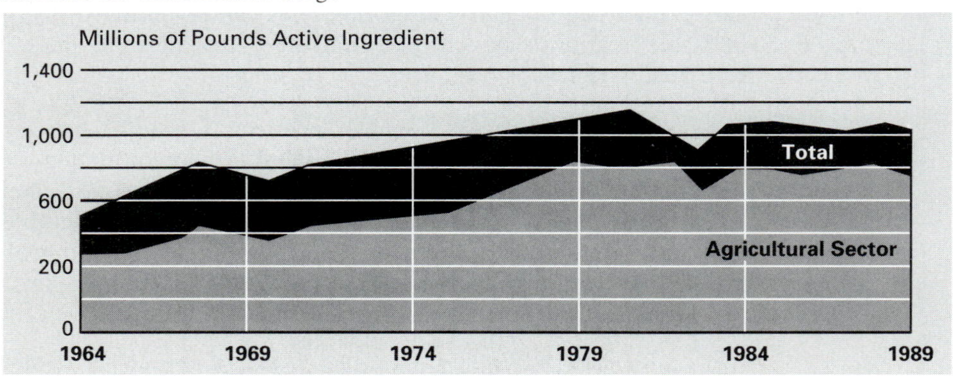

Source: United States Environmental Protection Agency, *Securing Our Legacy* (Washington, D.C.: EPA, 1992), p. 18.

also proposed a policy that would restrict the export of pesticides banned in the United States and is actively involved in related legislative efforts.[45]

Alternatives

Breaking the pesticide habit is difficult, but continuing the usage of pesticides is becoming more and more of a problem, not only because of the harmful effects of the chemicals used in pesticides but also because resistance to pesticides is growing. Species resistant to at least one or more pesticides meant to control them now number over 900, including 520 insects and mites, 150 plant diseases, and 113 weeds (Figure 8.2). In addition, at least 17 insect species are resistant to all major classes of insecticide, and several plant diseases are not affected by most fungicides used to control them. Developing resistance to pesticides is a natural evolutionary process, and resistant strains develop particularly quickly when pesticides are used to try and eliminate pests rather than just control them. If 99.9 percent of the insects in a field are killed, the survivors are a superstrain that can resist pesticides.[46]

Farmers can thus get on a pesticide treadmill. As pests develop resistance to the chemicals, higher doses and new products are necessary to achieve the same level of control. But as more and more pesticides are applied, they are less effective. Between 1945 and 1989, insecticide applications increased tenfold in the United States, but crop losses to insects nearly doubled. Overuse of pesticides can actually be counterproductive, as rice yields in Indonesia actually rose by 15 percent between 1987 and 1991 after the government banned the use of 57 insecticides and promoted alternative methods of pest control. Pesticide use dropped 65 percent and the government saved $120 million annually on pesticide subsidies.[47]

FIGURE 8.2 Pesticide-Resistant Species Since 1908

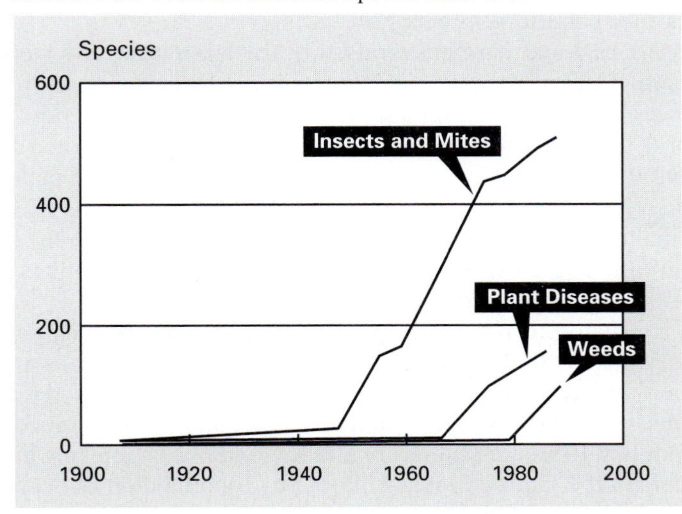

Source: From *Vital Signs 1994: The Trends That Are Shaping Our Future* by Lester R. Brown, Hal Kane, and David Malin Roodman, eds. Copyright © 1994 by Worldwatch Institute. Reprinted by permission of W. W. Norton & Company, Inc.

Those opposed to the widespread continued use of pesticides point out that there are many safer and more effective ways to control pests and weeds that deserve consideration. Some of these alternatives go under the general heading of modifying cultivation procedures. One of these methods is crop rotation, where the types of crops planted in fields are changed from year to year so that populations of pests that attack a particular crop don't have time to multiply to naturally uncontrollable sizes. Planting rows of hedges or trees in and around crop fields can act as barriers to invasions by pests and provide habitats for their natural enemies. Planting times can be adjusted so that most major insect pests starve to death before the crop is available. Crops can also be grown in areas where their major pests do not exist.[48]

Certain methods of biological control can also be used such as the introduction of various natural predators and pathogens, which are disease-causing bacteria and viruses. More that 300 biological pest control projects have been successful worldwide. Recently a tiny wasp was used to stop devastation of cassave crops in Africa by a mealybug. Other examples of biological control include the use of guard dogs to protect livestock from predators, using ducks to devour insects and slugs, using geese for weeding orchards and for controlling grass in gardens and nurseries, using birds and spiders to eat insects, and using allelopathic plants, which naturally produce chemicals that are toxic to their weed competitors or that repel or poison their insect pests.[49]

However, 10 to 20 years of research may be required to understand how a particular pest interacts with its various enemies and thus to determine the best biological control agent. Mass production of such agents is often difficult and farmers have found that they are slower to act and harder to apply than pesticides. Such agents must be protected from pesticides sprayed in adjacent fields, and there is the charge that some of these agents may also become pests in their own right. Some pest organisms can develop genetic resistance to viruses and bacterial agents used for biological control, and some may also devour other beneficial insects.[50]

Males of some insect pest species can be raised in the laboratory and sterilized by radiation or chemicals. They can then be released into the environment in large numbers to mate unsuccessfully with wild females to eventually reduce the pest population. This method works best if the females mate only once and if the infested area is relatively isolated so that it does not become repopulated with nonsterilized males and defeat the program. Chemical hormones can also be used to prevent insects from maturing completely and make it impossible for them to reproduce. But they must be applied at exactly the right time in the life cycle of the target insect and sometimes affect natural predators and other nonpest species.[51]

Food can also be irradiated to kill and prevent insects from reproducing in certain foods after harvest, which extends the shelf life of some perishable foods and destroys parasitic worms and bacteria. The FDA approved the use of low doses of ionizing radiation on spices, fruits, vegetables, and fresh pork in 1986, and in 1992 issued rules to cover irradiation of poultry. Irradiated foods are already sold in 33 countries, but surveys show that consumers in the United States will not buy food labeled as being irradiated because of fears about radiation exposure. Opponents say more studies are needed on potential harmful effects of the procedure. Even advocates worry that irradiation will give consumers a false sense of security, making them less careful about refrigerating and cooking foods properly. Thus there is some controversy over the use of this method.[52]

Many pest control experts believe that the best way to control crop pests is through using a carefully designed integrated pest management (IPM) program. In using this approach, each crop and its pests are evaluated as an ecological system and a program is developed that uses a variety of cultivation, biological, and chemical methods to control pests that threaten the crops. The overall aim of an IPM is not total eradication of the pests, but to keep the population of pests just below the level at which they cause economic loss. When this level is reached, biological and cultural controls are first applied. Small amounts of pesticides are applied only when absolutely necessary, and a variety of chemicals are used even then to retard development of genetic resistance.[53]

This approach allows farmers to escape from the pesticide treadmill and at the same time minimize hazards to human health, wildlife, and the environment in general. It has been estimated that farmers using these programs saved $579 million more than they would have by using traditional methods of pest control. These programs can (1) reduce inputs of fertilizer and irrigation water, (2) reduce preharvest pest-induced crop losses by 50 percent, (3) reduce pesticide use and control costs by 50 to 90 percent, and (4) increase crop yields and reduce crop production costs.[54]

However, use of IPM programs requires expert knowledge about each pest-crop situation and is slower acting and more labor intensive than the use of conventional pesticides. While long-term costs may be lower, initial costs may be higher than the use of pesticides. Methods that work in one area may not be applicable to another area with slightly different growing conditions, making each program more or less tailor-made to each situation. So far, this method has not been widely used in the United States because of many factors. Only about 1 percent of the Department of Agriculture's $1.6 billion research and extension budget is spent on integrated pest management programs. Pesticide companies, on the other hand, spend $1.7 billion annually in research and development worldwide. These IPM programs are strongly opposed by agricultural chemical companies that see little profit to be made from most alternative pest control methods.[55]

The term "sustainable agriculture" has taken on a new meaning in the current environment. Once the term referred to organic farming, which was a trendy program to grow crops without using synthetic chemicals. But, more recently the effort has come to include efforts to curb soil erosion by modifying plowing techniques and to protect water supplies by minimizing, if not eliminating, use of artificial fertilizers and pest controls. In addition to using new planting methods, farmers are also experimenting with novel ways to control pests without using chemicals. Thus the definition of sustainable agriculture has been considerably broadened so that it is acceptable to more people and has opened up an opportunity for dialogue for alternative farming methods.[56]

> "The 1990s are the beginning of the end of the chemical era," says Dave Dyer, executive director of American Farmland Trust. "Agrichemicals will shift from being the driving force to being the helping hand." The vision of such a future—one in which farmers cease to be among the nation's leading polluters of streams and groundwater, consumers no longer worry about eating an apple a day, the topsoil regenerates itself and the richest farmland on Earth becomes even richer—is too compelling to ignore.[57]

Cockroaches—A Case Study in Integrated Pest Management

Cockroaches carry viruses and bacteria that can cause hepatitis, polio, typhoid fever, plague, and salmonella. Attempts to control the cockroach consume one third of the pest control budget for urban sites and represent the largest expenditure for a single pest in U.S. homes and other establishments. Because of its growing resistance to many pesticides, however, the cockroach is not likely to be eliminated from homes and other buildings in the near future.

Both the professional pesticide applicator and the average consumer can control cockroaches by using the principles of Integrated Pest Management (IPM). A blend of old-fashioned practices and new technology, IPM is an ecological approach to pest management that takes into account the biology of the pest and its interaction with the environment. Although IPM may include the use of chemical pesticides, it considers all available options to achieve the greatest control with the least possible hazard.

Controlling cockroaches with IPM involves estimating the extent of the cockroach population, and then using a range of techniques to achieve tolerable levels. The basic control measure is to modify cockroach habitats by lowering the temperature, removing food, eliminating moisture, reducing clutter, and filling hiding spaces such as cracks and crevices. If these actions do not provide enough control, the appropriate pesticides may be used. Recent studies have shown that the most effective, least toxic, and least expensive method to control cockroaches is to apply 99 percent boric acid dust in cracks and crevices. The cockroaches ingest the powder while grooming themselves and die three to ten days later. (Because boric acid may be harmful if ingested by children, it should be used cautiously if children may be exposed.)

The IPM approach is being demonstrated for use in managing a variety of other pests, such as termites, grasshoppers, and aquatic weeds. Much of this work is done cooperatively with other federal and state agencies, pesticide user groups, universities, and the agricultural chemical industry. IPM is also viewed as a primary tool for managing pest populations that have become resistant to pesticides. Toward this end, we are working with the states to promote the use of innovative methods for dealing with pest resistance.

Source: United States Environmental Protection Agency, *Environmental Progress and Challenges: EPA's Update* (Washington, DC: U.S. Government Printing Office, 1988), p. 133.

The movement toward sustainable agriculture faces stiff resistance, however, from farm communities all over the nation and draws ridicule from mainstream agriculture. There is a lack of reliable research into low-chemical methods, and thus many farmers say the new techniques simply won't work on modern giant farms. These methods would raise production costs, say critics, and make it impossible for U.S. farmers to compete on world markets. Others see the movement as a step backward that could lower farm yields and income from agricultural activities, by replacing the mechanical and scientific advancements of the past 50 years with more sweat and a lower standard of living.[58]

Despite the facts that pesticide problems show no signs of going away and that more farmers will be forced to take up alternatives as older pesticides are banned or lose effectiveness and fewer new ones become available, official policies still do not acknowledge that a new era in farming is dawning. Despite success stories around the world and with ample reason to cut back on pesticide usage, farmers themselves have been slow to adopt sustainable practices. They see the illusion of a guarantee in pesticides that they can never get from other methods. Pesticides seem to have become as much a part of farming as seeds and fertilizer.[59]

TOXIC SUBSTANCES

The high standard of living that we enjoy in the United States would not be possible without the existence of the thousands of chemicals produced and used in various products. Most of these chemicals are not harmful if used properly, but others can be extremely detrimental to human health and the environment. They may cause health effects ranging from cancer to birth defects and may seriously degrade the environment. Even exposure to some of these chemicals in minute amounts may cause harm. Cancer, which is induced by toxic substances, may appear decades after the exposure and will usually be indistinguishable from a cancer caused by other means. Reducing or eliminating exposure to these harmful chemicals is one of the major goals of the regulatory process regarding toxic substances.

The health hazard posed by a chemical after it enters the environment depends on its toxicity and the extent humans are exposed to the chemical. Knowledge about the harmful effects of synthetic organic compounds has lagged far behind their introduction to the marketplace. While billions of dollars have been spent on product development and marketing, comparatively little has been spent on observing chemicals' interactions with living things and the environment. The vast majority of chemicals in use have not been fully tested for toxicity, which requires animal experiments that can take several years and can cost more than $500,000 per chemical.[60]

A study by the National Research Council (NRC) found no information at all of the possible toxic effects of more than 80 percent of the approximately 50,000 industrial chemicals used in the United States, a category that excludes pesticides, food additives, cosmetics, and drugs. There were many important unanswered questions about the remaining 20 percent. The study found that occupational exposure limits had been set for fewer than 700 of these 50,000 chemicals, and for those produced in amounts exceeding 1 million pounds a year, virtually no testing had been done on the potential for neurobehavioral damage, birth defects, or toxic effects that might span several generations by passing from parents to offspring.[61] In order to measure Americans' exposure to toxic chemicals, the NRC recommended that the EPA switch to testing blood rather than continue its fat-screening program, calling it outdated, underfunded, and flawed. Many of the new chemicals on the market are volatile organic chemicals that do not attach to fatty tissues, the NRC argued.[62]

Pesticides account for only a small share of the chemicals in common usage, yet they pose some of the greatest potential hazards. As mentioned in the previous section, they pose risks not only to farmworkers but also to the general population through residues on food and drinking water contamination. The relative threat posed by pesticide poisoning, food residues, and contaminated drinking water varies with the type of pesticide used and the care taken during application.[63] While the toxic effects of pesticides are better understood than those of industrial chemicals, there is still not enough information, according to some experts, to determine the health effects of more than 60 percent of the pesticides used in the country.[64]

Industrial wastes are another major source of toxic substances, as hazardous waste sites dot the countryside. Data on the generation and disposal of hazardous waste material are sketchier and more confusing than for pesticides, as countries apply different definitions to what is called "hazardous," "special," or simply "industrial" waste. This practice makes comparison between countries difficult if not impossible. But, most practices for disposing of hazardous waste still reflect the "out-of-sight, out-of-mind" mentality that has dominated waste disposal all over the world. This mentality exposes the public to unnecessary risks of contamination by toxic chemicals.[65]

There is also a growing concern about exposure to toxic chemicals in the home, and recent studies suggest that benign products found in and around the home emit their own stream of toxic fumes and chemicals. Many household products such as pesticides, particleboard, deodorizers, dry-cleaned clothes, and carpets may collectively pose more risk to human health and the environment than industrial chemical waste. Studies have shown that concentrations of contaminants, some of which caused cancer in rats and mice, were higher inside homes than outside, even when the homes were located in highly industrialized areas.[66]

Thus there is increasing concern about the risks to which humans are exposed by these toxic substances in our environment. More and different kinds of cancers such as breast, testicular, and prostate cancer are believed to be linked to chemical exposure that disrupts hormones in the body. Such chemicals are also believed to damage human reproductive systems by reducing sperm counts in men and causing uterine abnormalities and miscarriages in women. They may also affect immune systems, making humans susceptible to certain kinds of diseases. Chemicals are being blamed for many effects on the human body that can have long-term impacts on human health.[67]

But, not only human beings are affected by toxic chemicals. Wildlife is also exposed to the dangers of toxic substances. This exposure has implications for humans as well as for wildlife.[68] Researchers found that toxic compounds phased out in the 1970s, for example, continue to enter the waters of Lake Michigan as they leach from the soil or are carried by the wind. Such chemicals have tainted nine of ten lake trout, and have rendered one of four unfit for human consumption. Forty percent of chinook salmon in the lake are said to exceed public health standards for safe levels of PCBs, and also contain DDT, mercury, dieldrin, and other toxins.[69]

When PCBs were phased out in the 1970s, the levels of these compounds in fish in the Great Lakes were ten times higher than today. But there are still plenty of them in the lakes because old dump sites containing PCBs continue to leak the chemical into the lakes. Also, PCBs in bottom sediments are long-lived as they are only slowly converted to nontoxic compounds. Dredging to deepen ship channels through the lakes has stirred up such persistent chemicals trapped in these sediments. PCB concentrations have also increased in recent years because of the drop in lake levels.[70]

Regulation

There are 7 million known chemical compounds, 65,000 of which are in substantial commercial use and available on the market. Some 1,000 new chemicals are put into production each year and thus into the environment. Those chemicals that are harmful to human health and the environment must be identified and steps taken to reduce their associated risks. The EPA has a number of legislative tools to use in controlling exposure to toxic chemicals by regulating their release into the air, water, and land (Exhibit 8.2). This chapter focuses on the Toxic Substances Control Act, as most of the other acts have been or will be covered in other chapters.

Before the Toxic Substances Control Act (TSCA), previous laws that dealt with these substances authorized the government to act only after widespread exposure and possibly serious harm had already occurred. One major concept underlying TSCA is

EXHIBIT 8.2 Major Toxic Chemical Laws Administered by the EPA

Statute	*Provisions*
Toxic Substances Control Act	Requires that EPA be notified of any new chemical prior to its manufacture and authorizes EPA to regulate production, use, or disposal of a chemical.
Federal Insecticide, Fungicide and Rodenticide Act	Authorizes EPA to register all pesticides and specify the terms and conditions of their use, and remove unreasonably hazardous pesticides from the marketplace.
Federal Food, Drug and Cosmetic Act	Authorizes EPA in cooperation with FDA to establish tolerance levels for pesticide residues on food and food products.
Resource Conservation and Recovery Act	Authorizes EPA to identify hazardous wastes and regulate their generation, transportation, treatment, storage, and disposal.
Comprehensive Environmental Response, Compensation, and Liability Act	Requires EPA to designate hazardous substances that can present substantial danger and authorizes the cleanup of sites contaminated with such substances.
Clean Air Act	Authorizes EPA to set emission standards to limit the release of hazardous air pollutants.
Clean Water Act	Requires EPA to establish a list of toxic water pollutants and set standards.
Safe Drinking Water Act	Requires EPA to set drinking water standards to protect public health from hazardous substances.
Marine Protection Research and Sanctuaries Act	Regulates ocean dumping of toxic contaminants.
Asbestos School Hazard Act	Authorizes EPA to provide loans and grants to schools with financial need for abatement of severe asbestos hazards.
Asbestos Hazard Emergency Response Act	Requires EPA to establish a comprehensive regulatory framework for controlling asbestos hazards in schools.
Emergency Planning and Community Right-to-Know Act	Requires states to develop programs for responding to hazardous chemical releases and requires industries to report on the presence and release of certain hazardous substances.

Source: United States Environmental Protection Agency, *Environmental Progress and Challenges: EPA's Update* (Washington, D.C.: U.S. Government Printing Office, 1988), p. 113.

that the government has the authority to act before a substance can harm human health or the environment—the substance is, in effect, guilty until proven innocent. Under TSCA, the EPA reviews risk information on all new chemicals before they are manufactured or imported, and decides whether they should be admitted, controlled, or denied access to the marketplace. Because of TSCA, the entire chemical industry was put under comprehensive federal regulation for the first time, as the law applies to virtually every facet of the industry—product development, testing, manufacturing, distribution, use, and disposal. In addition, importers of chemical substances are treated as domestic manufacturers, thus extending the EPA's control to certain aspects of the international chemical trade.[71]

The initial impact of TSCA was in the area of inventory reporting. The act required the EPA to compile and publish an inventory of chemical substances manufactured, imported, or processed in the United States for commercial purposes. The inventory was compiled from reports that manufacturers, importers, processors, or users of chemical substances were required to prepare and submit to the agency.[72] The first inventory was published in 1979 and contained information on over 62,000 chemicals that came from manufacturers and importers and included production volume and plant location. In 1986, manufacturers and importers were required to report current data for a subset of substances on the inventory and update the information every four years thereafter.[73]

After publication of the initial inventory, the premanufacture provisions of TSCA went into effect. These provisions require a manufacturer who has developed a new chemical not on the inventory list to submit a notice to the EPA at least 90 days before beginning manufacture or importation of a new chemical substance for commercial purposes other than in small quantities solely for research and development. The information that has to be given to the EPA includes a description of the new chemical substance, the estimated total amount to be manufactured and processed, and other such information.[74]

In addition to this information, submitters must append any test data in their possession or control, and descriptions of other data concerning the health and environmental effects of the substance. The EPA encourages, but does not require, the submitter to follow the premanufacture testing guidelines the EPA has published. In any event, all test data are to be submitted regardless of their age, quality, or results. About 80 percent of the new chemicals received appear to present no unreasonable risks to human health or the environment, but the rest must go through a more detailed review. This review includes a structure-activity analysis where a chemical's physical and chemical behavior is predicted by comparing the chemical's molecular structure with that of other chemicals for which the behavior is already known.[75]

The administrator of the EPA has a number of options available after receipt of this information: These include extending the 90-day premanufacture review period for an additional 90 days for good cause, requiring additional testing of the substance, and initiating no action within the 90-day period because the chemical is deemed not to present a hazard to health or the environment. If a hazard is believed to exist, the administrator may issue a proposed order to take effect on the expiration of the notification period to prohibit or limit the manufacture, processing, distribution in commerce, use, or disposal of such substance or to prohibit or limit any combination of such activities.

If a total ban on the substance is not necessary, the administrator can issue further directives regarding regulation of the substance. Possibilities include setting concentration levels, limiting the use of the chemical, requiring warnings or instructions on its use, requiring public notice of risk or potential injury, and regulating methods of disposal. If the administrator has reason to believe the method of manufacture rather than the chemical itself is at fault, the manufacturer may be ordered to revise quality control procedures to the extent necessary to remedy whatever inadequacies are believed to exist.

In addition to these premanufacture notification provisions, other sections of TSCA deal with testing, evaluation, and control of existing substances. The act empowers the EPA administrator to require manufacturers or processors of potentially harmful chemicals already in use to conduct tests on these chemicals. The need for such testing must be based on the following criteria: (1) The chemical may present an unreasonable risk to health or the environment, or there may be substantial human or environmental exposure to the chemical; (2) there are insufficient data and experience for determining or predicting the health and environmental effects of the chemical; and (3) testing of the chemical is necessary to develop such data.[76]

The overall goal of the Existing Chemicals Program is to reduce unreasonable risks of injury to health or the environment from chemicals that are already in commerce. An interagency committee has been established by the act to assist the administrator to determine chemicals that should be tested, but the administrator's actions are not limited to these recommendations by the committee. This committee may designate, at any one time, up to 50 chemicals from its list of recommended substances for testing. Within one year, the administrator must either initiate testing requirements for these designated chemicals or publish in the Federal Register any reasons for not initiating such requirements.

The law also allows the EPA to require companies to submit unpublished health and safety data on a list of specified chemicals that are suspected of causing cancer or other health effects. These data are used to evaluate risks associated with exposure and to determine whether toxicity testing should be done if it has not been conducted.[77] Chemical manufacturers, processors, and distributors are also required to inform the EPA immediately when they obtain evidence that a chemical poses a substantial risk of harm to human health or the environment. Such notices may include unpublished toxicity and exposure studies and may lead to further action by the EPA or other agencies.[78]

The EPA is also authorized to monitor the exposure of humans and the environment to chemicals in order to identify potential hazards. Since 1960, for example, the agency has monitored the levels of PCBs and chlorinated pesticides such as DDT in humans, and developed improved methods for monitoring chemicals in human tissues and fluids such as a method to measure dioxin in fatty tissue. The agency also has conducted a number of chemical exposure studies in order to determine the risk humans may face from exposure to certain substances.[79]

By the end of 1987, the EPA had requested additional health and environmental testing by the manufacturers of 63 chemical groups for possible regulatory control. Congress and the EPA determined that several chemicals posed such a high risk that they needed to be regulated more stringently. In 1978, for example, the EPA instituted regulatory controls over the manufacture, use, and disposal of polychlorinated biphenyls

(PCBs) and banned the aerosol uses of chlorofluorocarbons (CFCs). In 1989, the agency banned the manufacture of most asbestos products.[80]

Results

Some of the actions taken under these provisions of TSCA have shown dramatic results. Restrictions on the use and disposal of PCBs, for example, have resulted in a significant decline of these residues in the environment, food, and human tissues. The number of individuals with high PCB levels has declined from over 8 percent to less than 1 percent of the population[81] (Figure 8.3). The term "PCB" refers to a widely used group of 209 different toxic, oily, synthetic chlorinated hydrocarbon compounds known as polychlorinated biphenyls. These substances were used in many commercial activities, especially as heat transfer fluids in electrical transformers and capacitors. They were also used as hydraulic fluids, lubricants, and dye carriers in carbonless copy paper, and in paints, inks, and dyes.

Over time, PCBs began to accumulate in the environment because of leaking electrical equipment and other sources. They eventually reached humans through the food chain and caused serious health problems in high concentrations.[82] The release of PCBs into the environment went largely unchecked from the late 1920s until the 1970s, but after that, the toxic effects of PCBs have been recognized. Tissue assays of more

FIGURE 8.3 PCB Levels

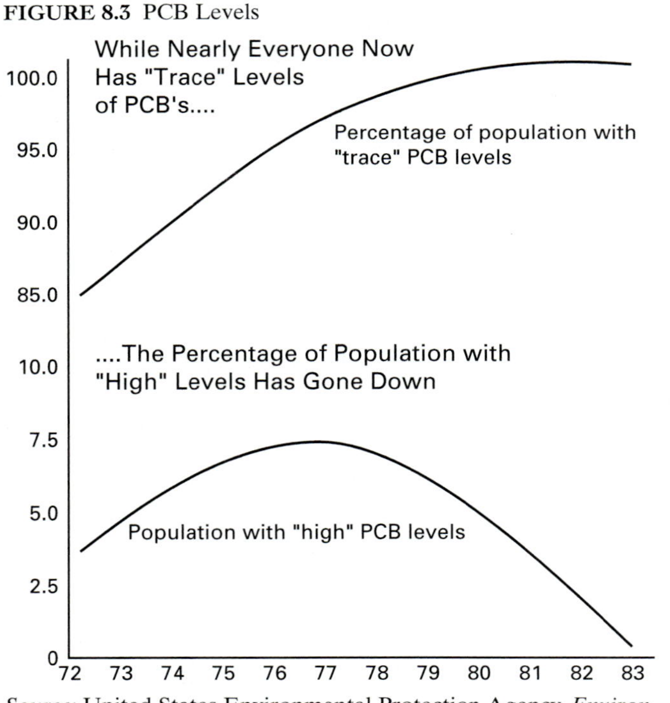

Source: United States Environmental Protection Agency, *Environmental Progress and Challenges: EPA's Update* (Washington, D.C.: U.S. Government Printing Office, 1988), p. 121.

than 4,000 samples from human beings in the 1970s showed that the entire U.S. population was carrying some burden of the substance. In 1973, the Food and Drug Administration (FDA) first established tolerances in PCBs for certain foods, and in 1979, those tolerances were lowered. These tolerances cover milk and manufactured dairy products, poultry and eggs, finished animal products and animal feed components, fish and shellfish, and infant and junior foods.[83]

In 1976, Congress directed the EPA to ban the manufacture, processing, distribution, and use of PCBs except in totally enclosed electrical equipment. The agency also imposed requirements to reduce the risk of PCB transformer fires, which can expose people in office buildings, apartment complexes, shopping malls, and other areas to severe risks. During a transformer fire, highly toxic furans and dioxins that are produced by the combustion of PCBs can spread throughout a building. The EPA has also established regulations for the safe disposal of PCBs and for the cleanup of PCB spills. New technologies, such as the use of incinerators and chemical and biological methods for detoxifying PCBs, have been promoted.[84]

Other actions to deal with toxic substances have been more controversial. Asbestos is another substance that was widely used for many purposes such as fireproofing and pipe and boiler insulation in schools and other buildings. Asbestos was often mixed with a cement-like material and sprayed on ceilings and other surfaces. Asbestos can break down into tiny fibers or dust that can be inhaled and lodged in lung tissue where it can cause lung cancer and asbestosis, a chronic scarring on the lungs that hinders breathing. Because of these harmful effects, which began to be discovered in the 1970s, asbestos use was sharply curtailed. In 1989 the EPA issued a ruling that aimed to eliminate almost all remaining uses of asbestos in the United States by 1997, such as usage in brake shoes and some construction materials. This ruling was later overturned by an appeals court that didn't challenge the EPA's scientific basis for claiming that asbestos posed a health risk, but faulted the agency for failing to consider less economically burdensome steps to reduce health risks from asbestos products.[85]

One of the major controversies concerning asbestos is the issue of removal of the substance from old buildings. Proponents of removal argued that damaged or crumbling asbestos materials in old buildings posed a danger to occupants of the building and maintenance workers. Thus it should be removed to prevent people from exposure to asbestos fibers that could be inhaled in these buildings. But some reports suggested that expensive remedial action to remove asbestos in commercial and public buildings was not warranted. These reports concluded that undisturbed asbestos in buildings in good repair are unlikely to produce airborne asbestos concentrations any higher than the levels found outside. Moreover, improper removal of the substance could actually increase such concentrations to dangerous levels.[86]

Dioxin has long been believed to be one of the most dangerous substances created by humans, and as a result reactions to the discovery of dioxin have been swift and sweeping. In 1982, for example, the town of Times Beach, Missouri, was discovered to be contaminated with dioxin from tainted oil sprayed on its roads and several horse arenas in the area. The government responded by evacuating the town, buying out the residents for about $40 million, and fencing the 480-acre area off completely. At the time, the government was largely applauded for acting so quickly and decisively, but in later years, it was accused of overreacting, as some research suggested that dioxin posed little or no threat to humans.[87]

Thus the government began to backtrack on dioxin and reevaluate its regulations of the substance.[88] Some evidence suggested that this reappraisal was as much a result of a well-financed public-relations campaign by the paper and chlorine industries as it was a result of new research. The trade associations of these two industries had aggressively promoted two pieces of evidence, which the industries themselves were involved in producing, suggesting that dioxin is less dangerous than previously thought. The reason for this effort is that chlorine made by some big chemical companies is used in bleaching pulp for paper, and some dioxin is released in the process. Stringent regulations could result in the purchase of very costly equipment to reduce the use of chlorine and dioxin emissions.[89]

In 1994, the EPA released a new study strengthening the link between dioxin and cancer as well as other problems such as harm to the reproductive and immune systems of humans. The study reaffirmed conclusions reached in 1985 that dioxin is a "probable human carcinogen," but stopped short of labeling the substance a known carcinogen. The report did not suggest specific regulations or policy changes, but these were expected to come after the comment period on the report was over and the agency issued its final assessment.[90] However, an independent panel of scientists chosen by the EPA to review its report accused the agency of overstating the risks of dioxin and that its conclusions were not scientifically defensible and thus it could not endorse the report as currently drafted.[91] Thus the controversy continues.

Lead became a concern in the early 1990s, particularly in regard to children. Studies showed that 3 million to 4 million American children, or about 1 out of every 6, under six years of age have lead poisoning. Only 7 percent of these children came from medium- and high-income families, but 25 percent of poor white children were affected and an incredible 55 percent of children coming from impoverished black families. Symptoms of lead poisoning include abdominal pains, muscular weakness and fatigue, nervous system disorders, high blood pressure, and even death in the most severe cases. Children are believed to be particularly sensitive to the toxic metal because their nervous systems and brains are still developing.[92]

The Centers for Disease Control, in response to this problem, lowered its threshold definition of dangerous blood levels of lead by 60 percent, to 10 micrograms a deciliter from 25 micrograms a deciliter. This standard is roughly equivalent to one part per 10 million. The government also launched an antilead campaign to warn Americans about the dangers of lead poisoning, a campaign the lead industry accused of relying on scare tactics and lacking useful information. Consumer groups also feared that the campaign would not reach low-income people whose children were most at risk.[93]

Sources of this lead include old and crumbling lead-based paint, airborne lead resulting from industrial processes, lead solder in old plumbing, and even though leaded gasoline has been largely phased out, the soil near major highways in particular is still contaminated. Potential solutions include (1) establishment of a national surveillance system for children with elevated blood lead levels; (2) elimination of leaded paint and contaminated dust in houses; (3) reduction of exposure to lead in water, food, air, soil, and places of play; and (4) an increase in community programs for the prevention of lead poisoning in children.[94]

About three quarters of the nation's housing, including most housing built prior to 1978, contains lead-based paint. In 1978, the Consumer Product Safety Commission banned the use of lead-based paint. In 1992, Congress passed a law requiring that purchasers and renters of pre-1978 housing be told of possible lead hazards starting in 1995, and calling for gradual abatement of federally owned and assisted housing starting with older buildings. A rule proposed in late 1994 implementing this legislation did not require dwellings be inspected for lead paint hazards prior to being sold or leased, but if the owners or landlords know of such hazards and don't disclose them, they could be sued along with their real-estate agents for damages. The rule was expected to cost about $75 million a year.[95]

Thousands of lawsuits are pending nationwide over this issue. New York City alone faces some 600 to 700 lawsuits on behalf of children alleging lead poisoning. The city owns some 45,000 housing units and thus is a landlord for many apartments containing lead-based paint and other violations. Settlements have averaged $500,000 for those cases that have been settled. And in 1995, a Manhattan federal court certified a class action suit against the Federal Home Loan Mortgage Corporation on behalf of children allegedly exposed to lead hazards in houses where the agency had been involved in the mortgage. This problem is thus likely to become the next major toxic torts litigation area after asbestos.[96]

The great difficulty in all these areas is establishing a firm linkage between exposure to a substance and adverse health effects such as cancer and reproductive disorders in order to make reasonably accurate risk assessments. This task requires a mix of epidemiological and toxicological evidence, but each method has its limitations. Epidemiological studies only measure harmful effects once they have occurred and provide no predictions for future effects given present exposures. Such studies are a blunt tool and often miss the small contribution of a pollutant among all the competing explanations for higher rates of particular diseases. Epidemiologists also find it difficult to assess the health effects of slight exposures to toxic substances.[97]

On the other hand, data from toxicological studies showing that a substance causes cancer in laboratory animals are generally not regarded as sufficient to prove carcinogenicity in humans. Massive doses of the substance are often used to ensure that any toxic effects that might exist are not missed, and health effects at these high levels are then extrapolated to effects at lower levels using mathematical models. But this procedure means that little direct information on health effects at low levels of exposure are collected. Most substances are also tested individually, but in real life, people are exposed to many chemicals at the same time. Such multiple exposures complicate risk assessment, as adding chemicals together can produce synergistic, additive, or antagonistic effects.[98]

There is a wide range of individual sensitivity to chemicals, and some evidence suggests that part of the population is more sensitive than the average person to the effects of a whole range of chemicals. Just as problematic is the fact that some portions of the population are exposed to more pollutants than others. All of these factors mean that a great deal of scientific uncertainty exists about the potency of substances such as those mentioned and no neat formulas for the health risks posed by these and other substances can be produced. Even the best studies and risk assessments are plagued by genuine controversy, and yet policy and regulations as well as the outcome of court cases depend on these studies and assessments.[99]

Other Efforts to Control Toxic Substances

After the Bhopal tragedy, community right-to-know bills were introduced into more states and both houses of Congress. Under Title III of the Superfund Amendments and Reauthorization Act (SARA) of 1986, also known as the Emergency Planning and

Community Right-to-Know Act, facilities that manufacture, process, or use any of 320 designated chemicals in greater than specified amounts, must report routine releases of those chemicals. These facilities are required to report on releases of toxic chemicals into the air, water, and land, and must also report on offsite transfers referring to transfers of wastes for treatment or disposal at a separate facility. Beginning in 1992, these reports must also contain detailed source reduction and recycling information as mandated by the Pollution Prevention Act.[100]

In addition to reporting routine releases, the law also provides for emergency notification of chemical accidents and releases, planning for chemical emergencies, and reporting of hazardous chemical inventories. With regard to accidental releases, the information that must be reported includes (1) the name of the chemical; (2) the location of the release; (3) whether the chemical is on the "extremely hazardous" list; (4) how much of the substance has been released; (5) the time and duration of the incident; (6) whether the chemical was released into the air, water, or soil, or some combination of the three; (7) known or anticipated health risks and necessary medical attention; (8) proper precautions, such as evacuation; and (9) a contact person at the facility.[101]

The EPA is required to make information from these reports available to the public. This Toxics Release Inventory, as it is called, is designed to assist citizen groups, local health officials, state environmental managers, and the EPA to identify and control toxic chemical problems. This information can aid communities to develop an emergency plan should an accidental release pose a serious threat to the community that requires drastic action (see box). One of the EPA's challenges is to interpret this

Required Elements of a Local Emergency Plan

An emergency plan must:

- Use the information provided by industry to identify the facilities and transportation routes where hazardous substances are present.
- Establish emergency response procedures, including evacuation plans, for dealing with accidental chemical releases.
- Set up notification procedures for those who will respond to an emergency.
- Establish methods for determining the occurrence and severity of a release and the areas and populations likely to be affected.
- Establish ways to notify the public of a release.
- Identify the emergency equipment available in the community, including equipment at facilities.
- Contain a program and schedules for training local emergency response and medical workers to respond to chemical emergencies.
- Establish methods and schedules for conducting "exercises" (simulations) to test elements of the emergency response plan.
- Designate a community coordinator and facility coordinators to carry out the plan.

Source: United States Environmental Protection Agency, *Chemicals in Your Community: A Guide to the Emergency Planning and Community Right-to-Know Act* (Washington, D.C.: U.S. Government Printing Office, 1988), p. 5.

information to help state and local officials evaluate and manage the risks posed by substances present in their communities.[102]

In 1993, 32 more chemicals were added to the list of chemicals that companies were required to report about including those regulated under the Resource Conservation and Recovery Act. Even more recently, 286 more were added, bringing the total to 654 chemicals and chemical categories. The initial list of 320 chemicals accounted for only a small percentage of the chemicals used in the country, so the EPA believed that the program should be extended to include additional chemicals that exhibited similar toxicity characteristics to those originally reported. Expansion of the list should give the public a more complete picture of toxic chemicals in their communities.[103]

Companies reporting under this law can, under very limited conditions, request that the identity of specific chemicals in their reports not be disclosed to the public. In order to protect a chemical's identity in this fashion, companies must be able to prove that the information has not been reported under any other environmental regulation, and that it is a legitimate trade secret; that is, that disclosure could damage the company's competitive position. If this protection is granted, the company must still report the chemical's identity to the EPA, but the agency will keep the original reports in a confidential file and "sanitized" versions, with the chemical name deleted, will be made available to the public. Health professionals can obtain access to trade secret chemical information if they need it to diagnose and treat patients or to do research.[104]

The law's purposes are to encourage and support emergency planning for responding to chemical accidents and to provide local governments and the public with information about possible chemical hazards in their communities. It requires that detailed information about the nature of hazardous substances in or near communities be made available to the public. This law is different from many other federal statutes, in that it does not preempt states or local communities from having more stringent or additional requirements. More than 30 states have such laws giving workers and citizens access to information about hazardous substances in their workplaces and communities.[105]

The law also focused industry's attention on pollution prevention and source reduction opportunities. Many companies took advantage of this opportunity to cut their emissions voluntarily. The day before the first TRI figures were released in 1987, the CEO of Monsanto publicly announced that the company would voluntarily reduce its toxic air pollution to only 10 percent of 1987 levels by 1992, and then continue to work toward the ultimate goal of zero emissions. Many thought this goal unattainable, because in the 1987 report the company indicated it had produced 61 million pounds of toxic air releases worldwide. But by 1992, the company's global emissions were actually cut by 92 percent compared with 1987 levels.[106] Thus the law has had a positive effect in getting companies to voluntarily reduce their toxic emissions, and these emissions have been reduced by approximately one half since the requirement went into effect.

In 1991, the EPA announced a 33/50 program whose goal was to reduce releases and transfers of 17 highly toxic, high-priority chemicals such as cadmium, mercury, lead, and benzene, 33 percent by the end of 1992 and 50 percent by 1995, using the 1988 TRI data as a baseline. This voluntary program attempts to create a partnership among government, industry, and the communities involved. In asking almost 6,000 companies to join the program, the EPA stressed the benefits of pollution prevention such as community health protection, competitive advantage from reducing product loss and waste disposal expense, potential avoidance of future liabilities and regulatory requirements, and improved community relations. Within one year, more than 700 companies had made explicit commitments to the program.[107]

Several other strategies are being pursued by the EPA including one that focuses agency attention on the highest risk toxic chemicals. Some 60,000 chemicals are being used in the United States, and most of these were in use before laws existed that required they be evaluated for health and environmental risks prior to being manufactured. The agency recently made a special effort to revitalize the review of these existing chemicals and make it a major priority. This strategy includes chemical screening that will be linked more directly to risk management. The EPA will be screening groups of chemicals, developing rules for these groups wherever possible. This procedure should greatly speed up the review process regarding existing chemicals.[108]

Under TCSA's information-gathering authority, the EPA was given a means to develop an integrated approach to the control and management of toxic chemicals all over the country. The agency shares such information among all its regulatory programs as well as with other agencies. An outreach service has been established to help the regional offices and states improve their risk assessments. This service is called the Chemical Assessment Desk and provides other parts of the agency with toxicity and risk information of the chemicals that have been reviewed in the toxic substances program. The EPA is also working with other countries to coordinate the gathering, testing, and evaluation of existing chemicals of common concern. This work is being done through the Organization for Economic Cooperation and Development, and through such efforts, it is hoped that a coordinated and comprehensive regulation of toxic chemicals can be developed.[109]

Courts have in some cases turned a more sympathetic ear to claims for compensation based on the fear of cancer. Residents of Hardeman County, Tennessee, discovered that their drinking wells had been contaminated with chemicals linked to cancer and other diseases. The Velsicol Chemical Corporation was blamed for the problem, as it has used a site in the area to dump waste from a nearby plant that manufactured pesticides. A group of 128 people sued the company in state court over alleged injuries such as liver and kidney damage. But the Hardeman County residents also demanded compensation for their fear of contracting cancer in the future.[110]

In earlier years, most courts would have dismissed such claims, but in August 1988, a federal appeals court approved the awarding of $207,000 for "cancerphobia" to five people serving as representatives of the class action. Awards for other class members were to be determined by a trial judge if the amounts are not settled out of court beforehand. Some lawyers claimed that the number of such claims has proliferated in the past two or three years, and cancerphobia charges are becoming routine in large toxic injury cases. And the courts, as the above case illustrates, are showing signs of going along with these cases and awarding damages.[111]

In 1990, attention turned to the effects of toxic chemicals on children. Environmental groups began to cite scientific evidence suggesting that the nation's 63 million youngsters are significantly more endangered by toxic chemicals than are adults. Children are believed to be more susceptible to toxins because the poisons quickly become more concentrated in small bodies. Their less developed detoxifying organs also raise the risk, as do rapidly developing cells, which increase the danger from carcinogens and neurotoxins. To protect children, environmentalists are asking the government to set new safety standards for pesticides, toxins, and air pollutants, taking children into account.[112]

Detoxifying the Environment

Most of the toxic chemicals released into the environment come from pesticide use in agriculture and from the disposal of hazardous industrial waste material. Strategies that reduce pesticide use as indicated in the first part of this chapter and that minimize industrial waste generation offer cost-effective approaches to lessening risks from toxic substances. The quick fixes of pesticide spraying and end-of-pipe pollution control must be replaced with new production systems aimed at reconciling economic profits and environmental protection. Some experts believe that with technologies and methods now available, pesticide use could probably be halved and the creation of industrial waste cut by a third or more over the next decade.[113]

Detoxifying the environment involves the development of policies and establishment of funding priorities that actively promote new methods of production in agriculture and industry. Greater public commitments to research and development are needed, some experts believe, in the areas of biological, cultural, and genetic methods of pest control. Developing countries, in particular, may need to stop subsidizing chemical pesticides so heavily, which only encourages farmers to apply more chemicals than is economically justified and increases all risks associated with toxic farm chemicals.[114]

With regard to disposal of industrial wastes, more research and development money must focus on waste-reducing technologies. Virtually no country, according to one expert, has yet designed an effective, long-term strategy to reduce industrial chemical wastes. By investing in waste reduction, governments can avoid future problems and costs arising from waste mismanagement, shortfalls in treatment capacity, and public opposition to siting new disposal facilities. Few developing countries have even established the basic foundations for a hazardous waste management system, and most have no regulations governing toxic waste disposal or facilities that are capable of adequately treating and disposing of such materials.[115]

Perhaps the most promising strategy is to make industries assume more responsibility for the societal costs and risks associated with hazardous substances. Some experts think this strategy is crucial to fostering a transition to safer chemicals and products. If industries all over the world, for example, had to prove that chemical substances were safe, and if they faced strict liability for damages caused from the manufacture, use, and disposal of their products, the risks may diminish throughout the chemical cycle. Such a process would weed out risky substances in industrial laboratories before they are introduced into the environment, rather than wait for a regulatory agency to correct abuses after many years of using the chemical.[116]

The Chemical Manufacturers Association (CMA) established a "Responsible Care" initiative to improve the performance of the industry health, safety, and the environment. This initiative was signed by over 170 leading chemical companies, representing more than 90 percent of the industrial chemical production in the country. The CMA promised to report progress in implementing the initiative, and asked residents who lived near chemical companies to be active participants in the initiative. The initiative was said to be the industry's way of making sure it was not part of the problem but part of the solution. The initiative listed the following principles.[117]

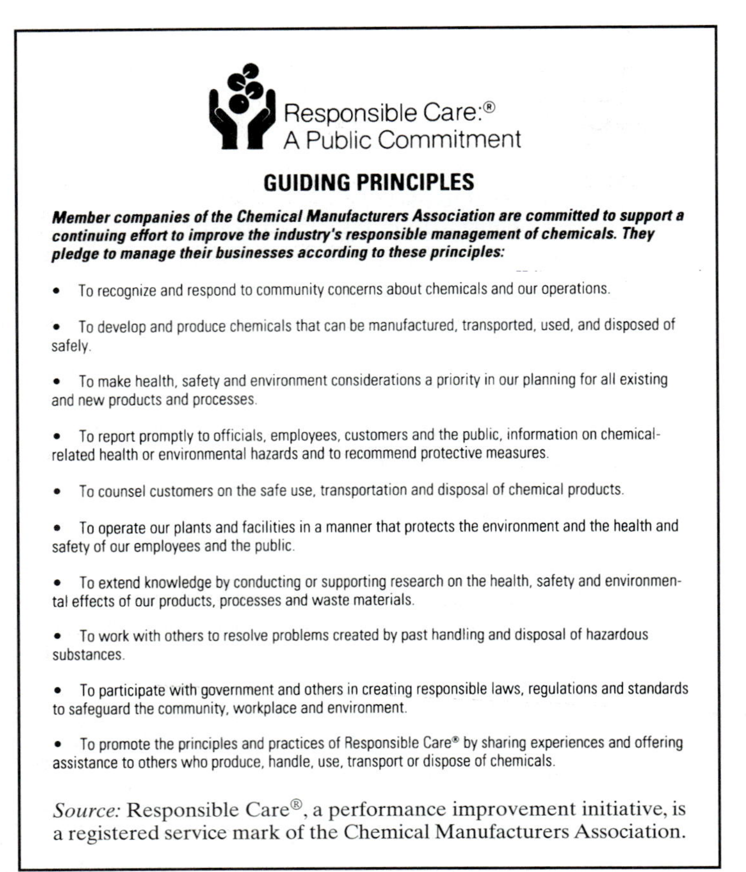
Responsible Care:®
A Public Commitment

GUIDING PRINCIPLES

Member companies of the Chemical Manufacturers Association are committed to support a continuing effort to improve the industry's responsible management of chemicals. They pledge to manage their businesses according to these principles:

- To recognize and respond to community concerns about chemicals and our operations.
- To develop and produce chemicals that can be manufactured, transported, used, and disposed of safely.
- To make health, safety and environment considerations a priority in our planning for all existing and new products and processes.
- To report promptly to officials, employees, customers and the public, information on chemical-related health or environmental hazards and to recommend protective measures.
- To counsel customers on the safe use, transportation and disposal of chemical products.
- To operate our plants and facilities in a manner that protects the environment and the health and safety of our employees and the public.
- To extend knowledge by conducting or supporting research on the health, safety and environmental effects of our products, processes and waste materials.
- To work with others to resolve problems created by past handling and disposal of hazardous substances.
- To participate with government and others in creating responsible laws, regulations and standards to safeguard the community, workplace and environment.
- To promote the principles and practices of Responsible Care® by sharing experiences and offering assistance to others who produce, handle, use, transport or dispose of chemicals.

Source: Responsible Care®, a performance improvement initiative, is a registered service mark of the Chemical Manufacturers Association.

There is also the possibility that many chemicals can be replaced with natural substances. The Danish company Novo Nordisk has developed more than 40 industrial enzymes to replace synthetic chemicals in many processes such as stonewashing jeans and ripening apples more quickly. These enzymes are much less harmful to the environment than the chemicals they replaced. These enzymes are natural catalysts that can speed up a chemical reaction without being consumed in the process. They are also biodegradable and require up to a third less energy than many synthetic chemicals. The business has been growing by 10 to 15 percent a year, and the company has emerged as a pioneer in what is called "green chemistry," the search for more benign substitutes for synthetic chemicals.[118]

Some suggest that society has begun to turn from an uncritical acceptance of a chemically driven culture. The focus of industry has begun to shift toward a fairly widespread adoption of the waste-management hierarchy of source reduction, recycling, treatment, and minimal residual storage and disposal. Perhaps the time has come, from the perspective of the environmental and health communities, to restrict the use of many toxic materials, and even prohibit some of them outright. While zero discharge may be unrealistic, this goal suggests the direction in which environmentalists are heading. Industry is approaching a similar position, but for largely different reasons. The high costs of waste disposal, combined with increased liability, make source reduction an economic imperative. Thus business is being led in the same directions as environmental and health advocates when it comes to control of toxic substances.[119]

Questions for Discussion

1. Who was one of the first people to point out the danger of pesticide usage? What particular pesticide did this person focus on and what dangers were highlighted? What was the effect of this action?
2. What threats do pesticides pose to human health and the environment? Describe the present regulatory system with regard to pesticides. Do you believe this system adequately protects human health and the environment? Why or why not? Who is most at risk from pesticides?
3. What alternatives to pesticides exist? Are these alternatives effective? What is an integrated pest management system? What are the advantages and disadvantages of this method? What does the future hold with regard to pesticide usage?
4. What threats do toxic substances pose to human health and the environment? Name some chemicals that you have heard about that are dangerous. What risks do these chemicals pose to human health and how have they been controlled?
5. Describe the Toxic Substances Control Act. What are its major provisions? How does the act work with respect to new and existing chemicals? What impacts has it had on the chemical industry? On the whole, do you think it is a good piece of legislation?
6. What are community right-to-know bills and what do they normally contain? What is a toxic release inventory, and what has been its effect on communities and on the chemical industry? Do these requirements help communities plan for emergency situations?
7. What new strategies are being pursued with respect to toxic chemicals? Do you believe these strategies hold much promise? What direction do courts seem to be taking? Is this a development that bears watching?
8. What long-term strategies make the best sense regarding toxic chemicals? What additional things should industry do to protect the public and the environment? Examine the responsible care initiative of the Chemical Manufacturers Association. What do you think of these principles? Can they be implemented?

Endnotes

1. G. Tyler Miller, *Living in the Environment,* 6th ed. (Belmont, CA: Wadsworth, 1990), p. 549.
2. *Your Guide to the U.S. Environmental Protection Agency,* OPA 212 (Washington, D.C.: Environmental Protection Agency, 1982), p. 9.
3. Gary Gardner, "Preserving Agricultural Resources," *State of the World 1996,* Lester R. Brown et al. (New York: W. W. Norton, 1996), p. 89.
4. Miller, *Living in the Environment,* p. 551–552.
5. Ibid., p. 550.
6. Rachel Carson, *Silent Spring* (Greenwich, CT: Fawcett, 1962). See also "Averting a Death Foretold," *Newsweek,* November 28, 1994, pp. 72–73.
7. Peter Weber, "A Place for Pesticides?" *World Watch,* May–June 1992, p. 19.
8. Miller, *Living in the Environment,* p. 554.
9. Gardner, "Preserving Agricultural Resources," p. 90.
10. "Farmers' Cancers Linked to Pesticides in Study," *Times-Picayune,* September 23, 1992, p. A5.

11. Miller, *Living in the Environment,* p. 554.
12. Ibid., p. 555. Some scientists argue that the food-pesticide cancer risk is overrated. While they acknowledge that the nation's food basket is teeming with potential carcinogens, they say that 99.9 percent are natural chemicals that foods develop as by-products of cooking or natural toxins that plants secrete in self-defense. See Marilyn Chase, "Scientists Term Food-Pesticide Cancer Risk Overrated," *Wall Street Journal,* October 9, 1992, p. B1.
13. Rita Beamish, "Pesticide–Breast Cancer Link to Be Investigated," *Times-Picayune,* October 22, 1993, p. A3.
14. Anastasia Toufexis, "Watch Those Vegetables, Ma," *Time,* March 6, 1989, p. 57.
15. Gisela Bolte, Dick Thompson, and Andrea Sachs, "Do You Dare to Eat a Peach?" *Time,* March 27, 1989, pp. 24–27.
16. "Revenge of the Apples," *Wall Street Journal,* December 17, 1990, p. A8. See also Reed Irvine and Joseph C. Goulden, "Apple Growers Sue CBS," *New Dimensions,* March 1991, p. 76. In order to avoid an Alar type scandal, banana importers agreed to stop using the pesticide aldicarb on bananas after unusually high pesticide residues in some fruit prompted California health officials to issue a health advisory for young children. See Maura Dolan, "Pesticide Use Stopped on Bananas," *Times-Picayune,* June 16, 1991, p. A16.
17. Amal Kumar Naj, "Insecticides Found to Be Traveling Throughout World," *Wall Street Journal,* September 29, 1995, p. B4.
18. Allen A. Boraiko, "The Pesticide Dilemma," *National Geographic,* 157, no. 2 (February 1980), 151. See also "Pesticides Leave Enduring Trail, Causing Disagreement," *Dallas Times Herald,* July 20, 1986, p. A53.
19. United States Environmental Protection Agency, *Meeting the Environmental Challenge* (Washington, D.C.: U.S. Government Printing Office, 1990), p. 19.
20. *1978 Report: Better Health and Regulatory Reform* (Washington, D.C.: Environmental Protection Agency, 1979), p. 29.
21. United States Environmental Protection Agency, *Environmental Progress and Challenges: EPA's Update* (Washington, D.C.: U.S. Government Printing Office, 1988), pp. 114–115.
22. EPA, *Meeting the Environmental Challenge,* p. 19.
23. EPA, *Environmental Progress and Challenges,* p. 129.
24. "Pesticide-icides," *Regulation,* Nos. 3–4, (1987), p. 9.
25. Ibid.
26. Ibid.
27. Scott Ferguson and Ed Gray, "1988 FIFRA Amendments: A Major Step in Pesticide Regulation," *Environmental Law Reporter,* 19 ELR, 2–89, pp. 10070–10082.
28. Sonia L. Nazario, "EPA Under Fire for Pesticide Standards," *Wall Street Journal,* February 17, 1989, p. B1. See also Jim Stiak, "Pesticides and Secret Agents," *Sierra,* May–June 1988, pp. 18–21.
29. EPA, *Environmental Progress and Challenges,* pp. 128–129.
30. United States Environmental Protection Agency, Office of Communications, Education, and Public Affairs, *Securing Our Legacy* (Washington, D.C.: EPA, 1992), p. 18. In 1991, the EPA was criticized in a report issued by the General Accounting Office for reacting too slowly in reviewing scientific studies needed to assess a pesticide's potential to contaminate groundwater. See "Report: EPA Lax Toward Pesticides in Water Supply," *Times-Picayune,* May 8, 1991, p. A8.
31. Ibid., p. 134. After several years of controversy, the state of Virginia took action in 1991 to ban the use of granular carbofuran, the first time anywhere a chemical had been banned solely because it imperiled wildlife. See Deana West, "Taking Aim at a Deadly Chemical," *National Wildlife,* June–July 1992, pp. 38–41.
32. Ted Williams, "Hard News on 'Soft' Pesticides," *Audubon,* March–April 1993, pp. 30–40.
33. EPA, *Environmental Progress and Challenges,* pp. 134–135.

34. Andy Pasztor and Barry Meier, "EPA Has Started to Crack Down on Use of Pesticides Thought to Be Hazardous," *Wall Street Journal,* December 6, 1984, p. 7.
35. David Holzman, "Farm Workers Reap Cancer Risks," *Insight,* September 4, 1989, 52–53.
36. "EPA Issues Rules on Pesticides Made by Biotechnology," *Wall Street Journal,* October 5, 1984, p. 37.
37. EPA, *Environmental Progress and Challenges,* p. 140.
38. Ibid., pp. 139–140. In 1991, the EPA registered for the first time two pesticides derived from biological organisms that were genetically engineered using recombinant DNA techniques. See EPA, *Securing Our Legacy,* p. 18.
39. Robert E. Taylor and Art Pine, "Science Academy Urges Major Revision in U.S. Standards on Pesticides in Food," *Wall Street Journal,* May 21, 1987, p. 28.
40. "Tough Pesticide Rules Planned," *Times-Picayune,* April 27, 1994, p. A6. In early 1995, however, the EPA agreed to ban dozens of widely used pesticides and comply with the Delaney Clause as part of a settlement of a lawsuit charging that the banned chemicals hurt farmworkers and consumers and that the government had broken the law by not pulling them from the market. Environmentalists charged that the EPA had never enforced the Delaney Clause, and sued the agency along with the state of California and farmworker advocates. The settlement required the EPA to phase out a group of 34 chemicals in two years from being used on processed foods, and within five years for usage directly on crops. Another 87 chemicals that the EPA lists as known carcinogens must be analyzed within five years for their possible presence in processed foods. See Charles McCoy, "EPA Agrees to Ban Pesticides, Comply with Rule in Food Act," *Wall Street Journal,* February 9, 1995, p. B11.
41. Timothy Noah and Scott Kilman, "New Pesticide Bill Is Expected to Make Foods Safer by Dumping Outdated Laws," *Wall Street Journal,* July 26, 1996, p. A4; Lawrence L. Knutson, "Clinton Signs Pesticide Bill," *Times-Picayune,* August 4, 1996, p. A3.
42. EPA, *Meeting the Environmental Challenge,* p. 20. Pesticide use, however, took a sharp upturn in 1994 and 1995, totaling 1.23 billion pounds in 1994, and 1.25 billion pounds in 1995, a significant increase from 1997 through 1993, when pesticide use held at or below 1.1 billion pounds a year. See Timothy Noah, "U.S. Pesticide Use Took Sharp Upturn in 1994 and 1995," *Wall Street Journal,* May 28, 1996, p. A28.
43. Jonathan Tolman, "Poisonous Runoff from Farm Subsidies," *Wall Street Journal,* September 8, 1995, p. A10.
44. See David Weir and Constance Matthlesson, "Will the Circle Be Unbroken?" *Mother Jones,* June 1989, pp. 20–27.
45. EPA, *Meeting the Environmental Challenge,* p. 20.
46. Peter Weber, "Resistance to Pesticides Growing," *Vital Signs 1994,* Lester R. Brown et al. (New York: W. W. Norton, 1994), p. 92.
47. Gardner, "Preserving Agricultural Resources," p. 89.
48. Miller, *Living in the Environment,* p. 558.
49. Ibid., p. 559.
50. Ibid., p. 560. See also Amal Kumar Naj, "Can Biotechnology Control Farm Pests?" *Wall Street Journal,* May 11, 1989, p. B1.
51. Miller, *Living in the Environment,* p. 561.
52. Ibid., p. 562. Richard Gibson and Rose Gutfeld, "Irradiated Food Shows New Signs of Life," *Wall Street Journal,* March 1, 1993, p. B1.
53. Ibid.
54. Ibid., p. 563.
55. Ibid. Weber, "A Place for Pesticides?" pp. 24–25.
56. J. Madeleine Nash, "It's Ugly, but It Works," *Time,* May 21, 1990, p. 30.
57. Jeanne McDermott, "Some Heartland Farmers Just Say No to Chemicals," *Smithsonian,* 21, no. 1 (April 1990), 127.

58. Sue Shellenbarger, "Back to the Future: A Movement to Farm Without Chemicals Makes Surprising Gains," *Wall Street Journal,* May 11, 1989, p. A1.
59. Weber, "A Place for Pesticides?" pp. 24–25.
60. Sandra Postel, "Controlling Toxic Chemicals," *State of the World 1988* (New York: Worldwatch Institute, 1988), p. 124.
61. Ann Misch, "Assessing Environmental Health Risks," *State of the World 1994,* Lester R. Brown et al. (New York: W. W. Norton, 1994), pp. 119–120.
62. Barbara Rosewicz, "U.S. Urged to Test Blood to Measure Toxic Chemicals," *Wall Street Journal,* May 7, 1991, p. B4.
63. Ibid., p. 121–122.
64. Misch, "Assessing Environmental Health Risks," p. 120.
65. Ibid., p. 120.
66. Kathy Boccella, "Toxic Dangers Lurking Underfoot Inside Home," *Times-Picayune,* April 16, 1995, p. A19.
67. Misch, "Assessing Environmental Health Risks," pp. 120–132; Amanda Spake, "Is the Modern World Giving Us Cancer?" *Health,* October 1995, pp. 53–56; Michael D. Lemonick, "Not So Fertile Ground," *Time,* September 19, 1994, pp. 68–70.
68. See Daniel Glick, "The Alarming Language of Pollution," *National Wildlife,* April–May 1995, pp. 38–45.
69. Peter Steinhart, "Innocent Victims of a Toxic World," *National Wildlife,* 28, no. 2 (February–March 1990), p. 21.
70. Ibid., p. 23.
71. EPA, *Better Health and Regulatory Reform,* p. 21.
72. United States Environmental Protection Agency, Office of Toxic Substances, *Reporting for the Chemical Substance Inventory* (Washington, D.C.: EPA, 1977), p. 1.
73. EPA, *Environmental Progress and Challenges,* p. 123.
74. Appendix I to Premanufacture Notification Draft Guideline, *The Chemical Reporter* (Washington D.C.: Bureau of National Affairs, 1978), p. 1124.
75. EPA, *Environmental Progress and Challenges,* p. 126.
76. Environmental Protection Agency, *The Toxic Substances Control Act,* 1976, p. 2.
77. EPA, *Environmental Progress and Challenges,* p. 123.
78. Ibid., p. 124.
79. Ibid.
80. EPA, *Meeting the Environmental Challenge,* p. 18.
81. EPA, *Environmental Progress and Challenges,* p. 116.
82. Ibid., p. 120.
83. Paul N. Cheremisinoff, "Focus on High Hazard Pollutants," *Pollution Engineering,* February 1990, p. 75.
84. EPA, *Environmental Progress and Challenges,* p. 120. See also Amal Kumar Naj, "Battle Against Toxic PCBs Gains Ground as Bacteria Are Found That Eat Them," *Wall Street Journal,* November 9, 1988, p. B4; Frank Edward Allen, "EPA Stumbles onto Promising Method for Destroying Large Amounts of PCBs," *Wall Street Journal,* March 21, 1991, p. A3.
85. Barbara Rosewicz, "Appeals Panel Rejects EPA Ban on Asbestos," *Wall Street Journal,* October 22, 1991, p. A3.
86. Mitchell Pacelle, "Asbestos Report Finds Cleanups Unwarranted," *Wall Street Journal,* September 25, 1991, p. B1; David Stipp, "Removing Asbestos Doesn't Guarantee Substance Is Gone," *Wall Street Journal,* March 22, 1993, p. A6.
87. Eric Felten, "The Times Beach Fiasco," *Insight,* August 12, 1991, pp. 12–19.
88. "EPA Will Take Another Look at Dioxin," *Times-Picayune,* April 11, 1991, p. A4; Keith Schneider, "Scientists Backtracking on Dioxin," *Times-Picayune,* August 18, 1991, p. A22.

89. Jeff Bailey, "How Two Industries Created a Fresh Spin on the Dioxin Debate," *Wall Street Journal,* February 20, 1992, p. A1.
90. Albert R. Karr, "New Report by EPA Strengthens Link Between Dioxin and Cancer, Other Ills," *Wall Street Journal,* September 13, 1994, p. B8; Gary Lee, "EPA Study Links Dioxin to Cancer," *Times-Picayune,* September 12, 1994, p. A4. See also Sharon Begley, "Don't Drink the Dioxin," *Newsweek,* September 19, 1994, p. 57.
91. Kathryn E. Kelly, "Cleaning Up EPA's Dioxin Mess," *Wall Street Journal,* June 29, 1995, p. A16.
92. Ann Blackman and Janice M. Horowitz, "Controlling a Childhood Menace," *Time,* February 25, 1991, pp. 68–69.
93. "Anti-Lead Campaign Draws Fire Inside and Outside the Industry," *Wall Street Journal,* November 25, 1992, p. B6.
94. Blackman and Horowitz, "Controlling a Childhood Menace," p. 69.
95. Timothy Noah, "Proposal Seeks New Warnings of Lead in Paint," *Wall Street Journal,* October 27, 1994, p. B6.
96. Amity Shlaes, "Hour of Lead," *Wall Street Journal,* December 20, 1995, p. A14.
97. Misch, "Assessing Environmental Health Risks," pp. 132–133.
98. Ibid., pp. 133–134. Some progress has been made in developing procedures to assess cumulative damages. See Peter Hong, "Do Two Pollutants Make You Sicker Than One?" *Business Week,* September 28, 1992, pp. 77–78.
99. Misch, "Assessing Environmental Health Risks," pp. 134–35. See also David Stipp, "How Sand on a Beach Came to Be Defined as Human Carcinogen," *Wall Street Journal,* March 22, 1993, p. A1.
100. United States Environmental Protection Agency, Office of Pollution Prevention and Toxics, *Expanding Community Right-to-Know: Recent Changes in the Toxics Release Inventory* (Washington, D.C.: EPA, 1995), p. 2.
101. United States Environmental Protection Agency, *Chemicals in Your Community: A Guide to the Emergency Planning and Community Right-to-Know Act* (Washington, D.C.: U.S. Government Printing Office, 1988), p. 7.
102. EPA, *Environmental Progress and Challenges,* p. 124.
103. EPA, *Expanding Community Right-to-Know,* p. 4.
104. EPA, *Chemicals in Your Community,* p. 14.
105. Ibid., pp. 2–4.
106. Ron Chepesiuk, "The Nutrasweet Smell of Environmental Respectability," *Tomorrow,* April–May 1995, pp. 18–20.
107. EPA, *Securing Our Legacy,* p. 22.
108. EPA, *Meeting the Environmental Challenge,* p. 18.
109. EPA, *Environmental Progress and Challenges,* p. 125.
110. Paul M. Barrett, "Courts Lend Sympathetic Ear to Claims for Compensation Based on Cancer Fears," *Wall Street Journal,* December 14, 1988, p. B1.
111. Ibid.
112. Sonia L. Nazario, "Children Become Centerpiece of Efforts to Set Tighter Restrictions on Pollutants," *Wall Street Journal,* October 15, 1990, p. B1.
113. Postel, "Controlling Toxic Chemicals," p. 118.
114. Ibid., pp. 133–134.
115. Ibid., pp. 134–135.
116. Ibid., p. 136.
117. "Handle with Responsible Care," *Wall Street Journal,* April 11, 1990, p. A9.
118. Julia Flynn, "Novo Nordisk's Mean Green Machine," *Business Week,* November 14, 1994, pp. 72–78.
119. Daniel Mazmanian and David Morell, "The Elusive Pursuit of Toxics Management," *The Public Interest,* No. 90 (Winter 1988), pp. 92–93.

Suggested Reading

Billings, Charlene W. *Pesticides: Necessary Risk.* New York: Enslow Publishers, 1992.

Bogard, William. *The Bhopal Tragedy: Language, Logic, and Politics in the Production of a Hazard.* Boulder, CO: Westview Press, 1989.

Bosso, Christopher John. *Pesticides and Politics: The Life Cycle of a Public Issue.* Pittsburgh, PA: University of Pittsburgh Press, 1987.

Carson, Rachel. *Silent Spring.* Greenwich, CT: Fawcett, 1962.

Crone, Hugh D. *Chemicals and Society.* Cambridge, MA: Cambridge University Press, 1986.

Dinham, Barbara. *Pesticide Hazard: A Global Health and Environmental Audit.* New York: Humanities, 1993.

Dover, Michael J. *A Better Mousetrap: Improving Pest Management for Agriculture.* Washington, D.C.: World Resources Institute, 1985.

Edelstein, Michael R. *Contaminated Communities: The Social and Psychological Impacts of Residential Toxic Exposure.* Boulder, CO: Westview Press, 1988.

Francis, Bettina M. *Toxic Substances in the Environment.* New York: Wiley, 1994.

Guerrero, Peter F. *Pesticides: Thirty Years Since Silent Spring.* New York: Diane Publishers, 1992.

Gustafson, David I. *Pesticides in Drinking Water.* New York: Van Nos Reinhold, 1993.

Harte, John, et al. *Toxics A to Z: A Guide to Everyday Pollution Hazards.* Berkeley: University of California Press, 1991.

Horn, D. J. *Ecological Approach to Pest Management.* New York: Guilford Press, 1988.

Kurzman, Dan. *A Killing Wind: Inside Union Carbide and the Bhopal Catastrophe.* New York: McGraw-Hill, 1987.

Marco, G. J., et al. *Silent Spring Revisited.* Washington, D.C.: American Chemical Society, 1987.

Patrick, David R., ed. *Toxic Air Pollution Handbook.* New York: Van Nos Reinhold, 1994.

Pimental, David, and Hugh Lehman, eds. *Pesticide Question: Environment, Economics, and Ethics.* New York: Chapman and Hall, 1992.

Postel, Sandra. *Defusing the Toxics Threat: Controlling Pesticides and Industrial Waste.* Washington, D.C.: Worldwatch Institute, 1987.

Scott, Ronald M. *Chemical Hazards in the Workplace.* New York: Lewis Publishers, 1989.

Sheehan, Helen E., and Richard P. Weeden. *Toxic Circles: Environmental Hazards from the Workplace into the Community.* Camden, NJ: Rutgers University Press, 1993.

Toxics in the Community. New York: Gordon Press, 1991.

United States Environmental Protection Agency. *Chemicals in Your Community: A Guide to the Emergency Planning and Community Right-to-Know Act.* Washington, D.C.: U.S. Government Printing Office, 1988).

Van den Bosch, Robert. *The Pesticide Conspiracy.* Berkeley: University of California Press, 1989.

Vighi, Marco, and Cozo Funari, eds. *Pesticide Risk in Groundwater.* New York: Lewis Publishers, 1995.

Yount, Lisa. *Pesticides.* New York: Lucent Books, 1995.

CHAPTER 9 Waste Disposal

Every American household generates more than a ton of rubbish per year on average. The amount of solid waste generated annually per person was 1,460 pounds in 1988 and is expected to rise to 1,744 pounds per person by the year 2010. Added to this solid waste produced by households is 60 million tons from commercial activities and another 90 million tons from industry, making a total of around 400 million tons of solid waste produced each year in American society. On top of this figure, there is a further 250 million tons of hazardous waste generated each year, which requires special treatment.[1]

Until the mid-1970s, this rubbish was largely handled by small firms or local government. Most of it was trucked to out-of-town sites and dumped. The dangerous waste was simply buried. Nobody really cared what happened to all this waste material as long as it was out of sight. However, times have changed. Disposal of these wastes in the modern world, especially those considered to be hazardous, is a costly and time-consuming business, requiring complex measures to control. Uncontrolled waste presents environmental and health risks that necessitate action to prevent degradation of water, soil, and air, and to protect human health.

Initial concerns regarding waste disposal focused on the fire hazards solid wastes posed and on the particulates generated by open burning. In the mid-1960s, for example, legislation was enacted to restrict open burning of garbage. In the latter half of the 1970s, attention focused on the problems negligent hazardous waste disposal was causing for the environment and human health because of leaching, contamination, corrosion, and poisoning of land, water, vegetation, and animals as well as human beings by toxic chemicals and heavy metals. Investigations disclosed that between 1950 and 1979, over 1.5 trillion pounds of hazardous wastes had been dumped in about 3,300 sites around the country. Fifty-three chemical companies in 1978 alone had dumped 132 billion pounds of industrial waste. Incidents such as Love Canal heightened public apprehension about the hazardous waste problem.[2]

As a result of concern about hazardous waste, which eventually came to pose a threat to human health and wildlife, federal and state governments have passed strict environmental standards related to disposal of hazardous waste. The solid waste problem has become the subject of increasing attention as more and more waste was generated and traditional disposal methods became unacceptable. These developments led to the formation of waste disposal companies who have the ability to meet new environmental standards requiring safer and costlier methods of disposal.[3] These companies generally use four methods to dispose of solid and hazardous waste safely.

1. ***Landfills.*** These burial sites for waste are high-tech operations lined with impermeable materials and constantly monitored. Trenches are built into the base of the landfills to collect noxious fluids that could leak out and contaminate drinking water. These are then pumped to the surface where they are made inert. Landfills that do not meet such stringent standards are being closed all over the country.
2. ***Incineration.*** There are over 100 incinerators in the United States today that burn waste material and in some cases generate energy from this burning. Hazardous waste is burned at very high temperatures in special incinerators. These incinerators must have technology to capture noxious fumes that are generated in the process of burning. Some of these incinerators have been in operation for several years and may need to be modernized.
3. ***Recycling.*** Some hazardous waste is recycled, but the problems with respect to recycling hazardous waste are even greater than the problems involved with recycling solid waste. Many areas have created something called a waste exchange to facilitate the exchange of both solid and hazardous waste and promote recycling of this material.
4. ***Storage.*** Some hazardous waste is simply stored in covered facilities that are not in close proximity to populated areas. Proper protection must be provided for leaking drums that might threaten groundwater supplies, and the stored material must be constantly monitored for leakage and other potential problems.

The costs of disposing of waste material has increased dramatically over the past several years. In 1978, for example, it cost about $2.50 to have a ton of hazardous waste dropped into a safe hole in the ground. In 1987 the cost of disposing of hazardous waste in this fashion ranged from $200 a ton upward. Burning the waste cost $50 a ton in 1978, while in 1987 fees ran at over $200 a ton and $2,000 a ton for really nasty waste. Garbage companies a decade ago charged only $3 a ton to get rid of common or garden rubbish. More recently, Long Island City, which ships out nearly all of its household rubbish, paid $130 a ton to have the stuff removed.[4]

Even though operating margins of the garbage companies have increased, making them quite profitable, the industry has problems with regulators and the public. The laws regulating disposal are inconsistent, laxly monitored, and seldom enforced, which creates distortions in the market. Definitions of hazardous waste differ and national laws on air, water, and soil pollution are seldom mutually coherent. Companies find it difficult to get permits for landfill sites and incinerators because of the "not in our backyard" syndrome. Nations and communities alike object to the importation of waste from other countries and communities.[5]

SOLID WASTE

Nonhazardous solid waste can be defined as any unwanted or discarded material that is not in liquid or gaseous form. These wastes include municipal garbage and industrial refuse as well as sewage, agricultural refuse, demolition wastes, and mining residues. About 98.5 percent of this nonhazardous solid waste comes from mining, oil, and natural gas production, and industrial activities. Mining waste is often left piled near mine sites where it can pollute the air, surface water, and groundwater. This waste material consists of overburden, which is the soil and rock cleared away before mining, and the tailings discarded during ore processing. Industrial waste consists of scrap metal, plastics, paper, fly ash from electrical power plants, and sludge from industrial waste treatment plants. Much of this waste is disposed of at the plant site where it is produced.[6]

Waste from homes and businesses in or near urban areas makes up the remaining 1.5 percent of nonhazardous solid waste produced in the country. Almost 160 million tons of this municipal solid waste are discarded in the United States every year, or about 400,000 tons every day. This is enough waste to fill the Astrodome in Houston more than twice daily for a year. This waste stream has increased 80 percent since 1960 and is expected to increase another 20 percent in the next 10 years. The average amount of municipal solid waste generated per person in the United States is about two to five times that in most other developed nations.[7]

Americans throw out almost four pounds of trash a day consisting mainly of paper products and yard wastes, which make up about 59 percent of all municipal solid waste[8] (Figure 9.1). Most of the rest consists of glass, plastic, aluminum, iron, steel, tin, and other nonrenewable mineral resources. Only about 10 percent of these potentially reusable resources are recycled in this country. The rest is hauled away and dumped or burned at a cost of almost $5 billion a year. Each year these wasted resources have enough aluminum in them to rebuild the country's entire commercial air fleet every three months, enough iron and steel to supply the nation's automobile companies, and enough wood and paper to heat 5 million homes for 200 years.[9]

Disposal Options

Disposal of this amount of solid waste poses problems for our society that will only grow in magnitude. There is a problem with space as urban land grows more expensive, making it too costly to open new landfills. A second problem concerns the leachate, which is the distillation of chemicals in the waste material that percolates into the soil

FIGURE 9.1 Composition of Municipal Waste

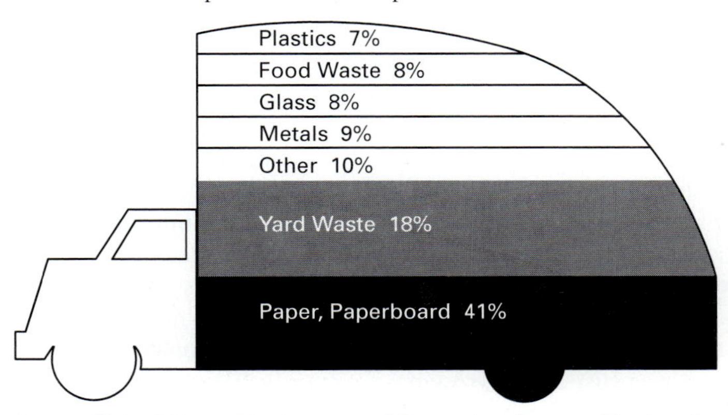

Source: United States Environmental Protection Agency, *Meeting the Environmental Challenge* (Washington, D.C.: U.S. Government Printing Office, 1990), p. 14.

and groundwater. Third is the problem of organic decomposition, which is unsightly and smelly, and a producer of methane gas as air and sunlight decompose some of the material. There is an ethical problem related to the problems we are creating for future generations who will have to deal with the trash we have disposed of in many cases, and who will have fewer resources available because we have wasted them. And finally there is a political problem, consisting of the negative response most people have toward the building of more landfills and incinerators. A closer look at each of the disposal methods will more clearly indicate the nature of the problem.

Landfills About 75 percent of this waste material is disposed of in landfills. Before the Resource Conservation and Recovery Act (RCRA) was passed in 1976, much of the solid waste in this country was discarded into open dumps or burned in crude incinerators. These open dumps created health hazards, polluted surface waste and groundwater, and often caught fire and filled the air with pollutants. The RCRA banned all open dumps and required landfills to be upgraded to sanitary landfills. These sanitary landfills are disposal sites in which wastes are spread out in layers, compacted, and covered with a fresh layer of soil after each day's activity.[10]

Open burning of solid waste is not allowed in these landfills, odor is seldom a problem, and rodents and insects cannot thrive underneath the layers of garbage. These landfills are also to be constructed so they reduce water pollution from runoff and leaching. Semi-impermeable clay under the landfill and a collection pipe, as well as a subterranean trench around the site, trap most of the leachate. Pumps then send it back to the top of the landfill where some of the chemicals are filtered out of the leachate. Instruments constantly monitor the leachate to make sure it is not leaking into groundwater. This kind of landfill can be put into operation quickly, generally has a low operating cost, and can handle massive amounts of solid waste. After it has been filled and allowed to settle for a few years, the land can often be used for some constructive purpose such as a park, a golf course, an athletic field, a wildlife area, or for some other similar purpose.[11]

Many people assume that the trash buried in a landfill eventually decomposes, but this does not happen, as most of the garbage in a landfill is virtually mummified. The layers of dirt covering the garbage and the landfill's bulk tend to keep air and rain from reaching the garbage, and as a result, oxygen-loving bacteria play only a limited role in transforming ordinary household garbage into decomposed earth. Other bacteria do digest biodegradable material, but this process takes a long time and may stop altogether under extremely dry conditions. The average landfill is often like a pyramid where the trash buried there lies embalmed.[12]

Research into actual landfills shows that some food debris and yard waste does degrade but at a very slow rate—only 25 percent in the first 15 years with little or no additional change for at least another 40 years. Most of the remaining trash seems to retain its original weight, volume, and form for at least four decades. Newspapers were found to be the largest single commodity in landfills, taking up as much as 16 percent of the volume of waste dumped at the average site. Most of these newspapers, no matter their age, come out in remarkably readable shape. Plastics constituted less a percentage of garbage than was suspected, growing from only 10 percent in the 1960s to 13 percent in later years. It was surmised that plastic products are made thinner and lighter as more of them are produced, allowing them to be squashed flat and thus take up less space.[13]

The problem with this method of waste disposal is that many of these landfills are close to overflowing—almost 75 percent of such landfills are expected to reach capacity in 10 years. In more than 40 states, rising land costs, technical risks, new EPA regulations, leachate problems, increasing space scarcity, and an escalating number of NIMBY (not in my back yard) movements are forcing the closing of landfills. Many of these landfills are small, however, and the lost space is offset to some extent by the creation of huge new landfills that can handle enormous amounts of garbage. For the past 20 years, the big garbage companies who manage these landfills have been gaining over half the market.[14]

Nonetheless, many communities where old landfills have reached capacity are having trouble siting new landfills. Some cities without enough landfill space are shipping their trash elsewhere. This problem is especially true in states like New Jersey, where only 11 landfills remained in 1991, down from 300 in the 1970s, and in New York, where landfill costs dwarf those in other areas. About 15 million tons of trash were shipped between states in 1989, with more than half of this coming from New York and New Jersey alone.[15]

In March 1987, the infamous garbage barge left Long Island City in New York loaded with 3,100 tons of refuse and illustrated the problem. The owners of the barge had initially hoped to dispose of its load in a landfill in Moorehead City, North Carolina, but because of public outrage, the barge continued on a two-month journey along the Atlantic Coast and into the Gulf of Mexico, only to be turned away at every port at which it landed. Eventually the barge had to return home to where it started, and its load of garbage was finally going to be incinerated. The incident received national news coverage and called the nation's attention to the solid waste disposal problem.[16]

States with large areas for landfills did not want to become the nation's garbage dump and started to restrict garbage imports from other states. Finally, the Supreme Court ruled in 1992 that such barriers violate constitutional protection of interstate commerce unless a state can show that out-of-state waste poses a unique danger.[17] In 1994, the Supreme Court struck down local laws that prevented haulers from shipping waste to other states where space is cheaper. Such flow-control laws were becoming common as municipal governments tried to ensure a steady stream of garbage for treatment facilities in which they had a financial interest.[18] Thus competition for garbage disposal has been encouraged.

Landfill operators are learning new strategies to locate dump sites. New technologies enable them to cut the dust, noise, and risk of groundwater contamination from landfills, and they are bringing trash and people together in pricy home developments. The notion that a garbage dump might make a welcome neighbor for even upscale homeowners is said to be taking hold nationwide. Landfill operators are offering financial incentives to potential homeowners, trading millions of dollars in home value guarantees for the right to build new dumps, expand old ones, or keep existing ones running. These efforts are helping to overcome the NIMBY syndrome.[19]

Incineration Alternative disposal methods include incineration, which can reduce garbage weight as much as 75 percent and reduce the volume by as much as 90 percent. Waste-to-energy incinerators also produce heat that can be sold to generate electricity, which helps to reduce the cost of incineration. These incinerators do not pollute groundwater and add little to the air pollution problem if equipped with effective pollution control devices. But many incinerators in the United States that were built in the 1960s and 1970s were equipped with ineffective pollution controls, which were poorly maintained. Controls proposed by the EPA in 1988 required all new incinerators to use the best available technology to remove 95 percent or more of the air pollutants generated in the incineration of solid wastes.[20]

Incineration, however, still leaves about 25 percent of the original waste material in the form of toxic ash residues. This ash, which is usually disposed of in ordinary landfills, can be contaminated with hazardous substances such as lead, cadmium, mercury, dioxins, and other toxic metals. When disposed of in ordinary landfills, such waste can pose problems for human health and the environment.[21] In 1994, the Supreme Court ruled that cities must follow federal hazardous waste rules when disposing of ash produced by municipal incinerators. It was expected that this decision would make municipal incinerators much less competitive with landfills in many parts of the country.[22]

Incineration competes with the long-term goal of reducing waste, as this method involves the commitment of a certain amount of garbage to be profitable, providing incentives to produce more rather than less garbage. The costs to the community of building an incinerator can be high as property values will go down if the facility is located near residential areas. The costs of building the facility itself can also be expensive, and then it will have to be mothballed after 20 to 30 years of useful life. All of these technological uncertainties discourage investment in new methods that might result in better incineration technology.

In 1991, about 16 percent of the nation's volume of trash was burned as companies and municipalities invested heavily in trash incineration. By the end of the century, incinerators are projected to be burning about 25 percent of the solid waste generated in the country. But many of these facilities may be halted by the NIMBY movement that makes it hard to find new sites for incinerators. People are concerned about the air pollution generated by incinerators and the ash disposal problem. Conservationsts also oppose dependence on incinerators because they encourage continuation of a throwaway approach that wastes resources. The use of incineration encourages people to continue tossing away paper, plastics, and other burnable materials rather than looking for ways to recycle and reuse these resources and to reduce waste production.[23]

Composting Since yard waste constitutes a significant percentage of municipal solid waste, these wastes can be decomposed in backyard compost bins and used in gardens and flower beds. Currently, only about 1 percent of the yard waste generated in the United States is composted, but this method of disposing of yard wastes has the potential of reducing the solid waste stream by almost 18 percent. Degradable solid waste from slaughterhouses, food-processing plants, and kitchens can also be composted, as can organic waste produced by animal feedlots, municipal sewage treatment plants, and various industries. Much of this material can be collected and degraded in large composting plants, and sold at a profit as a soil conditioner and fertilizer.[24]

One company believes that composting could handle up to 60 percent of the country's municipal solid waste, and is spending $20 million on research to develop disposable diapers that break down in composting systems. Disposable diapers pose a particular problem as they constitute a significant percentage of solid waste, and by holding out composting as a solution, the company hopes to encourage the construction of new garbage-handling systems. The commercial process advocated relies on the same biological degradation of material as in the backyard process. Such plants can turn garbage into compost under controlled temperatures in 3 to 14 days. After this process, the compost must cure for 30 to 180 days before it can be used for landscaping or farming.[25]

Recycling Another option is recycling, which is being pursued by an increasing number of communities. As of 1993, 40 states had comprehensive recycling programs, and the number of communities that had recycling programs jumped from 50 to over 4,000 in a three-year period. Curbside recycling programs now cover one third of all U.S. households, meaning that tens of millions of Americans now make a daily ritual of sorting their garbage for collection.[26] A study by the EPA showed that more than one fifth of all garbage in the country is transformed into new products, up from 17 percent in 1990, and that the amount of trash going to landfills and incinerators was being reduced because of these efforts.[27]

Resources can be recycled using high- or low-technology approaches. In the former approach, machines shred and separate urban waste to recover glass, iron, aluminum, and other useful materials. These materials are then sold to manufacturing industries as raw materials for the production process. The remaining trash, which consists of paper, plastics, and other combustible wastes, are either recycled themselves or incinerated. The heat produced by this incineration can be used to produce steam or electricity to run the recovery plant and for sale to industries and residential developments. Very few of these resource recovery plants are in existence today.[28]

The low-technology approach involves homes and businesses separating various kinds of waste and disposing of it in separate containers. Paper, glass, metals, and plastics are usually separated in this manner, and picked up by compartmentalized city collection trucks, private haulers, or volunteer recycling organizations and then sold to scrap dealers, compost plants, and manufacturers. This method is used to recycle most solid waste in this and other countries. Studies have shown that it takes the average American family only 16 minutes a week to separate trash in this manner.[29]

This low-technology approach to recycling produces little air and water pollution and has low start-up costs and relatively moderate operating costs. This approach also saves energy and provides jobs for unskilled workers. Recycling creates three to six times more jobs per unit of material than landfilling or incineration. Collecting and selling cans, paper, and other materials for recycling is an important source of income for many people all over the world, especially the homeless and the disadvantaged. In some developing nations, small armies of poor go through urban garbage disposal sites and remove paper, metals, and other items and sell them to factories to earn some money for themselves.[30]

Using scrap iron instead of iron ore to produce steel requires 65 percent less energy and 40 percent less water, and produces 85 percent less air pollution and 76 percent less water pollution. Recycling aluminum produces 95 percent less air pollution and 97 percent less water pollution, and, perhaps most importantly, requires 95 percent less energy than mining and processing bauxite. Increasing the recycling rate of paper is a key to preventing further clearing and degradation of forests and reducing the unnecessary waste of timber resources. Thus recycling has tremendous advantages when it comes to resource and energy conservation.[31]

As recycling programs increased across the country, the demand for recycled material did not keep pace, and recycled material started to collect as companies had trouble making money from recycling.[32] However, demand for recycled materials took off in 1994, and the piles of recycled materials that had been collecting began to disappear. Prices for paper, cardboard, and some plastics reached record highs, and the recycling movement was being transformed into something of a smoothly running commodity industry. Rather than having to pay for the programs because collection costs exceeded the value of recycled products, many cities began to break even or make money on the programs when all costs were taken into account. Large companies began to get more and more involved in processing recycled material, bringing new economies of scale and more cost-effective collection and processing technologies to recycling.[33]

Nearly 39 percent of all paper products consumed in the United States were recycled in 1992, up from 22 percent in 1970. The number of newsprint recycling plants in the United States grew from 9 to 29 from 1988 to 1992. Recycled paper is made into new paper, cereal boxes, toilet tissue, and bedding for farm animals. The recycling rate for aluminum cans was 64 percent in 1991 because of the huge saving involved in using recycled aluminum. But steel's recycling rate had increased to 25 percent and efforts are being made to increase this percentage. And glass, whose recycling rate was 20 to 30 percent, hoped to increase its market share because glass bottles, like aluminum cans, are completely recyclable.[34]

One of the most troublesome problems with recycling is the increasing use of plastic packaging, which is difficult to burn or recycle, and, because it is not biodegradable, will clog landfills for centuries. The problem with recycling plastics is that plastic isn't one thing, because there are literally thousands of polymers around, each with different properties. Six of these polymers account for 97 percent of postconsumer waste, but from the point of view of pure recycling, this is about five too many. When mixed, plastic waste results in a drab polymer pudding that can't be used for anything with demanding specifications.[35]

An effort to deal with this problem is the development of a material identification system to assist in the separation of plastic bottles and create a higher value for recycled material. Bottles are coded by the most widely used resins (Exhibit 9.1) for identification purposes. Engineers have developed machines that can read these markers as the containers whiz past on a conveyor belt and separate them automatically. They have also developed a process that can separate shredded mixed plastics by flotation so flakes of individual resin types can be skimmed off and collected separately.[36]

The amount of recycled plastic rose to only 3.5 percent in 1993 from 2.2 percent in 1990, a rather disappointing performance. However, in 1995, the demand for PET containers, which make up 95 percent of all plastic bottles, was growing by 21 percent a year, largely because of rising cotton prices, which drove foreign manufacturers to cheaper cotton substitutes. The demand for synthetic fibers, especially polyester, increased demand for used PET bottles, which can be recycled into polyester fibers to use in making clothes. Prices more than tripled in a year as 565 pounds of PET plastic containers were recycled when there was a demand for 800 million pounds.[37]

Some companies have made a switch from plastic containers to other paper-based packaging because of the recycling problem. In 1990, for example, McDonald's announced it was phasing out its foam sandwich containers, which accounted for about 75 percent of the foam packaging used by the company. Environmentalists had long

EXHIBIT 9.1 Plastic Container Code System for Plastic Bottles

PET	Polyethylene Terephthalate: Used for soda bottles because it holds carbonation. Recycled into new soda bottles, carpets, and synthetic textiles.
HDPE	High-Density Polyethylene: Used for milk, water, and detergents. Makes detergent bottles, recycling bins, and irrigation pipes.
PVC	Polyvinyl Chloride: Appears in some cooking oil bottles and in film for meat packaging. Can be recycled into pipe or fencing.
LDPE	Low-Density Polyethylene: Makes shopping bags and other film products and flexible margarine tubs. Can be used for new bags and film.
PP	Polypropylene: Found in yogurt cups and ketchup bottles. Can be recycled into auto parts, carpets, and textiles.
PS	Polystyrene: The clear version is used in salad takeout trays; the foamed version is used in food containers. Can be remade into office products, cafeteria trays, and videocassette cases.
No label	Includes all other resins and layered multimaterial. Includes products with a mixture of the other six resins. Rarely recycled.

complained that the polystyrene-based containers were difficult to recycle and did not degrade in a landfill, contributing to the waste disposal problem. The company had been engaged in an expensive campaign to show that foam containers could be recycled, sponsored by the chemical industry that manufactured polystyrene, but eventually gave in to the critics of the packaging. The paper wrap it now uses takes up less volume and weighs about a third of the old polystyrene container.[38]

The chemical industry is concerned to find ways to recycle plastics before the movement to ban or switch products as McDonald's did grows any larger. But the task isn't easy as researchers are trying a number of different approaches. Almost all the major chemical companies have plunged into recycling ventures. And one company claimed to have developed the first biodegradable plastic by using starch that can be derived from potatoes, corn, rice, or wheat. The material could be used, so it was claimed, to replace some of the nonbiodegradable plastic derived from petrochemicals.[39]

Attitudes of Americans will have to change to make recycling a more comprehensive approach to the waste disposal problem. The average American has been conditioned by advertising and example to a lifestyle that involves a throwaway attitude toward waste. Waste is something bad to be discarded and not seen as a resource. The out-of-sight, out-of-mind approach to waste is prevalent in this country. We emphasize making, using, and replacing more and more items to increase economic wealth regardless of the environmental costs to society. Because environmental costs are not reflected in market prices, consumers have little incentive to recycle and conserve renewable resources.[40]

The city of Seattle has instituted a successful recycling program that is fundamentally a program using market incentives to change consumers' behavior. Residents of the city are faced with a "pay-as-you-throw" system of garbage disposal. Homeowners pay $10.70 a month for a 19-gallon mini-can but pay $31.75 for three full-size, 32-gallon cans. Curbside recycling of plastic beverage bottles, glass, cans, newspapers, and other waste paper is free to homeowners. Four out of five households in the city recycle their trash, and 90 percent put out one can or less of garbage a week. While volunteer efforts have received a great deal of attention, it is clearly market incentives in the form of high incremental costs for additional garbage cans that have changed the behavior of Seattle residents.[41]

Other efforts have been made to increase the demand for recycled products. In 1993, the Clinton administration issued an executive order requiring that paper purchased by federal agencies meet strict recycling standards.[42] The state of Florida urged companies to set up joint ventures to purchase garbage from the state and in return, the state promised to buy back recycled goods from those same ventures. Other states developed programs to offer low-interest loans and grants to help businesses buy recycling machines or develop recycled products. Others offered tax credits for investing in machines that helped recycling efforts. Still others were willing to pay more for recycled goods or specified that a certain percentage of their purchases must involve recycled goods.[43]

Private business organizations also got in the act to promote recycling. Big paper buyers banded into groups such as the Recycled Paper Coalition to promote buying of recycled paper. The Environmental Defense Fund forged an alliance to harness the $1 billion annual paper-purchasing power of six companies including McDonald's to demand more recycled fibers in paper.[44] In 1993, the Buy Recycled Business Alliance had grown to nearly 500 large and small companies from just 25 the year before, and spent $10.5 billion on recycled products compared with about $2.7 billion the previous year. A major goal of the alliance has been to require member suppliers to provide them with higher volumes of recycled raw materials and greater recycled content in finished goods.[45]

Reusable Products The reuse of products involves using the same product over and over in its original form. Examples of reusable products are glass beverage bottles that can be collected, washed, and refilled by bottling companies. Up until 1975, most beverage containers in the United States were refillable glass containers. Today they make up only 15 percent of the market, while nonrefillable aluminum and plastic containers have taken over. Another product is cloth diapers that can be washed and reused to replace disposable diapers that cause such a serious landfill problem. Carrying lunches in lunch boxes that can be reused instead of in paper bags or baggies that are discarded is another example.[46]

Since one of the largest components of municipal solid waste is packaging, reusable packaging holds great promise to reduce the garbage disposal problem. Although containers and packaging account for 30 percent of the total waste stream by weight, they account for a larger percentage of the volume, as packaging is generally bulky. To make matters worse, a majority of packaging contains foil or plastics, making the material difficult, if not impossible, to recycle. The small juice box that has become so popular because of its size and convenience is made up of so many different materials it is not cost-efficient to try and separate the materials so they can be recycled. The package was banned in several states.[47]

Reuse of products extends resources and reduces energy use and pollution even more than recycling. Refillable glass bottles are the most energy-efficient beverage container on the market, as three times more energy is needed to crush and remelt a glass bottle to make a new one than is required to clean and refill a bottle. If reusable glass bottles replaced the 82 billion throwaway beverage cans produced in the United States every year, it is estimated that enough energy would be saved to supply the annual elec-

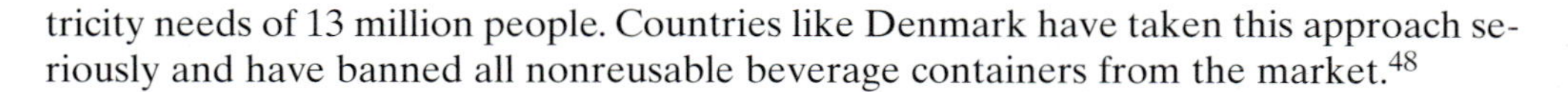

tricity needs of 13 million people. Countries like Denmark have taken this approach seriously and have banned all nonreusable beverage containers from the market.[48]

Source Reduction Perhaps the best method of dealing with the solid waste disposal problem is not to produce so much waste in the first place. Reducing unnecessary waste of nonrenewable mineral resources can extend supplies of these resources even more dramatically than recycling or reuse. This method generally saves more energy than recycling and reduces the environmental impact of disposing of waste material. Manufacturers can conserve resources by using less material per product. They can also make products that last longer, and abandon the planned obsolescence approach that may contribute to short-term profits but has long-term implications for resource availability. Products can also be designed so that they are easy and inexpensive to repair, so that their useful life may be extended.[49]

Several efforts have been made to cut the amount of waste produced. McDonald's, working with the Environmental Defense Fund, launched a major waste reduction effort in 1991 that affected its suppliers, workers, and even customers. The company announced a series of 42 initiatives that would cut the huge waste stream at its 11,000 restaurants by more than 80 percent in just a few years.[50] In 1992, the music business announced that it will eliminate the so-called long box in which CDs were sold that created an estimated 23 million pounds of garbage in 1990 alone.[51] And a coalition of northeastern governors issued a challenge to the nation's 200 largest makers of consumer packaging urging them to reduce the volumes of solid waste their products generate.[52]

Waste Exchange The EPA is in the process of building a nationwide database to facilitate waste exchange, and encourage companies to think of waste as a resource. While the idea is not necessarily new, as there are already some 18 regional nonprofit waste exchanges in the United States and Canada, the EPA's national focus may help to spread the use of exchange. As one example of how exchange can be useful, an upstate New York company used to spend $575 on hauling and "tipping fees" each time it dumped a truckload of scrap wood in a landfill hundreds of miles from its plant. Now for less than $200 it takes each load of scrap wood to a company in Pennsylvania that grinds it into an assortment of scented chips used as an air freshener. The EPA hopes to facilitate such exchanges with its computerized system that will be updated daily to end delays in getting rid of waste material.[53]

In 1990, companies listed about 8 million tons of materials with exchanges, and between 10 and 30 percent of that listed material eventually changed hands. That amount is miniscule against the total amount of waste produced in the country, but it is a promising beginning. The Pacific Materials Exchange based in Spokane, Washington, listed about 400 available waste items in 1992, including 60,000 pounds of pickle waste, 7 tons of rubber tires, and 55 gallons of vanilla. The Exchange was awarded a grant by the EPA to develop a computer network for the 23 waste exchanges that existed in North America at the time.[54]

Integrated Waste Management

The basic problem with waste disposal is cultural, in that societies generate and dispose of waste based on cultural values that are exemplified in lifestyles and methods of manufacture. Exhibit 9.2 shows three different approaches to the waste disposal problem that

EXHIBIT 9.2 Three Systems for Handling Discarded Materials

Item	*For a High-Waste Throwaway System*	*For a Moderate-Waste Resource Recovery and Recycling System*	*For a Low-Waste Sustainable Earth-System*
Glass bottles	Dump or bury	Grind and remelt; remanufacture; convert to building materials	Ban all nonreturnable bottles and reuse (not remelt and recycle) bottles
Bimetallic "tin" cans	Dump or bury	Sort, remelt	Limit or ban production; use returnable bottles
Aluminum cans	Dump or bury	Sort, remelt	Limit or ban production; use returnable bottles
Cars	Dump	Sort, remelt	Sort, remelt; tax cars lasting less than 15 years, weighing more than 818 kilograms (1,800 pounds) and getting less than 13 kilometers per liter (30 miles per gallon)
Metal objects	Dump or bury	Sort, remelt	Sort, remelt; tax items lasting less than 10 years
Tires	Dump, burn, or bury	Grind and revulcanize or use in road construction; incinerate to generate heat and electricity	Recap usable tires; tax or ban all tires not usable for at least 96,000 kilometers (60,000 miles)
Paper	Dump, burn, or bury	Incinerate to generate heat	Compost or recycle; tax all throwaway items; eliminate overpackaging
Plastics	Dump, burn, or bury	Incinerate to generate heat or electricity	Limit production; use returnable glass bottles instead of plastic containers; tax throwaway items and packaging
Yard wastes	Dump, burn, or bury	Incinerate to generate heat or electricity	Compost; return to soil as fertilizer; use as animal feed

Source: G. Tyler Miller, *Living in the Environment,* 6th ed. (Belmont, CA: Wadsworth, 1990), p. 370. © Wadsworth Publishing Co.

really involve three different cultures with regard to resource use and disposal of waste material. The sustainable approach would seem to make the most sense given the increasing costs of garbage disposal and the near exhaustion of some of our nonrenewable resources. The movement toward this kind of society, however, involves a change of basic values and notions of economic wealth that are not going to change overnight. But change of some sort seems inevitable given the dimensions of the waste disposal problem.

The ultimate disposal of solid waste involves a use of various methods that can be combined into an integrated management system that is designed to emphasize certain management practices that are consistent with each community's demography and waste stream characteristics. In such an integrated waste management system, each component is designed so that it complements, rather than competes, with the other components in the system. These components must be combined in a cost-effective manner that is beneficial to each community and meets the need for safe garbage disposal that will continue to be effective for future generations.[55]

For example, a small town in a rural area may depend completely on landfilling to dispose of its waste, while large cities may come to rely more heavily on recycling to reduce the burden on its landfills. Some cities utilize a complete integrated waste management system by stressing source reduction, having a mandatory recycling program, incinerating the remaining combustible trash, and landfilling the residual ash and any other items not suitable for incineration. Each community will need to tailor the system to meet its individual needs and take advantage of its resources.

The final report of the Municipal Solid Waste Task Force provided a valuable framework for apportioning municipal waste disposal responsibilities among federal, state, and local authorities. It emphasized voluntary efforts as opposed to increased federal regulation, and saw the EPA playing largely a role of information provider. The EPA could, according to this report, develop technical and educational guidance, data collection and research and development programs, and act as a national clearinghouse for this information. The task force envisioned a cooperative effort by all levels of government to track volumes and types of wastes, with the EPA taking the lead in conducting research and development in technical areas related to combustion, landfilling, recycling, and source reduction. Six national objectives were identified as part of a national agenda to solve the municipal solid waste problem.[56]

1. Increase the waste planning and management information (both technical and educational) available to states, local communities, waste handlers, citizens, and industry, and increase data collection for research and development.
2. Increase effective planning by waste handlers, local communities, and states.
3. Increase source reduction activities by the manufacturing industry, government, and citizens.
4. Increase recycling by government and by individual and corporate citizens.
5. Reduce risks from municipal solid waste combustion in order to protect human health and the environment.
6. Reduce risks from landfills in order to protect human health and the environment.[57]

To accomplish these objectives, the use of more market incentives has been encouraged. Perhaps the greatest "market failure" with respect to solid waste disposal is the distortion of market incentives that results from municipal systems that charge flat fees for garbage collection. Wastefulness is encouraged by systems that charge such a single fee regardless of the amount of trash put out by households. Consumers are not being made aware of the true disposal costs resulting from their buying and discarding habits. If consumers were charged per pound for the type of garbage disposed of, the proper signals would be sent so that their purchasing choices would be affected by disposal costs.[58]

HAZARDOUS WASTES

While initial concerns regarding waste disposal focused on the fire hazards posed by solid wastes, the disposal of hazardous wastes began to get more attention as dump sites were discovered all around the country that posed potential threats to groundwater supplies, as well as to wildlife and the environment in general.[59] There was little concern about this problem until 1977, when it was discovered that hazardous chemicals leaking from an abandoned waste dump in Niagara Falls, New York, were contaminating a suburban development. This development was called Love Canal, and the publicity surrounding this problem made the public at large and especially public and elected officials aware of the dangers that buried hazardous wastes could pose for human health and the environment. The incident made the headlines for several years, and the name of Love Canal became a household word that brought to mind untold dangers lurking beneath the surface in many parts of the country (see box).

Disposal of hazardous wastes is a costly and time-consuming business, requiring complex measures to control. Uncontrolled waste presents environmental and health risks that necessitate action to prevent degradation of water, soil, and air, and to protect human health. The total quantity of hazardous wastes produced in the country is difficult to estimate. The EPA estimated that 264 million tons are produced in this country each year, while the Office of Technology Assessment (OTA) had a much higher annual estimate of 400 million tons. About 95 percent of this waste was generated and either stored or treated on-site by large companies such as chemical manufacturers, petroleum companies, and other industrial facilities.[64]

Hazardous wastes have been defined as wastes that (1) cause or significantly contribute to an increase in mortality or in serious irreversible or incapacitating reversible illness; or (2) pose a substantial present or potential hazard to human health or the environment when improperly treated, stored, transported, or disposed of or otherwise managed.[65] They are wastes that cannot be managed by routine procedures, because if they are improperly managed, they can cause a threat to public health and the environment. Improper disposal of hazardous wastes has also been responsible for other kinds of environmental damage—fires, explosions, pollution of surface water and air—as well as posing serious threats to human health through poisoning via the food chain or through direct contact.[66]

Hazardous wastes include wastes that pose a fire hazard (ignitable), dissolve materials or are acidic (corrosive), are explosive (reactive), or otherwise pose dangers to human health and the environment (toxic) (Figure 9.2). The characteristics of ignitability include liquids with a flash point (the temperature at which a vapor easily ignites in air) of less than 140 degrees F, materials that burn so vigorously and persistently when ignited that they create a hazard, and ignitable compressed gases. Corrosive materials include aqueous wastes with a pH of less than or equal to 2.0 or greater than or equal to a pH of 12.5, and liquid wastes that corrode steel at a rate equal to or greater than 0.25 inch per year at a test temperature of 130 degrees F.[67]

Those materials that are reactive include materials that react violently with water, those that form potentially explosive mixtures when combined with water, and those that, when mixed with water, will generate toxic gases, fumes, or vapors in quantities sufficient to endanger human health or the environment, and materials that are capable of detonation or explosive reactions if subjected to a strong initiating source or if heated under confinement. Finally, toxic materials are those which have deleterious biological effects, such as poisoning, on human beings and other animal life.[68]

Hundreds of potentially dangerous substances can be found in hazardous waste, but as shown in Exhibit 9.3, the most common are few in number. While there are about 14,000 regulated producers of hazardous waste, by far the majority are chemical

Love Canal

The Love Canal incident provides a horror story related to the disposal of hazardous wastes. Love Canal, located in Niagara Falls, New York, was an uncompleted, abandoned nineteenth-century waterway. It had been used as an industrial dump site since the 1930s, and in 1947 was purchased by Hooker Chemical and Plastics Company to dispose of drums of toxic chemical wastes. The site was covered and sold to the Niagara Falls Board of Education in 1953, who proceeded to build an elementary school and a playing field on the site. Part of the site was also sold to a developer who built several hundred homes on the periphery of the old canal.[60]

In 1976, after some years of unusually heavy rains and snow, the chemicals began seeping into basements of the houses. The canal itself overflowed and chemicals that had leaked from the decayed drums entered the environment. In August 1978, the New York State Department of Health declared the Love Canal area "a grave and imminent peril" to the health of those living nearby. Investigations were conducted into complaints about an abnormal number of miscarriages, birth defects, cancer, and a variety of other illnesses. Eleven different actual or suspected carcinogens, including the dreaded dioxin, were found among the many chemicals leaching into the air, water, and soil. Air monitoring equipment found pollution levels ranging as high as 5,000 times the maximum safe level. Finally, President Carter declared Love Canal a disaster area, making federal disaster relief aid available to the residents, and signed an emergency order under which the federal government and New York State would share the cost of relocating the area families.[61]

By July 1979 (1) two hundred and sixty-three families had been evacuated, two hundred and thirty-six homes had been purchased by the state, a thousand additional families had been advised to leave their homes; (2) housing values had dropped to nil; (3) almost $27 million had been appropriated by municipal, state, and federal agencies for providing temporary housing, closing off the contaminated area, and containing the leachate (including digging a trench, installing a drain pipe to catch the leachate, and covering the canal with a clay cap to seal it); and (4) nine hundred notices of claims had been filed against Niagara Falls, Niagara County, and the Board of Education for a total of more than $3 billion in damages to health and property, and other suits had been filed against Hooker Chemical Company.[62]

In all, the damage suits against Hooker Chemical totaled more than $650 million filed by the state and the federal government alone. Private parties also sued Hooker for hundreds of millions of dollars in additional damages. In October of 1983, Hooker claimed that it had reached out-of-court settlements on most of the personal-injury claims brought by residents of the Love Canal area. The company said it had settled with 1,345 of the 1,431 original claimants who sought a total of $16 billion in compensatory and punitive damages. The claims were for personal injury, wrongful death, or property damage resulting from exposure to chemical residues. The company ended up paying about $5 million to $6 million with insurance companies paying about $25 million more.[63]

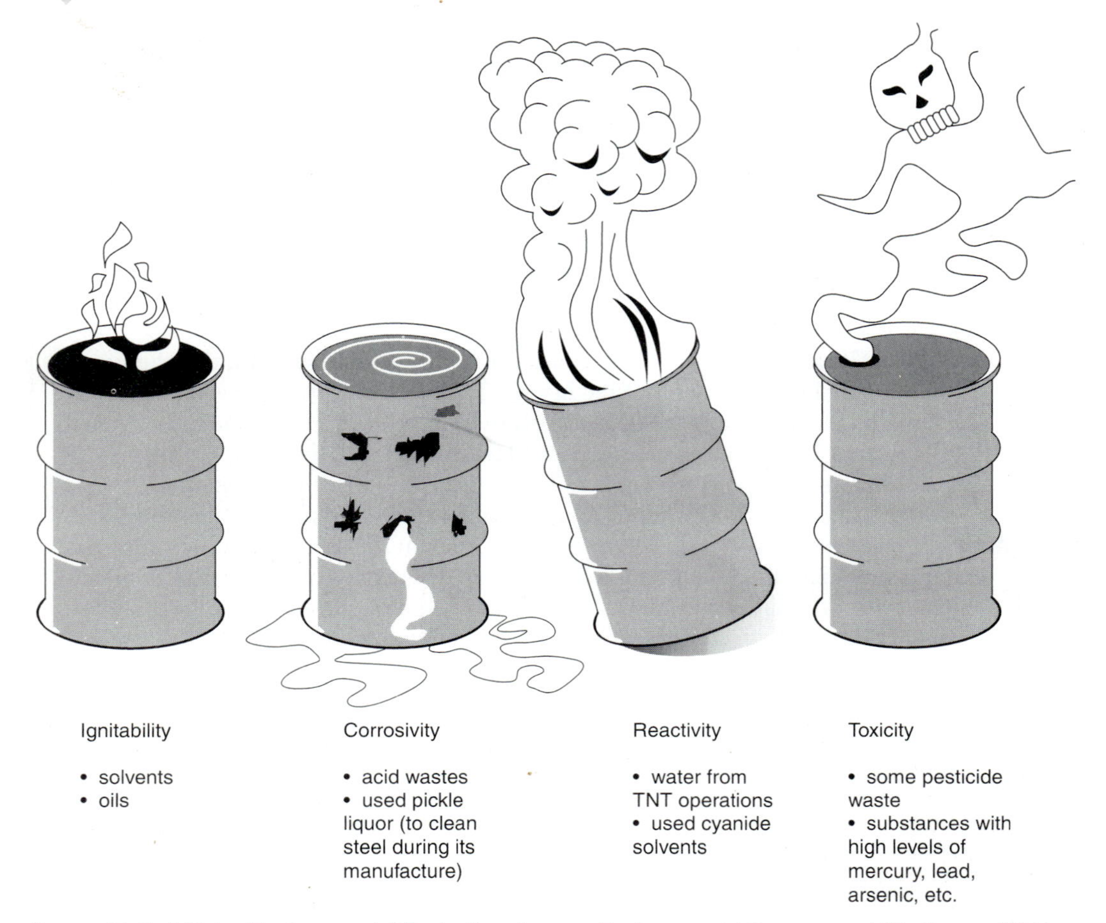

Source: United States Environmental Protection Agency, *Environmental Progress and Challenges: EPA's Update* (Washington, D.C.: U.S. Government Printing Office, 1988), p. 85.

FIGURE 9.2 Characteristics of Hazardous Waste Materials

manufacturers and allied industries. Ninety percent of the hazardous waste produced in the country comes from facilities that generate large quantities of more than 2,200 pounds per month. A much smaller amount comes from small quantity generators that produce between 220 and 2,200 pounds per month.[69]

The continuing problem of dealing with waste generation is complex and expensive. The EPA itself won't estimate what portion of the estimated 264 metric tons of hazardous waste generated annually is disposed of improperly, but some estimates state that one out of every seven companies producing toxic wastes may have dumped illegally at some time. Probably huge quantities of such wastes go into streams, pastures, or vacant lots, where the risk of human contamination is high and the chance of detection slim.[70] The usual way such waste was disposed of in the past was to put barrels of it on the back dock to be picked up by a waste hauler and disposed of somewhere out of mind and out of sight. No records were kept of what was in the barrels

EXHIBIT 9.3 Examples of Hazardous Waste Generated by Businesses and Industries

Waste Generators	*Waste Type*
Chemical Manufacturers	Strong Acids and Bases Spent Solvents Reactive Wastes
Vehicle Maintenance Shops	Heavy Metal Paint Wastes Ignitable Wastes Used Lead Acid Batteries Spent Solvents
Printing Industry	Heavy Metal Solutions Waste Inks Spent Solvents Spent Electroplating Wastes Ink Sludges Containing Heavy Metals
Leather Products Manufacturing	Waste Toluene and Benzene
Paper Industry	Paint Wastes Containing Heavy Metals Ignitable Solvents Strong Acids and Bases
Construction Industry	Ignitable Paint Wastes Spent Solvents Strong Acids and Bases
Cleaning Agents and Cosmetics Manufacturing	Heavy Metal Dusts Ignitable Wastes Flammable Solvents Strong Acids and Bases
Furniture and Wood Manufacturing and Refinishing	Ignitable Wastes Spent Solvents
Metal Manufacturing	Paint Wastes Containing Heavy Metals Strong Acids and Bases Cyanide Wastes Sludges Containing Heavy Metals

Source: United States Environmental Protection Agency, *Solving the Hazardous Waste Problem* (Washington, D.C.: EPA, 1986), p. 8.

and where they were finally dumped. All this has changed with the development of legislation and regulation to control the disposal of hazardous waste.

Regulation

Responsibility for control and eradication of hazardous waste disposal problems is lodged in the EPA's Office of Solid Waste and Emergency Response. This office implements two federal laws related to hazardous waste disposal, the Resource Conservation and Recovery Act (RCRA), which regulates current and future waste practices, and the Comprehensive Environmental Response, Compensation, and Liability Act (CERCLA), commonly called Superfund, which provides for cleaning up of old waste sites. Legislation thus focuses on preventing future contamination from improper waste disposal and the cleanup of existing waste sites where hazardous waste was disposed of improperly and poses a threat to human health and the environment.

The Resource Conservation and Recovery Act (RCRA)

RCRA, originally passed in 1976, controls the generation, transportation, storage, and disposal of wastes at existing or future waste facilities. Each year about 3,000 facilities manage 275 million metric tons of RCRA waste in the country. Specifically, the law provides for (1) federal classification of hazardous waste; (2) a "cradle-to-grave" manifest (tracking) system for waste material; (3) federal safeguard standards for generators and transporters, and for facilities that treat, store, or dispose of hazardous

FIGURE 9.3 The Hazardous Waste Manifest Trail

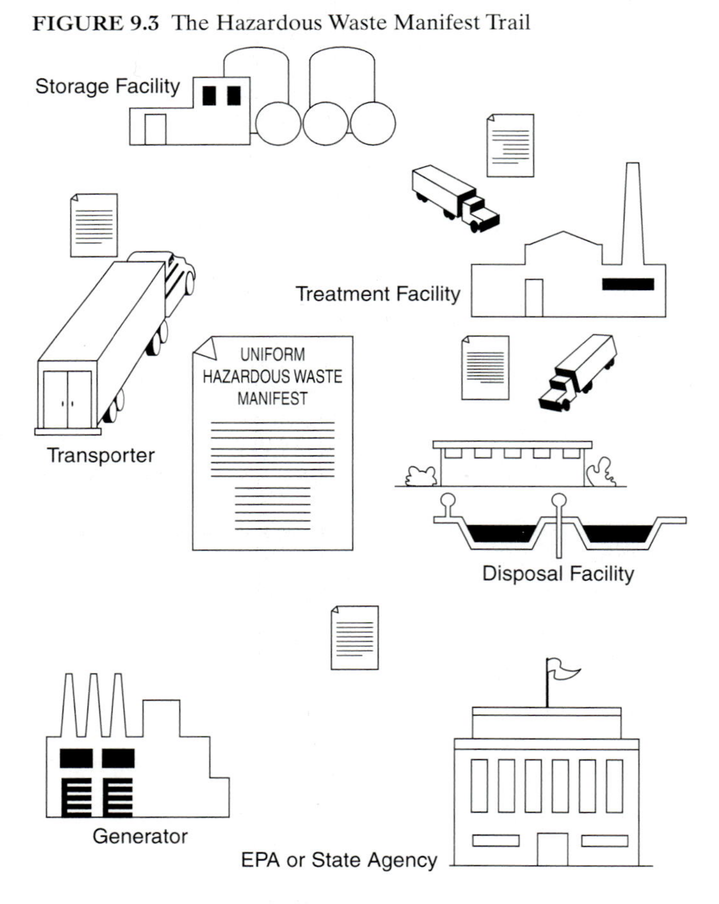

A one-page manifest must accompany every waste shipment. The resulting paper trail documents the waste's progress through treatment, storage and disposal. A missing form alerts the generator to investigate, which may mean calling in the state agency or EPA.

Note: a manifest is unnecessary for waste treated and disposed of at the point of generation.

Source: United States Environmental Protection Agency, *Environmental Progress and Challenges: EPA's Update* (Washington, D.C.: U.S. Government Printing Office, 1988), p. 88.

wastes; (4) enforcement of standards for facilities through a permitting system; and (5) authorization of state programs to replace federal programs.[71]

The basic purpose of RCRA is to protect groundwater from toxic pollution. The law provides cradle-to-grave control of hazardous waste material, from point of production through the point of disposal (Figure 9.3). Those who produce wastes have to obtain a permit to manage them on their own property. When shipping them to a treatment, storage, or disposal facility, they have to provide a manifest containing basic information about the waste material. All treatment, storage, and disposal operations are required to meet minimum standards to protect public health and the environment.[72]

Regulations to implement RCRA were developed in phases. The first phase included identification of solid wastes considered to be hazardous and the establishment of reporting and record keeping for the three categories of hazardous waste handlers: generators, transporters, and owners or operators of treatment, storage, and disposal (TSD) facilities (see box). In November 1980, these regulations became effective. By July 31, 1985, the EPA had identified 52,864 major generators of hazardous wastes, 12,343 transporters, and 4,961 TSD facilities.[73]

The second phase involved the development of technical standards related to the design and safe operation of the various types of treatment, storage, and disposal facilities. These standards serve as the basis for issuing permits to such facilities. Technical standards have been issued for incinerators, and for new and existing land disposal facilities, along with financial responsibility and liability, and insurance requirements for all facilities. Landfills, for example, must now include double liners, leachate detection and collection systems, and groundwater monitoring.

Congress intended the states to eventually assume responsibility for the RCRA hazardous waste program. The EPA is authorized to approve qualified state plans for

GENERATORS MUST

- Determine if waste is hazardous
- Apply for permit for waste handled on-site
- Originate and follow up manifest for waste moving off-site
- Keep records

TRANSPORTERS MUST

- Deliver hazardous waste to designated facility
- Carry manifest with the shipment
- Report and clean up spills

TSD FACILITIES MUST

- Apply for permit from EPA
- Meet Interim Status Standards until permit approved or denied
- Meet permit conditions after permit is issued

Source: The Resource Conservation and Recovery Act: What It Is; How It Works, SW-967 (Washington, D.C.: U.S. Environmental Protection Agency, 1983), pp. 5–6.

hazardous waste management. To receive final authorization to operate the entire RCRA program, states must adopt regulations fully "equivalent to" and "consistent with" federal standards. Mississippi became the first state to receive full authorization to operate its own program. States can be granted interim authorization by setting regulations that are "substantially equivalent to" EPA's regulations. All 50 states were expected to seek final authorization to manage their own hazardous waste programs.

Congress reauthorized RCRA in late 1984 (these revisions were called the 1984 Hazardous and Solid Waste Amendments, or HSWA), imposing new and far-reaching requirements on the 175,000 enterprises that generate small amounts of waste per month (between 220 and 2,200 pounds) and those that own or operate underground storage tanks. The former are called small quantity generators (SQGs), and rules implemented in 1985 and modified in 1986 require any business producing or using hazardous materials or chemicals to register with the agency, and to be able to prove that the wastes from these materials are being disposed of properly.[74]

The revisions also required treatment of all hazardous wastes to EPA specific levels or methods before they could be disposed of in landfills. Because of this "land ban," it was expected there would be more treatment of hazardous waste. For those wastes that are landfilled, the EPA established more stringent requirements for land disposal facilities, which was expected to reduce the number of landfills that were permitted. Finally, the revisions required facility owners to clean up leaks of waste material that occur at their facilities, and gave the EPA broadened authority to implement this requirement.[75]

The Comprehensive Environmental Response, Compensation, and Liability Act (CERCLA)

Commonly called Superfund, CERCLA provided money to the EPA and gave it authority to direct and oversee cleanup of old and abandoned waste sites that pose a threat to public health or the environment. The law provided funding for the government to clean up inactive waste sites where responsible parties cannot be found or where those responsible are unable or unwilling to perform the cleanup, and created liabilities for parties who were associated with waste sites, either to perform the cleanup or to reimburse the EPA for the cost of the cleanup. Superfund was first authorized in 1980 for $1.6 billion, with the amount of money dependent upon the size of the National Priority List, the extent of cleanup necessary, responsible party contributions, scope of the fund, and the amount of money the EPA could manage efficiently.

Superfund imposes liability on responsible parties for the costs of removal or remedial action, costs of response by other parties or entities, and for damage to, or destruction of, natural resources. The liability of the law is joint and several; that is, a single party may be held responsible for all cleanup costs even if other parties are involved. This might occur where other parties have disappeared or become insolvent, defunct, or bankrupt. Liability also attaches to a party without regard to fault or negligence. These costs can be high; for example, for a facility other than a vessel or vehicle, liability includes the total of all response costs plus up to $50 million for damages.[76]

The first phase of this effort was to conduct a nationwide inventory of such sites and establish priorities for cleanup. In ranking these dump sites, the EPA took five exposure pathways into account: (1) the population put at risk, (2) the hazard potential of substances at the sites, (3) the potential for contamination of drinking water, (4) the possibility of direct human contact, and (5) the potential for destruction of sensitive ecosystems. Once sites have been identified, the EPA can require owners of old or abandoned dumps to perform the cleanup work themselves, or, where this is not possible or where immediate action is needed, the EPA and the states can step in and do the cleanup.

Two general responses can be made to these hazardous waste sites. Removal actions are short-term actions that stabilize or clean up a hazardous site, and typically involve removing tanks or drums of hazardous substances from the surface, excavating contaminated soil, installing security measures at a site, or providing a temporary alternate source of drinking water to residents. Remedial actions include the study, design, and construction of longer-term actions aimed at a permanent remedy. These remedial actions can be taken only at sites on the National Priorities List (NPL), a list of the nation's most serious hazardous waste sites. Typical remedial actions include removing buried drums from a site, constructing underground walls to control the movement of groundwater, incinerating the waste material, or applying bioremediation techniques or other innovative technologies to contaminated areas[77] (see box).

Steps in Cleaning Up an Uncontrolled Waste Site

After someone alerts EPA about a potential problem site, what happens? If the site is found to present a release or threat of release to public health or the environment that must be addressed quickly, EPA may take emergency measures to remove the threat. These removal actions range from installing security fencing to digging up and removing wastes for safe disposal at a RCRA approved facility. Such actions may be taken at any site, not just those on the National Priority List (NPL). These actions can take place at any time during investigation or cleanup at a site when a determination is made that response should not be delayed.

1. Identification and Preliminary Assessment

If response can be delayed without endangering public health and the environment, we can take additional time to evaluate the site further. We collect all the available information on the site from our files, state and local records, and U.S. Geological Survey maps. We analyze the information to determine the size of the site, parties most likely to have used it, local hydrological and meteorological conditions, and the impact of the wastes on the environment.

2. Site Inspection

Inspectors then go to the site to collect sufficient information to rank its hazard potential. They look for evidence of hazardous waste, such as leaking drums and dead or discolored vegetation. They may take samples of soil or water. Inspectors analyze the ways hazardous materials could be polluting the environment, for example, through runoff into nearby streams. They also check to see if the public (especially children) have access to the site.

3. Ranking Sites for the National Priorities List

Sites are evaluated according to the type, quantity, and toxicity of wastes at the site, the number of people potentially exposed, the pathways of exposure, and the importance and vulnerability of the underlying ground-water supply. This information is used to determine the Hazard Ranking System score. If the score is 28.5 or above, the site may be proposed for listing on Superfund's National Priorities List. Each state may also propose one site for listing if it is the top priority site in the state.

4. Negotiating with Potentially Responsible Parties

After the parties potentially responsible for the contamination are identified, EPA notifies them of their potential liability. We then negotiate with them to reach an agreement to undertake the studies and subsequent cleanup actions needed at the site. If negotiations are not successful, EPA may use its enforcement authorities to require responsible parties to take action, or the Agency may choose to clean up the site and seek to recover costs at a later date.

5. Remedial Investigation

The objective for hazardous waste sites placed on the NPL is long-term cleanup. To select the cleanup strategy best suited to each unique site, a more extensive field study or remedial investigation is conducted by EPA, the state, or the responsible parties. This study includes extensive sampling and laboratory analyses to generate precise data on the types and quantities of wastes present at the site, the soil type and water drainage patterns, and resulting environmental or public health risks.

6. Feasibility Study and Cleanup

Cleanup actions must be tailored exactly to the needs of each individual site. The feasibility study analyzes those needs and evaluates alternative cleanup approaches on the basis of their relative effectiveness and cost. Remedial actions must use permanent solutions and alternative treatment to the maximum extent practicable. They may include technologies such as ground-water treatment or incineration.

7. Post-Cleanup Responsibilities

After cleanup, the state is responsible for any long-term operation and maintenance required to prevent future health hazards or environmental damage.

Source: United States Environmental Protection Agency, *Environmental Progress and Challenges, EPA's Update* (Washington, D.C.: U.S. Government Printing Office, 1988), p. 97.

The original Superfund authorization ran out in October 1985, and proposals were introduced into Congress to raise the amount of money in a reauthorization that ranged from $5.3 billion to $10 billion. Finally, in October 1986, a new $9 billion Superfund program, called the Superfund Amendments and Reauthorization Act of 1986 (SARA), was passed and signed into law by the President. The measure set more stringent toxic waste cleanup guidelines and permitted the formation of risk retention groups to offer pollution liability coverage. To fund the bill, petroleum taxes were increased to $2.75 billion with a higher burden placed on imports. Another $2.5 billion was raised from a broad-based corporate surtax levied at a rate of 0.12 percent on corporate alternative minimum taxable income exceeding $2 million. Other sources of funding included a $1.4 billion tax on chemical feedstocks, $1.25 billion from general revenues, and $0.6 billion from interest and recoveries from companies responsible for toxic dumps.[78]

In early 1989, a thorough review of the Superfund program was completed that introduced several new principles to guide implementation. An enforcement first policy was adopted to make those responsible for the problem pay for the cleanup. Another policy involved regular checks at all priority sites for imminent threats that needed quick action to protect human health and the environment. The EPA also began to deal with the worst problems at the worst sites first in order to concentrate its efforts on the most serious problems. And finally, the use of treatment technologies to reduce the volume and toxicity of wastes instead of simply containing contaminated materials was encouraged.[79]

By the end of 1991, according to official EPA figures, the agency had surveyed more than 30,000 potential Superfund sites and completed more than 2,700 emergency removal actions. Some 1,200 sites had been placed on the NPL, and 93 percent of these had remedial investigation or site work underway. By the end of 1991, surface cleanup had been completed at 196 sites and by March 15, 1992, all cleanup construction was complete at 71 sites. An additional 70 sites were expected to be added to the NPL list each year for the rest of the decade, and since it takes seven to ten years to clean up a site, studies were underway to shorten this time period so the EPA could meet its goal of cleaning up 650 sites by the end of FY 2000.[80]

With regard to its enforcement first policy, six out of ten designs for permanent cleanups and projects to implement them were done by responsible parties in 1991, up from 42 percent in 1989. In FY 1991, the EPA secured a record $1.4 billion in commitments to conduct site work from those responsible for hazardous waste pollution. This amount equaled the entire tax-financed 1991 Superfund budget, and brought the total obtained since Superfund began to $5.1 billion, three fourths of which was recovered under the enforcement first policy (Figure 9.4). Because the average cleanup cost is more than $25 million per site, cleanups financed by responsible parties make it possible to conduct more cleanups in the same time period.[81]

Despite these efforts, progress under Superfund is slow, as there is little incentive in the program for companies to develop new cleanup technologies and a great deal of incentive for companies to spend millions of dollars on lawyers to put off spending hundreds of millions on actual cleanup. For example, Shell Oil along with its insurers spent $40 million in legal fees before agreeing to spend several hundred million dollars to clean up a site near Denver. Five companies spent $16 million in legal fees before coming to an agreement to clean up a site near St. Louis that cost $14 million.[82]

Some studies indicated that nearly 90 percent of the money spent by insurers in Superfund claims went for legal and related costs rather than for cleaning up pollution, and total legal costs at the average site amounted to about 40 percent of cleanup costs. Litigation can proceed at three levels in these disputes over liability. The first level involves fights between the government and companies that might be liable for pollution at the site. The second level involves battles among companies that each may bear some

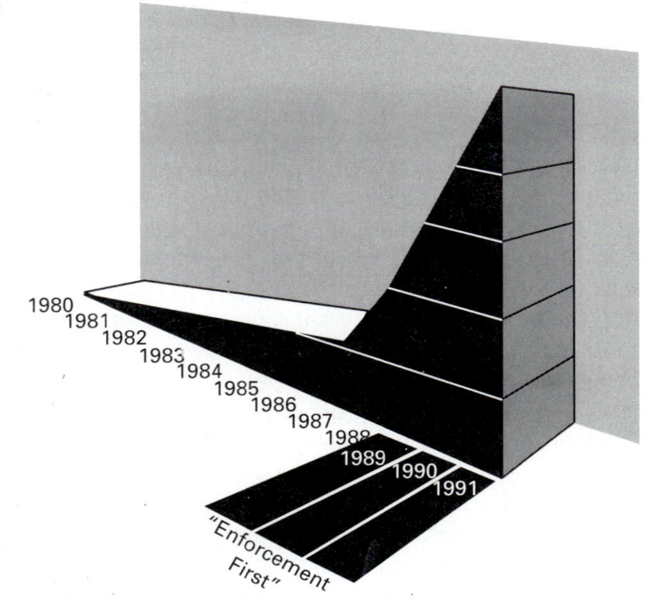

Source: United States Environmental Protection Agency, *Securing Our Legacy* (Washington, D.C.: EPA, 1992), p. 16.

FIGURE 9.4 Paying for Past Neglect

liability over exactly what liability each company has for the cleanup. And the third level concerns disputes between the insurance companies and their policyholders over what amount, if any, the insurers will pay for the cleanup. All of this legal maneuvering takes a considerable amount of time and money.[83]

Studies also showed that a great deal of time was involved in studying the site to determine the best treatment remedy. The Rand Corporation in 1993 determined that the average site took eight years of study before treatment actually begins.[84] Federal investigators found that contractors who do such cleanup work had billed the government for parties, air travel, and limousine services for employees' spouses, season baseball tickets, and club memberships. These expenses may be allowable, some investigators said, as morale boosters under federal regulations.[85]

More and more people came to believe the law needed major revision, and the Clinton administration in 1994 brokered an unprecendented compromise between business and environmentalists over how to rewrite the Superfund program. Instead of holding a company responsible for the full cost of a cleanup, the plan would have reduced litigation by requiring businesses that agree to a settlement to pay only their prorated share of cleanup expenses. The plan also called for less stringent cleanup, as polluters would only have to clean sites to meet standards that protect health and the environment rather than returning the site to a pristine state. And finally, if at least 85 percent of businesses with insurance coverage agreed not to sue their insurers, the insurers would pay $8.1 billion over 10 years into an Environmental Insurance Resolution Fund, which would be available to companies cited by the EPA to perform the cleanup.[86] The plan fell victim to election-year politics, however, and was not passed. Thus Superfund continues to limp along under the old rules.

The question of liability took on a new twist in the late 1990s, as businesses that purchased land had to take greater steps to protect themselves from being stuck with the bill for cleanup should that land later turn out to contain hazardous wastes. Insurance to cover this risk became increasing difficult to purchase as insurers found coverage of such risks difficult to quantify. Under Superfund, current owners of contaminated property can be held liable for cleanup even if the previous owners were to blame for the problem. Companies began to use environmental auditors in increasing numbers to check the land and any buildings that may be on the property for hazardous wastes before buying. Such auditing, however, was not foolproof and attracted many questionable people who wanted to cash in on the demand for their services.[87]

Also in 1990, a decision by a court of appeals expanded the notion of lender liability under Superfund law, when it ruled that a bank foreclosing on contaminated property could be held liable as an owner for cleanup of the site. The question before the court was what level of activity would expose a holder of a security interest to liability? The court answered that under Superfund a lender may be liable for the entire cost of the cleanup if its involvement with the borrower ever put it in a position to "affect hazardous waste disposal decisions" if it so chose. It did not matter to the court when the contamination took place or when the loan was made, even if these actions took place decades before the Superfund law was enacted. Because of joint and several liability, one bank could also be held liable for the entire cost of the cleanup if it was the only bank in the group that lent money that was still solvent; thus loans of only thousands of dollars could conceivably generate liabilities in the millions.[88]

Under pressure from banks after this ruling, the EPA issued rules emphasizing that lenders were broadly protected from liability. They would be permitted to engage in a variety of activities including monitoring and protecting security interests, providing financial advice to borrowers, requiring the cleanup of a facility during the life of a loan, and even foreclosing on the property, all without incurring Superfund liability.[89] However, an appeals court in Washington said the agency had gone too far, and that Congress had not intended to give the EPA authority to define liability for a class of potential defendants. Thus the rules issued by the EPA to protect banks and savings institutions from cleanup liability were invalidated.[90]

Because of the huge sums often involved in liability for hazardous waste cleanup, the issue of disclosure of such potential liability in financial statements is of concern. The Securities and Exchange Commission (SEC) mounted a disclosure study in 1988 that was intended to develop disclosure guidelines with respect to liability for cleanup, but any flagrant violations it discovered would be referred to the proper authorities for enforcement action. Companies claimed they would not know the full extent of their liabilities for several years, because of litigation with insurance companies and other potentially responsible parties. But the SEC believed that companies that had potential liability at several sites could at least disclose the minimum cost of cleanup at these sites.[91]

Meanwhile, as of October 1995, Superfund spending topped $26 billion after having been first estimated to cost only $5 billion and take five years to complete its cleanup efforts. An additional $9.5 billion was committed to cleanups bringing total Superfund resources to $35.6 billion. Only 82 sites of the 1,290 on the NPL were deleted from the list, most involving little or no cleanup activities. Construction of cleanup facilities has been completed at an additional 223 sites that are undergoing monitoring and maintenance before they can be eligible for removal from the list. For every site that has made it onto the list, there are 10 potential sites waiting in the wings. Thus some $30 billion has been spent over 15 years and less than 5 percent of the job is finished[92] (Table 9.1).

Management of Hazardous Waste

The National Academy of Sciences outlined three ways to deal with hazardous waste: (1) waste prevention by waste reduction, recycling, and reuse; (2) conversion to less hazardous or nonhazardous material; and (3) perpetual storage (Figure 9.5). The first approach is considered by many to be the most desirable, as the best way to deal with waste of any kind is simply not to produce so much in the first place that has to be disposed of in some fashion. The goal of waste prevention is to reduce the amount of waste produced by modifying industrial or other processes and by reusing or recycling the hazardous wastes that are produced.[93]

Companies that have tried to reduce their wastes often find that waste reduction and pollution prevention save them money. The Minnesota Mining and Manufacturing Company (3M) has had a waste reduction program in place since 1975, and claimed that its waste production was cut in half and that this reduction saved it $300 million annually. Many firms have little incentive to reduce their wastes, however, because of

TABLE 9.1 Hazardous Waste Sites on the National Priority List, 1994

[Includes both proposed and final sites listed on the National Priorities List for the Superfund program as authorized by the Comprehensive Environmental Response, Compensation, and Liability Act of 1980 and the Superfund Amendments and Authorization Act of 1986]

State	*Total Sites*	*Rank*	*Percent Distribution*	*Federal*	*Non-Federal*
Total	1,296	(X)	(X)	160	1,136
United States	**1,283**	**(X)**	**100.0**	**158**	**1,125**
Alabama	13	28	1.0	3	10
Alaska	8	42	0.6	6	2
Arizona	10	36	0.8	3	7
Arkansas	12	32	0.9	—	12
California	96	3	7.5	23	73
Colorado	18	22	1.4	3	15
Connecticut	16	25	1.2	1	15
Delaware	19	20	1.5	1	18
District of Columbia	—	(X)	—	—	—
Florida	58	6	4.5	5	53
Georgia	13	28	1.0	2	11
Hawaii	4	46	0.3	3	1
Idaho	10	37	0.8	2	8
Illinois	37	11	2.9	4	33
Indiana	33	12	2.6	—	33
Iowa	19	20	1.5	1	18

TABLE 9.1 *(Continued)*

State	*Total Sites*	*Rank*	*Percent Distribution*	*Federal*	*Non-Federal*
Kansas	10	37	0.8	1	9
Kentucky	20	19	1.6	1	19
Lousiana	14	27	1.1	1	13
Maine	10	37	0.8	3	7
Maryland	13	28	1.0	4	9
Massachusetts	30	13	2.3	8	22
Michigan	77	5	6.0	1	76
Minnesota	41	8	3.2	3	38
Mississippi	5	45	0.4	—	5
Missouri	23	17	1.8	3	20
Montana	9	41	0.7	—	9
Nebraska	10	37	0.8	1	9
Nevada	1	50	0.1	—	1
New Hampshire	17	24	1.3	1	16
New Jersey	108	1	8.4	6	102
New Mexico	11	34	0.9	2	9
New York	85	4	6.6	4	81
North Carolina	23	17	1.8	2	21
North Dakota	2	49	0.2	—	2
Ohio	38	10	3.0	5	33
Oklahoma	11	35	0.9	1	10
Oregon	13	28	1.0	2	11
Pennsylvania	102	2	8.0	6	96
Rhode Island	12	32	0.9	2	10
South Carolina	26	15	2.0	2	24
South Dakota	4	46	0.3	1	3
Tennessee	18	22	1.4	4	14
Texas	30	13	2.3	4	26
Utah	16	25	1.2	4	12
Vermont	8	42	0.6	—	8
Virginia	25	16	1.9	6	19
Washington	56	7	4.4	20	36
West Virginia	6	44	0.5	2	4
Wisconsin	40	9	3.1	—	40
Wyoming	3	48	0.2	1	2
Guam	2	(X)	(X)	1	1
Puerto Rico	9	(X)	(X)	1	8
Virgin Islands	2	(X)	(X)	—	2

— Represents zero. X Not applicable.

Source: U.S. Bureau of the Census, *Statistical Abstract of the United States: 1995,* 115th ed. (Washington, D.C.: U.S. Government Printing Office, 1995), p. 237.

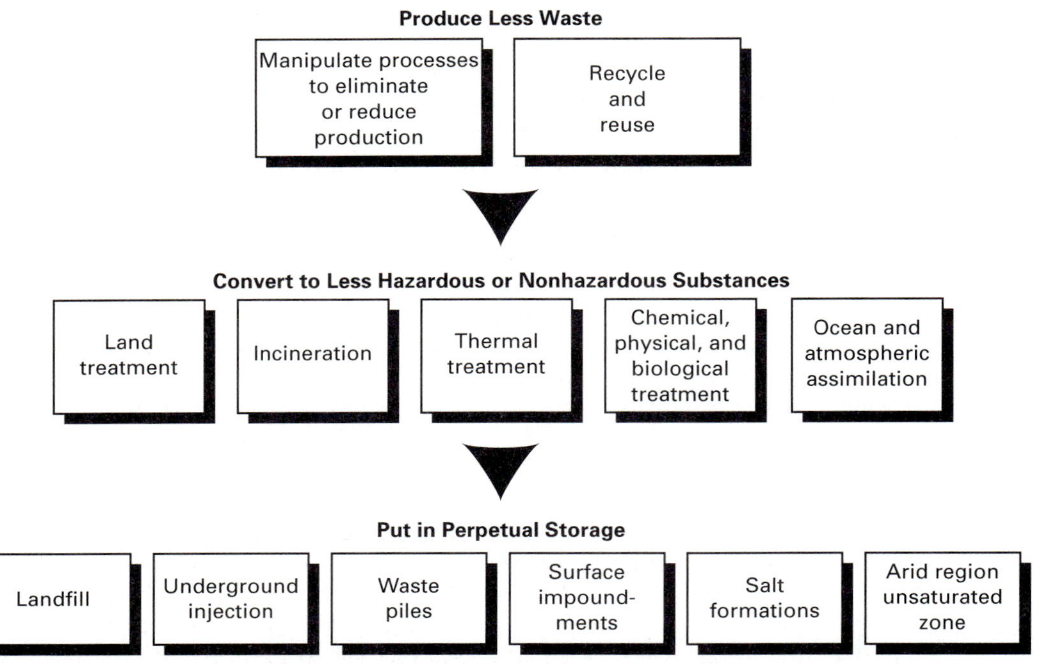

Source: G. Tyler Miller, *Living in the Environment,* 6th ed. (Belmont, CA: Wadsworth, 1990), p. 475. © Wadsworth Publishing Co.

FIGURE 9.5 Options for Dealing with Hazardous Waste

the small percentage that waste disposal amounts to in comparison to the total value of the products they produce. Some states have tried to encourage waste reduction by offering technical assistance, providing a database of information about waste material, and providing matching grants to large and small companies and communities wanting to implement waste reduction projects.[94]

According to EPA estimates, at least 20 percent of the hazardous waste generated in the United States could be recycled, reused, or exchanged so that one industry's waste becomes another's raw material, yet only about 5 percent of such waste is managed in this manner. Various waste exchanges have developed around the country whose purpose is to reduce the amount of waste that has to be disposed of in some manner. By publishing information about available waste streams, companies are made aware of what's out there that can be useful to them. The Pollution Prevention Act of 1990 requires the chemical industry to report all SARA emissions prior to their treatment, recycle, or disposal, and thus adds to the need for an information resource like a waste exchange. A waste exchange is not a broker nor does it get involved in a potential transaction. Rather it is a means to bring companies together who otherwise might remain unaware of each other.[95]

Conversion of hazardous wastes to less hazardous or nonhazardous materials could include spreading degradable wastes on the land to let them degrade over time, burning them on land or at sea in specially designed incinerators, thermally decomposing them, or treating them chemically or physically. Over 60 percent of the hazardous wastes generated in this country could be incinerated, according to the EPA, but this is an expensive method of waste disposal, and has waste disposal problems of its own because of the toxic waste residue. Much of the waste material generated in some foreign countries is detoxified in large treatment plants, but these facilities have not been used to dispose of much of the waste generated in this country.[96]

Bioremediation techniques began to offer some promise in dealing with the hazardous waste problem, as the costs of disposal mounted. This technique involves the use of bacteria to get rid of toxic components of hazardous waste and leave behind harmless substances. Bioremediation takes longer than more conventional cleanup methods, and may thus be inappropriate where fast action is needed, but the technique does have a decided cost advantage. In addition to being under attack as potentially unsafe, incineration can cost up to $1,000 per ton, while bioremediation costs less than $100 a ton in some cases. Use of this process may also be safer, as bioremediation eliminates the problem rather than relocate it to somewhere else. Before it can be used on a large scale, however, several problems must be solved.[97]

The last phase of hazardous waste management involves placing these wastes in perpetual storage. Such storage methods can involve the use of secured landfills or underground vaults, which should be located in a geologically and environmentally secure place that is carefully monitored for leaks. The costs of using this method are high, and the toxic waste disposal and management industry has been expanding by 20 to 30 percent per year to handle waste material. Much waste material is also injected into deep wells that are theoretically drilled beneath aquifers that are tapped for drinking water and irrigation. While these disposal methods are safer than they were in previous years, they are still prone to leakage and other dangers.[98]

Finally, the development of an international hazardous waste trade must be mentioned. In order to save money and avoid regulatory hurdles, cities and waste disposal companies in the United States and other industrialized countries ship large amounts of hazardous waste to other countries. Many of these legal exports from the United States go to Canada and Mexico. To ship waste to these and other countries, all U.S. companies have to do is to notify the EPA of their intent to ship this material, get written permission from the recipient country, and file an annual report with the agency.[99] Such lax provisions encourage countries such as the United States to transport its hazardous wastes problems to other countries.

These legal shipments of hazardous wastes may be only the tip of the "sludgeberg," as it is called, as there is evidence mounting of a growing trade in illegal shipments of hazardous wastes between countries. Waste disposal firms can charge high prices for picking up hazardous wastes from companies that generate them, and if they can dispose of them legally or illegally in other countries at low cost, they can make large profits. Most of the recipient countries, however, are beginning to realize that importation of hazardous waste can threaten the health and environment of their country and weaken long-term economic growth. Many are adopting a "not in our country" (NIOC) attitude similar to the NIMBY phenomenon in this country.[100]

In late 1990, it was reported that millions of tons of toxic waste were being dumped in Latin America each year, leaving poisonous residues that will endanger lives for decades. Scientists in Brazil called the dumping of hazardous waste in their country an illicit trade that is shrouded in secrecy and often done by small and unregistered companies. This dumping was devastating the environment and said to cause cancer, birth defects, nerve damage, and blood disorders. According to Greenpeace, Latin America made a perfect dumping ground as there was lots of space, loads of corrupt inspectors, and widespread ignorance of the problem.[101] By 1995, China with its cheap labor, vast area, and spotty legal enforcement had become the world's chief dumping ground. Some of this material can be reclaimed, but the rest has to be incinerated or dumped along with tons of useless trash sold as recyclable material by garbage dealers.[102]

Leaders from 116 countries drafted an international treaty in 1989 that was designed to help control the export of hazardous waste material. The treaty would ban such exports unless the government of the receiving country gives prior written permission to receive the waste material.[103] In 1994, industrial nations who were members of the Organization for Economic Cooperation and Development agreed to stop dumping their toxic wastes in poor countries. The ban on waste exports for incineration or burial was made effective immediately, and exports of wastes for recycling will be illegal as of December 31, 1997, in order to close the loophole of exporting waste under the guise of recycling. Some countries including the United States had wanted a more limited ban that would have allowed them to continue exports for recycling subject to approval from the importing nation.[104]

Questions for Discussion

1. Describe the solid waste problem. How is most of this waste disposed of currently? What is the problem with this method of disposal? What happens to waste material using this method?
2. What problems exist with incineration as a method of waste disposal? Would you recommend this method as a viable one to deal with the solid waste problem? Why or why not? What can be done to improve incineration?
3. Is recycling the answer to the solid waste problem? What types of recycling exist? What can be done to develop more recycling programs across the country? What kinds of materials are a particular problem when it comes to recycling?
4. What is an integrated waste-management system? What incentives can be used to dispose of waste more effectively? Is it realistic to think that waste will be seen as a resource in our country?
5. What characteristics does hazardous waste possess? What are some of the most common hazardous wastes? Why are they so difficult to dispose of appropriately?
6. Describe the Resource Conservation and Recovery Act. What are its major provisions? What is cradle-to-the-grave control? How is this accomplished? What responsibilities does business have under this system?
7. Describe the Superfund program. To what problem is it directed? What is joint and several liability? Why was the law constructed in this manner? Does this provision pose a particular problem for business? What new developments are taking place with respect to liability?
8. What options exist for dealing with hazardous waste? What is the best approach in your opinion? What are some companies doing to reduce the waste they produce? Do waste reduction techniques have widespread applicability?

Endnotes

1. "The Garbage Industry: Where There's Muck There's High Technology," *The Economist,* April 8, 1989, p. 23; John E. Young, "Tossing the Throwaway Habit," *World Watch,* May–June 1991, p. 28.

2. Samuel Epstein et al., *Hazardous Wastes in America* (San Francisco: Sierra, 1982), p. 303.
3. "The Garbage Industry," p. 23.
4. Ibid., p. 24.
5. Ibid., pp. 24–26.
6. G. Tyler Miller, *Living in the Environment,* 6th ed. (Belmont, CA: Wadsworth, 1990), p. 351.
7. United States Environmental Protection Agency, *Environmental Progress and Challenges: EPA's Update* (Washington, D.C.: U.S. Government Printing Office, 1988), p. 87.
8. Ibid.
9. Miller, *Living in the Environment,* p. 351.
10. Ibid., p. 359.
11. Miller, *Living in the Environment,* p. 359.
12. Richard Wolkomir, "I Learned That It Just Keeps Getting Deeper," *Smithsonian,* 21, no. 1 (April 1990), 152.
13. Dan Grossman and Seth Shulman, "Down in the Dumps," *Discover,* April 1990, pp. 39–40.
14. EPA, *Environmental Progress and Challenges,* p. 87; Jeff Bailey, "Economics of Trash Shift as Cities Learn Dumps Aren't So Full," *Wall Street Journal,* June 2, 1992, p. A1.
15. Rose Gutfeld, "Smelly Imports: As Eastern Landfills Reach Capacity, States Send Garbage West," *Wall Street Journal,* April 26, 1991, p. A1.
16. Terri Thompson and Mimi Bluestone, "Garbage: It Isn't the Other Guy's Problem Anymore," *Business Week,* May 25, 1987, p. 150.
17. Paul M. Barrett, "Justices Reject States' Efforts to Bar Waste," *Wall Street Journal,* June 2, 1992, p. A3.
18. Paul M. Barrett and Jeff Bailey, "Garbage Haulers Win Big Victory in High Court," *Wall Street Journal,* May 17, 1994, p. A3.
19. Steve Mills, "Neighbors Welcome Landfills," *Times-Picayune,* September 11, 1994, p. A8.
20. Miller, *Living in the Environment,* p. 361.
21. Ibid.
22. Paul M. Barrett, "Incinerators Dealt Big Blow by High Court," *Wall Street Journal,* May 3, 1994, p. A3.
23. Miller, *Living in the Environment,* p. 361.
24. Ibid., pp. 361–362.
25. Zachary Schiller, "Turning Pampers into Plant Food?" *Business Week,* October 22, 1990, p. 38.
26. Bruce Van Voorst, "Recycling: Stalled at Curbside," *Time,* October 18, 1993, pp. 78–80.
27. David Kalish, "Recycling Making a Dent in Landfill-Bound Waste," *Times-Picayune,* January 7, 1995, p. C8.
28. Miller, *Living in the Environment,* p. 363.
29. Ibid., p. 365.
30. Ibid. See also Gene Kramer, "Recycling's New Payoff May Be in Jobs," *Times-Picayune,* October 3, 1994, p. A3.
31. Ibid. See also Dana Milbank, "Aluminum's Envious Rivals Turn Green, Rush to Show They, Too, Are Recyclable," *Wall Street Journal,* September 18, 1991, p. B1.
32. Van Voorst, "Recycling: Stalled at Curbside," pp. 78–80; Bruce Van Voorst, "The Recycling Bottleneck," *Time,* September 14, 1992, pp. 52–54.
33. David Stipp, "Cities Couldn't Give Away Their Trash; Now They Get Top Dollar from Recyclers," *Wall Street Journal,* September 19, 1994, p. B1. See Jeff Bailey, "Waste of a Sort: Curbside Recycling Comforts the Soul, but Benefits Are Scant," *Wall Street Journal,* January 19, 1995, p. A1, for a more pessimistic view of recycling.
34. Milbank, "Aluminum's Envious Rivals Turn Green," p. B1.
35. JoAnn Gutin, "Here's Another Look at Plastics," *E Magazine,* May–June 1994, pp. 30–31.
36. Ibid., p. 32.

37. Stephanie Anderson Forest, "There's Gold in Those Mills of Soda Bottles," *Business Week,* September 11, 1995, p. 48.
38. Scott Kilman, "McDonald's to Drop Plastic Foam Boxes in Favor of High-Tech Paper Packaging," *Wall Street Journal,* October 2, 1990, p. A3.
39. Michael Waldholz, "New Plastic Is Promoted as a Natural," *Wall Street Journal,* January 24, 1990, p. B1. Scientists are also designing a plastic that can be made without toxic by-products and recycled with help from an enzyme. See Joan E. Rigdon, "Scientists Find Key to a 'Greener' Plastic," *Wall Street Journal,* August 26, 1993, p. B8.
40. Miller, *Living in the Environment,* p. 368.
41. Kenneth Chilton, *Talking Trash: Municipal Solid Waste Mismanagement* (St. Louis, MO: Washington University Center for the Study of American Business, 1990), p. 19. See also Randolph B. Smith, "Aided by Volunteers, Seattle Demonstrates Recycling Can Work," *Wall Street Journal,* July 19, 1990, p. A1. By 1993, 42 percent of the city's trash was going into recycling bins, and 90 percent of all single-family homes were participating. But the program has hit some snags. See Bill Richards, "Recycling in Seattle Sets National Standard but Is Hitting Snags," *Wall Street Journal,* August 3, 1993, p. A1.
42. Timothy Noah, "Clinton Orders Agencies to Buy Recycled Paper," *Wall Street Journal,* October 21, 1993, p. A16.
43. Laurie M. Grossman, "Florida to Buy Back Its Recycled Waste," *Wall Street Journal,* June 19, 1992, p. B1.
44. Mary Beth Regan, "How Much Green in 'Green' Paper?" *Business Week,* November 1, 1993, pp. 60–61.
45. Anita Sharpe, "Business Group Quadruples Outlay on Recycled Items," *Wall Street Journal,* October 13, 1993, p. B8.
46. Miller, *Living in the Environment,* p. 369.
47. Gary McWilliams, "The Big Brouhaha Over the Little Juice Box," *Business Week,* September 17, 1990, p. 36; David Stipp, "Lunch-Box Staple Runs Afoul of Activists," *Wall Street Journal,* March 14, 1991, p. B1. In 1994, Maine repealed its ban on the drink boxes. See Dave Kansas, "Maine Repeals U.S.'s Only Ban on Drink Boxes," *Wall Street Journal,* April 7, 1994, p. B1.
48. Miller, *Living in the Environment,* p. 369.
49. Ibid., p. 370.
50. Frank Edward Allen, "McDonald's Launches Plan to Cut Waste," *Wall Street Journal,* April 17, 1991, p. B1.
51. Meg Cox, "Music Firms Try Out 'Green' CD Boxes," *Wall Street Journal,* July 25, 1991, p. B1; Meg Cox, "CD Marketers Will Eliminate Paper Packaging," *Wall Street Journal,* February 28, 1992, p. B1.
52. Frank Edward Allen, "Governors Groups to Urge 200 Firms to Cut Waste by Using Less Packaging," *Wall Street Journal,* March 15, 1991, p. A4.
53. Evan I. Schwartz, "A Data Base That Truly Is Garbage In, Garbage Out," *Business Week,* September 17, 1990, p. 92.
54. David Kalish, "Talking Trash," *Times-Picayune,* July 5, 1992, p. F1.
55. Chilton, *Talking Trash,* pp. 7–8.
56. Ibid., p. 18.
57. EPA Municipal Solid Waste Task Force, *The Solid Waste Dilemma: An Agenda for Action* (Washington, D.C.: U.S. Environmental Protection Agency, 1989), pp. 24–25.
58. Chilton, *Talking Trash,* pp. 19–20.
59. Epstein et al., *Hazardous Wastes in America,* p. 303.
60. U.S. Council on Environmental Quality, *Environmental Quality 1979* (Washington, D.C.: U.S. Government Printing Office, 1979), p. 177.

61. Ibid., pp. 176–177.
62. Ibid., p. 177.
63. "Occidental Settles Most Injury Claims in Love Canal Case," *Wall Street Journal,* October 11, 1983, p. 3.
64. Miller, *Living in the Environment,* pp. 474–475.
65. United States Office of Research and Development, *Controlling Hazardous Waste Research Summary* (Washington, D.C.: Environmental Protection Agency, 1980), p. 4.
66. *The Resource Conservation and Recovery Act: What It Is; How It Works,* SW-967 (Washington, D.C.: Environmental Protection Agency, 1983), p. 3.
67. Benjamin A. Goldman, James A. Hulme, and Cameron Johnson, *Hazardous Waste Management: Reducing the Risk* (Washington, D.C.: Island Press, 1986), p. 20.
68. Ibid.
69. EPA, *Environmental Progress and Challenges,* p. 80.
70. Barry Meier, "Dirty Job: Against Heavy Odds, EPA Tries to Convict Polluters and Dumpers," *Wall Street Journal,* January 7, 1985, p. 1.
71. Ronald J. Penoyer, *Reforming Regulation of Hazardous Waste* (St. Louis, MO: Washington University Center for the Study of American Business, 1985), p. 1. Subtitle D of RCRA covers solid wastes that are primarily nonhazardous. The solid waste disposal facilities regulated under this subtitle fall into four general categories: landfills, surface impoundments, land application facilities, and waste piles. The requirements for these facilites are designed to (1) protect endangered species, food crops, groundwater, air, and floodplains; (2) prevent disease transmission; and (3) ensure the safety of employees and nearby residents. Some of these solid waste disposal facilities also accept household hazardous waste and waste from small-quantity hazardous waste generators. See United States Environmental Protection Agency, *Solving the Hazardous Waste Problem* (Washington, D.C.: EPA, 1986), pp. 25–26.
72. United States Environmental Protection Agency, *1978 Report: Health and Regulatory Reform* (Washington, D.C.: EPA, 1979), p. 16.
73. *The Resource Conservation and Recovery Act,* SW-967, p. 5; *The New RCRA: A Fact Book* (Washington, D.C.: Environmental Protection Agency, 1985), p. 4.
74. *The New RCRA,* pp. 1–2.
75. EPA, *Environmental Progress and Challenges,* pp. 88–89.
76. Penoyer, *Reforming Regulation of Hazardous Waste,* p. 17.
77. United States Environmental Protection Agency, Solid Waste and Emergency Response, *The Superfund Program: Ten Years of Progress* (Washington, D.C.: EPA, 1991), p. 2.
78. Alexander & Alexander Government and Industry Affairs Office, *Washington News,* 4, no. 11 (October 31, 1986), 1–2. See also Superfund Amendments of 1986, P.L. 99–499.
79. United States Environmental Protection Agency, Office of Communications, Education, and Public Affairs, *Securing Our Legacy* (Washington, D.C.: EPA, 1992), pp. 16–17.
80. Ibid., p. 17.
81. Ibid.
82. Amal Kumar Naj, "How to Clean Up Superfund's Act," *Wall Street Journal,* September 15, 1988, p. 26. See also Christopher Elias, "Waste Site Cleanup Liability May Be Hazardous to Lenders," *Insight,* November 13, 1989, pp. 42–44.
83. Jonathan M. Moses, "Insurer Payouts Over Superfund Flow to Lawyers," *Wall Street Journal,* April 24, 1992, p. B3.
84. Bob Sablatura, "Much Spent on Superfund with Little to Show for It," *Times-Picayune,* October 29, 1995, p. A18.
85. H. Josef Hebert, "Superfund Contractors Splurge on Limos, Cabins," *Times-Picayune,* April 10, 1992, p. A6.

86. Mary Beth Regan, "Toxic Turnabout?" *Business Week,* April 25, 1994, p. 30–31; Timothy Noah, "Disputes End Revision Effort for Superfund," *Wall Street Journal,* October 6, 1994, p. B6. Efforts continue to revise Superfund legislation. See John J. Fialka, "Senate GOP Leaders Offer Bill to Limit Liabilities of Some in Superfund Suits," *Wall Street Journal,* January 23, 1997, p. A4; Johh J. Fialka, "Superfund Ensnares Thousands of Small Firms in a Legal Nightmare, Fueling Overhaul Drive," *Wall Street Journal,* March 19, 1997, p. A20.
87. Eric Felter, "Toxic Risks Keep Buyers Guessing," *Insight,* August 13, 1990, p. 41. See also Robert Tomsho, "Gumshoes Help Companies Cut Dump Cleanup Bills," *Wall Street Journal,* November 5, 1991, p. B1.
88. Dennis R. Connolly, "Superfund Whacks the Banks," *Wall Street Journal,* August 28, 1990, p. A10. See also Christopher Elias, "Waste Site Cleanup Liability May Be Hazardous to Lenders," *Insight,* November 13, 1989, pp. 42–44.
89. Amy Dockser Marcus and Ellen Joan Pollock, "EPA Plans Rule to Curb Liability on Loans to Owners of Waste Sites," *Wall Street Journal,* February 14, 1991, p. B5; Barbara Rosewicz, "Bank Liability Could Be Cut for Cleanups," *Wall Street Journal,* June 6, 1991, p. A3.
90. Richard B. Schmitt, "Appeals Court Invalidates EPA Rules Shielding Lenders from Superfund Law," *Wall Street Journal,* February 10, 1994, p. B2.
91. Amal Kumar Naj, "See No Evil: Can $100 Billion Have No Material Effect On Balance Sheets?" *Wall Street Journal,* May 11, 1988, p. 1.
92. Sablatura, "Much Spent on Superfund," p. A18.
93. Miller, *Living in the Environment,* p. 475. Because of success in reducing the amounts of waste generated, business decreased substantially at hazard waste dumps in 1993, as volumes fell by 15 percent and prices decreased by about 33 percent from the year before. And 1994 was expected to be an even worse year. See Jeff Bailey, "Slump at Hazardous-Waste Dumps Raises Concerns," *Wall Street Journal,* August 5, 1994, p. B3.
94. Ibid., pp. 475–476.
95. Mindy Brodhead-Averitt, "Trash or Treasure?" *Business and Industry Coordinator Magazine,* November–December 1991, p. 1.
96. Miller, *Living in the Environment,* p. 475.
97. Robert D. Hof, "The Tiniest Toxic Avengers," *Business Week,* June 4, 1990, pp. 96–98. See also Susan Chollar, "The Poison Eaters," *Discover,* April 1990, pp. 76–78; Richard Lipkin, "Biological Warfare on Toxic Waste," *Insight,* June 19, 1989, pp. 50–51. See also United States Environmental Protection Agency, Office of Research and Development, *Bioremediation: Innovative Pollution Treatment Technology* (Washington, D.C.: EPA, 1993). There is some evidence that plants might also be useful in waste cleanup. See Jerry E. Bishop, "Pollution Fighters Hope a Humble Weed Will Help Reclaim Contaminated Soil," *Wall Street Journal,* August 7, 1995, p. B1; John Carey, "Can Flowers Cleanse the Earth?" *Business Week,* February 19, 1996, p. 54.
98. Miller, *Living in the Environment,* p. 477.
99. Ibid., p. 478.
100. Ibid.
101. Todd Lewan, "Officials: Waste Producers Dump on Latin America," *Times-Picayune,* December 11, 1990, p. A17.
102. Craig S. Smith, "China Becomes Industrial Nations' Most Favored Dump," *Wall Street Journal,* October 9, 1995, p. B1.
103. Miller, *Living in the Environment,* p. 478.
104. Clare Nullis, "Wealthy Nations Ban Dumping in Poor Countries," *Times-Picayune,* March 26, 1994, p. A23.

Suggested Reading

British Medical Association Staff. *Hazardous Waste and Human Health.* New York: Oxford University Press, 1991.

Chilton, Kenneth. *Talking Trash: Municipal Solid Waste Mismanagement.* St. Louis, MO: Washington University Center for the Study of American Business, 1990.

Conway, R. A., et al., eds. *Hazardous and Industrial Solid Waste Minimization.* New York: ASTM, 1989.

Curlee, T. Randall. *Waste-to-Energy in the United States: A Social and Economic Assessment.* New York: Greenwood, 1994.

EPA Municipal Solid Waste Task Force. *The Solid Waste Dilemma: An Agenda for Action.* Washington, D.C.: U.S. Environmental Protection Agency, 1989).

Erwin, Lewis, and L. Hall Healey, Jr. *Packaging and Solid Waste: Management Strategies.* New York: AMACOM, 1990.

Foden, Charles, and Jack Weddell. *Hazardous Materials Emergency Action Data.* New York: Lewis Publications, 1991.

Fortuna, Richard, and David J. Lennett. *Hazardous Waste Regulation: The New Era.* New York: McGraw-Hill, 1987.

Gano, Lila. *Hazardous Waste.* New York: Lucent Books, 1991.

Gibbs, Lois. *The Love Canal: My Story.* Albany, NY: State University of New York Press, 1982.

Harris, Christopher, et al. *Hazardous Wastes: Confronting the Challenge.* New York: Greenwood, 1987.

Harris, Christopher, and Scott A. Henry. *Hazardous Chemicals and the Right to Know.* New York: Executive Publishers, 1994.

Head, J. J., ed. *Solid Waste and Recycling.* Raleigh: Carolina Biological, 1992.

Hester, R. E., and R. M. Harrison, eds. *Waste Incineration and the Environment.* New York: CRC Press, 1994.

Jackman, Alan P., and Roger L. Powell. *Hazardous Waste Treatment Technologies.* New York: Noyes, 1991.

James, Jody, ed. *Environmental Awareness: Solid Waste.* New York: Bancroft-Sage, 1991.

James, Jody, ed. *Environmental Awareness: Toxic Waste.* New York: Bancroft-Sage, 1991.

Jamishidi, Mohammad, et al. *Waste Management: From Risk to Remediation.* Upper Saddle River, NJ: Prentice Hall, 1994.

Kharbands, O. F., and E. A. Stallworthy. *Waste Management: Towards a Sustainable Society.* New York: Greenwood, 1990.

LaGrega, Michael. *Hazardous Waste Management.* New York: McGraw-Hill, 1994.

Legett, Jeremy. *Waste War.* New York: Marshall Cavendish, 1991.

Muir, Warren, and Joanna Underwood. *Promoting Hazardous Waste Reduction.* New York: INFORM, 1987.

Office of Technology Assessment. *From Pollution to Prevention: A Progress Report on Waste Reduction.* Washington, D.C.: U.S. Government Printing Office, 1987.

Palmer, Joy. *Waste and Recycling.* London: Trafalgar, 1995.

Peck, Dennis L., ed., *Psychosocial Effects of Hazardous Toxic Waste Disposal on Communities.* Springfield, IL: Thomas, 1989.

Pollution Engineering Staff. *Hazardous Waste and Toxic Substances Laws and Regulations.* New York: Gulf Publications, 1994.

Rhyner, Charles R., et al. *Waste Management and Resource Recovery.* New York: Lewis Publications, 1995.

Solid Waste Dilemma: An Agenda for Action. New York: Diane Publishers, 1993.

Tesar, Jenny E. *Waste Crisis.* New York: Facts on File, 1991.

Wentz, C. A. *Hazardous Waste Management,* 2nd ed. New York: McGraw-Hill, 1995.

CHAPTER 10

Deforestation and Species Decimation

The sustainable management of forest resources and the plant and animal species they contain is one of the most pressing resource management problems facing the world today. While potentially a renewable resource, forests are being mismanaged all over the world and are threatened by human activities related to harvesting the wood resources forests provide or cutting them down in the interests of agricultural or urban development. The destruction of these forests destroys the habitats of thousands of wildlife species, which are then threatened with extinction. The world simply must come to grips with what is happening to these forests and begin to develop a sustainable approach to forest management that will conserve these resources and the species they support for present as well as future generations.

Forests have many resources that are useful for human purposes. They supply humans with lumber for building houses and other buildings, wood that can be used for fuel, pulpwood used in making paper, medicines that can cure diseases, and many other products that are worth over $150 billion a year worldwide. Many forests are also places where mining is done, where some livestock is grazed, and where humans enjoy the recreational aspects of forest lands and seek solitude and a chance to enjoy the beauty of such areas. Worldwide, about half the timber cut each year is used as fuel for heating and cooking, especially in developing nations, one third is converted into lumber and other wood products used in building, and one sixth is used as pulp in making a variety of paper products.[1]

But forests also have vital ecological functions that most people are not aware of, which makes them vulnerable to being cut down for human purposes. Many forests are watersheds that absorb, hold, and gradually release water that recharges springs, streams, and aquifers. In performing this function, forests regulate the flow of water from mountains that receive a great deal of snowfall to croplands and urban areas, and help control soil erosion, the severity of flooding, and the amount of sediment washing into rivers, lakes, and reservoirs. Forests also provide habitats for a larger number of wildlife species than any other part of the natural world, making them the world's major reservoir of biological diversity. They also absorb the noise associated with an urban world and thus provide solitude, and in some cases even absorb air pollutants.[2]

Forests also play a vital role in global carbon and oxygen cycles through photosynthesis in which trees remove carbon dioxide and add oxygen to the world's air supply.

When trees are harvested and burned, as they are in the Amazon rain forest, the carbon they contain is released into the atmosphere as carbon dioxide, and the destruction of tree cover also leads to the release of some of the carbon stored in the exposed soil in which the trees grow. Thus deforestation creates a double adverse effect in increasing carbon dioxide levels worldwide by removing some carbon dioxide–absorbing capacity and adding to carbon dioxide levels at the same time. Large-scale deforestation thus contributes to global warming, which, as described in an earlier chapter, may alter global climate, food production, and sea levels all over the world.[3]

DEFORESTATION

Since 1950, deforestation around the world has accelerated. Vast tracts of forest have disappeared from Japan, the Philippines, the mainland of Southeast Asia, much of Central America, western North America, eastern South America, the Indian subcontinent, and sub-Saharan as well as the horn of Africa. Forest clearing is beginning to take place in Siberia and the Canadian north as well as Alaska. Fires have destroyed many hectares of trees in the Amazon Basin, and trees in Central Europe are dying because of pollution. Thus forests are under threat as never before.[4]

Trees still cover 26 percent of land area worldwide, and three fourths of the original forest area still has some tree cover, according to some estimates. But just one third of the original total, which amounts to about 12 percent of the land area worldwide, retains a mantle of intact forest ecosystems. The rest is said to consist of biologically impoverished stands of commercial timber and fragmented regrowth. This kind of deforestation has resulted in a surge of economic growth in come countries, but has also caused catastrophic losses to human welfare and ecological health worldwide.[5]

Deforestation is the total conversion of this forest to other uses in which no forest remains. This practice must be distinguished from depletion, which is caused by logging or local usage and removal of some trees, where the forest in some sense is left standing. The fundamental causes of this deforestation are poverty, overpopulation, insecure land tenure, misguided government policies, inequitable distribution of land and wealth, the need to put land to more intensive uses, and other such reasons. More proximate causes are said to be agriculture, cattle ranching, mining, building of hydroelectric dams with resultant flooding, encroachment, and logging.[6]

The International Hardwood Products Association in Alexandria, Virginia, has stated that subsistence farmers in developing countries are responsible for 80 percent of tropical deforestation.[7] However, the Food and Agricultural Organization of the United Nations estimates that half the forest cleared in the tropics each year is for shifting cultivation by landless farmers. As the number of subsistence farmers grows, more land is cleared, thus making for something of a vicious circle that is difficult to arrest. Clearing forest land for permanent agriculture is the most important cause of deforestation.[8]

Rapid population growth and poverty push landless people to clear and cultivate forest land to grow crops and to cut trees for fuelwood. In some countries, wealthy landowners influence the government to encourage landless peasants to clear tropical forests for cultivation because this helps defuse political pressures for equitable land distribution. Thus the government has built roads and other projects to open up these lands to settlement. When they arrive in the rain forest, settlers clear patches of forest to grow enough food on which to survive by using slash-and-burn cultivation. Many of the nutrients in the soil come from the trees themselves, and when these nutrients are exhausted, the farmers move on to another patch of rain forest.

Policies of developing nations and international aid-lending agencies have encouraged this kind of deforestation. The mass migration of the urban poor to tropical forests would not be possible without highway, logging, mining, ranching, and dam-building projects that open up these usually inaccessible areas. Many of these projects have been financed by loans from the World Bank and other international lending agencies whose policies are greatly influenced by developing nations. It was not until 1987 that the World Bank set up a department to review the environmental impacts of the projects it was supporting.[9]

Reliance of the rural poor on wood as a source for energy is the second most important cause of tropical rain forest deforestation. More than one-half the inhabitants of developing nations, or 2.5 million people, do not have adequate fuelwood for cooking and heat and must continually destroy rain forests for these purposes. The women in these countries spend much of their time gathering wood, as the poorest people in developing nations depend most directly on the rain forests for fuelwood. Many countries suffering from fuelwood shortages have inadequate forestry policies and budgets, and are planting 10 to 20 times fewer trees than needed to offset forest losses and meet increased demands for fuelwood and other forest products.[10]

Commercial logging for export is not the primary cause of deforestation on a global basis; however, it can be an important problem in certain supplier countries. Nigeria, for example, was a leading exporter of tropical logs in 1960, but by 1985, its forests had been so depleted that it became a net importer of forest products. Malaysia cut down its forests three or four times faster than they were being replenished during the 1980s, and it has lost half of its rain forests in a 20-year period.

Damage to the forest during logging may be caused by damage to seedlings and residual trees. Erosion of soil particularly in mountainous areas and an increased susceptibility to wildfire can cause further damage to the residual forest. Logging roads may be used by hunters, timber poachers, or farmers who move in and clear the remaining forest. Nonetheless, in some regions with low population density and limited forest conversion, the economic gains from logging encourage land clearing and road construction.[11]

Governments of developing countries encourage the exploitation of forest resources to pay off foreign debts and allow logging without forest management or reforestation. These governments take the long-term benefits of their forests for granted, and sell logging rights too cheaply to exploit short-term benefits. Logging concessions are not usually granted on a competitive basis and turn over too frequently for the forests to regenerate. Internal political instability and corruption can keep logging concessions from developing programs for reforestation. There are thus many political problems that do not encourage management of forests for the long-term benefit of the country as a whole.[12]

Tropical Rain Forests

The different kinds of forests throughout the world call for different kinds of management practices. Tropical rain forests are scattered in an uneven green belt that lies roughly between the Tropic of Cancer and the Tropic of Capricorn. Rain forests grow in regions where at least four inches of rain falls monthly, where the mean monthly temperature exceeds 75 degrees F, and where frost never occurs. Not all the rain forests in the world are tropical, witness the Olympic Peninsula of Washington and the Tongass National Forest in Alaska. But these forests are much less diverse and contain far fewer species of trees and other forms of life than the tropical rain forests.[13]

Tropical rain forests are broad-leafed woodlands that have at least one hundred inches of rain each year, making them very wet and humid places. There are two types of tropical rain forests. The first is the closed forest where trees and undergrowth combine to cover the ground, making the terrain very difficult to traverse. Closed humid forests are located in high rainfall regions of the Amazon Basin in Brazil and on the islands of Southeast Asia. Open forests, on the other hand, are formations with continued grass cover, such as the savannah woodlands of Africa, which are considered to be open tropical forests.[14]

Rain forests once covered 5 billion acres of land in the tropics, but only half of the original rain forests exist today, and even these are rapidly disappearing. They still cover about 5 to 7 percent of the earth's surface, a land area about three-quarters the size of the lower United States. But during the 1980s alone, about 8 percent of the tropical rain forests were lost worldwide, a decline from 1,910 million hectares in 1980 to 1,756 million hectares in 1990 (a hectare is approximately equivalent to three acres of land surface). During this period, these rain forests were cleared at an average annual rate of 15.4 million hectares, an area almost twice the size of Austria. Some estimates indicated that this destruction was taking place at the rate of close to 40 hectares (100 acres) per minute. If deforestation were to continue at this rate, almost all tropical forests would be gone or severely depleted in just 30 years.[15]

These rain forests are located in three main geographic areas including Asia, Latin America, and Africa. During the 1980s, Asia had 16 tropical countries that covered 336.5 million hectares, and tropical forests covered 305.5 million of these hectares. Some 1.82 million hectares of these forests were being cleared each year during this period. There were 23 tropical countries in South and Central America with a total land mass of 895.7 billion hectares, including 678.7 million hectares of tropical rain forest. These forests were being destroyed at the rate of 4.12 million hectares each year. Finally, Africa had 37 tropical countries covering 703.1 million hectares. Forests cover 216.6 million hectares with 1.33 million hectares being cleared annually.[16]

The primary causes of this deforestation vary by region. In Asia, the forests are threatened by commercial logging and agricultural expansion. The main pressures to cut down the rain forest in Africa come from fuelwood collection, overgrazing of cattle, and logging. In Latin America, forest clearance is associated with cattle ranching, population resettlement schemes, and major development projects such as hydroelectric dams and highways. Commercial logging is also a factor in this deforestation. Globally, some 5.9 million hectares are logged annually in the tropics, including 4.9 million hectares in primary forests. These forests supply about 30 percent of the world's log exports, 12 percent of sawnwood exports, and 60 percent of plywood and veneer exports.[17]

Tropical rain forests are home to from 5 million to 10 million species of plants and animals. At least 50 percent, and perhaps even as high as 90 percent, of all plant and animal species exist in tropical rain forests, yet these rain forests cover only a small percentage of the earth's surface. They thus contain the richest diversity of life found anywhere on earth. It is estimated that two thirds of the plant and animal species in a tropical rain forest are in the canopy, and when all the trees are removed destroying the canopy, the habitat of these plants and animals is also destroyed.[18]

A report issued by the U.S. National Academy of Sciences claims that a typical 4-square-mile patch of rain forest may contain 750 species of trees, 125 kinds of mammals, 400 types of birds, 100 of reptiles, and 60 of amphibians. Each type of tree in the rain forest may support more than 400 insect species. Actually the number of insects are so great that they can only be guessed at, but one hectare may contain as many as 42,000 species. Some of these species are of such huge size as to stagger the imagination. It has been claimed that some lily pads are 3 feet or more across, some butterflies have 8-inch wingspans, and some fish can grow to more than 7 feet long.[19] Tropical forests thus represent the most biologically diverse communities and the the most complex systems known in the universe.

> No other habitat on earth contains such a profusion or weight of plant life per hectare. Under the tropical sun, moreover, everything grows at astonishing speed. The rainforest produces new vegetable tissues faster than any other community on land. But death, too, is ever-present. A smell of decay hangs in the air and underfoot is a thin layer of debris. The forest's dynamism is fuelled by the speed of decay in the hot, damp, atmosphere, which acts as an incubator for scavenging and digesting organisms, and by the powerful flow of nutrient-carrying water from the ground to the canopy, drawn by the suction of evaporation from its myriad leaves. The rainforest is thus in dynamic balance, at a hectic rate of turnover.[20]

Many drugs that are now prescribed in developed countries owe at least part of their potency to chemicals from wild plants from the rain forests. Aspirin is made according to a chemical "blueprint" supplied by a compound extracted from the leaves of tropical willow trees. Quinine, used to combat malaria, comes from the bark of a tree in South America. The rosy periwinkle, which has been used by tribal healers for generations, supplies vital materials for drugs effective against Hodgkins disease, leukemia, and several other cancers. The U.S. National Cancer Institute has identified more than 1,400 tropical forest plants with the potential to fight cancer.[21]

Sales of drugs with active ingredients derived from rain forests have been estimated to be greater than $100 billion per year. Yet only 1 percent of plants in the tropical rain forests that have been identified have even been analyzed as to their potential use for curing diseases. There is the possibility that a newly discovered species of roundworm might produce an antibiotic of extraordinary power. An unnamed moth contains a substance that blocks viruses in a manner never envisioned by molecular biologists. An obscure herb could be the source of a blackfly repellant. Millions of years of evolution through natural selection have made these organisms that exist in rain forests superhuman chemists in developing substances that may defeat many biological problems that undermine human health.[22]

These forests thus provide a habitat that encourages and supports a biological diversity found nowhere else on earth. As such, they provide a genetic resource of untold value. Although the tropical rain forests are thus rich in resources, they are also very fragile and have great difficulty recovering from severe or repeated human disturbances. This fragility is due not only to the scattered distribution of most species and their ecological requirements but also to their interdependencies. The soil in most tropical forests is poor and easily eroded when the forest cover is removed. One billion people worldwide are affected by floods, fuelwood shortages, and soil degradation caused by tropical deforestation. Many of these people will starve to death during the next 30 years if deforestation of the rain forest continues.[23]

The land beneath the forest is usually poor and yields crops for only two or three seasons. Nutrients are contained in soil eroded by rain, and when the land is depleted, the farmers clear more land by burning rain forests. The sun bakes the exposed soil, clouds stop forming over the barren land, and rain patterns change. If the entire forest were destroyed, the land would die, and a reforested Amazon is highly unlikely to rise from the ashes. Temperatures would rise and precipitation levels would fall, severely hindering the growth of new rain forest. There would be a decrease in local evaporation, which provides one-half the water in the Amazon Basin. Eventually, these changes can convert a diverse tropical forest into a sparse grassland or even desert.[24]

> The entire rainforest community, which supports so much life, is held in a fine and delicate balance. As each component dies, its nutrients are recycled by a whole community of decomposing organisms and are then reabsorbed by the plants to provide a new life for the forest. In the rainforests, this system of recycling has evolved over millions of years to become supremely efficient; it is only this which enables their poor soils to support a paradoxically luxuriant growth. The rich vegetation therefore gives a false impression of the soil's potential; for once the trees have gone, the land quickly becomes unproductive. Agriculturalists and cattle ranchers learn this to their cost, as the land they use gradually deteriorates. In the search for new, more fertile soils, they clear ever more forest land, contributing to the destruction already wreaked by development schemes and exploitation for timber and minerals.[25]

It is now known that the rain forest is an incredibly efficient ecosystem that wastes little energy or matter that is essential to its survival. Because the soil is poor, the forest functions like a delicately balanced organism that recycles most of its nutrients and much of its moisture. Water evaporates from the upper leaves of trees, cooling them in the process as they collect the intense sunlight. This evaporated moisture is gathered into clouds, which return the moisture to the system in torrential rainfall. Dead animals and plants decay quickly, returning essential nutrients from the soil back to growing plants. Virtually no decaying matter seeps into the region's rivers because of efficient recycling.[26]

Tropical rain forests are a renewable resource if managed properly with long-term interests in mind and on a sustainable basis. The concept of sustainable yield management if implemented appropriately can be used to achieve a balance between the harvest and growth of a forest, which theoretically can provide timber products indefinitely. Sustainable harvesting should not diminish the benefits to future generations, and requires the protection of soil, water, wildlife, and timber resources in perpetuity. The mix of benefits and products from rain forests may vary from region to region, but in the aggregate these forests must be as useful to future generations as the primary forests are to current generations.[27]

The Smithsonian Institution in conjunction with the International Hardwood Products Association, has developed a consensus regarding the principles that should govern tropical rain forest management. These principles recognize rain forests as dynamic systems that provide a variety of goods and services that are useful for humans. Choices must be made as to how these goods and services are going to be utilized, but these choices must be made intelligently with long-term goals of sustainability providing a framework for decision making. These principles include the following.

- Forests are dynamic. If not utilized, trees are continually dying and being replaced. Sustainable utilization systems can be devised and are viable so long as they mimic natural forest dynamics and work within the nutrient limitations of the ecosystem. The problems arise not so much in devising such management systems, but in making them work under the prevailing socio-economic conditions.
- Forests provide a wide variety of goods and services. Non-timber products, such as fruits, nuts, rubber, and rattan, often harvested by forest-dwelling people, provide economic benefits that can be more valuable than timber removal. Economic analysis of the value of timber harvesting should take into account the potential threat to other major opportunities if log extraction damages production of these products.
- Timber should realize its true value. Where timber prices do not fully cover the replacement and environmental costs, there is no incentive to restore the forest so that it can be sustained at its original level. Governments commonly collect far too little forest revenue to reinvest in replacement efforts that incorporate all of the affected environmental factors. Where the logger has no incentive to utilize low grade and waste products, maximum utilization will not occur, and sustainability could be endangered by high-grading and forest fires (fueled by dry matter left behind after logging operations).
- Fiscal and financial incentives for conservation and regeneration must be provided. In the same way that logging should bear its environmental costs, conservation, regeneration, and reforestation should enjoy profits from the environmental benefits they convey to the society at large.
- Re-entering a forest to cut (re-cutting) too frequently is a common problem, which inevitably degrades the forest. Re-cutting may occur in order to satisfy the annual volume a logger is obliged to cut, or to take advantage of remaining stands of commercial value if the logger's concession is too short to ensure him the access to the next harvest. Re-cutting often takes place before there has been adequate re-growth. Frequent re-cutting compounds any damage from the first cut, thereby slowing down regeneration.
- Timber concession agreements are currently much too short in duration. Most are not more than 20 to 30 years. As a result, the concessionaire does not have practical reason to invest in the long-term future of the concession. Concession contracts should extend over at least two cutting cycles, and should be subject to periodic review for compliance with good management practices.[28]

The primary forests, which are those essentially unmodified by recent human intervention, can best be sustained by the preservation in perpetuity of tracts of such forests, which are representative of diverse environments. Once sufficient areas of these primary forests are secured, other primary forest areas may be subjected to sustainable utilization practices that preserve the structure and function of the ecosystem. Secondary forests, those left after exploitation of marketable timber and regrowth after deforestation and agricultural abandonment, can be managed with the goal of accelerating the rate of growth to reach maximum sustainable production, but all resource values such as water, soil, and wildlife must be considered in making decisions. An increasing proportion of tropical lands that are now deforested or inadequately stocked, must be planted with useful trees to meet future needs for industrial forest products and to reduce the pressure on primary forests. These are called tree plantations and require continuous monitoring and flexible management.[29]

In July 1989, leaders of the United States, Italy, France, West Germany, Great Britain, Canada, and Japan met in Paris for a summit on environmental issues. These countries endorsed the concept of writing off debts of developing countries in return for preservation of vanishing tropical rain forests.[30] Most major banks in the developed world hold large amounts of hard currency debt from many of these developing countries, but it is unlikely that this debt will be paid in full by these countries. If conservation organizations acquire title to this debt, they may be able to negotiate with debtor countries to obtain repayment in local currency and use the proceeds for conservation of the rain forests. Several of these debt-for-nature swaps have been worked out in recent years, and while the results have been meager compared with the overall problems, they are still substantial[31] (Table 10.1).

Some companies have discovered that selling goods with nature in mind can be a sound business practice, and believe that harvesting rain forests for fruits, nuts, essences, and oils, instead of destroying them through slash-and-burn farming and ranching, can provide the incentive to ensure that the forests survive, and have tried to promote this strategy in their products. Ben & Jerry's Rainforest Crunch ice cream has been available since October 1989, and has proved popular with consumers. Community Products Inc. expected to sell $1.5 million of its Brazil nut-and-cashew brittle. This candy, packaged in colorful tins that tout the environmental integrity of its ingredients, is sold in department stores and by mail order for $12 a pound.[32]

Interest in this idea of sustainable harvesting has grown in recent years, encouraged by studies such as one that appeared in the British journal *Nature,* which reported that a hectare of land harvested in such a fashion could generate an annual income of almost $700, while a plantation or cattle ranch on the same parcel of land would yield only $150 over the same time period. To promote the harvesting of these products, Brazil established four large "extractive reserves" in which logging is banned and long-term harvesting rights are given to rubber trappers and gatherers of nuts and fruits.

TABLE 10.1 Commercial Debt-for-Nature Swaps

Country	*Funds for Conservation*	*Face Value of Debt*	*Cost*
Bolivia	$250,000	$650,000	$100,000
Costa Rica	$42,872,904	$79,853,631	$12,515,474
Dominican Rep	$582,000	$582,000	$116,400
Ecuador	$10,000,000	$10,000,000	$1,422,750
Guatemala	$90,000	$100,000	$75,000
Jamaica	$437,000	$437,000	$300,000
Madagascar	$3,149,229	$3,149,229	$1,455,268
Mexico	$250,000	$250,000	$180,000
Nigeria	$93,446	$149,800	$64,788
Philippines	$10,622,354	$11,440,000	$5,638,750
Poland	$50,000	$50,000	$11,500
Zambia	$2,270,000	$2,270,000	$454,000
Total (1/92)	$70,666,933	$108,931,660	$22,333,930

Source: Huntington Williams III, "Banking on the Future," *Nature Conservancy,* May–June 1992, p. 27.

Several more of these kinds of reserves are under consideration, as the idea is beginning to catch on that nontimber forest products offer better livelihoods for many forest dwellers that timber extraction or slash-and-burn farming.[33]

The country of Brazil became the target for worldwide criticism in the way it was allowing its rain forests to be destroyed in the interests of industrial development. Environmental activists from around the world charged that the government's policies are resulting in the wanton destruction of Brazil's rain forest, its wildlife, and its native peoples, and are also endangering the world environment. Fires set by ranchers and homesteaders in the Amazon region were said to be spewing 7 percent of the world's carbon dioxide into the atmosphere, contributing to the global warming. The government called much of this criticism unjust and cruel, but sought to stem some of the outcry by announcing a new environmental program to slow rain forest destruction.[34]

Provisions of the plan included establishing a five-year, $100 million program to zone the Amazon region for agriculture, mining, and other uses. Raw-timber exports and tax incentives to cattle ranchers were to be suspended, at least temporarily. The production and sale of toxic chemicals used in mining and agriculture was to be regulated. The plan also called for the creation of 7 million acres of new national parkland, and involved studying a possible expansion of the areas set aside for use of the country's 220,000 remaining native people.

The issue over rain forest preservation in Brazil, however, was framed as a battle between developed and developing nations. The developed world sees the Amazon as a vast nature reserve that holds cures for illnesses and guarantees the climate, while the Brazilian government sees the Amazon as containing resources to be exploited and providing space for its fast-growing population. Many Brazilians resent outside interference in their affairs, and see debt-for-nature swaps as just one more way for outside people to meddle in the country's affairs. Such strident nationalism drew a negative reaction from critics of the government, who called the program nothing more than an attempt to appease foreign criticism, thus limiting its success.[35]

Besides these efforts, there has been one major attempt to develop a comprehensive plan to save the rain forests on a worldwide basis. A Tropical Forestry Action Plan was prepared over several years with input from governments, forestry agencies, agencies of the United Nations, and nongovernmental organizations to help tropical countries come to grips with deforestation. Each of these nations was to come up with a formal proposal for managing and protecting its forests, with the help of international agencies, and T.F.A.P. would channel $8 billion in aid over a five-year period to implement those programs. The effort was launched with assurances that mistakes of the past, which included duplication of effort, rip-offs by contractors, consultants, and corrupt officials, and a tendency to promote the donor's priorities at the expense of developing countries, would not be repeated.

The specific purposes of the plan, which covered all tropical forests, was to improve the lives of rural people, to increase food production, to improve methods of shifting agriculture, to ensure the sustainable use of forests, to increase supplies of fuelwood and the efficiency of its use, and to expand income and employment opportunities.[36] The plan had five components: (1) forestry in land use, which hoped to make better use of forest lands to ensure sustainable practices in forestry; (2) development of a sustainable logging industry to improve management of tropical logging; (3) resolving the fuelwood crisis through conservation measures, use of alternative fuels, and development of new fuelwood resources; (4) conservation of forest species and ecosystems; and (5) building knowledge and expertise in forest management.[37]

Regarding success of the plan, a report issued in 1991 was not very complimentary about the plan as it had been implemented up to that point. Countries had been chosen for aid because of their ability to digest large amounts of money rather than by the size of their uncut tropical forests. Many of the proposed action plans involved opening of previously pristine forests for exploitation. For example, in Cameroon the T.F.A.P. proposed construction of a 370-mile road through virgin rain forest to open it up to development. Apparently the plan was also sold in different ways to rich and poor nations. In industrial nations the plan was touted as a way to save the forests, while in developing nations, the plan was presented as one more source of funding for traditional forestry projects.[38]

The most serious problem with the plan, however, seemed to be it's based on a flawed premise. The plan was founded on the assumption that tropical forests can be harvested and managed without damaging the ecosystem, yet many doubted there was evidence enough to support this assumption. Since we know so little about the intricate codependencies that tie the myriad species of plants, animals, and insects of these forests into a working system, some biologists wonder whether tropical forestry is sustainable at any commercial level. Under threat of a funding cutoff from sponsoring organizations, the United Nation's Food and Agriculture Organization, which was principally responsible for administering the plan, ceded control of the program to an outside governing council and agreed to participate in the program's redesign.[39]

None of these efforts seem to have been successful to halt rain forest destruction. From June to November of 1995, tens of thousands of fires were burning over the 2 million square miles of the Brazilian rain forest. As Brazil's economy rebounded, many farmers and ranchers started to expand their operations again into further and further reaches of the rain forest. Most of this burning was illegal, but the government seemed powerless to halt the destruction. After several years of delay, the government finally came up with $2.4 million to fund a detailed survey of the damage that has been done to the rain forest in the last three years.[40]

For the immediate future, it seems that large quantities of valuable lumber will continue to be destroyed, depriving nations and communities of opportunities to realize the full economic benefit from forest industries' development and trade. More watersheds will be degraded. There will be widespread soil erosion, continued siltation of rivers and dams, and further loss of agriculture land causing food shortages. Flooding may result in further unnecessary loss of life in some countries that have no forests of their own, but suffer because of deforestation that takes place many miles from the country. All of these problems call for further action to protect the rain forests from further needless destruction.

If destruction of the rain forests continues much longer, the damage to this special kind of ecosystem may be irreversible. The impact of deforestation on regional climates could change the character of the rain forests and lead to even greater extinctions of plant and animal species than is taking place at present. The poor of these regions would then have to endure even more misery than they are at present. The people in the rest of the world also have a stake in what happens to the rain forests, as the world needs the rain forests as a functioning system. This is a more important issue than the issue of who owns the rain forests and thus has responsibility for them. In the final analysis, the responsibility for saving the rain forests belongs to everyone rather than just a few nations.[41]

Old-Growth Forests

Old-growth or virgin forests are those that contain massive trees that are hundreds and thousands of years old, such as the great stands of Douglas fir, giant sequoia, and coastal redwoods in the western part of the United States. These forests are thick with trees and vegetation and generally have a greater diversity of plant and animal life than secondary forests. The latter are, as the name secondary implies, stands of trees that have grown up after the virgin forest was cut down, and result from secondary ecological succession. Many forests in the United States and other temperate areas such as Europe are secondary forests that grew after the logging of virgin forests or the abandonment of agricultural lands.[42] Old growth is described by the Wilderness Society in the following manner.

> Old growth . . . is distinguished from young and mature forest by its vigorous diversity of tree species and sizes, including massive Douglas-firs up to nearly three hundred feet tall and thirty feet or more in circumference at the base. Beneath these aspiring giants grow smaller trees—western hemlock, red cedar, bigleaf maple, and other shade-tolerant species. The trees together form a canopy of several layers that diffuses light into an ample soft radiance; here and there shafts of sun penetrate directly to the forest floor. Because shrubs, herbs, and seedling trees are usually sparse and patchy, the forest seems spacious, deep to the eye. Everywhere, more various even than their living progeny, are the generations of trees gone by—the standing dead with dry needles still intact, limbless snags, rotting stubs coated with moss, and downed logs and limbs of all sizes and states of decay.[43]

Logged areas in the Northwest were replanted haphazardly or not at all for the first half of this century, as little thought was given to conservation of these resources. During the last 40 years, replanting efforts have intensified and sustained yield has at least developed as a goal of forest management. While rotation cycles vary from forest to forest, and even from site to site, planted trees in this area of the country generally require at least 70 years to reach harvestable proportions. As these secondary crops mature on private lands, timber companies are pressing for higher and higher annual cuts on the national forests and other public lands that contain practically all that remains of the old-growth forest in the lower 48 states.[44]

Conservationists and other environmentalists are fighting to save the rapidly disappearing old-growth stands in national and state forests in the Pacific Northwest, and have been embroiled in controversy for several years. Timber companies want to clear-cut these stands and replace them with tree farms. Only 10 to 15 percent of the country's old-growth forests are left, according to some estimates, and at present cutting rates most of these irreplacable forests will be gone within 15 to 20 years. People who want to stop this deforestation believe that the total destruction of these forests will go down in history as one of the great ecological crimes of the century. They also point out that it is hypocritical for the United States to pressure Brazil not to cut down its tropical rain forests, while continuing cutting down its remaining old-growth forests.[45]

The current plight of the old-growth forests had its origins in the late 1940s, when a postwar housing boom resulted in increased cutting of trees on private lands. The logging industry was thus forced to turn to public lands, including the old-growth forests that were prized because of the high quality and quantity of their timber. The National Forest Service and the Bureau of Land Management cooperated in selling rights to new tracts of forest. This policy, when combined with modern logging machinery that makes cutting on mountain slopes easier, has put old-growth forests are risk.[46] Current pressures have increased because of the profitable trade in logs to foreign countries, which take logs from private lands. As log exports have increased, federal forests, from which logs can't be exported, are left as the last remaining large source of wood and wood products for the domestic market.[47]

Though forbidden to do so by the National Forest Management Act of 1976, timber in virtually every national forest in Washington and Oregon has been sold at a level above a sustained yield level, somewhere around 23 percent above, according to Forest Service data for twelve of the forests.[48] Since World War II, these areas have been logged primarily in a dispersed clear-cut system that checkerboards the landscape with 25- to 50-acre clearings, leaving patches of forest of about the same size or somewhat larger. This practice was implemented to minimize damage to the forest, but the result has been to chop it into pieces, which damages its capacity as a wildlife habitat and jeopardizes the remaining trees themselves.[49]

Clear-cutting removes all trees from a given area in a single cutting to establish a new, even-aged stand or tree farm. The clear-cut area may consist of a whole stand of trees, a group, a strip, or a series of patches. After all the trees are cut, the site is either naturally reforested from seed released by the harvest, or is reseeded by foresters with genetically superior seedlings raised in a nursery. Clear-cutting increases the volume of timber harvested per acre, reduces the need for road building, and shortens the time needed to establish a new stand of trees. Timber companies prefer this method of tree harvesting because it requires less skill and planning than other harvesting methods and gives them a greater return.[50]

But clear-cutting can also lead to severe soil erosion if done on steeply sloped land, sediment water pollution, flooding from melting snow and heavy rains, and landslides. It also leaves ugly, unnatural forest openings that take decades to regenerate themselves. The number and types of wildlife habitats are also reduced and thus the biological diversity of the forests is affected. Once a site has been clear-cut, it is hard to break the cycle and wait 100 to 400 years for an uneven-aged stand to regrow through secondary ecological succession.[51]

Selective cutting involves cutting down intermediate-aged or mature trees either singly or in small groups. This practice reduces crowding, encourages the growth of younger trees, and maintains an uneven-aged stand with trees of great variety. Given enough time, the stand will regenerate itself. If done properly, selective cutting helps protect the site from soil erosion. This method is favored by those who wish to use forests for multiple purposes and wish to preserve biological diversity. But, selective cutting is also much more costly and, unless the value of the trees is high, is not economically profitable.[52]

The controversy again centers on the ecological versus the economic functions of the old-growth forest. Where is a proper balance to be found between maintaining some old-growth forest for its ecological value and cutting it down to exploit its economic

potential? Is it necessary to preserve some of these forests or can they all be cut down without any serious consequences? How much should be preserved and on what ethical principles can such a decision be made? These questions continue to surface in the controversy that seemingly has no successful resolution to satisfy all parties.

To stop continued clear-cutting, the northern spotted owl entered the fray, as studies indicated that populations of the owl were declining rapidly as logging shrank and fragmented its habitat. In fact, some studies showed that it was poised on the brink of population collapse. The Forest Service was charged by law with the perpetuation of vertebrate species native to the national forests, and thus had to come up with some kind of plan to preserve the species. It was well aware that if the owl made the endangered species list, it could virtually shut down logging in the Northwest. Thus initially, the Forest Service proposed setting aside 550 spotted owl habitat areas in Oregon and Washington and 200 more in the Douglas fir zone of northwestern California, each area containing 2,200 acres of old-growth forest.[53]

While the controversy centered on the owls, what was really at stake was the health and survival of the entire Douglas fir ecosystem of the Northwest, the natural ecological system on which owls and watersheds as well as the region's human economy all depend. The owls are an indicator of the plight of that entire ecosystem, but it will accomplish little, so said some environmentalists, to set aside preserves for them while the forest all around these areas remains available for harvesting. If the ecosystem at large is destroyed, so will be the owls along with many other forms of wildlife. Thus the Forest Service's plan was criticized on the basis that the acreage was too small and the number of areas too few to ensure the owl's survival. [54]

In April 1989, the U.S. Fish and Wildlife Service recommended that the spotted owl be placed on the endangered species list, which would protect it from further logging. Industry officials predicted an economic apocalypse that would cost the region 132,000 jobs and $3 billion in local payroll, not to mention the $1.6 billion the federal government would have received from selling logging rights in national forests. Environmentalists disputed these figures, calling them wild exaggerations and filed three separate lawsuits against the various federal agencies that control forest and wildlife policies in areas that contain old-growth forests.[55]

A new concern was introduced in 1990, when environmental groups and cancer researchers asked for federal protection for the Pacific yew, a tree with bark that provides a scarce new cancer-fighting drug. In 1960, the National Cancer Institute had already found the first indications that an extract of the yew's bark could be used against ovarian as well as other cancers in women. The active chemical compound in this bark was later isolated and called taxol, but since it takes the bark from three trees to produce enough taxol for one cancer victim, the substance is in very short supply. Efforts to synthesize the drug have not worked. The petition filed by these groups asked the Interior Department to list the Pacific yew as threatened under the Endangered Species Act because its habitat is the old-growth forest. The tree is apparently one of the slowest-growing species in the world, and thrives in the shady undergrowth of ancient pine trees.[56]

In response to these pressures, the Forest Service unveiled a new master plan that called for a substantial reduction of logging in national forests, to an average 10.8 billion board feet by 1995 from the current average annual cut of 12.2 billion board feet. The plan would halve clear-cutting in national forests and would put greater emphasis on wildlife protection and recreation. The timber industry quickly predicted a 20 percent increase in the cost of lumber because of this change, and a consequent rise in housing prices putting thousands of people out of the market. "We are squeezing 65,000 families out of affordable housing," said one industry spokesperson, "to provide bird houses for spotted owls."[57]

Finally, the owl was brought under protection of the Endangered Species Act by being listed in June 1990, along with a comprehensive logging plan announced by the Bush administration. The plan was said to save 450 pairs of owls, which the administration claimed was 125 more than under a plan proposed by an interagency scientific panel in the spring. Logging would be reduced in 1991 to 750 million to 800 million board feet from about 950 billion board feet in 1990. Some 1,000 jobs would be lost under the plan, as compared with 7,600 under the earlier proposal. The administration also proposed to set up an interagency task force to devise a forest management plan for fiscal 1991 that it said would be designed to protect the owl while mitigating economic dislocation.[58]

Under pressure from a federal district judge, the U.S. Fish and Wildlife Service eventually designated 11.6 million acres of old-growth forest as "critical habitat" for the owl, either placing it off limits to loggers or otherwise restricting cutting. Added to the more than 3 million acres already off limits, the total acreage affected would have been an area the size of Massachusetts and Vermont, by far the largest amount of land placed off limits to industry.[59] According to the timber industry, the move would put 100,000 people out of work, an estimate that was challenged by environmentalists. The controversy was again seen as owls versus jobs, but as some articles pointed out, the issue is much more complicated.[60]

The owl became an issue in the 1992 presidential campaign, and after Clinton was elected, plans were made to hold a timber summit of all interested parties. It was estimated that the summit would attract 300,000 people including loggers, environmentalists, journalists, and other parties. The outcome of the summit was a plan that tried to strike a balance between the needs of nature and the needs of man, but that seemed to please neither side in the debate. The administration plan decreed that the amount of logging on federal land would be sharply reduced, allowing only 1.2 billion board feet to be harvested annually, compared with a peak of 5 billion board feet in the 1980s, and offered a $1.2 billion aid package for worker retraining and community investment.[61]

Loggers were shocked by the reduction and claimed that 85,000 jobs would be lost and the timber-dependent town devastated. But since areas of old growth outside of wilderness areas and parks would be vulnerable, environmentalists were not pleased either. They did not like the notion that any additional logging of old growth should be allowed, warning that it could push endangered species to extinction and imperil one the the country's most vital natural habitats. Lawmakers from the Northwest also attacked the plan calling it a sellout.[62] Such reaction only shows how difficult it is to resolve issues of this kind to everyone's satisfaction.

In 1995, the new Republican Congress found a way to open up more federal land to logging. It passed an Emergency Salvage Timber Sale Program which President Clinton signed into law as part of a wider budget-cutting package. The law suspends existing environmental laws such as the Endangered Species Act as they pertain to salvage logging. Environmentalists feared that the program had far-reaching implications beyond merely thinning dead or dying trees, because the program also allowed for cutting "associated trees" considered in imminent danger of mortality or infection, which could refer to every tree in the forest. They were also concerned about the creation of new roads to facilitate logging, as once roads were built for salvage operations, whole areas would be opened up to exploitation.[63]

This controversy is much more than just a battle between a particular type of owl and an industry. It is really a conflict between different philosophies regarding the place of nature and the relationship of human beings to their natural environment. Fundamental questions are involved in the debate about the future of the environment and the priorities to be established. Are the old-growth forests, and by extension, nature itself, there for humans to use and exploit, or are they to be preserved in their natural state? How much wilderness does the country need, and how is this need to be balanced against the need for lumber to build housing? How much human discomfort in the form of job change and other dislocations can be justified in the name of conservation or preservation?[64]

Sooner or later the timber industry, Forest Service, and the public at large are going to have to come to grips with the reality that we can't continue to cut more wood than the land is capable of reproducing, or we will exhaust timber resources for future generations. At the present time, according to some experts, too many trees are falling, millworkers are losing their jobs, spotted owls and other animals face possible extinction, and the land itself is slipping away into rivers and streams. Losses are high for all parties concerned. Rapid growth of demand for logs that depletes the forest must give way to slower growth and eventually some kind of equilibrium where the forest's wealth is conserved and carefully recycled. "Each member of the community is supported by what the land in its health allows, a diverse and vigorous commonwealth sustaining itself through time."[65]

Forest Management

Such controversies over old-growth and tropical rain forests show how difficult it is to resolve conflicting demands on resources. The need for preserving these forests because of their ecological value is important. But they are also useful for timber, medicines, ranching, nuts and other such products, as well as for recreation. In addition, cutting down the forests affects other wildlife such as salmon and other fish that are affected when rivers are silted because of runoff of clear-cut areas during heavy rainstorms. When there are so many interests involved, how can decisions be made to balance these interests or establish priorities, and who should make these decisions?

There is some concern that centralized management is part of the problem, and that the rights of indigenous peoples who have inhabited and conserved forests since time immemorial need to be recognized. Local tribes who depend on the forest are generally powerless in national politics and their rights are recognized only in the breach. Yet history shows, according to some observers, that indigenous people have saved more tropical forests than all the world's conservation groups put together. They are a vast storehouse of knowledge regarding how the rain forest works and practices that are sustainable, since their existence depends on preservation of the forest, and yet this knowledge is lost as these peoples are destroyed or moved out of the way in the never-ending quest for development.[66]

Large organizations act to maximize their budgets, and the Forest Service is no exception. It favors commodity extraction over nonconsumptive uses, because it retains most of the proceeds from timber sales. It does not benefit from camping, fishing, hiking, or other such uses, nor does it benefit from saving the fish and wildlife that depend on its forests for existence. To counter these incentives, user fees might be charged for every use of the forest and its budget should come exclusively from these fees, which might result in a change in priorities. Communities could also be more involved in management of local forests.[67]

Allowing local countries and peoples to share in the profits when drugs are discovered is another way to promote preservation of forests. The giant drug firm Merck & Company signed a contract with a Costa Rican organization called the National Institute for Biodiversity, which gave the company access to the country's estimated 12,000 species of plants and 300,000 species of insects in exchange for sharing the profits from any drug discoveries. It was expected that the country would eventually make more money from these sales than from its traditional products of bananas and coffee, giving it an incentive to preserve the rain forest. Even so, the agreement did not recognize the rights of local tribes and assumed the state could use the resources as it decided. Not a single country in the world extends intellectual property rights to indigenous knowledge, treating it as if it were a national resource to be exploited.[68]

Wood certification programs are cropping up around the world. These programs typically involved some process to label wood products as having been produced under conditions of sustainability, which can then be taken into account by purchasers of these products. This process is similar to green marketing efforts of companies and retailers. Such labeling is supposed to work by creating a market demand for products from forests that are managed in a sustainable fashion. The criteria developed by the Forest Stewardship Council are the most widely used for this purpose. They are said to be consistent but yet flexible enough to recognize country differences[69] (Exhibit 10.1). Forest management operations should follow these principles.

EXHIBIT 10.1 The Forest Stewardship Principles

1. Respect the laws of the country it [forest management] is operating in and international treaties and agreements to which the country adheres.
2. Define and legally establish long-term tenure and rights to use land and forest resources.
3. Respect the legal and customary rights of indigenous peoples to own, use and manage their lands and resources.
4. Support the social and economic well-being of forest workers and local communities.
5. Encourage the optimal use of forest products and services to ensure economic viability.
6. Maintain the critical ecological functions of the forest and minimize adverse impacts on biological diversity, water resources soils, non-timber resources and ecosystems and landscapes.
7. Write, implement and keep up-to-date an appropriate management plan.
8. Conduct regular monitoring that assesses the condition of the forest, the yields of forest products, the chain of custody and management operations.
9. Not replace natural forests by tree plantations. Plantations should complement natural forests and reduce pressures on them.

Source: Peter Knight, "Save a Forest—Use Bar Codes," *Tomorrow,* June–July 1995, p. 69.

Some experts argue that none of these schemes will work to reduce demand for wood products, which they argue is necessary given the strain the forests are currently under around the world. Nor do they address the other driving forces of deforestation. The basic problem is that the prices of forestry products do not reflect the true costs of cutting down forests and externalize these costs on others. The price of wood does not reflect the costs of flooding or destruction of habitat for other wildlife. Nor does it reflect the carbon storage function of the forest, which has been estimated at $3,000 in net present value per hectare for some forests. Deforestation also destroys nontimber forest products that do not register fully in the money economy. Scenery with real economic value for the tourism industry is also destroyed. Many economic services provided by forests are not factored into the prices of wood products[70] (Exhibit 10.2).

EXHIBIT 10.2 Economic Services Provided by Intact Forest Ecosystems

Service	*Economic Importance*
Gene pool	Forests contain a diversity of species, habitats, and genes that is probably their most valuable asset; it is also the most difficult to measure. They provide the gene pool that can protect commercial plant strains against pests and changing conditions of climate and soil and can provide the raw material for breeding higher-yielding strains. The wild relatives of avocado, banana, cashew, cacao, cinnamon, coconut, coffee, grapefruit, lemon, paprika, oil palm, rubber, and vanilla—exports of which were worth more than $20 billion in 1991—are found in tropical forests.
Water	Forests absorb rainwater and release it gradually into streams, preventing flooding and extending water availability into dry months when it is most needed. Some 40 percent of Third World farmers depend on forested watersheds for water to irrigate crops or water livestock. In India, forests provide water regulation and flood control valued at $72 billion per year.
Watershed	Forests keep soil from eroding into rivers. Siltation of reservoirs costs the world economy about $6 billion per year in lost hydroelectricity and irrigation water.
Fisheries	Forests protect fisheries in rivers, lakes, estuaries, and coastal waters. Three fourths of fish sold in the markets of Manaus, Brazil, are nurtured in seasonally flooded *varzea* forests, where they feed on fruits and plants. The viability of 112 stocks of salmon and other fish in the Pacific Northwest depends on natural, old-growth forests; the region's salmon fishery is a $1-billion industry.
Climate	Forests stabilize climate. Tropical deforestation releases the greenhouse gases carbon dioxide, methane. and nitrous oxide, and accounts for 25 percent of the net warming effect of all greenhouse gas emissions. Replacing the carbon storage function of all tropical forests would cost an estimated $3.7 trillion—equal to the gross national product of Japan.
Recreation	Forests serve people directly for recreation. The U.S. Forest Service calculates that in eight of its nine administrative regions, the recreation, fish, wildlife, and other nonextractive benefits of national forests are more valuable than timber, grazing, mining, and other commodities.

Source: From *State of the World 1994: A Worldwatch Institute Report on Progress Toward a Sustainable Society* by Lester R. Brown, et al., eds.

Few attempts have been made to calculate the prices of forest products under full ecological pricing schemes, but nature-blind pricing is said to bias production toward high-impact technology and boost consumption to unsustainable levels by encouraging inefficiency. The wood products industry is largely stuck in an old commodity mind-set, favoring quantity over quality and volume over value. Yet enormous opportunities may exist for cost-effective savings through efficiency improvements. Ecological pricing would accelerate technological advances and discourage waste of wood products. It would encourage recycling of paper products by making virgin timber more expensive.[71]

Moving toward ecological pricing of timber products means a change in government policies. The first order of business is for governments to stop subsidizing deforestation. Once this is done, then ecological pricing can be implemented. The most powerful means government has at its disposal to accomplish this goal is to use taxes to make forest products reflect their true cost of production. These taxes could be used to reduce high-impact uses of forests. Governments would also have to impose tariffs on goods from countries that did not employ ecological pricing. The imposition of these measures may require a political will that seems to be lacking in most countries; however, some experts see this kind of pricing as critical to a long-term solution to the forest management problem.[72]

SPECIES DECIMATION

The issue of preserving species of plants and animals goes to the heart of one's approach to the environment, and raises many fundamental ethical questions. If one believes that the environment has no value apart from its usefulness to human beings, then the conservation approach makes the best sense. Plant and animal resources must be conserved so that they are not decimated more than necessary to serve human needs for this and future generations. Arguments to slow deforestation because species are being destroyed that may prove to have some medical benefit to humans are based on this view of the environment. If, however, one believes that plants and animals deserve to exist in their own right because they are part of nature, then preservation is the right approach, which means preserving old-growth and tropical forests and the habitats of the species in them because they are a part of nature and deserve to be preserved for their own sakes, not for what benefits they may be able to provide human beings.

The loss of biological diversity through species decimation is the most important process of environmental change, because it is the only process that is wholly irreversible. Once a species is totally eliminated and there are none of that species left, or at least not enough to continue reproduction, that species is gone forever. The chances that the same environmental conditions will ever exist so that the same species will evolve at some future time is for all practical purposes nonexistent. Yet humans take this biological wealth for granted, perhaps because we have always been surrounded by so many plant and animal species that we have not noticed any decimation. Biological wealth is taken less seriously than material or cultural wealth even though biota is part of a country's heritage. Plants and animals are the product of millions of years of evolution centered on that particular place, and hence as much of a reason for national concern as the particularities of language and culture.[73]

Some scientists claim that humans have reduced biological diversity to its lowest level since the end of the Mesozoic era 65 million years ago, when the dinosaurs disappeared. It is the first time in history that plant communities, which anchor ecosystems and maintain the habitability of the earth, will also be devastated. Given the interdependence of human beings and the other species that inhabit the earth, it is remarkable that the task of studying biodiversity is still at an early stage.[74] Some estimates state that the earth is populated with 5 million to 30 million different wild species of plants and animals, and that so far scientists have identified only about 1.8 million species, and two thirds of these are insects[75] (Table 10.2).

Speciation is a process that separates genetic variations into distinct units that are called species. During the process of speciation, the original population of organisms with similar genes, called a gene pool, is divided into two or more gene pools. Each of these new gene pools acquires a unique set of characteristics through mutation and selection. Speciation follows most commonly from the physical division of a gene pool as the separation inhibits interbreeding between individuals in the two populations.[76] These species are like a giant genetic library of successful survival strategies that have developed over billions of years.

The biological diversity found in these species is the foundation for the services the ecosystem provides and upon which we and other species depend for our existence. They are also the source of future biological evolution and genetic engineering. As more and more species are eliminated, the evolutionary pattern for future species is affected, and biodiversity is reduced for future generations. Because the fate of every

TABLE 10.2 Known and Estimated Diversity of Life on Earth

Form of Life	*Known Species*	*Estimated Total Species*
Insects and Other Invertebrates	989,761	30 million insect species, extrapolated from surveys in forest canopy in Panama; most believed unique to tropical forests.
Vascular Plants	248,400	At least 10 to 15 percent of all plants are believed undiscovered.
Fungi and Algae	73,900	Not available.
Microorganisms	36,600	Not available.
Fishes	19,056	21,000, assuming that 10 percent of fish remain undiscovered; the Amazon and Orinoco Rivers alone may account for 2,000 additional species.
Birds	9,040	Known species probably account for 98 percent of all birds.
Reptiles and Amphibians	8,962	Known species of reptiles, amphibians, and mammals probably constitute over 95 percent of total diversity.
Mammals	4,000	
Misc. Chordates[1]	1,273	Not available.
Total	1,390,992	10 million species considered a conservative count; if insect estimates are accurate, the total exceeds 30 million.

[1]Animals with a dorsal nerve chord but lacking a bony spine.

Source: From *State of the World 1988: A Worldwatch Institute Report on Progress Toward a Sustainable Society*, by Lester R. Brown, et al., eds. Copyright © 1988 by the Worldwatch Institute. Reprinted by permission of W.W. Norton & Company, Inc.

species is to become extinct eventually, there must be an influx of new species for life to continue. Thus human beings are destroying the future every time species are eliminated, and reducing the number and kind of plant and animal species that are going to be evolving for future generations.

Whatever the absolute numbers, and the range is wide as indicated above, there seems to be widespread agreement among scientists that more than half the species on earth live in moist tropical forests. Some scientists believe that as much as 90 percent of plant and animal species may be found in these forests.[77] There are two overwhelming species-rich groups in these forests: the arthropods (especially insects) and the flowering plants, which are concentrated in these forests as nowhere else on earth. The fragile superstructure of species build up when the environment remains stable enough to support their evolution during long periods of time, but such communities can be easily destroyed by relatively minor disturbances in the physical environment.[78]

The rain forest is an interdependent community where neither animals nor plants can survive without the other, and all are part of an elaborate network of interactions that weaves through every part of the forest. Over time, species adapt to one another with each lineage refining its ecological niche from generation to generation. The requirements a species has for survival become ever more distinct over time, and animals and plants in the forest become increasingly successful survivors, evolving ever more efficient at avoiding exploitation. As species become more specialized, they leave room for others, and so give rise to new species with narrower ecological niches. This makes for an ever more increasing biodiversity.[79]

Humans, however, are one predator that can upset the delicate balance of the rain forest and destroy it and all the species it contains. Human activity has already had a devastating effect on species diversity, and the rate of human extinctions is accelerating. It is estimated the rain forest clearing alone eliminates 2 to 3 percent of all the species in the forest every year. The global loss could be as much as from 4,000 to 6,000 species per year. Some experts estimate that this rate is on the order of 10,000 times greater than the naturally occurring background extinction rate that existed prior to the appearance of human beings.[80] Perhaps as much as 25 percent of the earth's species will be lost by the year 2050, and even this estimate may be conservative. What seems clear is that the earth will witness the extinction of millions of species before the end of the century.[81]

Habitat destruction seems to be the prime mover behind species extinction, as some 67 percent of all endangered, vulnerable, and rare species of vertebrates are threatened by habitat degradation or destruction. If a habitat is reduced 90 percent in an area, roughly one half of the species in the area will be lost. The big animals and the animals at the top of the food chain go first because it takes a lot of territory to support a meateater. If the country is carved up with settlements and clear-cutting, the remaining wilderness becomes a series of ecological islands, too small to support the full range of biological diversity that existed previously.[82]

Species in Danger

There are many species throughout the world that are in danger of becoming extinct, including some surprises. While sharks strike terror into the hearts of most people, several species are under threat because of commercial shark fishing. Shark fins in particular are in demand in Asia where they are used to make the delicacy of shark-fin soup.

Fishing for shark meat was also encouraged to prevent overfishing of swordfish and other species. Most species of shark reproduce only once every two years, and infant-mortality rates exceed 50 percent. Thus it is difficult for sharks to maintain their numbers as they increasingly come under pressure because of overfishing.[83]

Whales came under such a threat of extinction that the International Whaling Commission instituted a worldwide ban on hunting whales for profit in 1986. Some whales were still allowed to be hunted for scientific purposes. Nonetheless, Norway decided to violate the ban in 1993 and began hunting the minke whale, arguing that a careful harvest of relatively plentiful species like the minke is harmless. Despite being branded an international outlaw and threatened with an economic backlash, the country insisted that its whaling would continue. This action set off a debate about the value of whales and whether they are somehow special and should be considered something more than just another animal.[84]

Tigers are another animal that stimulates fear in people because of its size and strength. Yet the tiger in Asia is a vanishing breed, disappearing faster than any other large mammal with the possible exception of the rhinoceros. It is estimated that no more than 5,000 to 7,500 of the creatures remain on earth, a population decline of about 95 percent in this century. Where they once rambled across most of Asia, they are now confined to small shrinking pockets of their forest habitat. The tiger is under great demand for its skin, its bones that are ground up into potions to heal diseases like rheumatism, and its other body parts. The tiger is also under threat because of human encroachment on its habitat.[85]

The salmon of the Pacific Northwest are vanishing, plunging from a population of around 100 million in the 1850s to about 15 million at present. Some estimates put the current salmon population at much lower levels. According to one estimate, 107 separate salmon stocks, populations that were born in and return to spawn in a particular stream, are extinct and an additional 89 are close to passing into history. The major culprit for salmon losses are said to be the 18 huge dams that provide the region with federally subsidized hydroelectric power. These dams block the upstream migration of salmon to their spawning grounds. Salmon streams are also affected by soil runoff from areas that have been clear-cut during heavy rains.[86]

Birds are flying into trouble, as they are in decline all over the world. According to a report by BirdLife International, a Cambridge, England–based conservation group, 70 percent of the world's 9,600 bird species are in decline, and some 1,000 species are threatened with extinction in the near future. Millions of wild birds around the world are being trapped, shot, cooked, plucked, starved, poisoned, driven out of their nests, covered with oil, or illegally caged (Exhibit 10.3). The clearing of forests has led to declines of 250 bird species that breed in North America and winter to the south. Depletion of wetlands in North America has caused duck breeds to decline by 30 percent since 1955, while the use of DDT in some parts of the world has caused bird's eggshells to become too thin, resulting in loss of young. Oil spills have killed thousands of waterfowl.[87]

Many of these and other animals are under threat because of the illicit trade in endangered species. As species become more and more rare, their price goes up on world markets. Only about 1,000 giant pandas remain in the world, driving their price up as high as $112,000, giving a tremendous incentive for poachers to capture and sell the few that remain in the wild. Elephants are hunted with assault rifles for their ivory

EXHIBIT 10.3 Why Birds Are Disappearing

Threats	*Consequences—A Few Examples*
Habitat Loss	
Deforestation	At least half of the 250 bird species that breed in North America and winter to the south have declined in recent years.
Forest Fragmentation	In North America, parasitic cowbirds infest more than half of other songbirds' nests in many areas where clearings have been cut into forests for roads or developments.
Overgrazing and Plowing of Grassland	Once-abundant species such as Eurasia's bustards and North America's prairie chickens have all but disappeared.
Desertification	Whitethroat warblers that breed in Britain but winter in Africa have declined by 75 percent in 27 years.
Draining of Wetlands	North America's most common ducks have declined by 30 percent in 27 years.
Pesticides	Widespread use of DDT in Africa has caused birds' eggshells to become too thin, leading to declines in populations of at least six birds-of-prey. Use of carbofuran on Virginia fields left tens of thousands of dead birds.
Chemical Contamination	
Acidification of streams	Disappearance of fish has destroyed food supply for loons on northern lakes, and local extinction of the dipper in Wales.
Use of lead shot and sinkers	Consumption of lead scattered by hunters and fishers has caused widespread lead poisoning of swans, geese, ducks, and loons.
Oil spills	Persian Gulf War oil field spills attracted and killed thousands of waterbirds. Exxon Valdez spill killed 300,000 birds.
Exotic Species	Introduction of brown tree snake on Guam extinguished four of the island's five endemic bird species. Roaming house cats in Victoria, Australia, kill 13 million small animals per year, including members of 67 bird species.
Overhunting	In Italy alone, 50 million songbirds end up on dinner plates each year.

Source: Howard Youth, "Flying into Trouble," *World Watch,* January–February 1994, p. 19. © Worldwatch Institute.

tusks that at one time sold for $150 a pound.[88] Poachers also haunt nearly half of America's 366 park areas, supplying animal parts to illegal traffickers who operate in several states. The illegal killing of these animals is a $200-million-a-year business, with as much as $100 million of that amount going for medicinal purposes.[89]

Overfishing

The United Nations Food and Agriculture Organization has estimated that one third of the 200 fisheries it monitors are depleted or being overfished. Some 40 percent of species the United States monitors are also depleted or overfished, and an additional 42 percent are estimated to have lower populations than needed to produce the best catches. As a result, six countries agreed to stop pollack fishing for two years in the Bering Sea. The 1995 quotas for Atlantic tuna has been cut 50 percent from 1993 levels. Salmon fishing was banned in 1994 off the Pacific Northwest. And also in 1994, rules took effect to stop overfishing off New England.[90]

Fish catches have declined in most fishing regions around the world (Table 10.3). The catch has fallen in all but two of the world's fifteen major fishing regions, and in four of them it has shrunk by more than 30 percent. The basic problem seems to be an expanded capacity to catch fish and government subsidies that have lured more people and boats into the business even after the point of diminishing returns has been reached. Between 1970 and 1990, the world's fishing fleet doubled from 585,000 to 1.2 million large boats. Today the industry is said to have double the capacity needed to make the annual catch. This growing disparity between the capacity of the world's fishing fleets and the limits of the oceans has already translated into job losses, as in the last few years more than 100,000 fishers have lost their source of income. One hundred times that number could be out of work in the coming decades.[91]

Other factors contributing to the decline of fish populations include the catching of younger fish to meet increasing demand, which depletes the number of adults of spawning age, a self-defeating strategy. The explosion of people living near seacoasts

TABLE 10.3 Change in Catch for Major Marine Fishing Regions

Region	*Peak Year*	*Peak Catch*	*1992 Catch*	*Change*[1]
Atlantic Ocean		(million tons)	(million tons)	(percent)
Northwest	1973	4.4	2.6	–42
Northeast[2]	1976	13.2	11.1	–16
West Central	1984	2.6	1.7	–36
East Central	1990	4.1	3.3	–20
Southwest	1987	2.4	2.1	–11
Southeast[2]	1973	3.1	1.5	–53
Mediterranean and Black Seas[2]	1988	2.1	1.6	–25
Pacific Ocean				
Northwest	1988	26.4	23.8	–10
Northeast[2]	1987	3.4	3.1	– 9
West Central	1991	7.8	7.6	– 2
East Central	1981	1.9	1.3	–31
Southwest	1991	1.1	1.1	– 2
Southeast	1989	15.3	13.9	– 9
Indian Ocean				
Western	still rising		3.7	+ 6[3]
Eastern	still rising		3.3	+ 5[3]

[1]Percentages were calculated before rounding off catch figures. [2]Rebounding from a larger decline. [3]Average annual growth since 1988.

Source: From *State of the World 1995: A Worldwatch Institute Report on Progress Toward a Sustainable Society* by Lester R. Brown, et al., eds. Copyright © 1994 by Worldwatch Institute. Reprinted by permission of W. W. Norton, Inc.

has damaged fish habitats and spawning grounds in estuaries. The use of sophisticated technology to locate fish and the building of floating factories to process them has made fishing more efficient. Larger and more effective nets have increased the bycatch, a term referring to unwanted fish that are thrown back dead into the ocean. Some 80 percent of the tonnage hauled in by shrimpers consists of this bycatch.[92]

The stakes are high, as fish are a key source of protein for nearly 20 percent of the world's people. In Asia alone, an estimated 1 billion people rely on fish as their primary source of animal protein. The same is true of many people in island nations and the coastal states of Africa. While income generated from marine fishing generates only 1 percent of total income worldwide, fishing and related industries added $18.5 billion to the U.S. gross domestic product in 1992. In Southeast Asia, more than 5 million people fish full-time, contributing some $6.6 billion toward the region's national income. Around the world, some 200 million people depend on fishing and fish-related industries for their livelihoods.[93]

The Georges Bank off the coast of New England was once said to be so abundant with fish that a sailor could walk from one ship to another on the backs of cod that filled the water. But during the 1980s, the seemingly inexhaustible fish stocks of the region began to disappear, and by the mid-1990s, the area was all but shut down and the New England fleet was rusting and rotting at dockside. Hopefully, the fish will return if their spawning grounds are protected, but in the meantime, the federal government helped the thousands of people thrown out of work with programs to buy out boat owners and retrain fishers for new careers. If the fish do return, a much smaller fleet will be chasing them.[94]

A similar situation exists in the Gulf of Mexico, where certain fish populations such as the redfish and red snapper have been depleted. The red snapper in particular has received a great deal of attention. Trawling and overfishing sent the catch of snapper plummeting between 1981 and 1987, when the combined commercial and recreational catch fell from 16.9 million pounds to 5.1 million pounds, a decline of nearly 70 percent. Some experts estimate that it will take until 2019 to rebuild the snapper population to a sustainable level, but only if a solid plan is put in place immediately. Direct overfishing of the snapper is already tightly controlled, so attention has turned to find ways to reduce the bycatch from shrimp trawlers, where it is estimated that 85 percent of all baby snappers are killed each year.[95]

Managing fish populations on a sustainable basis is a difficult and complex process. Generally, fishery managers attempt to maximize the catch of fleets in any given year while maintaining a large enough population of fish to support fishing in future years. That level is called "maximum sustainable yield" to use the jargon of the procedure. But it is easy to miscount fish populations as they tend to move around and are sometimes concentrated in certain regions of the ocean. In addition, politics often intervenes and pressures are placed on managers to err in the direction of keeping fishers working and allowing fishing to continue when it should be stopped.[96]

Protecting Species

According to one expert, creating parks and reserves free from human interferences has long been considered to be the key to conserving plants and animals. Consistent with this strategy, some 425 million hectares of land in some 3,500 areas worldwide enjoy some degree of protection. These areas are set aside to protect an intact example of each of the earth's ecological zones and to reconcile their preservation with the economic needs of surrounding communities. But designating parks, which some think is a static solution to a dynamic problem, is no longer enough to avert a mass extinction of species. According to some estimates, as much as 1.3 billion hectares would have to be set aside to conserve representative samples of all the earth's ecosystems.[97]

Many of these parks and reserves are too small to maintain population sufficient to ensure species survival. Fragmenting ecosystems, whether done though clear-cutting or cutting roads through rain forest, is destructive of wildlife, which need larger habitats to survive. Many of the parks in the United States, for example, do not encompass an entire ecosystem that is necessary to sustain wildlife. Studies have shown that the smallest parks have lost the greatest share of species, but that even very large parks such as Rocky Mountain and Yosemite have lost between a quarter and a third of the native wildlife. These parks do not encompass enough area to preserve the habitat of many species that reside in these areas, but extension of the parks to encompass them will meet with severe resistance.

Biologists estimate that as many as 2,000 species of mammals, reptiles, and birds will have to be bred in capacity to escape extinction as natural ecosystems are cleared and fragmented. Zoos have become something of an ark in sustaining animals for which wild habitat is no longer sufficient until such time when human demands on the biosphere stabilize and the animals can be reintroduced into their natural habitat. Zoos are involved in genetic management of increasing numbers of vulnerable species. But zoos are limited by budgets, and probably can afford no more than 900 species—less than half the 2,000 species in these groups that face extinction. And they can do almost nothing for the hundreds of thousands of insects and invertebrates threatened with extinction. In addition, they are under attack by animal rights activists.[98]

Likewise, botanical gardens could back up ecological restoration by maintaining threatened plant species and restoring them to natural settings when appropriate. But conserving the full genetic range of threatened plant species in these gardens alone is unattainable. The Office of Technology Assessment warns that cultivating sufficient populations of plants to maintain diversity is unrealistic, even though it may be theoretically possible for the botanic gardens of the world to grow the estimated 25,000 to 40,000 flowering plants that are threatened. Protecting a diversity of plant life, the agency states, will depend on maintaining them in their natural state.[99]

Another way to protect species—the method used by environmentalists in the old-growth controversy—is to protect them more directly through legislation. The real issue in the old-growth controversy is protection of the trees, but because trees do not have standing before the law, the owl is used as a surrogate. It is considered to be an endangered species, and thus protected by law, giving the environmentalists a way to stop the logging of old-growth forests. Because its habitat is essential to its survival, this is a reason to stop the logging of these forests. In order for the owl to survive, its habitat has to be protected.

An endangered species is one having so few individual survivors that the species could soon become extinct over all or most of its natural range. Species that meet this definition are given protection by the Office of Endangered Species of the U.S. Department of the Interior's Fish and Wildlife Service. This office had listed 919 species as endangered or threatened at the beginning of 1995, with an additional 4,001 under review. Of these, about 2,100 to 2,600 are likely to be listed eventually. About 3,000 of the 25,000 plant species in the United States have been identified as endangered by scientists, but only 204 of these plants are under federal protection. It was predicted that by the year 2000, up to 700 native plant species may become extinct in the country.[100]

The Endangered Species Act of 1973 was enacted to "provide for the conservation, protection, and propagation of species and subspecies of fish and wildlife that are presently threatened with extinction or likely within the foreseeable future to become threatened with extinction." The act makes it illegal for the United States to import or to carry on trade in any product made from an endangered species unless it is used for an approved scientific purpose or to enhance the survival of the species. The law also provides protection for endangered and threatened species in the United States and abroad by making it illegal to in any way affect the life and health of species that are considered to be endangered.

The law gives authorization to the National Marine Fisheries Service of the Department of Commerce to identify and list endangered and threatened marine species. The Fish and Wildlife Service in the Department of the Interior identifies and lists all other endangered and threatened species in the country. When listed as endangered, a species cannot be hunted, killed, collected, or injured in the United States. It is unlawful for any person to take any endangered species, where taking is defined to include "any action to harass, harm, pursue, hunt, wound . . . or attempt to engage in such conduct."[101] Any decision to add a species to or remove one from the list must be based on biological considerations rather than economic considerations.

Once a species is listed as endangered or threatened, a plan is supposed to be developed for its recovery. The law requires the appropriate agencies to employ all methods and procedures needed to return the listed species to a point where it no longer needs protection. It is not enough for agencies to simply avoid activities that impact negatively on the endangered or threatened species; they must also be responsible to "continuously develop programs which positively affect rare plants and animals and which will bring them to the point where they can be taken off the list of threatened and endangered species."[102]

According to one source, of the roughly 600 species that have been listed as threatened or endangered over the past two decades, only seven have become extinct. Nine species have recovered sufficiently to be removed from the list and many others have made remarkable recoveries. These include animals like the American alligator, the California gray whale, the peregine falcon, and the bald eagle.[103] The latter was reclassified in 1994 from an endangered species to the less urgent category of threatened in all but three of the lower 48 states. The eagle population, which numbered in the tens of thousands in the 1800s, plummeted to 417 pairs in 1963 when it was listed as endangered. Since then the number of eagles has risen steadily, reaching 4,016 pairs in 1994 plus several thousand juveniles across the lower 48 states.[104]

However, research has shown that efforts to promote the survival of about 400 endangered species nationwide costs taxpayers at least $884.2 million. This figure did not take into account money spent by the private sector or declines in land value, an issue that has become more important as more landowners have been affected by efforts to protect species. The figure included only funds set aside by the federal government to restore to health species listed as endangered in the 20 years of the act's existence. Many individual animals have had large sums spent on their behalf (Table 10.4). Such information has given opponents of the law adequate incentive to try and change the law to

TABLE 10.4 Endangered Species and Their Costs

Species	Cost
1. Atlantic Green Turtle	$88,236,000
2. Loggerhead Turtle	85,947,000
3. Blunt-Nosed Leopard Lizard	70,252,000
4. Kemp's Ridley Sea Turtle	63,600,000
5. Colorado Squawfish	57,770,000
6. Humpback Chub	57,700,000
7. Bonytail Chub	57,700,000
8. Razorback Sucker	57,700,000
9. Black-Capped Vireo	53,538,000
10. Swamp Pink	29,026,000

Source: National Wilderness Institute, as reported by Michele Kay, "Group: Saving Animals from Extinction Costly," *Times-Picayune,* March 24, 1994, p. A10.

protect property rights and use different kinds of reward systems to encourage protection of endangered species.[105]

There are many unanswered questions regarding preservation of species. Do we have a moral obligation to save all species currently in existence? If not, how do we make distinctions between those saved and those allowed to become extinct? Can we identify keystone species so that these can be preserved? Proposals have been made to preserve only the most biologically unique groups of organisms, as the federal government does not have the resources to protect every single subspecies, variety, or distinct population of species.[106] But who decides what species are truly unique and what criteria are used to make this distinction? These questions arise because a growing human population is crowding out more and more plants and animals who share the same limited space and resources.

Meanwhile, inadequate funding and staffing, which reflect political priorities, have resulted in a hopeless backlog of work for those agencies responsible for implementing legislation to protect species.[107] Because of delays, some plant and animal species become extinct before they can be listed.[108] The lack of vigorous enforcement reflects uncertainty about how much of the natural world we really want to preserve and how many species can be preserved.[109] While the law is the most powerful tool the nation has for preserving biological diversity, it is not enough, according to some critics. If we are serious about sharing the earth with other biological life forms, "we need to elaborate on the original articulation of the right of other forms of life to exist, and take our commitment to preserving sustainable ecosystems more seriously."[110]

There are said to be serious procedural problems with the way the act works. If anything is done at all, it is done after a species may have reached the point of no return. The chances of successfully restoring a species are reduced if we wait until a species is almost gone before taking action. The minimum size required for a species to exist indefinitely without active help by humans is considerably larger than that needed for mere survival over a few generations. We need to be thinking, according to some experts, more proactively about the preservation of entire systems rather than just individual species. What is really needed is a better understanding of ecosystem dynamics and the role certain species play in the whole system.[111]

It is recommended that a strategy encompassing both preservation and active ecosystem restoration where possible is needed to minimize the global extinction crisis. Severely degraded tropical land need not be written off as a total biological loss, as some biologists believe that nearly all the land deforested so far in the Amazon has the capacity to regenerate. This potential for regeneration is based on studies of how natural ecosystems repair themselves. Such restoration aims to reestablish viable native communities of plants and animals, but this restoration cannot be done haphazardly. Advocates of restoration maintain that the successful conservation of biological diversity depends less on keeping humans out of fragile ecosystems than on making sure they do the right things when they are there.[112]

Every species distinction diminishes humanity, as every organism contains on the order of 1 million to 10 million bits of information in its genetic code. This code has been hammered into existence by an astronomical number of mutations and episodes of natural selection that has taken place over the course of thousands or even millions of years of evolution. For most people, the power of evolution by natural selection may be too great to conceive, let alone duplicate. Species diversity, the world's available gene pool, is one of our planet's most important and irreplaceable resources, as each species represents a unique combination of traits, each one of which is an evolutionary solution to biological problems.[113]

> The loss of biological diversity does not mean the disappearance of a few familiar "showcase" species, but rather the loss of complex interwoven systems of plants and animals that make up the Earth's ecosystems. The loss of biological diversity represents a moral and ethical catastrophe of unprecedented proportions involving natural wonders which have evolved over hundreds of millions of years, and which constitute a priceless resource largely untapped and little understood by mankind. The extinction crisis should give us a new awareness that the living things with which we share the planet are not only a source of beauty, wonder, and joy, but are integral to our very survival.[114]

As species are decimated largely as a result of habitat destruction, the capacity for natural genetic regeneration is greatly reduced, causing, as one scientist puts it, the death of birth. The science of genetic engineering does not make new genes, but depends on rearrangement of existing genes. From this perspective, the ultimate importance of tropical forests is in their genetic stock, from which incalculable and inconceivable benefits may be derived. With each species lost, the potential growth of the life sciences and certain possibilities for genetic engineering are forever curtailed and impoverished. As one scientist put it: "If we permit the loss of the rainforests, and with them a major portion of biological diversity, it might with justice be viewed as one of the greatest acts of desecration in human history."[115] The same could be said for the loss of species and the destruction of habitats all over the world, including the spotted owl and old-growth forests.

Questions for Discussion

1. What resources do forests provide for human purposes? What ecological functions do they also perform? Is there a conflict between these roles? How can this conflict, if it exists, be mitigated? What role do forests play in the global carbon and oxygen cycles of the world?
2. What is a tropical rain forest? Where are they located? How much land did they once cover compared to what they cover today? How many species do they contain? Is the soil that supports the

rain forest rich in nutrients? How is the rain forest able to sustain itself? What keeps it going? What is the rain forest's source of energy?

3. What are the fundamental causes of deforestation? How does government policy encourage such practices? What do you think of the principles stated in the chapter regarding rain forest management? Do you think they are feasible from an economic standpoint? Do they recognize the ecological functions of the rain forest?
4. What are old-growth forests? Where are they located? What pressures are being put upon them? Is it important to preserve them for their ecological functions? Do they have intrinsic value? What role did the northern spotted owl play in the controversy over old-growth forests? What is at stake in this controversy? What is the owl's current status? What is the current state of affairs with respect to old-growth forests?
5. What is different about species decimation with regard to other forms of ecological destruction? How much biological diversity presently exists when compared with other historical time periods? Does this concern you personally? Where do most of the world's species live? Why? How many are being lost every year? What value do these species have for humans? Do they have a right to exist irrespective of their instrumental value?
6. What can be done to protect and preserve species? What is an endangered species? How does this process of protection work? What problems exist with the legislation and its implementation? Should the legislation be revised, and if so, in what manner? What other strategies are most likely to be effective in preserving species?

Endnotes

1. G. Tyler Miller, *Living in the Environment,* 6th ed. (Belmont, CA: Wadsworth, 1990), p. 285.
2. Ibid., p. 286–287.
3. Ibid., p. 285.
4. Alan Thein Durning, "Redesigning the Forest Economy," *State of the World 1994,* Lester R. Brown et al. (New York: W. W. Norton, 1994), p. 23. See Also Eugene Linden, "The Tortured Land," *Time,* September 4, 1995, pp. 42–53.
5. Ibid., p. 24.
6. The Smithsonian Institution and International Hardwood Products Association, *Tropical Forestry Workshop: Consensus Statement on Commercial Forestry Sustained Yield Management and Tropical Forests,* October 1989, p. 2.
7. International Hardwood Products Association, *World's Tropical Forests: A Renewable Resource,* undated, p. 1.
8. United Nations Food and Agricultural Organization, *The Tropical Forestry Action Plan.*
9. Miller, *Living in the Environment,* p. 289.
10. Ibid., pp. 291–292.
11. *Tropical Forestry Workshop,* p. 2.
12. Lisa L. Lyles, "A Long and Rocky Road to Reversing Rainforest Destruction," *Nature,* June 9, 1988, p. 491. A recent example is provided in Guyana and Suriname where governments have opened huge tracts of forest for logging by timber and trading companies. Economic hardship and the lure of logging revenue combined to make the rain forest more valuable for exploitation. The countries, desperate for quick cash, granted huge concessions to Asian logging consortiums getting but a pittance in return, according to some critics. See Eugene Linden, "Chain Saws Invade Eden," *Time,* August 29, 1994, pp. 58–59.
13. Peter H. Raven, "Endangered Realm," *The Emerald Realm: Earth's Precious Rain Forests,* Martha E. Christian, ed. (Washington, D.C.: National Geographic Society, 1990), p. 10.
14. Miller, *Living in the Environment,* p. 284.
15. Ibid., p. 287; Anjali Acharya, "Tropical Forests Vanishing," *Vital Signs 1995,* Lester R. Brown et al. (New York: W. W. Norton, 1995), p. 116.

16. World Wildlife Fund, Position Paper No. 3, *Tropical Forest Conservation,* August 1989. There is some controversy over the extent of forest loss around the world. See Richard Monastersky, "The Deforestation Debate," *Science News,* July 10, 1993, pp. 26–27.
17. Acharya, "Tropical Forests Vanishing," p. 116.
18. Tom Lovejoy, "Infinite Variety—A Rich Diversity of Life," *The Rainforests: A Celebration,* Lisa Silcock, ed. (San Francisco: Chronicle Books, 1990), p. 36.
19. Eugene Linden, "Playing with Fire," *Time,* September 18, 1989, p. 77.
20. Julian Caldecott, "The Rainforest—An Overview," *The Rainforests: A Celebration,* Lisa Silcock, ed. (San Francisco: Chronicle Books, 1990), pp. 14–15.
21. World Wildlife Fund, *Tropical Forest Conservation.*
22. Durning, "Redesigning the Forest Economy," p. 28.
23. Joanne Omang, "In the Tropics, Still Rolling Back the Rain Forest Primeval," *Smithsonian,* 17, no. 12 (March 1987), 56–67.
24. "Amazon Forest Unlikely to Rise from Ashes," *Science News,* Vol. 137 (March 17, 1990), p. 164.
25. Ghillean T. Prance, "Introduction," *The Rainforests: A Celebration,* Silcock, ed., p. 9.
26. Linden, "Playing with Fire," p. 78.
27. *Tropical Forestry Workshop,* p. 3.
28. Ibid., pp. 4–5.
29. Ibid., pp. 3–4.
30. Gilbert M. Grosvenor, "Environmental Promises," *The National Geographic Society,* November 1989, p. 536.
31. See Huntington Williams III, "Banking on the Future," *Nature Conservancy,* May–June 1992, pp. 23–27; "Mexico Signs 'Debt for Nature' Deal," *Times-Picayune,* February 26, 1991, p. A10.
32. Susan Dillingham, "From the Rain Forest to the Shelves," *Insight,* June 4, 1990, p. 41.
33. Ibid. See also Thomas A. Carr, Heather L. Pedersen, and Sunder Ramaswamy, "Rain Forest Entrepreneurs," *Environment,* 35, no. 7 (September 1993), 12–15.
34. Michael S. Serrill, "A Dubious Plan for the Amazon," *Time,* April 17, 1989, p. 67.
35. Ibid.
36. *The Tropical Forestry Action Plan.*
37. Ibid.
38. Eugene Linden, "Good Intentions, Woeful Results," *Time,* April 1, 1991, pp. 48–49.
39. Ibid., p. 49.
40. Mac Margolis, "Taking Two Steps Back," *Newsweek,* January 8, 1996, p. 52.
41. Linden, "Playing with Fire," p. 85.
42. Miller, *Living in the Environment,* p. 184.
43. John Daniel, "The Long Dance of the Trees," *Wilderness,* Spring 1988, p. 21.
44. Ibid., p. 23.
45. Miller, *Living in the Environment,* pp. 303–304.
46. Michael D. Lemonick, "Showdown in the Treetops," *Time,* August 28, 1989, pp. 58–59.
47. Charles McCoy, "Spotted Owl's Fate Puts Timber Policy, Northwest Logging Jobs at Loggerheads," *Wall Street Journal,* April 27, 1989, p. A6.
48. Daniel, "The Long Dance of the Trees," p. 23. Also see Christopher Elias, "Forest Service's Eager Beavers Draw Fire with Timber Sales," *Insight,* October 17, 1988, pp. 38–41.
49. Ibid., p. 27.
50. Miller, *Living in the Environment,* p. 298.
51. Ibid., pp. 298–299.
52. Ibid., pp. 296–297.
53. Daniel, "The Long Dance of the Trees, p. 27.
54. Ibid.

55. McCoy, "Spotted Owl's Fate," p. A6.
56. "Cancer Researchers Join Tree Battle," *Times-Picayune,* September 20, 1990, p. A10. See also Marilyn Chase, "Clashing Priorities: Cancer Drug May Save Many Human Lives—At Cost of Rare Trees," *Wall Street Journal,* April 9, 1991, p. A1. While no one was able to develop a synthetic version, developers of taxol began to contract with European countries to derive the drug from the needles and twigs of more plentiful yew species there. Robert L. Jackson, "FDA Is Urged to Approve Taxol for Use Against Ovarian Cancer," *Times-Picayune,* November 18, 1992, p. F10.
57. "The Year of the Deal: 23rd Environmental Quality Index," *National Wildlife,* 29, no. 2 (February–March, 1991), 37.
58. Rose Gutfeld, "U.S. Unveils Plan to Save Spotted Owl, but Some See Strategy as Short-Sighted," *Wall Street Journal,* June 27, 1990, p. A6.
59. Charles McCoy, "U.S. Prepares a Plan Protecting Owl Beyond Logging Critics' Expectations," *Wall Street Journal,* April 26, 1991, p. A1; Charles McCoy, "U.S. Will Designate 11.6 Million Acres In Northwest to Protect Owls, Tall Trees," *Wall Street Journal,* April 29, 1991, p. B2.
60. Peter Hong, "Tree-Huggers vs. Jobs: It's Not That Simple," *Business Week,* October 19, 1992, pp. 108–109.
61. Ted Gup, "It's Nature, Stupid," *Time,* July 12, 1993, pp. 38–40.
62. Ibid.
63. Eamon Lynch, "Logging Without Laws," *Audubon,* January–February 1996, pp. 14–18. See also Kathie Durbin, "The 'Timber Salvage' Scam," *The Amicus Journal,* 17, no. 3 (Fall 1995), 29–31.
64. Ted Gup, "Owl vs. Man," *Time,* June 25, 1990, p. 58.
65. Daniel, "The Long Dance of the Trees," p. 33. Other areas of old growth are also under threat such as the Tongass National Forest in Alaska. See Martha Brant, "The Alaskan Assault," *Newsweek,* October 2, 1995, p. 44.
66. Durning, "Redesigning the Forest Economy," p. 26. See also Mark Schleifstein, "Natives May Show Way to Saving Rain Forests," *Times-Picayune,* December 13, 1994, p. B3.
67. Durning, "Redesigning the Forest Economy," p. 27.
68. Ibid., pp. 28–29. See also Jed Horne, "Drug Giant Banks on Rain Forest Deal, *Times-Picayune,* August 16, 1992, p. A1; Eugene Linden, "Lost Tribes, Lost Knowledge," *Time,* September 23, 1991, pp. 46–55.
69. Peter Knight, "Save a Forest—Use Bar-Codes," *Tomorrow,* May–June, 1995, pp. 68–69; Martin Wright, "A Greener Forest Just Around the Bend," *Tomorrow,* May–June, 1995, pp. 46–48.
70. Durning, "Redesigning the Forest Economy," pp. 32–34.
71. Ibid., pp. 35–36.
72. Ibid., p. 37.
73. Edward O. Wilson, "Threats to Biodiversity," *Scientific American,* 261, no. 3 (September 1989), 108.
74. Ibid., p. 110.
75. Miller, *Living in the Environment,* p. 318.
76. Otto T. Solbrig, "The Origin and Function of Biodiversity," *Environment,* 33, no. 5 (June 1991), 19.
77. Lovejoy, "Infinite Variety," p. 36.
78. Wilson, "Threats to Biodiversity," p. 110.
79. Caldecott, "The Rainforest—An Overview," p. 14.
80. Wilson, "Threats to Biodiversity," p. 112.
81. Sierra Club, *The Extinction Crisis,* June 1987, p. 1.
82. David Holzman, "Species Extinction Mires Ecosystem," *Insight,* March 26, 1990, p. 52.

83. Philip Elmer-Dewitt, "Are Sharks Becoming Extinct?" *Time,* March 4, 1991, p. 67. See also Vernon Church, "Danger: No Sharks!" *Newsweek,* December 14, 1992, pp. 64–65.
84. Michael D. Lemonick, "The Hunt, the Furor," *Time,* August 2, 1993, pp. 42–45. See also Eugene Linden, "Sharpening the Harpoons," *Time,* May 24, 1993, pp. 56–57.
85. Eugene Linden, "Tigers on the Brink," *Time,* March 28, 1994, pp. 44–51.
86. Sharon Begley, "Better Red Than Dead," *Newsweek,* December 12, 1994, pp. 79–80; Charles McCoy, "Salmon Advocates Win First Big Case Against the Government," *Wall Street Journal,* March 30, 1994, p. B1.
87. David Briscoe, "World Could Go the Way of the Birds, Experts Warn," *Times-Picayune,* January 25, 1994, p. A5; Howard Youth, "Flying into Trouble," *World Watch,* January–February 1994, pp. 10–19.
88. Andrea Sachs, "A Grisly and Illicit Trade," *Time,* April 8, 1991, pp. 67–68; Sharon Begley, "Killed by Kindness," *Newsweek,* April 12, 1993, pp. 50–56.
89. David Van Biema, "The Killing Fields," *Time,* August 22, 1994, pp. 36–37.
90. Emily Smith, "Not 'So Many Fish in the Sea,' " *Business Week,* July 4, 1994, pp. 62–64. See also Michael D. Lemonick, "Too Few Fish in the Sea," *Time,* April 4, 1994, pp. 70–71.
91. Peter Weber, "Protecting Oceanic Fisheries and Jobs," *State of the World 1995,* Lester R. Brown et al. (New York: W. W. Norton, 1995), pp. 21–23.
92. Smith, "Not 'So Many Fish in the Sea,' " pp. 62–64.
93. Weber, "Protecting Oceanic Fisheries and Jobs," p. 23.
94. John McQuaid, "Scientists Confounded by Nature, Politics," *Times-Picayune,* March 28, 1996, p. A10.
95. John McQuaid, "Managers Maneuver at Cliff's Edge," *Times-Picayune,* March 27, 1996, p. A6. See also Mark Schleifstein, "Red Snapper Catch Could Grow," *Times-Picayune,* November 16, 1995, p. B4.
96. McQuaid, "Scientists Confounded by Nature, Politics," p. A8. Proposals have been made to tax the amount of fish caught as a better way of managing the resource. See Gary S. Becker, "How to Scuttle Overfishing? Tax the Catch," *Business Week,* September 18, 1995, p. 30.
97. Edward C. Wolf, "Avoiding a Mass Extinction of Species," *State of the World 1988* (Washington, D.C.: Worldwatch Institute, 1988), p. 102.
98. Ibid., p. 117. See also Jesse Birnbaum, "Just Too Beastly for Words," *Time,* June 24, 1991, p. 60.
99. Ibid.
100. Miller, *Living in the Environment,* p. 321; Thomas Lambert, *The Endangered Species Act: A Train Wreck Ahead* (St Louis, MO: Washington University Center for the Study of American Business, 1995), p. 20. A report issued in early 1996 by the Nature Conservancy said that mammals and birds are doing relatively well compared with other groups, but that a high proportion of flowering plants and freshwater species are in trouble. Of the 20,481 species examined, about two thirds are secure or apparently secure, 1.3 percent are extinct or possibly extinct, 6.5 percent are critically imperiled, 8.9 percent are imperiled, and 15 percent are considered vulnerable. This study was said to be the most comprehensive assessment to date of the state of American fauna and flora. See William Dicke, "Many Animals, Plants in Peril, Survey Finds," *Times-Picayune,* January 2, 1996, p. A4.
101. Thomas France and Jack Tuholske, "Stay the Hand: New Directions for the Endangered Species Act," *The Public Land Law Review,* Spring 1986, p. 15.
102. Ibid., p. 4.
103. John C. Sawhill, "Saving Endangered Species Doesn't Endanger Economy," *Wall Street Journal,* February 20, 1992, p. A19.
104. Michael D. Lemonick, "Winged Victory," *Time,* July 11, 1994, p. 53; Gary Lee, "Eagle Populations Soar," *Times-Picayune,* June 30, 1994, p. A1.
105. Michele Kay, "Group: Saving Animals from Extinction Costly," *Times-Picayune,* March 24, 1994, p. A10.

106. Lambert, *The Endangered Species Act,* pp. 13–14.
107. See John Lancaster, "Endangered Species Act Isn't Working Smoothly," *Times-Picayune,* October 31, 1990, p. C11.
108. See Cass Peterson and Philip J. Hilts, "Waiting in Line for Protection, Endangered Species Dying," *Houston Chronicle,* September 21, 1987, Sec. 7, p. 4.
109. See Fiona Sunquist, "Should We Put Them All Back?" *International Wildlife,* September–October, 1993, pp. 34–40.
110. Sierra Club, *Endangered but Not Yet Protected,* Fall 1986, p. 4.
111. M. John Faybee, "A Hard Act To Follow," *Backpacker,* October 1990, p. 89.
112. Wolf, "Avoiding a Mass Extinction of Species," p. 110.
113. Wilson, "Threats to Biodiversity," p. 114.
114. Sierra Club, *The Extinction Crisis,* p. 1.
115. Lovejoy, "Infinite Variety," p. 38.

Suggested Reading

Amelung, Torsten, and Markum Diehl. *Deforestation of Tropical Rain Forests: Economic Change and Impact on Development.* Ann Arbor: University of Michigan Press, 1992.

Berkmuller, Klaus. *Environmental Education About the Rain Forest.* Washington, D.C.: Island Press, 1992.

Botkin, Daniel. *Forest Dynamics: An Ecological Model.* New York: Oxford University Press, 1993.

Bullock, Charles S., III, ed. *Forest Resource Policy.* New York: Wiley, 1993.

Bunce, R. G., and D. C. Howard, eds. *Species Dispersal in Agricultural Habitats.* New York: Wiley, 1994.

Caufield, Catherine. *In the Rainforest.* New York: Alfred A. Knopf, 1985.

Christian, Martha E., ed. *The Emerald Realm: Earth's Precious Rain Forests.* Washington, D.C.: National Geographic Society, 1990.

Erwin, Keith. *Fragile Majesty: The Battle for North America's Last Great Forest.* Seattle, WA: Mountaineer's Books, 1989.

Gillis, Malcolm, and Robert Repetto. *Deforestation and Government Policy.* New York: ICS Press, 1988.

Gradwohl, Judith, and Russell Greenberg. *Saving the Tropical Forests.* Washington, D.C.: Island Press, 1988.

Hazlewood, Peter T. *Cutting Our Losses: Policy Reform to Sustain Tropical Forest Resources.* Washington, D.C.: World Resources Institute, 1989.

International Hardwood Products Association. *World's Tropical Forests: A Renewable Resource,* undated.

Kelly, David. *Secrets of the Old Growth Forest.* Layton, UT: Gibbs Smith, 1988.

Norse, Elliott. *Ancient Forests of the Pacific Northwest.* Covelo, CA: Island Press, 1989.

Perry, David A. *Forest Ecosystems.* Baltimore: Johns Hopkins, 1995.

Postel, Sandra, and Lori Heise. *Reforesting the Earth.* Washington, D.C.: Worldwatch Institute, 1988.

Repetto, Robert. *The Forest for the Trees: Government Policies and the Misuse of Forest Resources.* Washington, D.C.: World Resources Institute, 1989.

Ricklefs, Robert E., and Dolph Schluter, eds. *Species Diversity in Ecological Communities.* Chicago: University of Chicago Press, 1993.

Rosenzweig, Michael L. *Species Diversity in Space and Time.* Cambridge: Cambridge University Press, 1995.

Sadler, Tony. *Forests and Their Environment.* Cambridge: Cambridge University Press, 1994.

Shoumatoff, Alex. *The World Is Burning.* Boston: Little, Brown, 1990.

Silcock, Lisa, ed. *The Rainforests: A Celebration.* San Francisco: Chronicle Books, 1990.

The Smithsonian Institution and International Hardwood Products Association. *Tropical Forestry Workshop: Consensus Statement on Commercial Forestry Sustained Yield Management and Tropical Forests,* October 1989.

Spurr, Stephen H., and Burton V. Burns. *Forest Ecology,* 3rd ed. New York: Kreiger, 1992.

Ward, Peter. *The End of Evolution.* New York: Bantam Books, 1994.

Wiese, Robert J., and Michael Hutchins. *Species Survival Plans: Strategies for Wildlife Conservation.* Washington, D.C.: American Zoo and Aquarium, 1994.

Wilderness Society. *Ancient Forests: A Threatened Heritage.* Washington, D.C.: Wilderness Society, 1989.

Williams, Michael. *Americans and Their Forests.* New York: Cambridge University Press, 1989.

Wolf, Edward C. "Avoiding a Mass Extinction of Species." *State of the World 1988.* Washington, D.C.: Worldwatch Institute, 1988.

CHAPTER 11

Coastal Erosion and Wetlands Protection

The coastlines of the United States are important for reasons that have to do protection of an ecosystem where plant and animal life are abundant. The coastal zone is considered to be the relatively warm, nutrient-rich, shallow water that extends from the high-tide mark on land to the edge of the continental shelf. While this area represents only 10 percent of the total ocean area, it is home to 90 percent of all ocean plant and animal life and provides resources for most of the major commercial marine fisheries. It is one of the earth's most important ecosystems and thus needs to be protected from environmental degradation.[1]

Several different kinds of habitats are located in the coastal zone, including estuaries, which are coastal areas where freshwater from rivers, streams, and land runoff mixes with salty seawater. These estuaries provide aquatic habitats that have a lower average salinity than the water of the open ocean, making them unique places for the development of aquatic life forms. Along with inland swamps and marshes and tropical rain forests, these estuaries produce more plant biomass per square meter each year than any other ecosystem in the world.[2]

Land that is flooded all or part of the year with freshwater or salt water is called a wetland, and wetlands that extend inland from estuaries and are covered all or part of the year with salt water are known as coastal wetlands. These nutrient-rich areas are among the world's most productive ecosystems. Many people view these areas as desolate and worthless; they are often regarded as wastelands, as sources for mosquitoes, flies, and unpleasant odors. Wetlands are places to be avoided, or better yet, developed, as many people think they should be dredged and filled in to be used for housing or commercial developments or used as depositories for human-generated waste materials.[3]

However, these coastal wetlands provide human beings with a remarkable variety of benefits. They serve as a spawning and nursery ground for many species of marine fish and shellfish, thus ultimately providing seafood. They are also breeding grounds and habitats for waterfowl and other kinds of wildlife. Some spend their entire life in wetlands and other wildlife use them primarily as nesting, feeding, or resting grounds. Estuaries and coastal wetlands are spawning grounds for 70 percent of the country's seafood, including shrimp, salmon, oysters, clams, and haddock. This fishing industry is a $15-billion-a-year endeavor and provides jobs for millions of people.[4]

Coastal wetlands also dilute and act as filters for large amounts of waterborne pollutants, and thus help to protect the quality of adjacent waters used for swimming, fishing, and habitats for wildlife. Wetlands remove nutrients such as nitrogen and phosphorus and thus help prevent eutrophication of waters. They also filter harmful chemicals such as pesticides and heavy metals, and trap suspended sediments that produce turbidity (cloudiness) in water. Due to their position between upland and deep water, wetlands can also intercept surface-water runoff from large land areas before it reaches open water, and thus filter that water to improve its quality.[5]

Estuaries and coastal wetlands also protect population centers in coastal areas by absorbing damaging waves caused by tropical storms and hurricanes, and act as giant sponges to absorb floodwaters. Wetlands vegetation can also reduce shoreline erosion by absorbing and dissipating wave energy and encouraging the deposition of suspended sediments. Plants in wetland areas are important in protecting against erosion because they increase the durability of the sediment by binding soil with their roots. They also dampen wave action in the area and reduce current velocity through friction. Many states are recommending the planting of wetland vegetation to control shoreline erosion in coastal areas.[6]

Inland wetlands are those areas located away from coastal areas that are covered with freshwater all or part of the year. They include bogs, marshes, swamps, and mudflats. Wet tundra is also considered to be a wetland, as 58 percent of the state of Alaska is covered with this kind of wetland. Wetlands are most common on floodplains along rivers and streams, in isolated depressions surrounded by dry land, and along the margins of lakes and ponds. Certain wetland types are common in certain regions, such as the pocosins of North Carolina, bogs and fens of the northeastern and north-central states, inland saline and alkaline marshes and riparian wetlands of the arid and semiarid West, prairie potholes of Minnesota and the Dakotas, and the cypress-gum swamps of the South.[7] Figure 11.1 shows the relative abundance of wetlands in the United States by state.

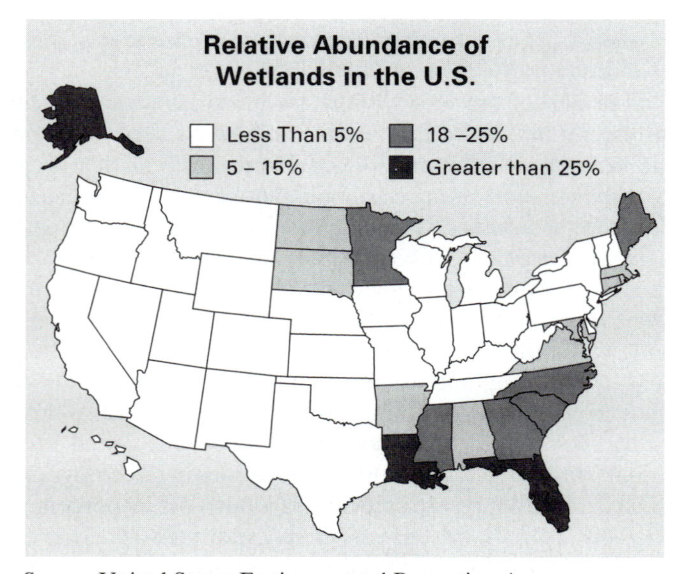

Source: United States Environmental Protection Agency, Communications, Education, and Public Affairs Office, *Securing Our Legacy* (Washington, D.C.: EPA, 1992), p. 31.

FIGURE 11.1 Relative Abundance of Wetlands in the United States

These inland wetlands are also important ecosystems that are rapidly being destroyed and degraded by human activity. They provide habitats for a variety of fish, waterfowl, and other wildlife. Most freshwater fish feed upon wetland-produced food and use wetlands as nursery grounds. Most of the important recreational fish spawn in wetlands, and a great variety of birds use the wetlands to raise their young. Wetlands also help reduce the frequency, level, and velocity of floods and riverbank erosion. They act as natural sponges that absorb flooding waters and help protect adjacent and downstream property from flood damage. Wetlands upstream of urban areas are especially valuable because urban development increases the rate and volume of surface-water runoff, thus greatly increasing the risk of flood damage. Many wetlands also help recharge groundwater aquifers by holding water and allowing it to infiltrate the ground. Wetlands are also important in growing certain crops such as blueberries, cranberries, and rice, which is used to feed half the world's people.[8]

COASTAL EROSION AND RISING SEA LEVELS

The erosion of coastlines, beaches, and barrier islands has accelerated over the past 10 years as a result of rising sea levels. This erosion has become particularly prevalent on the world's sandy coastlines, at least 70 percent of which have retreated during the past few decades.[9] Some 86 percent of California's 1,100 miles of exposed Pacific shoreline is receding at an average rate of between 6 inches and 2 feet per year, and Monterey Bay, south of San Francisco, loses as much as 5 to 15 feet annually. Cape Shoalwater, Washington, which is about 70 miles west of Olympia, has been eroding at the rate of more than 100 feet a year since the turn of the century. In North Carolina, erosion cuts into beachfront property as much as 60 feet in some places, and the Cape Hatteras lighthouse may soon be surrounded by the ocean.[10]

Such coastal erosion is only one of the natural processes that have altered the world's shorelines ever since the oceans were first formed. The scouring action of waves and the pounding of storms, as well as the rise and fall of ocean levels, have changed coastlines over the centuries. These changes threaten developments that have been built on coastal areas as Americans have moved to be closer to the ocean and beaches of the country. Receding coastlines also threaten the survival of shore-dwelling wildlife such as sea turtles, which need beaches to lay their eggs.[11]

While sea level fluctuations are part of a natural cycle, over the past 100 years, the ocean has risen more than a foot, a rate faster than at any time in the past millennium. Some scientists believe that these natural changes are magnified by a fundamental change in the world's climate caused by the greenhouse effect mentioned in an earlier chapter. Some projection of sea level rise over the next 40 to 50 years suggest that most recreational beaches in developed areas could be eliminated unless protective measures are taken. Such increased erosion will decrease natural storm barriers, and moderate storms could be turned into catastrophic ones because of the loss of protection. Further sea level rises will also permanently affect freshwater supplies, and large cities around the world could be threatened by saltwater intrusion.[12]

For most of recorded history, the sea level has changed slowly, which has fostered the development of a social order based on its relative constancy. But if global warming is a continuing phenomenon, the earth's temperature will be radically altered, and an accompanying sea level rise represents an environmental threat of unprecedented proportion. Higher global temperatures can alter sea levels in four ways: (1) density can decrease through the warming and subsequent expansion of seawater, which increases volume, (2) melting of alpine glaciers, (3) net increases in water as the fringes of polar glaciers melt, and (4) by more ice being discharged from ice caps into the oceans.[13]

The rate of thermal expansion depends on how quickly ocean volume responds to rising atmospheric temperatures, how fast surface layers warm, and how rapidly the warming reaches deeper water masses. The rate of sea level change expected on a global level in the forseeable future is unprecedented on a human time scale. With today's level of population and investment in coastal areas, the world has much more to lose from sea level rise than ever before. Only 30 countries in the entire world are completely landlocked, and while only 3 percent of the world's land area is at risk, this area encompasses one third of global cropland and is home to a billion people. Countless billions of dollars' worth of property in coastal towns, cities, and ports will be threatened and problems will occur with natural and artificial drainage, and with saltwater intrusion into rivers and aquifers.[14]

According to estimates by the EPA, erosion, inundation, and saltwater intrusion could reduce the area of present-day coastal wetlands in the United States up to 80 percent if current projections of the global sea level rises are realized. The extent of wetland loss will depend on the degree to which coastal towns and villages seek to protect their shorelines. Some 46 percent of all U.S. wetlands would be lost under a one-meter rise if shorelines were allowed to retreat naturally. The loss of up to 80 percent of these wetlands is envisioned under a more rapid rate of rise. The economic and ecological cost of such a loss has not been calculated for the United States, let alone for the rest of the world.[15]

Communities have but two choices when faced with such sea level rises: They can either retreat from the shore or fend off the sea by building jetties, seawalls, groins, and bulkheads to hold back the ocean. The price tag attached to some of these options may be higher than even developed nations can afford, especially when long-term ecological damage that these structures themselves can cause is taken into account. Preliminary estimates by the EPA for the cost of holding back the sea from U.S. shores ranges from $32 billion to $309 billion for a one-half-meter to two-meter rise in sea levels. This cost does not include money needed for repairing or replacing infrastructure.[16]

Legal definitions of private property and of who is responsible for compensation in the event of such disasters are already being debated. If the sea level continues to rise, pushing up the cost of adaption, these issues will likely become part of an increasingly acrimonious debate over property rights and individual interests versus those of society at large.[17] In poorer countries, evacuation and abandonment of coastal areas may be the only option, and as millions of people are displaced by rising sea levels and move inland, there will be increased competition with those already living in these areas for scarce food, water, and land. This competition may spur regional clashes and increase international tensions.[18]

Numerous measures can be taken by governments and private citizens to deal with this problem. Governments can begin to limit coastal development by ensuring that private owners bear more of the costs of settling in coastal areas. Restricting shoreline development in this country has fallen largely to individual states. Since 1971, 29 of 30 states with coastlines have adopted coastal zone management programs. In North Carolina, for example, developers cannot build large projects any closer than 120 feet from the first line of dunes. The state also outlaws permanent seawalls and manmade barriers. Florida controls seaside construction by requiring approval by the governor and state cabinet for any new building closer than about 300 feet to the water's edge. But a major problem with state regulation is the lack of coordination.[19]

The simplest and most effective response to the problem would be to prevent people from living near oceans. The nonprofit Nature Conservancy encourages this strategy by buying threatened coastal areas and refusing to have them developed. In the 1980s, this organization's 32 separate purchases in 8 states has sheltered more than 250,000 acres, including 13 barrier islands off the coast of Virginia that it bought for $10 million. But this policy is not likely to work on a comprehensive or long-term basis, as the pressures for living close to the ocean are too great for too many people.[20]

WETLANDS DESTRUCTION

Once there were over 200 million acres of wetlands in the lower 48 states, but by the mid-1970s, only 99 million acres remained. The average rate of wetland loss from 1955 to 1975 was 458,000 acres per year, with 440,000 acres of this loss inland wetlands, and 18,000 acres coastal wetlands. Currently, about 56 percent of the original coastal and inland wetlands areas in the lower 48 states have been destroyed, with about 80 percent of this loss due to draining and clearing of wetlands for agricultural purposes. Agricultural activities have had the greatest impact on forested wetlands, inland marshes, and wet meadows. Urban development was the major cause of coastal wetland losses outside of Louisiana, with submergence of Louisiana's coastal waters by rising Gulf waters as the leading factor behind wetland loss in that state. In addition to direct physical destruction, these habitats are also threatened indirectly by chemical contamination and other pollution[21] (Exhibit 11.1).

EXHIBIT 11.1 Major Causes of Wetland Loss and Degradation

Human Impacts	*Natural Threats*
Drainage	Erosion
Dredging and stream channelization	Subsidence
Deposition of fill material	Sea level rise
Diking and damming	Droughts
Tilling for crop production	Hurricanes and other storms
Grazing by domesticated animals	Overgrazing by wildlife
Discharge of pollutants	
Mining	
Alteration of hydrology	

Source: United States Environmental Protection Agency, *America's Wetlands: Our Vital Link Between Land and Water* (Washington, D.C.: Office of Wetlands Protection, 1988), p. 6.

Nearly half of the estuaries and coastal wetlands in the country have been destroyed or damaged because of these multiple uses, which puts these areas under great ecological stress. Extensive losses of wetlands has taken place in Louisiana, Mississippi, Arkansas, North Dakota, South Dakota, Nebraska, Florida, and Texas. The loss of these wetlands has greatly diminished our nation's wetland resources and reduced the benefits that these wetlands provided for the country's population. Water quality has been adversely affected in many parts of the country because of this loss, and damages from floods have increased. In addition, the country has seen a decline in waterfowl populations in recent years, which in part is the result of wetland destruction.[22]

Since the mid-1970s, America's waterfowl population has been in decline. Certain key species among the duck population are said to be in serious decline and have been so for at least a decade. The major part of the problem in this decline seems to be the loss of wetlands where waterfowl breed and nest their young. The best nesting grounds in North America, say some wildlife experts, are in a hilly region called the coteau that stretches across parts of South and North Dakota and on into Canada. While this region was once a landscape of wetlands, prairies, and grassy hills, it is now mostly farmland. Thus the loss of habitat in which to breed has resulted in a drop in the number of nests and the number of eggs laid, as well as the number of hatchlings that survive. This loss of habitat is compounded by the increased presence of contaminants in those breeding grounds that are left from the use of agricultural chemicals.[23]

Most of the damage has been done by dredging and filling operations, and contamination by waste material. These wastes contaminate the water and make it unfit for swimming and poison fish and shellfish to the point where they are inedible. About half of the country's estuaries and coastal wetlands remain undeveloped, but with the exception of Alaska, about 70 percent of the country's shoreline is privately owned, and many private owners find it hard to resist lucrative offers from developers. Some of these lands have been purchased by federal and state governments and by private conservation agencies to protect them from development.[24]

Coastal areas of the country are under particular threat because over half of the U.S. population lives along the coastlines. These coastal areas are also the sites of large numbers of motels, hotels, condominiums, beach cottages, and other real-estate developments. Most of the country's rivers have been dammed or diverted to meet energy, irrigation, and flood control needs, and these activities have changed the normal flow of freshwater into coastal wetlands and estuaries and modified their ability to function as a viable ecosystem. A great part of the problem lies in ignorance about the valuable ecological function these areas provide, as estuaries and coastal wetlands are one of the most productive and important natural ecosystems in the world. But they are also among the most intensely populated and stressed ecosystems, as human activities are increasingly impairing or destroying some of the more important services these ecosystems provide.[25]

Inland wetlands are often dredged or filled in and used as croplands and sites for urban and industrial development. As is the case with coastal wetlands, these areas are often considered to be wastelands because people do not understand their ecological importance. This destruction has greatly reduced the habitat of birds and other wildlife that live on or near these ecosystems, and has threatened some species with extinction. Better management practices are necessary to prevent further loss of these wetlands and to restore them to their original condition.[26]

THE EVERGLADES

The Everglades is said to be dying because of urban development and pollution. Today the Everglades, or what is left of it, is surrounded by an urban population of 4.5 million people and is polluted by sugarcane farms that have been carved out of its northern reaches. The Everglades was one of the largest wetlands systems in the world supporting a rich diversity of wildlife. But the population of wading birds has dropped from more than 2.5 million in the 1930s to about 250,000 in recent years. Some 13 animals in the Everglades are now on the endangered species list. Half of the original Everglades has been lost to development over the past several decades.[27]

The water that naturally replenishes the wetlands once came from Lake Okeechobee in a shallow sheet 50 miles wide, moving slowly south through the Everglades before emptying into the Bay of Florida. But, since the mid-1960s, the lake overflow has been channeled through a massive flood control project that can direct water to urban centers and agricultural interests as well as reduce the flooding of cattle pastures and farmland. The Corps of Engineers took the twists and turns out of the Kissimmee River and straightened it as well as draining 45,000 acres of wetlands, much of which was converted into pasture. This change in water flow patterns has affected the ecology of the Everglades forever and began to kill the unique ecosystem by greatly reducing the amount of water flowing through the system.[28]

Runoff from farmlands has also changed the Everglades, as high levels of phosphates and nitrates have transformed more than 20,000 acres of grass into cattails. These intruders absorb oxygen from the marsh and suffocate aquatic life at the bottom of the food chain. On shallow ponds and canals, algae have grown so thick that the sun is blocked from underwater plants.[29] Other chemicals have also entered the Everglades system and threaten both plants and wildlife. Runoff from the Miami metropolitan area has also increased nutrients and pollutants in the ecosystem.[30]

Protecting the Everglades is more than just a matter of protecting wildlife and a certain kind of ecosystem. The fate of the cities in South Florida may be closely tied up with what happens in the Everglades, as the water supply could dry up if the Everglades becomes a sea of cattails that do not hold and purify water. The Everglades replenishes the aquifer from which the cities of Miami and surrounding areas draw their water supply, and if this aquifer dries up, the sunny subtropical paradise of South Florida could become a barren wasteland. This prophecy became all too true in 1990, as rainfall was less than normal, creating a severe water shortage in South Florida that necessitated drastic measures as far as cutting water usage was concerned.[31]

Efforts are being made to restore the Everglades to something of their original condition. The Everglades Forever Act of 1994 passed by the state of Florida requires the building of artificial marshes to strain out the excess phosphorus, nitrogen, and other chemicals applied to the sugarcane farms located between the Everglades federal and state parks and Lake Okeechobee. The Corps of Engineers has also been reintroducing some of the the bends and twists in the river and undoing its channeling projects in order to increased the flow of water to the Everglades and provide additional wetlands for fish and nesting of birds.[32]

In early 1996, the Clinton administration announced a $1.5 billion plan to restore the natural water system to the Everglades with the federal government picking up half the tab. The project was aimed at reversing a half century of government policy that drained water that would have gone to the Everglades to increase coastal development and the sugarcane industry. Under the plan, the sugarcane industry would be required to sell 100,000 acres adjacent to the Everglades, which constitutes one fifth of its land, that would be allowed to go wild in an effort to restore water flow and replenish underground aquifers. The industry would also have to pay a penny-a-pound tax, amounting to about $245 million to support restoration efforts. The ultimate cost could be as high as $10 billion for engineering all the solutions that would allow the Everglades to coexist with agriculture and a growing human population.[33]

LOUISIANA'S WETLANDS

What is happening to wetlands and the coastline in Louisiana is one of the most serious problems regarding coastal erosion and wetlands protection facing the nation. What is done in that region will have a bearing on what happens to this resource across the nation, as Louisiana contains 40 percent of the nation's coastal marshes, but even more importantly, some 80 percent of the nation's coastal wetlands loss is taking place in that state. Each year these coastal marshes produce a commercial fish and shellfish harvest that was worth $806.2 million in 1994, and with the exception of Alaska, provide the largest finfish catch for sport fishers, an industry with an estimated total economic output of $3.5 billion annually.[34]

This region contains an ecosystem that supports over 30 percent of the nation's fisheries and 22 percent of the nation's oil and gas production. These coastal marshes also buffer destructive tidal surges caused by hurricanes and tropical storms and reduce flood damage to agricultural areas and population centers. They trap and hold freshwater that is a major water supply source for coastal communities, agriculture, and industry. These coastal marshes prevent salt water from intruding into these coastal freshwater supplies. These marshlands also provide a feeding, spawning, and nursery ground for a wealth of fish, shellfish, and wildlife that is almost as varied as that found in tropical rain forests.[35]

The wetlands, estuaries, and barrier beaches and islands of coastal Louisiana are at the southern end of the major waterflow migration route in the United States, and hunting in this flyway is valued at $58 million annually. Nearly 4 million ducks and geese, which is more than 66 percent of the water flow that use the flyway, find a winter haven in the coastal wetlands. These coastal marshes are thus ideal for sport fishing, hunting, and water-oriented recreation, and the out-of-pocket expenses of those people who use these areas for recreation exceed $337 million annually.[36]

In past decades, many storms and hurricanes have wreaked their havoc on the Louisiana coast. In the last 40 or 50 years, several hurricanes hit the coast causing more than $2 billion worth of damage. The coastal marshes in Louisiana were very helpful in keeping the losses down, and it has been suggested that had the coastal marshes not offered a buffer to coastal towns and cities the losses would have been at least 10 times greater, or some $20 billion in the coastal parishes.[37] If these estimates are true, than the coastal wetlands have at least an $18 billion value as a storm buffer alone. This estimate does not include the value of human lives saved as a result of the buffering effect.

Each year, about 40 to 60 square miles of these coastal wetlands disappear forever. Since the turn of the century, the state has lost more than a million acres of wetlands, an area 25 to 50 percent greater than the state of Rhode Island. While the state still has 5,156 square miles of saltwater, brackish, and freshwater wetlands remaining, the Corps of Engineers estimates that between the present time and the year 2040, another 1 million acres of wetlands will be lost to open water. This means that by the year 2040, a total of 2.4 million acres of wetlands will have been lost or converted to other uses.[38]

If this rate of loss is not reduced or arrested, the Gulf shoreline will advance inland as much as 33 miles in some parts of the state, jeopardizing federal, state, local, and private investments. Municipal and industrial water supplies will be threatened by saltwater intrusion. Coastal communities and cities such as New Orleans will be more vulnerable to hurricane tidal surges and flooding from tropical storms. Since 60 to 75 percent of the population in Louisiana live within 50 miles of the coast, many of these people may be forced to move inland in the not-too-distant future. And valuable fish and wildlife and recreational resources will be lost and difficult to restore.[39]

Causes of Destruction

These is no single cause of this wetland loss in the state of Louisiana, but ten major factors have been identified (Figure 11.2). These factors interact with each other, intensifying the impact of each single factor, thus producing a synergistic effect. Each of these factors accounts for a different percentage of the total loss, but flood protection in the

FIGURE 11.2 Major Causes of Wetland Loss

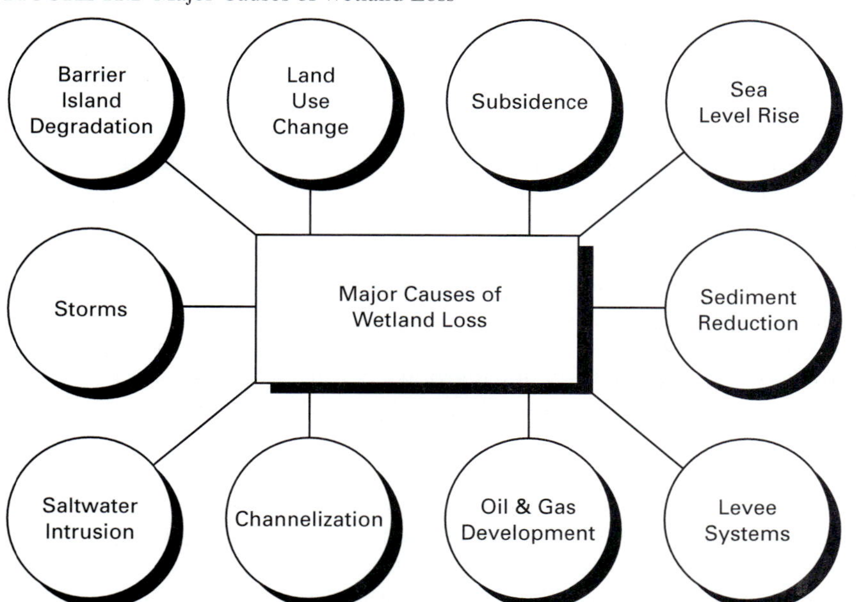

Source: U.S. Army Corp of Engineers in cooperation with the State of Louisiana, *Crisis on Louisiana's Coast . . . America's Loss,* undated, p. 5.

lower Mississippi Valley and economic developments in the Louisiana coastal wetlands have been responsible for a major part of wetland loss and consequent coastal erosion.[40]

The Louisiana coast stretches for about 40,000 miles of tidal shoreline and involves four different types of marshes: saline, fresh, brackish, and intermediate. Those marshes close to the salty Gulf waters tend to be saline or brackish marshes, and those farther inland from the Gulf and nearer to freshwater are the fresh or intermediate marshes.[41] All of these marshes were created over a period of some 7,000 to 10,000 years as the Mississippi River went through its "delta switching" phases. The river's mouth would switch to and fro like a hose, and in the process create new marshes. During one period, the river would flow in one direction spewing out millions upon millions of gallons of silty water, which would settle and form a marsh. Then gradually, the river would switch in another direction to form another marsh.[42]

The Mississippi draws its water supply from springs, lakes, rivers and ditches of 36 states in the country and from 4 Canadian provinces.[43] Thus the river covers a huge drainage area, and the soil from this area has been deposited along the coast of Louisiana for thousands of years. However, the river has also flooded vast areas of the country, especially in its lower reaches. In 1927, a so-called great flood caused major devastation along the lower reaches of the river. Protecting property along the lower reaches of the Mississippi River and its tributaries from the devastation of flooding has been a national imperative since that time. In order to control this flooding, the nation has invested $5.9 billion in building the world's largest flood control project. This project, called the Mississippi River and Tributaries Project (MR&T), is said to have prevented $111.3 billion of flood damage over the course of its operation.[44]

This flood control project, however, has confined the river between its banks along a good part of its course through southern Louisiana, and thus has changed the natural ecological functions the river performed in replenishing wetlands. The river would overflow its banks every spring, flooding the adjacent wetlands with nutrient- and sediment-rich water, which would build and sustain the diversity of the marshland. Since the levees were built, about the only water that flows into the wetlands is rainfall. The Mississippi carries 183 million tons of sediment down to the Gulf of Mexico each year, where it is dropping off the edge of the continental shelf into deep waters instead of building new wetlands[45] (Figure 11.3).

Before the levees, the Mississippi River delivered approximately 300 million tons of sediment to coastal Louisiana annually. The amount of sediment that is not reaching the marshes because of the levee system varies from 0.7 to 2.6 tons per day at a flow of 700,000 cubic feet per second (cfs). An average discharge of 450,000 cfs results in about 1.8 tons being carried for each unit of discharge. Estimating the rate per day at 810,000 tons and the weight per cubic yard at 1.5 tons, amounts to about 200 million cubic yards per year being lost to the ocean where it does not do any good and is lost forever.[46]

With no flood of freshwater each year to push back intruding salt water, the wetlands that cannot tolerate salt are being killed and replaced with open water ponds, which increase the interface between open water and wetlands, causing ever more erosion of coastal wetlands. Without the annual flow of freshwater containing enriching nutrients and sediment, many of the wetlands along the Gulf of Mexico are sinking out of sight. The river has not been allowed for many decades to do its job of replenishing existing wetlands and building new ones to increase coastal areas.[47]

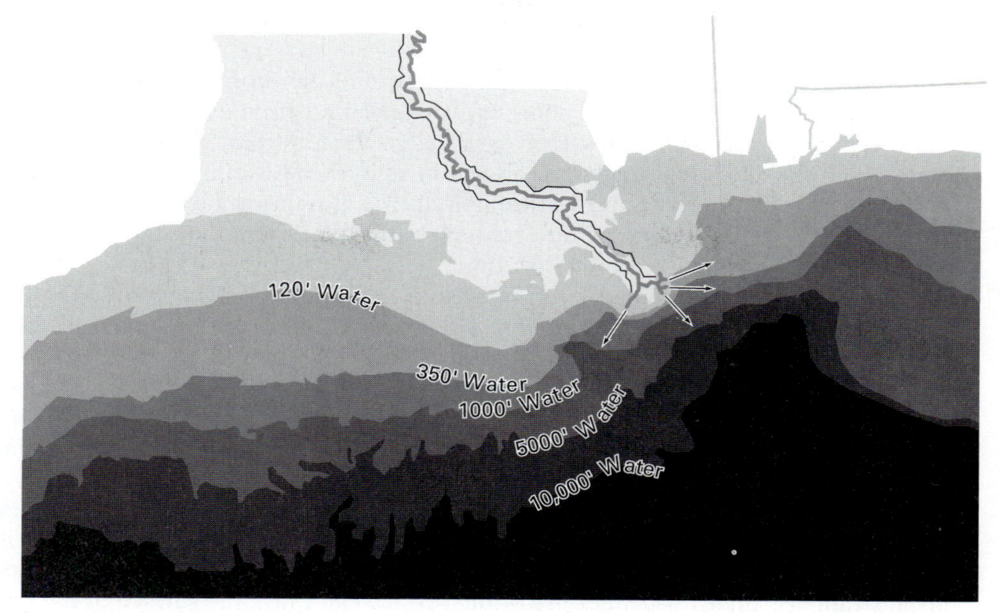

Source: The Louisiana Land and Exploration Company, *Louisiana's National Treasure,* undated, p. 13.

FIGURE 11.3 Mississippi Sediments Discharged into Deep Water

Economic development activities that have taken place in coastal wetlands interact with and intensify these natural processes. There has been a good deal of leveeing, channelization, oil exploration, and agricultural, urban, and industrial development in these coastal areas to accelerate the rate of loss. The wetlands area is laced with 8,200 miles of navigation, drainage, and petroleum access canals, which interrupt water and sediment flow over the wetlands. These canals also segment the wetlands and expose them to further erosion through wave action and other human activities. Salt water also flows into these areas and causes even greater erosion.[48]

Nature itself is responsible for some of the loss, as the long-term influences of subsidence, sea level rise, saltwater intrusion, and erosion have been felt by the wetlands. There have been significant changes due to natural forces in the relative land and water surface elevations. Subsidence causes the wetlands to sink an estimated 1.8 feet per century, and rising sea levels cover the wetlands with an additional one-half foot of water per century. These factors intensify saltwater intrusion and erosion and accelerate the conversion of wetlands to open water. Other minor losses occur as a result of storm-generated waves along the state's 40,000 miles of tidal shoreline and barrier islands.[49]

Three basic forms of wave erosion contribute to destruction of the wetlands. Human activities such as jetty construction and sand mining have contributed to shoreline erosion, which has proceeded at the rate of 33 feet per year along much of the barrier coast. These high rates of erosion threatened established development along the coast and also reduced Louisiana's first line of defense against incoming storm surges. If the beaches and marshes continue to disappear, cities such as New

Orleans will be subjected to much higher storm surges and direct wave attack during severe storms.

The second form of wave erosion is lake and bayshore erosion. The physical process where waves erode the shoreline also occurs within estuaries and along the shores of larger lakes and bays. Unlike waves in the Gulf that originate at some distance from the shore, wave generation within estuaries is localized and is caused primarily by prevailing winds and surface boat traffic. Bayshores facing prevailing winds appear to be the most vulnerable to erosion, but studies claim that the factors causing this erosion vary from location to location.

The third and last type of wave erosion is along canals and bayous. These natural and manmade waterways are widening as a result of bank erosion. Wind is less important in this type of erosion, while boat wakes and tidal energy seem to be the most important factors. Several studies have documented canal widening from ship traffic, while tidal energy in estuaries has contributed to barrier island erosion. It is highly likely that boat wake erosion in canals and bayous is more destructive to adjacent marshes than tidal erosion.

Subsidence or sinking of the wetlands happens when they are no longer replenished by waterborne sediments and become very compacted and sink under the open water. Subsidence is also caused by subsurface withdrawal of oil and gas, as withdrawal from strata less than 10,000 feet below the surface can cause the surface of the land to subside. Dewatering is also a factor in subsidence as when the water table is lowered due to drainage activities, the dewatered upper soils are subject to soil shrinkage, wind erosion, and biochemical oxidation. Urban expansion and agricultural drainage and flood control have led to extensive localized surface subsidence.

Because of these factors, the Louisiana coast began to erode, salt water crept farther inland, and plant life, unable to live in salty waters, began to die. Waste from drilling operations and chemical runoff from cropland also contributed to this process. As plant life died, so did its rootmat, and as the rootmat died, the soil began to wash away with the waves. Slowly but surely the coastal wetlands began to die and erode. The marshes were no longer populated with dense rushes of alligator weed, water hyacinths, and pickerelweed. Fresh marshes turned into saline or brackish marshes, and the typical vegetation became oystergrass, saltgrass, and winegrass.[50]

The Impact

Continued loss of the coastal wetlands will threaten most of the investment the state and the nation have made along the coast and will severely diminish related job opportunities. If another 1 million acres of wetlands are lost between now and the year 2040, commercial fish and wildlife harvests will be down to about 70 percent of the present harvest, having a national impact as about 46 percent of the nation's annual shrimp harvest comes from the Gulf waters of Louisiana's coastal marshes. An estimated 155 miles of banks in portions of major waterways built by federal and state governments, including the Gulf Intracoastal Waterway and the Mississippi River-Gulf Outlet, will be affected. The banks of these waterways will be lost to open water, requiring increased efforts to maintain.[51]

About 55 miles of hurricane protection levees and floodwalls will have to be shielded from erosion and enlarged to maintain the current level of protection, and in some locations, new projects will have to be mounted to replace the natural protection provided by the coastal wetlands. Nearly 100 miles of federal and state highways, about 27 miles of railroad tracks, and 1,570 miles of oil and gas pipelines, and 383 miles of gas, water, electric, power, and telephone lines will have to be relocated. And, it is estimated that about 1,800 businesses, residences, camps, schools, storage tanks, electric power substations, water control structures, and gasoline pumping stations will have to be protected or relocated. All of this additional protection and relocation because of wetlands loss will cost billions of dollars.[52]

Solutions

For the past several decades, the problems have been studied by state and federal agencies, as well as by private business organizations and environmental groups, probing for reasons for the wetlands loss and potential solutions. Each of these efforts has shed more light on the total problem and has confirmed that a piecemeal approach to the problems is not likely to be effective. This research has resulted in a much clearer understanding of the unique dynamics in the coastal area and has identified three ways that can used to approach wetland loss in the coastal regions.[53]

The first approach is wetland preservation, through restoring and maintaining the barrier shore and islands off the coast of the state, and building freshwater diversion structures in key places along the Mississippi. The former will keep wave action from further eroding the wetlands and help control further saltwater intrusion. The latter goal involves diverting freshwater from the river and allowing controlled flooding in areas where appropriate to enable the river to replenish the wetlands as it naturally did before the levees were built to prevent flooding.[54]

The second approach is to replace lost wetlands through the diversion of sediment-laden waters through outlets in the riverbanks and the disposal of sediment dredged from navigation channels into open water areas so new marshland can develop. Throughout the wetlands of coastal Louisiana are 2,800 miles of canals and rivers, which are constantly dredged to remain deep enough for navigation. Depositing this sediment back into the wetlands is a common and fruitful practice of the Corps of Engineers. And the third approach is to control development in coastal wetlands so that such growth will not further destroy wetlands and add to the problem. Another reason for controlling growth is to reduce the costs associated with relocation should this become necessary.[55]

An option with a high priority has been the restoration of Louisiana's barrier islands. These islands are the first line of defense against the sea and storm surges from hurricanes, and efforts have been made to raise the surfaces of these islands and close their breaches. Restoration of these islands will also limit wave erosion of interior marshes and help to prevent increases in the salinity of the bays behind them, limiting tidal mixing of the high-salinity waters of the open ocean. The ability of these islands to curtail wetland loss is limited, however, and these islands will not prevent wetlands from being submerged as the sea level rises nor will they prevent marsh erosion along canals and other waterways.

The diversion process plans to redirect water from the Mississippi River at three points in order to retard saltwater intrusion. It is hoped that reducing saltwater intrusion will slow down the loss of marshland and improve fish and wildlife productivity.

This diversion plan has been designed by the Corps of Engineers and will be funded by a yearly dedication of at least $5 million from Louisiana's mineral revenues. The dedication of these funds was done by Louisiana voters as they overwhelmingly approved establishing a Wetlands Conservation and Restoration Trust Fund, with the amount dedicated each year varying according to the price of oil but not exceeding $25 million. It is hoped that matching funds will be available from the federal government.[56]

Management of the marshes involves a variety of activities including regulation of the flow of water in and out of marsh systems, with the general goal of controlling water levels in the marshes. The size of these operations can vary from several acres to around 5,000 acres, with the larger wetland tracts being partitioned into smaller units. Water flow is usually regulated by a system of dikes or levees and some form of water control structure. Most of these plans incorporate features that prevent inflow of excess salt water and regulate the outputs or inputs of freshwater until a desired water level or salinity is reached. One of the most important advantages to this procedure is that private landowners can implement these measures themselves, and economic incentives are strong for these landowners to manage their marshes in their long-term interests.[57]

Some experts say that the growth of marsh management during the past several years and the expected use of similar projects during the next 20 years, will result in a total of 1,200,000 acres or about a third of Louisiana's wetlands being under some form of marsh management. Marsh management projects, however, often pit conservation and wildlife interests against each other. Some interests want to maximize the kinds of marsh grasses that entice ducks, while others want to regulate water levels to maximize the ability of fish to find protection during their growing years and move in and out of the marsh area at will. These two objectives are seldom compatible.[58]

A big job in saving the wetlands is one of educating the public as to the problem and what can be done. Projects involving citizens in cleanup can do a good deal to expose the public to the problem of wetland erosion. In two projects called Coastweeks and Beachsweep, over 3,400 volunteers turned out from across the state to clean 67 miles of coastal beaches. These efforts were sponsored in part by oil companies, state and federal agencies, and environmental groups. Government agencies recognize that a major part of any effort to save the wetlands will involve public understanding and citizen support behind action to establish a solution.[59]

A report complied by a technical subcommittee of the Gulf of Mexico Program issued at the end of 1990, indicated that between 1955 and 1975, freshwater and saltwater marshes in Louisiana were being converted to open water at the rate of about 50 miles a year, but since then the rate has slowed considerably to about half that amount. Some experts noted that this reduction of loss corresponded with the reduced rate of dredging that had been allowed and that tighter regulations on this activity were beginning to have an effect. Others argued, however, that the figures merely showed that the more fragile parts of the marshes had already disappeared, and what was left was the tougher, interior wetlands that could be expected to erode at a slower rate.[60]

Specifically, the study showed that in 1984, the state of Louisiana lost 26,860 acres of wetlands, but by 1990 that loss had slowed to 15,400 acres. Up to 87 percent of wetlands loss in the southern United States in the 1960s and 1970 was due to conversion of land to agricultural uses, but that much of the wetlands along the Gulf Coast was also giving way to open water. The cost of a acre of wetlands varies. The salt marshes could be replaced at a cost of about $10,000 per acre, but the forested swamp would cost $50,000 an acre or more to restore. The report also indicated that a 1 percent loss of wetlands in the Gulf amounted to a 1 percent loss in the shrimp harvest.[61]

At the end of 1990, a new bill was making its way though the federal government that was eventually passed as the Coastal Wetlands Planning, Protection, and Restoration Act. The act provides $50 million a year in combined federal and state spending for wetlands restoration and protection projects approved by a task force of federal agencies and the state. From the beginning, however, implementation of the legislation has been bogged down in bureaucratic problems and disputes, and efforts have largely focused on small projects aimed at making sure each part of the coast received some of the money. These projects will restore only a small percentage of the state's lost wetlands. The really big projects that would have more of an impact are awaiting funding from Congress, which is an iffy prospect in the era of budget cutting.[62]

ACTIONS TO PROTECT WETLANDS

Various approaches to wetlands protection can be taken by governments and private parties. Governments and private conservation organizations can purchase wetlands or easements on wetland areas and establish wildlife refuges, sanctuaries, or conservation areas. This preserves the wetlands in their natural state and protects them from development activities that would destroy their functions. The important thing in using this method is that once these areas have been set aside, they remain in that state and do not become subject to lobbying efforts by private citizens who want to change their classification to use the wetlands for their own purposes.[63]

Governments can also provide economic incentives to private landowners and industry to promote wetland preservation. As one example of this method, landowners who sell or donate wetlands to a government agency or qualified conservation organization can claim the value of the land as a charitable deduction. Governments can also create economic disincentives to wetland destruction, such as the "swampbuster" provisions written into legislation mentioned in a later paragraph. The intent of these disincentives is to discourage the further conversion of wetlands to farmland or other destructive actions.[64]

Regulation of wetlands is also important. Section 404 of the Clean Water Act establishes the major federal program under which activities in wetlands are regulated. This law requires a permit from the Army Corps of Engineers in order to discharge dredged or fill material into wetlands. Failure to obtain a permit or to comply with the terms of the permit can result in civil and/or criminal penalties. The Corps evaluates permit applications based on (1) regulations developed by the EPA in conjunction with the Corps that set the environmental criteria for permitting projects in wetland areas, and (2) factors to determine if the project is in the public interest. In addition to federal regulations, many states now have laws to regulate activities in wetlands.[65]

In 1986, the EPA established an Office of Wetlands Protection charged with providing leadership for a broad-based national effort for protecting the nation's wetland resources. In 1989, a Wetlands Action Plan was initiated with a goal of no net loss of wetlands over the short-term, and a gain in the quantity and quality of wetlands over the long-term. To achieve the no-net-loss goal, the EPA increased enforcement of federal restrictions on activities that destroyed or degraded wetlands. The EPA also provided guidance and support to state and local governments for wetlands protection, and worked with other federal agencies whose activities had an impact of wetlands, increasing public awareness of the ecological functions of wetlands, and conducting research to fill gaps in science to support wetland decisions.[66]

The EPA has become a center of wetlands expertise, according to its 1990 report, by providing more research, training, and communication on wetlands management. It is helping the states build comprehensive wetlands programs that incorporate both regulatory and nonregulatory approaches to wetlands protection. The EPA and other federal agencies involved in wetlands management are also developing better ways to monitor the health and extent of the nation's wetland resources. These agencies want to improve coastal water monitoring and to increase the number of estuarine/marine sanctuaries, protected refuges, reserves, and parks to preserve coastal areas and wetlands.[67]

The EPA also provided grants under this program to help states and tribes improve their ability to protect wetlands. Funding for this purpose grew from $1 million in 1990 to $5 million in 1991 to $8.5 million in 1992. In 1991, the EPA issued 60 such grants to 40 states, seven Indian tribes, and one territory. These grants enabled states and tribes to improve protection for their wetlands through conservation programs and other approaches that were tailored to local needs and situations.[68]

One of the most important areas to management is to reduce the pollution of coastal waters. The sources of pollution of these areas are shown in Figure 11.4. Nonpoint sources of pollution are a particular problem, and programs to control this pollution need to be strengthened in all coastal counties. Controls on point source discharges of toxics, nutrients, and other pollutants also need to be tightened to restore coastal water quality. Raw sewage flows from combined sanitary-storm sewers is a

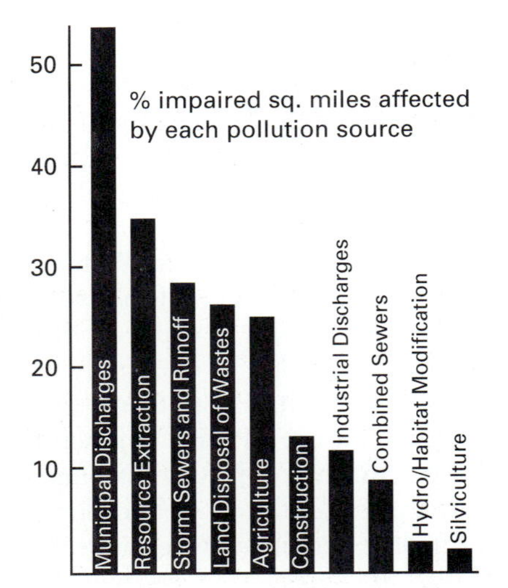

Source: United States Environmental Protection Agency, *Meeting the Environmental Challenge* (Washington, D.C.: U.S. Government Printing Office, 1990), p. 5.

FIGURE 11.4 Sources of Pollution in Estuaries

problem that is especially severe in many older seaboard cities that have combined systems. The EPA is requiring storm-water discharge permits for large cities in all coastal counties and is helping smaller municipalities with storm-water problems.[69]

Other laws, such as the Coastal Barriers Resource Act of 1982, also help to protect coastal zones from further destruction. This act prohibits most new federal expenditures and financial assistance for development of offshore barrier islands, which helps protect their important wetland resources. Of greater significance for inland wetlands is the "swampbuster" provision of the Food Security Act of 1985, which seeks to discourage the further conversion of wetlands for agricultural purposes by making any person who produces crops on wetlands converted to this purpose after December 23, 1985, ineligible for most federal farm benefits.[70]

There are many private protection strategies in addition to government actions. Because individual landowners and corporations own many of the nation's wetlands, they are in a key position to determine the fate of wetland properties under their control. Whether or not they actually own wetlands, citizens can help protect wetlands by supporting any number of wetland conservation activities. There are many opportunities for private citizens, corporations, and government agencies to work together to slow the rate of wetland loss and improve the quality of our remaining wetlands. Some of these options include the following.

- Rather than drain or fill wetlands, seek compatible uses involving minimal wetland alteration, such as waterfowl production, fur harvest, hay and forage, wild rice production, hunting and trapping leases, and selective timber harvest.
- Select upland sites for development projects rather than wetlands, and avoid wetland alteration or degradation during project construction.
- Donate wetlands, or funds for purchasing wetlands, to private or public conservation agencies.
- Maintain wetlands and adjacent buffer strips as open space.
- Construct ponds in uplands and manage for wetland and aquatic species.
- Support various wetland conservation initiatives by public agencies and private organizations.
- Participate in the Clean Water Act Section 404 program by reviewing public notices and, in appropriate cases, commenting on permit applications.[71]

In 1989, President Bush announced a no-net-loss policy with respect to wetland protection. This policy represented Bush's commitment to prevent further destruction of America's wetlands. But the policy never meant protecting wetlands at all costs, as in those cases where "unavoidable" damage was not possible to prevent, developers could offset the loss by restoring or creating roughly the same number of wetlands acres somewhere else. Also, critics believed that several exemptions built into the policy watered down the entire approach. For one thing, mitigation of damage would not be required if the amount of wetlands lost was considered "insignificant," and the mitigation process may not be practicable where there is a high proportion of land that is wetlands. This latter exemption was said to be directed at stripping protection from the arctic tundra areas, which are all wetlands.[72]

These exemptions were seen as huge loopholes through which developers and industry could escape. The words "unavoidable" and "insignificant" could be interpreted in many different ways, and the implementation of these words depended on who was doing the defining in a specific situation. In response to some of these criticisms, federal agencies issued a new definition of wetlands in 1990 that caused further controversy. The definition contained three tests to determine whether a territory qualifies as wetlands and thus comes under the no-net-loss policy: The territory under consideration must have the hydric soil typically found in wet areas, wetlands vegetation, and some wetness, even if the area is saturated for only one week during the growing season.[73]

These characteristics must be present under normal conditions, meaning that if a wetlands area were converted to cropland decades ago, it may still have the potential to revert back to its original wetlands state if left alone. This definition was expected to tighten the 404 program, under which regulation of dredging and filling in wetlands is provided. While the "swampbuster" bill mentioned above denied benefits to farmers who drain virgin wetlands, this new definition took aim at already converted cropland. Thus many millions of acres of farmland in some states became wetlands overnight because of this definition.[74]

Farmers were in an uproar, as they accused the new definition of being so expansive that vast chunks of farmland would be put under federal regulation for the first time. This definition would affect the cranberry bogs of Massachusetts, the rice paddies of Arkansas, and the sugar fields of Louisiana, as many farmers may need to get a permit from some federal agency before they could work their land. This new definition also discouraged some banks from lending to farmers because of the uncertain designation of their land, which in most cases was used as collateral for the loan. And it was expected to impair farmers' ability to sell excess acreage to developers. The definition thus introduced a new and divisive element into the debate over national wetlands policy and the implementation of the no-net-loss policy.[75]

After months of internal debate new revisions were proposed that would narrow the technical definition of wetland to mean land that is significantly wetter than under the original definition. The revisions would give wetlands protection to land with 15 straight days of standing water and 21 days of surface saturation. And with respect to wetlands vegetation, the revisions established a ranking system rating each species' preference for wet soil on a scale of 1 to 5, considering a territory wetlands only if the average of all plant life there was less than three. The revision would also permit more extensive "mitigation banking," which requires landowners to restore lost wetlands or create new ones in exchange for destroying existing wetlands.[76] This revision came under considerable criticism, and to defuse these attacks, the administration proposed increased federal spending to restore wetlands.[77]

Another bill introduced in 1991 would have drastically altered the way permits are issued for wetlands destruction. At the present time, the EPA can veto any permit issued by the Corps of Engineers for development if it determines the environmental damage would be too great. The new bill would have eliminated the EPA's authority over wetlands development and would also have required the federal government to buy all of the prime wetlands it believes should be protected. The EPA believed the bill would create a system with much more impetus to develop wetlands rather than protect them. This bill was introduced as part of the controversy over the definition of wetlands issued in 1990 that sparked such controversy and caused many in Congress to abandon the existing wetlands regulatory system.[78]

With the election of a Republican Congress in 1994, efforts to reduce wetlands protection center on rewriting the Clean Water Act with new definitions and regulations for wetlands that would require an area to be submerged for at least 21 days and remove protection for wetlands smaller than one acre. Critics of this definition argue that these provisions would deal a death blow to waterfowl production and open up many more acres of wetlands to developers and agricultural interests.[79] Measures have also been introduced under the general category of takings legislation mentioned in an earlier chapter that would compensate citizens for property values diminished by regulations, and thus would effectively halt enforcement of many existing regulations because of budget considerations.

Much of the controversy involves seasonal wetlands that are saturated only at certain times of the year, which environmentalists say are of great ecological importance. While these wetlands frequently are away from coastal areas in upland areas, they also are often the beginning of a river's watershed and provide storage for floodwaters as well as a habitat for migrating birds and other wildlife. Environmentalists believe that a scientific definition of how the wetlands fit into a particular watershed or how they are used by wildlife should be used. Political definitions greatly reduce the amount of wetlands protected and the protection they provide for wildlife and the public.[80]

In late 1996, the U.S. Army Corps of Engineers was preparing rules that would make it harder for landowners to get permission to develop federally protected wetlands. The current rule called Nationwide 26, because it is part of a group of regulations that apply everywhere in the country, allows owners of small plots of land to bypass the normal permit process that requires approval from several federal agencies. The proposed changes would phase out this exemption over a period of 18 to 24 months, at which time landowners wanting to change a federally classified wetland would have to undergo the normal permitting process. This move was expected to intensify the clash between the administration and the Republican-controlled Congress, who sharply disagreed on wetlands protection, and increase efforts to pass takings legislation to compensate small landowners for the lost economic potential of their property.[81]

Protection of wetlands is an important national goal, and more thought must be given to the development of strategies that are fair and just and yet accomplish the goal of saving this essential part of our environment. As with any other part of the environment, once wetlands are lost, it is difficult and expensive to restore them to their original condition, so the best strategy is to take steps to protect what is left and keep them in their original condition. This is essentially the goal behind the no-net-loss wetlands policy, although the question of how this policy is to be implemented is still up for debate. Perhaps the following quote from the EPA sums up the importance of wetlands to the nation and to future generations.

> Wetlands are an important part of our national heritage. Our economic well-being and our quality of life are largely dependent on our nation's wealth of natural resources, and wetlands are the vital link between our land and water resources. As wetlands are lost, the remaining wetlands become even more valuable. We have already lost over half of our nation's wetlands since America was first settled. We must now take positive steps to protect wetlands to ensure that the values they provide will be preserved for future generations.[82]

And so the wetlands controversy continues and is subject to political and economic pressures. Something like sustainable development is probably the most sensible approach to take with respect to the wetlands, for if they continue to erode, there are not going to be enough wetlands left to develop, and those states like Louisiana that have a large proportion of the nations's wetlands will be impoverished. The situation is serious as many people's lives are at stake and calls for an increased sensitivity of business and government leaders as well as citizens as to the unique ecology of the region and the importance of wetlands to the life of the region. There are lessons to be learned here that are vital to all regions of the country as to the importance of the ecological functions of areas that were formally considered to be wastelands that were ripe for development.

Questions for Discussion

1. What is happening to coastlines and sea levels around the country? Why is this happening? What choices do communities near the ocean have if these conditions worsen? What is being done about the problem? What strategies do you recommend?
2. What are coastal wetlands? What benefits do coastal wetlands provide for humans? What ecological functions do they perform? Why do many people regard these areas as worthless?
3. What are inland wetlands? Give some examples of the different types of inland wetlands around the country. What ecological functions do they perform? Were you aware of these functions before reading this chapter?
4. What are the major causes of wetlands destruction and degradation? What has been the extent of destruction across the country? Why are the wetlands a prime area for development?
5. Describe what is happening to the Everglades. Is the fate of Florida cities like Miami tied up with what happens there? In what ways? What can be done about the problem?
6. What is happening to coastal wetlands in Louisiana? What services do these wetlands provide for the state and country? How will these functions be affected if coastal erosion continues? What are the causes of this problem? Are some of these causes a surprise? Do they show what harm can be done through ignorance of ecology?
7. What are the various actions that can be taken to preserve wetlands? What is the government doing? What actions can private parties take to deal with the problem? What actions do you think will be most effective?
8. Describe the controversy over the no-net-loss policy with respect to wetlands. How are wetlands defined? What problems does this definition cause? What is the current status of this controversy? Are wetlands being adequately protected?

Endnotes

1. G. Tyler, *Living in the Environment,* 6th ed. (Belmont, CA: Wadsworth, 1990), p. 120.
2. Ibid.
3. United States Environmental Protection Agency, *America's Wetlands: Our Vital Link Between Land and Water* (Washington, D.C.: Office of Wetlands Protection, 1988), p. 1.
4. Miller, *Living in the Environment,* p. 122.
5. EPA, *America's Wetlands,* p. 4.
6. United States Environmental Protection Agency, *Environmental Progress and Challenges: EPA's Update* (Washington, D.C.: U.S. Government Printing Office, 1988), p. 60.
7. EPA, *America's Wetlands,* p. 2.
8. Ibid., p. 4.

9. Jodi L. Jacobson, "Holding Back the Sea," *State of the World 1990* (Washington, D.C.: Worldwatch Institute, 1990), p. 86.
10. Michael D. Lemonick, "Shrinking Shores," *Time,* August 10, 1987, pp. 40–41.
11. Ibid., p. 41.
12. Jacobson, "Holding Back the Sea," pp. 86–87.
13. Ibid., p. 81.
14. Ibid., pp. 83–84.
15. Ibid., p. 85.
16. Ibid., p. 92.
17. Ibid., p. 93.
18. Ibid., p. 95.
19. Lemonick, "Shrinking Shores," p. 47.
20. Ibid.
21. EPA, *America's Wetlands,* p. 6.
22. Ibid.
23. Richard Lipkin, "Plans That Are Habit-Forming," *Insight,* November 6, 1989, p. 56.
24. Miller, *Living in the Environment,* p. 129.
25. Ibid.
26. Ibid., p. 128.
27. James Carney, "Last Gasp for the Everglades," *Time,* September 25, 1989, p. 26.
28. Mark Schleifstein, "Everglades' Surrender Is Slow, Sad," *Times-Picayune,* March 26, 1006, p. A7.
29. Carney, "Last Gasp for the Everglades," pp. 26–27.
30. Paul A. Fresty, "Humans Prove Deadly to Once-Thriving Everglades," *Times-Picayune,* March 26, 1996, p. A7.
31. David Snyder, "Florida Drains Hopes, Aquifers Dry," *Times-Picayune,* March 4, 1990, p. A1.
32. Schleifstein, "Everglades' Surrender," p. A7.
33. Peter Katel, "Letting the Water Run into 'Big Sugar's' Bowl," *Newsweek,* March 4, 1996, p. 56. See also Timothy Noah, "Clinton Plans to Tax Sugar Growers in Florida to Fund Everglades Project," *Wall Street Journal,* February 16, 1996, p. B12; "Everglades Rehab Endorsed," *Times-Picayune,* February 20, 1996, p. A-3.
34. Mark Schleifstein, "Sinking Treasure," *Times-Picayune,* March 26, 1996, pp. A1, A4, A10.
35. United States Army Corps of Engineers, *Crisis on Louisiana's Coast . . . America's Loss* (New Orleans, LA: New Orleans District, 1989), p. 1.
36. Ibid., p. 2.
37. Interview with John Wong, Manager, Environmental Division, C. H. Fenstermaker and Associates, Inc., Civil Engineers, Land Surveyors, and Environmental Consultants, New Orleans, Louisiana, as quoted in Stewart T. Gibert, "Management of an Asset: The Louisiana Delta Wetlands," unpublished paper, College of Business Administration, Loyola University, New Orleans, Louisiana, 1990, p. 9.
38. Corps of Engineers, *Crisis on Louisiana's Coast,* p. 4. Schleifstein, "Sinking Treasure," p. A4
39. Ibid.
40. Ibid.
41. Robert Chabrelk and Greg Linscombe, *Vegetative Type Map of the Louisiana Coastal Marshes* (Baton Rouge, LA: Louisiana Department of Wildlife and Fisheries, 1987), pp. 23–24.
42. Coalition to Restore Coastal Louisiana, *Coastal Louisiana: Here Today, Gone Tomorrow?* (Baton Rouge, LA: Coalition, 1989), pp. 21–22.
43. Paul G. Kemp, Remarks made to a class on Environmental Issues for Management, College of Business Administration, Loyola University, New Orleans, Louisiana, Spring semester, 1990.
44. Corps of Engineers, *Crisis on Louisiana's Coast,* p. 6.
45. Ibid., pp. 5–6.

46. United States Army Corps of Engineers, "Atchafalaya Outlet," Mississippi River Commission pamphlet, p. 2.
47. Corps of Engineers, *Crisis on Louisiana's Coast,* p. 6.
48. Ibid., p. 8.
49. Ibid., p. 9.
50. Interview with Mike Windham, a refuge project manager with the Louisiana Wildlife and Fisheries, New Orleans District, as quoted in Gibert, "Management of an Asset: The Louisiana Delta Wetlands," unpublished paper, p. 14. See Peter Annin, "The Rat That Ate Louisiana," *Newsweek,* March 8, 1993, p. 67, for a story about the way in which nutria are also destroying wetlands in Louisiana.
51. Corps of Engineers, *Crisis on Louisiana's Coast,* p. 9.
52. Ibid.
53. Ibid., p. 10.
54. Ibid.
55. Ibid., p. 11.
56. Louisiana Department of Natural Resources, *Creation of Wetlands Trust Fund Approved,* October 1989, p. 3. See also Mark Schleifstein, "Saving the Coast," *Times-Picayune,* May 8, 1994, p. A16.
57. Corps of Engineers, *Crisis on Louisiana's Coast,* pp. 11–12.
58. Mark Schleifstein, "Conflicting Interests Squeeze Marshes," *Times-Picayune,* March 26, 1996, p. A5.
59. Gibert, "Management of an Asset," pp. 45–46.
60. Christopher Cooper, "La. Wetlands Loss Reported Slowing," *Times-Picayune,* December 5, 1990, p. B3.
61. Ibid.
62. Schleifstein, "Sinking Treasure," p. A10.
63. EPA, *America's Wetlands,* p. 8. The Nature Conservancy has protected more than 114,000 acres of Louisiana's wetlands by establishing a system of nine private nature preserves. See Nancy Jo Craig, "Saving Our Natural Resources: A Race Against Time," *Times-Picayune,* April 19, 1991, p. B7.
64. Ibid.
65. Ibid.
66. United States Environmental Protection Agency, *Meeting the Environmental Challenge* (Washington, D.C.: U.S. Government Printing Office, 1990), p. 5.
67. Ibid.
68. United States Environmental Protection Agency, Communications, Education, and Public Affairs, *Securing Our Legacy* (Washington, D.C.: EPA, 1992, p. 31.
69. EPA, *Meeting the Environmental Challenge,* p. 5.
70. EPA, *Environmental Progress and Challenges,* p. 61.
71. EPA, *America's Wetlands,* pp. 8–9.
72. Bob Marshall, "Bush Lets Industry Shove Wetlands," *Times-Picayune,* February 18, 1990, p. C16.
73. Rick Raber, "New Definition of Wetlands Bogs Down Some La. Farmers," *Times-Picayune,* April 21, 1990, p. A1.
74. Ibid.
75. Ibid.
76. Rick Raber, "White House Proposal Relaxes Wetlands Rules," *Times-Picayune,* August 3, 1991, p. A1; Michael D. Lemonick, "War Over the Wetlands," *Time,* August 26, 1991, p. 53.
77. See "Scientists Condemn Wetlands Proposal," *Times-Picayune,* October 9, 1991, p. A5; Bruce Alpert, "Wetlands Revisions Under Fire," *Times-Picayune,* May 15, 1991, p. A6; Bruce Alpert, "Bush to Push Wetland Spending," *Times-Picayune,* January 8, 1992, p. A2.

78. James O'Byrne, "EPA Chief: Bill Will Bog Down Wetlands Work," *Times-Picayune,* April 11, 1991, p. B3. See also James Gill, "Selling the State Down the River," *Times-Picayune,* April 14, 1991, p. B7.
79. Bob Marshall, "Fouling Up a Good Situation," *Times-Picayune,* May 28, 1995, p. C18.
80. Schleifstein, "Conflicting Interests Squeeze Marshes," p. A5.
81. John McQuaid, "Stricter Wetland Waiver Coming," *Times-Picayune,* December 9, 1996, p. A1.
82. EPA, *America's Wetlands,* p. 30.

Suggested Reading

Bingham, Gail, ed. *Issues in Wetlands Protection.* Washington, D.C.: Conservation Foundation, 1990.

Bolton, H. Suzanne, and Orville T. Magoon, eds. *Coastal Wetlands.* New York: American Society of Civil Engineering, 1991.

Burke, David G., et al. *Protecting National Wetlands.* Washington, D.C.: National Planning Association, 1989.

Charbreck, Robert H. *Coastal Marshes: Ecology and Wildlife Management.* Minneapolis: University of Minnesota Press, 1988.

Christie, Donna R. *Coastal and Ocean Management Law in a Nutshell.* St. Paul, MN: West, 1994.

Coalition to Restore Coastal Louisiana. *Coastal Louisiana: Here Today, Gone Tomorrow?* Baton Rouge, LA: Coalition, 1989.

Daiber, Franklin C. *Conservation of Tidal Marshes.* New York: Van Nostrand Reinhold, 1986.

Dennison, Mark S., and James F. Berry. *Wetlands: A Guide to Science, Law, and Technology.* New York: Noyes, 1993.

Dugan, Patrick. *Wetland Conservation: A Review of Current Issues and Required Action.* Washington, D.C.: Island Press, 1990.

Dugan Patrick, ed. *Wetlands: A World Conservation Atlas.* New York: Oxford University Press, 1993.

Holing, Dwight. *Coastal Alert: Ecosystems, Energy, and Offshore Oil Drilling.* Washington, D.C.: Island Press, 1990.

Kelley, Joseph T., et al. *Living with the Louisiana Shore.* Durham, NC: Duke University Press, 1984.

Kusler, Jon, and Mary Kentula. *Wetland Creation and Restoration: The State of the Science.* Washington, D.C.: Island Press, 1990.

Mitsch, William J. *Wetlands,* 2nd ed. New York: Van Nostrand Reinhold, 1993.

National Wildlife Federation. *Status Report of Our Nation's Wetlands.* Washington, D.C.: NWF, 1987.

Niering, William A. *Wetlands of North America.* New York: Thomson-Grant, 1991.

Pilkey, Orin H., Jr., and William J. Neal, eds. *Living with the Shore.* Durham, NC: Duke University Press, 1987.

Platt, Rutherford H., et al. *Coastal Erosion: Has Retreat Sounded?* New York: Natural Hazards, 1992.

Seoderi, Paul F. *Wetlands Protection: The Role of Economics.* Washington, D.C.: Environmental Law Institute, 1990.

Silverberg, Steven M., and Mark Dennison. *Wetlands and Coastal-Zone Regulation and Compliance.* New York: Wiley, 1993.

Simon, Anne W. *The Thin Edge: Coast and Man in Crisis.* New York: Harper & Row, 1978.

United States Army Corps of Engineers. *Crisis on Louisiana's Coast . . . America's Loss.* New Orleans, LA: New Orleans District, 1989.

United States Environmental Protection Agency. *America's Wetlands: Our Vital Link Between Land and Water.* Washington, D.C.: Office of Wetlands Protection, 1988.

Viles, Heather, and Tom Spencer. *Coastal Problems.* New York: Routledge, Chapman and Hall, 1995.

Whigham, Dennis F., ed. *Wetlands of the World.* Netherlands: Kluwer Academic Press, 1993.

Whigham, Dennis F., et al., eds. *Wetland Ecology and Management: Case Studies.* Netherlands: Kluwer Academic Press, 1990.

Williams, Michael, ed. *Wetlands: A Threatened Landscape.* New York: Blackwell, 1993.

CHAPTER 12

Management Theory and the Environment

Management theory reflects the dualism and anthropocentric approach to nature that is found in the rest of American society. Until recently, there have been few attempts to make the environment a central part of such theoretical efforts. While there has been a considerable amount of theoretical work accomplished in the last several decades to expand the notion of corporate responsibility and incorporate a social dimension into management theory based on the notion that the corporation is a multipurpose institution with both economic and social responsibilities, most if not all of this work does not even incorporate the environmental dimension as a central part of its theoretical structure. By and large, management theory does not deal with nature, and makes no mention of managing nature as a commons in the interests of society as a whole.

ECONOMIC FUNDAMENTALISM

Economic fundamentalism refers to the belief that the market can solve all the problems in society that are worth solving. Those problems that cannot be solved through markets are not worth worrying about, as they are externalities that can be either ignored or dealt with in something of a cursory fashion. People who are economic fundamentalists believe that the market can be utilized to allocate resources to deal with all the problems that society faces and that value questions that fall outside the market are not important enough to warrant much attention.

This mechanistic view of reality is borrowed from the worldview that emerged out of the Middle Ages when science began to get a foothold in Western society. The universe was looked at in mechanistic terms, as scientists searched for laws and mechanisms that would explain the movement of the planets and other bodies in the solar system. This worldview also was transferred to the social sciences, most particularly economics, which came to look at the economy in mechanistic terms and viewed market economies as mechanisms to allocate scarce resources to their most productive uses. The forces of supply and demand came to dominate thinking about the economy and became the primary way to explain how the system worked to allocate resources.

People were dehumanized in this view and became merely factors of production as land, labor, and capital were considered to be the major inputs to the process. Resources were allocated to their best uses through competition, which was seen to be a regulator of the whole process. Competition assured that consumers had a choice among competing products and were not the victims of a monopolist who could charge what he or she wanted. The system promoted self-interest as a universal motivating principle and thus was based on egoism to motivate people to participate in the system. But competition between competing egos would prevent any one ego from getting too large and dominating the rest. Thus the system provided a mechanism to balance competing egos against one another in order to keep things under control.

This pursuit of self-interest balanced in a competitive system assured that the society would be better off from the standpoint of material wealth than some alternative system. The system was given an ultimate utilitarian justification from an ethical standpoint, as individual egoism was swallowed up in utilitarianism, so to speak. People were not encouraged to develop a sense of social or community responsibility that would relate to marketplace behavior, as the system itself would assure that community interests were adhered to in marketplace transactions. People were thus encouraged to pursue their individual self-interest without any kind of external controls on their behavior that might promote the welfare of the community. People were let off the hook, so to speak, as the system itself would take care of any ethical concerns related to broader responsibilities to society.

External controls are vigorously opposed by economic fundamentalists, as the best government is the one that governs least. The government has certain functions to perform, like providing for the national defense and some kind of legal framework where disputes can be resolved, but it should not regulate the market system, and above all, not engage in any kind of long-range planning for the economy as a whole. The market mechanism would work perfectly if it were left alone and not interfered with by government or any other body. All such interference could do would be to promote inefficiencies and lead to a misallocation of resources. Thus economic fundamentalists fight government intervention at every corner, and oppose any positive functions of government beyond those mentioned.

What this system means as far as value is concerned, is that those things that cannot be traded on the market where the value can be determined by the forces of supply and demand, really have no value worth considering. The value of raising children, for example, which most societies consider to be one of the most important functions, cannot be valued on the market and is hence not considered as a part of the gross national product of the country. The environment also has no value in and of itself, as environmental deterioration is not factored into marketplace transactions, nor is resource depletion a part of our accounting system at the national level.

This adherence to economics as the basis of value makes for a certain kind of society. American society is a money-oriented culture, where everything has a price, everything is valued in economic terms, and the bottom line of everything that counts is money. Many, if not most, of the social and ethical problems facing our society are symptomatic of a much larger problem facing American society, a problem related to the pervasiveness of economic values in our society. By this is meant the pursuit of money and economic fortune, the pursuit of economic power and fame, the pursuit of economic gain as an end in itself and as the primary goal for which one ought to strive.

These values have pervaded every aspect of our lives and corrupted many of our institutions that were created to pursue other values to maintain something of a pluralistic balance in our society.

The main thing that counts in our society is money, and anything that can't be valued in monetary terms doesn't count. Why can't rain forests be saved? Because they have no economic value in their natural state. They have value only when the trees are cut down and sold on the market or when the ground underneath is used for grazing cattle or plowed and used to grow crops. The fact that rain forests may be vital to the preservation of the human race is of no consequence until that function can somehow be translated into economic value, which can then be taken into account in marketplace transactions.

This phenomenon stems from a fundamental flaw in the philosophy of John Locke, who is given credit for much of the thinking regarding property rights and the operation of markets. Something in a state of nature has no economic value and is of no utility to the human race. It is only when land, for example, is taken out of a state of nature and someone is given property rights to use that land for growing crops or for exploitation of minerals that the land then has economic value and can enter into marketplace transactions. In its natural state, the land is valueless.

The incentive structure of American society is also hooked into this economic value system. Doing good or doing wrong have little value in and of themselves, but have importance only as they entail fairly substantial economic consequences. Economic incentives are the most powerful ones, and to get corporations to dispose of hazardous wastes properly, for example, fairly substantial economic penalties or rewards have to be imposed in order for anything to get done in this regard. There is little incentive and a great deal of economic disincentive for corporations to voluntarily dispose of hazardous wastes properly or deal constructively with a host of other environmental problems they help to create.

Before the development of regulations, the traditional way to handle hazardous waste was to put barrels of the stuff on the back dock to be picked up by some waste hauler and disposed of somehow. Corporations neither kept records nor cared where the stuff was finally dumped, even though they were in the best position to know what hazards the substances posed to human health and the environment. But they operated in kind of an out-of-sight, out-of-mind mentality, and had no economic incentive to act responsibly. Consequently, the nation is stuck with having to clean up existing dumpsites all over the country with a program called Superfund, which is itself relatively ineffective.

Most environmental problems cannot be responded to in a market economy. Where there is no money to be made, incentives are lacking. If any one company in a given industry does choose to clean up its environmental problems, it will only place itself at a competitive disadvantage, because those costs will have to be reflected somewhere, most likely in the price of its products. Some corporations have found that they can save money by reducing their waste, for example, and have developed waste reduction programs that they claim have saved them millions of dollars. Some of these programs are described in the next chapter. But, the point is that unless economic incentives are built into the problem, such as in waste reduction, companies have no incentive to deal with environmental problems unless forced to do so by laws and regulations. And such incentives are not present across the board in all environmental problems.

Consumers have no incentive to voluntarily purchase automobiles with pollution control equipment even if the market offered such a choice. The impact that one car would make on the pollution problem is infinitesimal and can't be measured. Thus one would be spending money and not getting anything in return. There are not enough consumers that will act irrationally in an economic sense to make enough of these kinds of choices to make a difference. Even if they did, then a rational consumer would have an incentive to be a free rider, and enjoy clean air at someone else's expense.

When economic values dominate as they do in the marketplace, there are no incentives to respond to problems such as environmental degradation. Such a response has had to be made by the government, which can underwrite the rules by which all the players have to abide. Then corporations and consumers can be coerced into giving adherence to another set of values. But regulation has it own problems, not the least of which is the extensive enforcement mechanism that has to be developed to assure that everyone is in compliance with regulations. Placing all the moral freight on the regulatory process is asking that process to do more than it can deliver. But since the market allows corporations to abdicate environmental responsibility, the society really has no good alternative.

Some call this inability of market systems to respond to environmental pollution and degradation as market failure, but to use this term is not entirely accurate. Market systems were not designed to factor in environmental costs, and it is not fair to blame the system for not doing something for which it was not designed. Property rights are not appropriately assigned with regard to the environment, and nature often lacks a discrete owner to look after its interests. The rights of nature can be violated by market exchanges, and as a common property resource, nature can be overused and degraded as it is subject to the tragedy of the commons.

CORPORATE SOCIAL RESPONSIBILITY

The social environment was given increasing attention during the 1960s and 1970s by business corporations and schools of business and management. While the concept of corporate social responsibility may have had its origins in the 1930s as some scholars suggest, the concept really came into its own, so to speak, during the 1960s as a response to the changing social values of society.[1] Business executives began to talk about the social responsibilities of business and develop specific social programs in response to problems of a social, rather than economic, nature. Schools of business and management implemented new courses in business and society or in the social responsibilities of business.

Attempts to broaden the notion of corporate responsibility have largely rested on the doctrine of social responsibility or some variation thereof. While various definitions of social responsibility have been advocated, there seem to be five key elements in most, if not all, of these definitions. These elements are that (1) corporations have responsibilities that go beyond the production of goods and services at a profit; (2) these responsibilities involve helping to solve important social problems especially those they have helped create; (3) corporations have a broader constituency than stockholders alone as incorporated in the notion of stakeholders; (4) corporations relate to society and indeed have impacts that go beyond simple marketplace transactions; and (5) corporations serve a wider range of human values than can be captured by a sole focus on economic values.

The problem facing modern management theorists who accept the fact that a divergence often exists between social performance and marketplace behavior and one cannot be subsumed under the other, is how to connect social responsibilities with management behavior in such a way that they are not peripheral to mainstream business concerns. This is a most difficult task and is one that has not yet been successfully accomplished. While there is a good deal of evidence that the language of social responsibility and social responsiveness is now an accepted part of management theory and practice, such concepts remain more or less peripheral concerns that have not replaced the traditional economic responsibilities of business.

> In short, it appears that many of the changes that were anticipated in the Sixties and Seventies have in fact taken place, but more by absorption and adaptation than by replacement . . . Contemporary managers are more aware of the impact of business activity on the physical and social environment and the quality of life, but production, marketing, and finance remain their primary concerns. And in academe, the social issues/public policy area is well established as an essential element in a sophisticated and comprehensive management program—and a very popular theme in executive education—but it does not pervade the entire curriculum in the way that some of us once thought it might.[2]

Part of the problem is with the nature of social responsibility doctrine itself. The word "doctrine" is used consciously, as it seems that social responsibility was more of a doctrine than a serious theory of the corporation. Scholars and executives who advocated social responsibility seemed to do so more as an article of faith than as a theoretical paradigm that could bid for serious attention and begin to compete with the economic theory of the firm for a hold on the thinking of scholars and business executives. The debate about social responsibility was held largely on moral grounds: that corporations should balance responsibility with power, that they should be socially responsible out of a sense of enlightened or long-run self-interest, that they could avoid the haunting specter of government regulation by being socially responsible, that they needed to be responsive to social values as society changed in order to remain a viable institution, or that they could gain a better public image by being socially responsible.

These arguments were more in the nature of "ought" statements relative to how corporations should behave in order to create a good society. But they were never incorporated into a comprehensive theory that encompassed social as well as economic responsibilities. The result is that by and large the economic theory of the firm has survived intact from attacks mounted by social responsibility advocates and has not been replaced by any new theories or ways of thinking about the firm and its responsibilities. The bottom line of corporate organizations as well as for the nation as a whole is still economic in nature.

In recent years, there has been an explosion of theoretical work related in one way or another to this notion of social responsibility. Three major theoretical approaches—stakeholder theory, normative theory, and social contract theory—have been developed to provide a descriptive and analytical framework to aid in understanding how business and society relate to each other and how the responsibilities

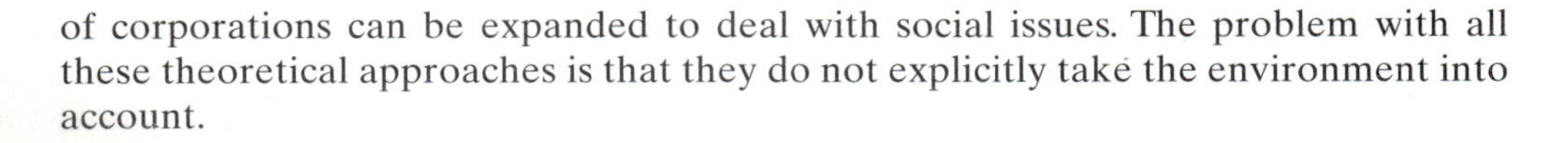
of corporations can be expanded to deal with social issues. The problem with all these theoretical approaches is that they do not explicitly take the environment into account.

Stakeholder Theory

Freeman is usually given credit for doing the seminal work on the stakeholder concept, even though Abrams urged business leaders to pay attention to their corporate constituents, a theme that was continued by the Committee for Economic Development 20 years later.[3] Since Freeman's work, however, the stakeholder concept has been widely employed to describe and analyze the corporation's relationship to society.[4] At least one conference has been held that dealt exclusively with the concept and was reported in a major journal.[5]

While each scholar may define the concept somewhat differently, each version generally stands for the same principle, namely that corporations should heed the needs, interests, and influence of those affected by their policies and operations.[6] A typical definition is that of Carroll, which holds that a stakeholder may be thought of as "any individual or group who can affect or is affected by the actions, decisions, policies, practices, or goals of the organization."[7] A stakeholder then is an individual or group that has some kind of stake in what business does and may also affect the organization in some fashion.

The typical stakeholders are considered to be consumers, suppliers, government, competitors, communities, employees, and of course, stockholders, although the stakeholder map of any given corporation with respect to a given issue can become quite complicated. [8] Stakeholder management involves taking the interests and concerns of these various groups and individuals into account in arriving at a management decision, so that they all are satisfied at least to some extent, or at least that the most important stakeholders with regard to any given issue are satisfied.

The problem with stakeholder theory is that with rare exception, it never takes the environment into account. The typical stakeholder map never includes plants or animals or nature in general, even though each can be enormously affected by corporate policies and activities, as the issue chapters in this book amply illustrate. Stakeholder theory is largely anthropocentric in nature as only humans count. The nonhuman or natural world has no stake and no interests that need to be taken into account.

Normative Theory: Ethics and Values

The normative approach to the field is called business ethics where the concern is to hold managers and their firms responsible for implementing ethical principles in their organizations and using moral reasoning in their decisions, in the formulation of policies and strategies, and in the general direction of their enterprises. The usual approach to ethical theory in the business ethics literature is to present either in cursory form or sometimes in greater detail the theory of utilitarianism based on the writings of Bentham and Mill as representative of a more general class of teleological ethics. Kantian ethical theory related to the categorical imperative as representative of the deontological approach to ethical decision making is also presented. Most business ethics scholars then proceed to consider certain notions of justice, usually the egalitarianism of John Rawls and the opposing libertarianism of Robert Nozick. They also generally include a discussion of rights and, at times, some variation of virtue theory.

What we are left with given these different theories is kind of an ethical smorgasbord where one has various theories from which to choose that will hopefully shed some light on the ethical problems under consideration and lead to a justifiable decision. But we are never told to any extent exactly how we are to decide which theory to apply in a given situation, what guidelines we are to use in applying these different theories, what criteria determine which theory is best for a given problem, and what to do if the application of different theories results in totally different courses of action.[9] This litany of conflicting theories and principles, all of which were initially meant as a universal approach to ethical problems, give conflicting signals to people in positions of responsibility in business and other organizations.

There was hope that a direct concern with ethics and values might triumph over the dominance of economic values where other approaches had failed. By establishing a broader framework of social values it was hoped that a more firm theoretical base could be built on which to rest the case for social responsibilities. In other words, economic responsibilities would then have to be seen as part of a much more comprehensive value framework that incorporated social responsibilities as an integral part of the system. Thus some scholars talked about a culture of ethics that embraces the most fundamental moral principles of humankind.[10]

One of the most ambitious efforts to construct an alternative paradigm based on ethical considerations is Amitai Etzioni's attempt to question the standard rationalist model at the root of economic theory.[11] Claiming that he does not advocate abandoning neoclassical economics, Etzioni tries to provide a broader framework for the integration of economics into a more comprehensive moral system and what he claims is a more realistic view of human relations and society. By developing the I & We paradigm, Etzioni anchors the self-interest of the individual I within a broader social collectivity called the We, where moral commitments to the development of a truly human community are paramount. He tried to show how moral concerns are evident in most areas of economic behavior.

But Etzioni never mentions the environment or includes it in his broader moral framework. The I & We paradigm leaves out animals and plants and other aspects of nature and is again anthropocentric in nature. The same is true of the other ethical theories mentioned previously. Utilitarianism, for example, leads to an analysis of good and bad consequences from a human perspective, and benefit-cost analysis based on this theory uses economic value to make its deliberations. Consequently, the field of environmental ethics, which is a separate field from business ethics, has not utilized these theories to any great extent, and has had to base its concerns about the environment on other theories related to intrinsic value and other such concepts as described in an earlier chapter.

Social Contract Theory

The emerging social responsibilities of business have also been expressed in terms of a changing contract between business and society that reflected changing expectations regarding the social performance of business.[12] The old contract between business and society was based on the view that economic growth was the source of all progress, social as well as economic. The engine providing this economic growth was considered to be the drive for profits by competitive private enterprise. The basic mission of business was thus to produce goods and services at a profit, and in doing this business was making its maximum contribution to society and being socially responsible.[13]

This changing contract between business and society was based on the view that the single-minded pursuit of economic growth produced detrimental side effects that imposed social costs on society. The pursuit of economic growth, some believed, did not necessarily lead automatically to social progress, but instead led to a deteriorating physical environment, an unsafe workplace, discrimination against certain groups in society, and other social problems. This contract between business and society involved the reduction of these social costs of business through impressing upon business the idea that it has an obligation to work for social as well as economic betterment. This idea was expressed by the Committee for Economic Development as follows:

> Today it is clear that the terms of the contract between society and business are, in fact, changing in substantial and important ways. Business is being asked to assume broader responsibilities to society than ever before and to serve a wider range of human values. Business enterprises, in effect, are being asked to contribute more to the quality of American life than just supplying quantities of goods and services.[14]

The changing terms of the contract are found in the laws and regulations that society has established as the legal framework within which a business must operate and through shared understandings that prevail as to each group's expectations of the other.[15] The social contract is a set of two-way understandings that characterize the relationship between business and society and the changes in this contract that have taken place over the past several decades are a direct outgrowth of the increased importance of the social environment of business. The "rules of the game" have been changed, particularly through the laws and regulations that have been passed relating to social issues such as pollution and discrimination.

More recent formulations of the social contract notion have emphasized a broad range of responsibilities related to consumers and employees and the responsibilities of multinational corporations to home and host countries.[16] Social contract theory has even more recently been used to bridge the gap between two streams of business ethics research that have been developed in the last 15 years, the empirical and the normative.[17]

Thus social contract theory has a long and rich history in business and society thought. But, the problem is that it is again profoundly anthropocentric and dualistic in nature. The social contract is always a contract between human beings; the natural world is never included in the bargain. Obviously, a social contract cannot include nature in a practical sense, as nature cannot speak for itself, at least in human language, and pursue its own interests in such an arrangement. But even when the social contract involves an intellectual exercise or thought experiment, nature is still not included.

John Rawls, for example, uses the concept of a veil of ignorance to provide a basis for a social contract, that in an original position where one did not know whether he or she was black or white, rich or poor, healthy or handicapped, young or old, people would agree on certain principles of justice that provided for the needs of the disadvantaged, because one might indeed turn up in this position once one stepped through the veil of ignorance and took his or her place in society.[18] This is obviously an intellectual exercise that cannot take place in practice. But then why not also consider that one might turn up as a tree or an animal, or perhaps even a rock. What kind of a contract would one want to make then to protect one's interests or right to exist? Again nature doesn't count and is not even a part of theoretical considerations with respect to the social contract.

THE GREENING OF MANAGEMENT THEORY

Some very recent attempts have been made to develop new theories of management and the corporation that do take the environment into account and make it a central part of the theoretical structure. These theories are very creative attempts to incorporate environmental concerns into management theory and recognize the void created when the environment is left out of consideration. They thus give us some idea of what management theories might look like that incorporate the environment in their theoretical structure.

Ecocentric Management

Some of these theoretical attempts focus on ecocentric management as opposed to anthropocentric management. Paul Shrivastava, for example, argues that industrial societies have focused primarily on the creation of economic wealth through technological expansion. Postindustrialization, however, is centered on the risks that accompany wealth creation and distribution, where risk is defined as a systematic way of dealing with the hazards and insecurities induced and introduced by modernization.[19]

The excessive production of hazards and ecologically unsustainable production or consumption of natural resources are the root cause of these risks associated with modernization. In a postindustrial society, risk is the central organizing concept, but science does not provide a single, unambiguous consensual answer to risk questions. Scientific rationality is no longer an adequate arbiter of risk disputes, as scientific interpretation of environmental risks are moderated by their respective political interests.[20]

The risks of modernization are rooted in ecologically destructive industrialization. Public perceptions of risk are heightened by frequent high-profile industrial accidents. To deal with this situation, corporations must better manage risk variables such as pollution, waste, and safety. The traditional management paradigm, which included a denatured view of the environment, a bias toward production and consumption, a financial risk bias, and anthropocentrism, must be abandoned in favor of an ecocentric management paradigm[21] (Exhibit 12.1).

Placing nature at the center of management/organizational concerns is the hallmark of an ecocentric management paradigm. Such a paradigm aims at creating sustainable economic development and improving the quality of life worldwide for all organizational stakeholders. Ecocentric management is based on biocentric or ecocentric values rather than solely economic values, the development of environmentally friendly products and packaging, a change from the concern to dominate and exploit nature to one of learning to live in harmony with nature, and full accounting for the social and environmental costs of production.[22]

Ecocentric management involves viewing the industrial system in cyclical terms consistent with the way natural ecosystems function rather than in linear fashion. An industrial ecosystem consists of a network of organizations that jointly seek to minimize

EXHIBIT 12.1 Traditional versus Ecocentric Management

Traditional Management	*Ecocentric Management*
Goals:	
Economic growth & profits	Sustainability and quality of life
Shareholder wealth	Stakeholder welfare
Values:	
Anthropocentric	Biocentric or Ecocentric
Rationality and packaged knowledge	Intuition and understanding
Patriarchal values	Postpatriarchal feminist values
Products:	
Designed for function, style, & price	Designed for the environment
Wasteful packaging	Environment friendly
Production System:	
Energy & resource intensive	Low energy & resource use
Technical efficiency	Environmental efficiency
Organization:	
Hierarchical structure	Nonhierarchical structure
Top-down decision making	Participative decision making
Centralized authority	Decentralized authority
High-income differentials	Low-income differentials
Environment:	
Domination over nature	Harmony with nature
Environment managed as a resource	Resources regarded as strictly finite
Pollution and waste are externalities	Pollution/waste elimination and management
Business Functions:	
Marketing aims at increasing consumption	Marketing for consumer education
Finance aims at short-term profit maximization	Finance aims at long-term sustainable growth
Accounting focuses on conventional costs	Accounting focuses on environmental costs
Human resource management aims at increasing labor productivity	Human resource management aims to make work meaningful & the workplace safe/healthy

Source: Paul Shrivastava, "Ecocentric Management for a Risk Society," *Academy of Management Review,* 20, no. 1 (January 1995), 131.

environmental degradation by using each other's waste and by-products, and by sharing and minimizing the use of natural resources. [23] It is an attempt to base an industrial system on ecosystem principles and emulate the recycling processes of nature, rather than design an industrial system on the linear principle where an inexhaustible supply of virgin resources is assumed along with bottomless sinks in which to dispose of waste material.

Stakeholder Theory

Attempts have also been made to expand the stakeholder concept to include the nonhuman dimension of nature. Mark Starik, for example, has also noted that stakeholders have been described almost exclusively in human terms. Much, if not most, of the literature employing the concept of stakeholder management has not explicitly identified either nonhuman nature or entities within the natural environment as stakeholders. He mentions several definitions of the term *stakeholder* and shows how the natural environment, its elements, processes, ecosystems, and nonhuman life forms have been excluded.[24]

There are several reasons for this omission, including the following: (1) Until very recently, most references to the business environment in management literature did not include the natural environment as a relevant business environment, but instead identified only economic, political, sociocultural, and technological business environments; (2) the stakeholder concept has been discussed almost exclusively in traditional political-economic terms, and since only humans have been perceived to possess and exercise political-economic power and legitimacy, only humans have been considered stakeholders, either as individuals or as groups. Nonhuman entities have not possessed political-economic voice, and, therefore, have not been able to identify nor assert their "stakes," whatever these may have been; and (3) stakeholder designation has not been perceived as necessary, since other stakeholders of organizations have attempted to play appropriate stand-in roles for the natural environment.[25]

Starik goes on to argue that nonhuman nature should be included as a stakeholder of organizations. As the natural environment becomes more important, and is seen to set certain parameters for business actions and is itself affected by those actions, managers are obliged to understand their firm's relationship with the natural environment, and including nonhuman entities as stakeholders can aid in this process. The nonnatural environment may also be considered as a nontraditional political entity, as nature has a voice that business and others would do well to heed and appreciate. The natural environment has amply demonstrated its physical powers of destruction in relation to organizational assets during natural disasters that increasingly is becoming of concern to insurance companies. And finally, human stakeholders standing in for the natural environment are necessary, but not sufficient to protect nonhuman nature's stakes. Adding nonhuman environmental stakeholders to the concept would make an organization's stakeholder map more nearly complete for total environmental problem identification, analysis, evaluation, and resolution.[26]

Within the stakeholder management process, stakeholder relationships are the primary entities that are managed. If an entity is not considered a stakeholder, relationships with it are not likely to be managed or to secure substantial management attention. Recognizing the natural environment as one or more stakeholders would therefore elevate it to the level to which managerial attention can be directed. The more particularized an organization treats its stakeholders, the better it can manage these relationships through the identification of and adaptation to specialized stakeholder needs and capacities.[27]

Sustainability

The concept of sustainability has been given a good deal of attention in recent years as mentioned in an earlier chapter, and has recently been applied to corporations in an attempt to broaden management theory to include the natural environment. Gladwin, Kennelly, and Krause, for example, state that most management theorizing and research continues to proceed as if organizations lack biophysical foundations, as organic and biotic limits in the natural world are excluded from the realm of organizational science.[28] To overcome this deficiency, they propose use of the sustainability concept to provide management theorists with an opportunity for reflection and reframing theories to include the natural environment.[29]

The authors admit at the outset that sustainable development has been defined in various ways since the core idea was promulgated by the Brundtland Commission, as a brief review of the literature on the subject will show (Exhibit 12.2). Yet they find certain principal components that these conceptions share. Sustainable development, they suggest, is a process of achieving human development in an inclusive, connected, equitable, prudent, and secure manner. Inclusiveness implies human development over time and space, as sustainability embraces both environmental and human systems, both near and far, in both the present and the future. Connectivity entails an embrace of ecological, social, and economic interdependence. Equity suggests intergenerational, intragenerational, and interspecies fairness. Prudence connotes duties of care and prevention: technologically, scientifically, and politically. And finally, security demands safety from chronic threats and protection from harmful disruption.[30]

The authors then go on to appraise certain paradigms in view of these components of sustainable development. They find the technocratic paradigm lacking because it represses many critical components bearing upon life support systems, fractures or severs the connections that sustainability requires, fails to deal adequately with equity considerations, and gives rise to risks and imbalances that threaten the future of the entire human community.[31] Likewise the ecocentric paradigm, despite its attractive ideology and admirable intent, is beset by internal contradictions and fails to truly integrate culture and nature.[32]

The sustaincentric paradigm, they argue, represents an emergent synthesis and is an attempt at a higher and deeper integration. Within this paradigm, economic and human activities are inextricably linked with natural systems, yet humans are neither totally disengaged from nor totally immersed in the rest of nature. The moral monism of both technocentrism and ecocentrism is rejected in favor of moral pluralism. The paradigm also accepts that material and energy growth are bounded by ecological and entropic limits and that growth cannot go on forever in a closed system. Sustaincentrism, however, offers a vision of development that is both people centered, concentrating on improvement of the human condition, and conservation based, maintaining the variety and integrity of nonhuman nature.[33]

The task ahead, the authors conclude, is one of reintegration—to reconceive their domain as one of organization-in-full community, both social and ecological. The sustaincentrism paradigm represents a tentative step toward the development of management theory as if sustainability matters and opens up a debate on the role of human organizations in our whole earth. Indeed, as ecological limits are approached and rules of distribution are enforced, societal goals are likely to shift from growth to development, and organizational incentive systems will shift in emphasis from quantity to quality. Organizations in harmony with sustainability will increase the quality of life in equitable ways that maintain or reduce energy/matter throughput. But such organizations cannot grow indefinitely,and assumptions of infinite growth must be removed from theories of strategy and organization. Sustainability implies theorizing about qualitative improvement in the absence of quantitative expansion, and recog-

EXHIBIT 12.2 Representative Conceptions of Sustainable Development

To maximize simultaneously the biological system goals (genetic diversity, resilience, biological productivity), economic system goals (satisfaction of basic needs, enhancement of equity, increasing useful goods and services), and social system goals (cultural diversity, institutional sustainability, social justice, participation) (Barbier, 1987:103).

Improving the quality of human life while living within the carrying capacity of supporting ecosystems (The World Conservation Union, United Nations Environment Programme & Worldwide Fund for Nature, 1991: 10).

Sustainability is a relationship between dynamic human economic systems and larger dynamic, but normally slower-changing ecological systems, in which (a) human life can continue indefinitely, (b) human individuals can flourish, and (c) human cultures can develop; but in which effects of human activities remain within bounds, so as not to destroy the diversity, complexity, and function of the ecological life support system (Costanza, Daly, & Bartholomew, 1991: 8).

A sustainable society is one that can persist over generations, one that is far-seeing enough, flexible enough, and wise enough not to undermine either its physical or its social systems of support (Meadows, Meadows, & Randers, 1992: 209).

Sustainability is an economic state where the demands placed upon the environment by people and commerce can be met without reducing the capacity of the environment to provide for future generations. It can also be expressed as . . . leave the world better than you found it, take no more than you need, try not to harm life or the environment, and make amends if you do (Hawken, 1993: 139).

Our vision is of a life-sustaining earth. We are committed to the achievement of a dignified, peaceful, and equitable existence. We believe a sustainable United States will have an economy that equitably provides opportunities for satisfying livelihoods and a safe, healthy, high quality of life for current and future generations. Our nation will protect its environment, its natural resource base, and the functions and viability of natural systems on which all life depends (U.S. President's Council on Sustainable Development, 1994: 1).

Sustainability is a participatory process that creates and pursues a vision of community that respects and makes prudent use of all its resources—natural, human, human-created, social, cultural, scientific, etc. Sustainability seeks to ensure, to the degree possible, that present generations attain a high degree of economic security and can realize democracy and popular participation in control of their communities, while maintaining the integrity of the ecological systems upon which all life and all production depends, and while assuming responsibility to future generations to provide them with the where-with-all for their vision, hoping that they have the wisdom and intelligence to use what is provided in an appropriate manner (Viederman, 1994: 5).

Source: Thomas N. Gladwin, James J. Kennelly, and Tara-Shelomith Krause, "Shifting Paradigms for Sustainable Development: Implications for Management Theory and Research," *Academy of Management Review,* 20, no. 4 (October 1995), 877.

nizes that while organizations cannot grow indefinitely, they can continue to develop indefinitely.[34]

Starik and Rands also believe that sustainability is a critical emerging management concept, but that the Brundtland Commission definition of sustainable development as "development that meets the needs of the present without compromising the ability of future generations to meet their own needs" is problematic. They argue that among the specific concepts that need to be included in a more comprehensive definition are irreplaceability, biodiversity, carrying capacity, socioeconomic and ecological system resilience, and futurity.[35] To deal with these deficiencies, they propose the following definition.

> Ecological sustainability is the ability of one or more entities, either individually or collectively, to exist and flourish (either unchanged or in evolved forms) for lengthy timeframes, in such a manner that the existence and flourishing of other collectivities of entities is permitted at related levels and in related systems.[36]

The authors then present a multilevel analyses of sustainability, which they argue is necessary to develop a theory of ecologically sustainable organizations (Exhibit 12.3). The factors at each level must be examined, and the interactions between these factors highlighted, as the nature of these interactions forms the foundation for organizations' ecological sustainability and provides the context within which all other organizational activities and relationships must be understood and assessed.[37] Ecologically sustainable organizations must be developed quickly and establish relationships with the multiple levels and systems that may themselves be moving toward sustainability.[38]

Finally, the Business Charter for Sustainable Development spells out on a more practical level certain principles believed to be important for a corporate organization to follow in order to become a sustainable organization. These principles cover the management or product/service design, operations management, precautionary management, employee/customer education, contractor/supplier management, and technology transfer (Exhibit 12.4). This charter was developed by a task force established by the International Chamber of Commerce, and was formally launched in April 1991 at the Second World Industry Conference on Environmental Management.

Resource-Based Theory

One of the most comprehensive attempts to develop a theory of the firm based on environmental considerations is the attempt to insert the natural environment into the resource-based view of the firm, which takes the perspective that valuable, costly to copy firm resources and capabilities provide the key sources of sustainable competitive advantage. For this view to remain relevant, says Stuart Hart, it must internalize the challenges created by the natural environment.[39] "Strategists and organizational theorists must begin to grasp how environmentally oriented resources and capabilities can yield sustainable sources of competitive advantage."[40]

Thus Hart develops a natural-resource-based theory of the firm to accomplish these objectives. The conceptual framework for this theory is composed of three interconnected strategies: pollution prevention, product stewardship, and sustainable development (Exhibit 12.5). Pollution prevention involves the minimization or elimination of emissions, effluents, and waste from corporate operations and can result in significant savings for some companies and a cost advantage relative to competitors. Product stewardship entails integrating environmental concerns into product design and development processes as firms are driven to minimize the life-cycle environmental costs of

EXHIBIT 12.3 Multilevel-Relationship-Induced Characteristics of Ecologically Sustainable Organizations

ECOLOGICAL LEVEL

Utilization of natural resource inputs at sustainable rates
Processes designed for maximization of conservation and minimization of waste
Development of goods and services for sustainable use and disposal/recycling
Generation of only assimilable outputs, which are ecologically useful or neutral
Effective mechanisms for sensing, interpreting, and responding to natural feedback
Promotion of values of environmental protection, sensitivity, and performance
Development of principles, strategies, and practices for ecosystem viability

INDIVIDUAL LEVEL

Inclusion of sustainability considerations in job design, selection, and training
Promotion of sustainability-oriented innovation by systems and structures
Reinforcement of a sustainability orientation by cultural artifacts

ORGANIZATIONAL LEVEL

Initiation of and involvement in environmental partnerships
Absence of targeted protests by environmental activists
Utilization of environmental conflict-resolution practices
Participation in industrial ecology and other waste-exchange arrangements
Allocation of extensive resources to interorganizational ecological cooperation

POLITICAL-ECONOMIC LEVEL

Encouragement of pro-sustainability legislation
Promotion of market-based environmental policy approaches
Encouragement and development of full-environmental-cost accounting mechanisms
Promotion of peak organization support for sustainable public policy
Promotion of peak organization sustainability-oriented self-regulatory programs
Participation in peak organizations specializing in promoting sustainability
Opposition to anti-sustainability and/or promotion of pro-sustainability subsidies

SOCIAL-CULTURAL LEVEL

Involvement with social-cultural elements to advance sustainability values
Involvement in educational institutions' environmental literacy efforts
Provision of environmental information to various media
Dissemination of sustainability information from culturally diverse stakeholders
Attention to environmental stewardship values of organizational members

Source: Mark Starik and Gordon P. Rands, "Weaving an Integrated Web: Multilevel and Multisystem Perspectives of Ecologically Sustainable Organizations," *Academy of Management Review,* 20, no. 4 (October 1995), 916.

EXHIBIT 12.4 The Business Charter for Sustainable Development

1. ***Corporate Priority:*** To recognize environmental management as among the highest corporate priorities and as a key determinant to sustainable development; to establish policies, programs, and practices for conducting operations in an environmentally sustainable manner.
2. ***Integrated Management:*** To integrate these policies, programs, and practices fully into each business as an essential element of management in all its functions.
3. ***Process of Improvement:*** To continue to improve corporate policies, programs, and environmental performance, taking into account technological developments, scientific understanding, consumer needs, and community expectations, with legal regulations as a starting point; and to apply the same environmental criteria internationally.
4. ***Employee Education:*** To educate, train, and motivate employees to conduct their activities in an environmentally responsible manner.
5. ***Prior Assessment:*** To assess environmental impacts before starting a new activity or project and before decommissioning a facility or leaving a site.
6. ***Products and Services:*** To develop and provide products or services that have no undue environmental impacts and are safe in their intended use, that are efficient in their consumption of energy and natural resources, and can be recycled, reused, or disposed of safely.
7. ***Customer Advice:*** To advise, and where relevant, to educate customers, distributors, and the public in the safe use, transportation, storage, and disposal of products provided; and to apply similar considerations to the provision of services.
8. ***Facilities and Operations:*** To develop, design, and operate facilities and conduct activities. taking into consideration the efficient use of energy and materials, the sustainable use of renewable resources, the minimization of adverse environmental impact and waste generation, and the safe and responsible disposal of residual waste.
9. ***Research:*** To conduct or support research on the environmental impacts of raw materials, products, processes, emissions, and wastes associated with the enterprise and on the means of minimizing such adverse impacts.
10. ***Precautionary Approach:*** To modify the manufacture, marketing, or use of products or services or the conduct of activities, consistent with scientific and technical understanding, to prevent serious and irreversible environmental degradation.
11. ***Contractors and Suppliers:*** To promote the adoption of these principles by contractors acting on behalf of the enterprise, encouraging and, where appropriate, requesting improvements in their practices to make them consistent with those of the enterprise; and to encourage wider adoption of these principles by suppliers.
12. ***Emergency Preparedness:*** To develop and maintain, where significant hazards exist, emergency preparedness plans in conjunction with the emergency services, relevant authorities, and the local community, recognizing potential boundary impacts.
13. ***Transfer of Technology:*** To contribute to the transfer of environmentally sound technology and management methods throughout the industrial and public sectors.
14. ***Contributing to the Common Effort:*** To contribute to the development of public policy and to business, government, and intergovernmental programs and educational initiatives that will enhance environmental awareness and protection.

EXHIBIT 12.4 *(Continued)*

15. ***Openness to Concerns:*** To foster openness and dialogue with employees and the public, anticipating and responding to their concerns about the potential hazards and impacts of operations, products, wastes, or services, including those of transboundary or global significance.
16. ***Compliance and Reporting:*** To measure environmental performance; to conduct regular environmental audits and assessments of compliance with company requirements, legal requirements, and these principles; and periodically to provide appropriate information to the board of directors, the shareholders, the employees, the authorities, and the public.

Source: Jan-Olaf Willums and Ulrich Goluke, *From Idea to Action: Business and Sustainable Development* (Oslo, Norway: ICC Publishing and Ad Notam Gyldendal, 1992), pp. 356–358.

their product systems. Through product stewardship, firms can exit environmentally hazardous businesses, redesign existing product systems to reduce liability, and develop new products with lower life-cycle costs. Finally, sustainable development dictates that efforts be made to sever the negative links between environment and economic activity in developing countries.[41]

Hart contends that his natural-resource-based view of the firm and its three interrelated strategies provide a conceptual framework for incorporating environmental considerations into strategic management. But he is also concerned that the framework have predictive and ultimately, normative value. To address these concerns, he develops a theoretical framework around two major themes: (1) the linkage between the natural-resource-based view and sustained competitive advantage, and (2) the interconnections among the three strategies.[42] In doing this he develops certain propositions to guide empirical research.[43] The natural-resource-based view of the firm, he argues, "opens up a whole new area of inquiry and suggests many productive avenues for research over the next decade.[44]

EXHIBIT 12.5 A Natural-Resource-Based View: Conceptual Framework

Strategic Capability	*Environmental Driving Force*	*Key Resource*	*Competitive Advantage*
Pollution Prevention	Minimize emissions, effluents, & waste	Continuous improvement	Lower Costs
Product Stewardship	Minimize life-cycle cost of products	Stakeholder integration	Preempt competitors
Sustainable Development	Minimize environmental burden of firm growth and development	Shared vision	Future position

Source: Stuart L. Hart, "A Natural-Resource-Based View of the Firm," *Academy of Management Review,* 20, no. 4 (October 1995), 992.

FUTURE DIRECTIONS

So where does this leave us with respect to the central question of management theory and its relation to the natural environment? The previously described attempts to broaden management theory away from economic fundamentalism by developing notions of corporate social responsibility and prescribing a corresponding set of social responsibilities for corporations suffer from the same problems as the traditional economic approach to management and organizations. They are profoundly anthropocentric, that is to say, human-centered, and reflect the dualism of Western culture with respect to its approach to nature, the traditional view that humans stand over against nature and are somehow apart from nature. These flaws leave them much less than satisfactory in a world besieged by environmental problems on a scale that was not imagined possible just a few years ago when environmental problems were seen as a quality-of-life issue, not the survival issue they are today.

In order to abandon anthropocentrism we must understand its origins, according to Purser, Park, and Montuori in a recent article. Organizational researchers need a better appreciation of the historical dimensions of anthropocentrism in modern thought before they can clearly differentiate between anthropocentric and ecocentric theories. They argue that three specific areas can be attributed to the consolidation and perpetuation of anthropocentric thought in the modern world: a linear perspective vision, a camera theory of knowledge, and the social construction of a "human-nature" dualism.[45]

The linear perspective is a way of knowing the world "where the observer views the landscape as if he or she were gazing through a window or lens of a camera," a perspective that was developed as an artistic technique during the Renaissance period.[46] This linear perspective became a crucial artistic and scientific tool, according to Purser, Park, and Montuori, and was "a precursor to scientific conceptualizations of the environment, with the world seen as a distant spectacle and the viewer as an immobile spectator, a precursor of the view that humans could locate themselves at the apex and center of the natural world through the detached inquiry Decartes would later make the crux of his method."[47] This disembodied way of knowing became dominant and privileged in our society and little emphasis was placed on the actual experiences of the people involved.

The camera theory of knowledge involves scientific abstraction, where the external world is regarded as under the domain of natural laws that are discovered by accurately observing systemic regularities in nature. Certain knowledge of nature could be obtained through accurate representation of objects with a minimum of interference from the observer, who is nothing more than a spectator observing a world that has become a spectacle or an object of investigation. Obtaining "true" knowledge of nature involves the use of adequate methods that can accurately measure nature's primary characteristics. The observer becomes detached, disembodied, and neutral, a mere recorder of events.[48]

Finally, human–nature dualism involves the construction of a moral hierarchy that assumes humans are above or apart from other, more lowly creatures. This categorical separation was necessary, so claim the authors, in order to support the claim that humans were morally superior to nonhumans, thus providing a moral justification for the domination of nature and the denial that nature has any inherent worth.[49] This dualism if reflected in the rejection of Darwin by many religious groups, because if Darwin is taken seriously, humans must admit that they are organically related to nature, and in some real biological sense, they are part of nature.[50]

These three pillars led to the consolidation of an anthropocentric worldview, which provided the foundation for knowing and managing nature by using the scientific method, where nature is seen as an assemblage of things that obey immutable mathematical laws. Science helps to uncover and use these laws to human advantage. But the dominance of this scientific conception of nature has been a major cause of estrangement from nature. This positivistic approach to reality has been socially reproduced in organizational theory and management practice, which traditionally excluded nature from consideration. Nature was there simply to be scientifically studied and manipulated in the interests of human progress.[51]

Couple this anthropocentric orientation with the egocentric orientation of traditional economic theory and one has a powerful paradigm where organizational self-interest is supported at the expense of the environment. With anthropocentrism and egoism as dominant value orientations, business organizations will pursue an economically advantageous course of action when confronted with a choice between environmental preservation or economic development. In the absence of government regulation, they will seek the most cost-effective solution to environmental problems, which may put the surrounding ecosystem as well as people dependent on that ecosystem at risk.[52]

Operating within this paradigm, organizations will pursue environmental reforms only if it is in their self-interest. They will draw boundaries around a narrow definition of themselves and attempt to advance their self-interest within this narrow domain. The natural environment is seen as something outside organizational concern unless it can serve utilitarian purposes. The environment is subordinated to the needs of the organization and economic system for profits and growth, and adequate consideration is not given to how activities or the organization will affect circular flows and exchange of materials within ecosystems. The organization always believes that economic growth is always possible and that new technology will ensure a perpetual and inexhaustible source of natural resources.[53]

Simply tinkering with the existing system and tacking the environment onto current theories based on traditional paradigms may not work to mitigate the environmental problems we face. The environmental management approach, which basically does not question traditional paradigms, is at least a more enlightened approach in that it takes environmental impacts into account and tries to mitigate these impacts by adjusting organizational behavior accordingly. But it again shares the assumptions mentioned above in that it is based on the belief that nature can still be managed in the interests of growth and continued development by taking certain environmental factors into account in the development of products and production as well as strategies. In other words, we may have to give some consideration to nature in order to get on with the real business of producing more and more goods and services.

The notion of sustainability may also be of questionable value in developing adequate environmental policies and practices. The notion shares traditional assumptions about growth and development. It is based on the assumption that economic growth can continue to be accomplished as long as it is sustainable. Most definitions of sustainability reflect an anthropocentric bias as the unit of sustainability is the human being even if it refers to future generations. Sustainable development operates within a utilitarian framework seeking the greatest good for the greatest numbers of people including future generations by reducing waste and inefficiency in the exploitation and consumption of nonrenewable natural resources and hopes to ensure a maximum sustainable yield of renewable resources. But it does not question the wisdom of continued economic growth.[54]

It seems that some entirely new understanding of the business organization and the system within which it functions is going to have to emerge, something along the lines of an ecosystem paradigm. But without an understanding of the ecosystem concept and a realization that business organizations are part of these complex natural systems, it is going to be difficult to develop such a paradigm. This calls for education of business leaders and students about ecology and environmental philosophies that can expand their understanding of the organization and its responsibilities to the natural environment.

> Clearly, the development of an ecocentric organization paradigm will require a better understanding of scientific ecological concepts, environmental philosophies, and the sociopolitical implications of expanding moral obligations to ecosystems. And only by enlarging the focal setting to a more contextual and ecocentric view of organizations will organizational theorists and management practitioners begin to seriously consider questions regarding an organization's impact on the natural environment.[55]

The notion of responsibility within an ecocentric paradigm, according to Purser, Park, and Montuori, is based on efforts to maintain, preserve, or restore the health of ecosystems. An ecosystem's integrity is intact to the extent it has the ability to maintain itself and the capacity for self-renewal. Ecosystem health has to do with maintaining "a balanced, integrated, adaptive community of organisms having a species composition, diversity, and functional organization comparable to the natural habitat of the region."[56] Healthy ecosystems do not require constant repair, upkeep, and management.[57] The health and integrity of a modified ecosystem can be determined by comparing it with that of a pristine counterpart.

This perspective requires a shift in values, values that allow cultural and economic systems to flourish within operating limits that are fitted to support ecosystem integrity. Primary emphasis is placed on the valuation of ecosystem integrity, and cultural and economic development is acceptable so long as ecological integrity or ecosystem health are sustainable. Thus the focus of this perspective is on ecological sustainability rather than sustainable development, which is a much different way of looking at the issue of sustainability.[58]

The ecocentric perspective involves an ethical holism, where each species and biological organism is seen to depend upon a web of relationships, and the integrity of an ecosystem is dependent upon the function, role, and operations of these various species and organisms interacting in mutually beneficial ways. This ethics does not focus on the study of species isolated from their habitat, and it decenters the privileged position of humans as the locus of value. This perspective amounts to a fundamental ethical shift, where human beings are not above and apart from the environment as conquerors, but are seen as members of a larger natural community with constraints on their behavior in the interests of the whole.[59]

The ecocentric responsibility paradigm is thus concerned with the sustainability of both natural and cultural values, and policy options are based on finding the best sustainable match between the requirements of sociotechnical organizations and nat-

ural environments. Ecological learning is also necessary for organizational theorists as there will most likely be no one right way to organize for ecological sustainability. Experimentation and continued theory development will be required in the development of the ecocentric paradigm. But both theorists and practitioners must confront the issue that their current thinking about organization-environment relationships may be dysfunctional.[60]

Finally, economic democracy must become a priority to reverse patterns of exploitation with regard to nature. Democratic coexistence must be emphasized where the preservation of the other is a condition for the preservation of the self. As more and more species become endangered, for example, the human being itself becomes more and more endangered. Advances in organizational theory and practice should focus on the design and development of ecologically democratic organizational forms and focus on alternatives to social engineering and authoritarian technocratic solutions to environmental problems. Ecological democracy will require new forms of self-management at the domain level, where organizations belonging to that domain can maximize collaboration with the environment, lower resistance to change in the interests of the whole, and undertake innovative approaches toward ecocentric organization development.[61]

Questions for Discussion

1. What are the important characteristics of economic fundamentalism? How does economic fundamentalism deal with the environment? Are any incentives provided within this system for environmental responsibility?
2. Describe stakeholder theory. How does it broaden the notion of corporate responsibility? How does stakeholder theory typically treat the environment? Is there any possibility of including the natural environment within the stakeholder framework?
3. What is social contract theory? Who or what is typically a part of the contract between business and society? Is nature typically a part of the contract? Is it possible to incorporate nature into social contract theory?
4. What are some of the different ways of understanding the concept of sustainable development? What would a sustainable corporation look like and what practices would it adopt? Is sustainability a viable concept for greening of management theory? What assumptions does it share with traditional management theory?
5. Describe the natural-resource-based theory of the firm. Does it bring the natural environment into theory as a central component? What implications does this approach have for strategy?
6. What is ecocentric management? How does it differ from an anthropocentric approach to organizations? What are the roots of anthropocentrism in modern organizational thinking? How can this be overcome by the notion of ecocentric responsibility?

Endnotes

1. See William C. Frederick, "From CSR1 to CSR2: The Maturing of Business and Society Thought," Graduate School of Business, University of Pittsburgh, 1978, Working Paper No. 279, p. 1.
2. Lee E. Preston, "Social Issues in Management: An Evolutionary Perspective," *Papers Dedicated to the Development of Modern Management,* Daniel A. Wren, ed. The Academy of Management, 1986, p. 56.
3. R. Edward Freeman, *Strategic Management: A Stakeholder Approach* (Boston: Pitman, 1984). See also F. Abrams, "Management's Responsibilities in a Complex World," *Harvard*

Business Review, 24, no. 1 (1951), 29–34; Committee for Economic Development, *Social Responsibilities of Business Corporations* (New York: CED, 1971).

4. See Abass F. Alkhafaji, *A Stakeholder Approach to Corporate Governance: Managing in a Dynamic Environment* (New York: Quorum Books, 1989); John W. Anderson, *Corporate Social Responsibility* (New York: Quorum Books, 1989); John J. Brummer, *Corporate Responsibility and Legitimacy: An Interdisciplinary Analysis* (New York: Greenwood Press, 1991); Kenneth E. Goodpaster, "Business Ethics and Stakeholder Analysis," *Business Ethics Quarterly,* 1, no. 1 (1991), 53–73.
5. Max Clarkson et al., "The Toronto Conference: Reflections on Stakeholder Theory," *Business & Society,* Vol. 33 (1994), pp. 83–131.
6. William C. Frederick, "Social Issues in Management: Coming of Age or Prematurely Gray?" Paper presented to the Doctoral Consortium of the Social Issues in Management Division, The Academy of Management, Las Vegas, Nevada, August 1992, p. 5.
7. Archie B. Carroll, *Business and Society: Ethics and Stakeholder Management,* 3rd ed. (Cincinnati: Southwestern, 1996), p. 60.
8. Ibid., pp. 69–73.
9. Tom Donaldson and Particia Werhane, for example, after presenting the theories of consequentialism, deontology, and what they call human nature ethics—which seems to be a variation of virtue ethics—state: "Indeed, the three methods of moral reasoning are sufficiently broad that each is applicable to the full range of problems confronting human moral experience. The question of which method, if any, is superior to the others much be left for another time. The intention of this essay is not to substitute for a thorough study of traditional ethical theories—something for which there is no substitute—but to introduce the reader to basic modes of ethical reasoning that will help to analyze the ethical problems in business that arise in the remainder of the book." Tom Donaldson and Particia Werhane, *Ethical Issues in Business: A Philosophical Approach,* 4th ed. (Upper Saddle River, NJ: Prentice Hall, 1993), p. 17.
10. William C. Frederick, "Toward CSR3: Why Ethical Analysis is Indispensable and Unavoidable in Corporate Affairs," *California Management Review,* XXVII, no. 2 (Winter 1986), 136.
11. Amitai Etzioni, *The Moral Dimension: Toward a New Economics* (New York: Free Press, 1988).
12. See Melvin Anshen, *Managing the Socially Responsible Corporation* (New York: Macmillan, 1974).
13. Milton Friedman, "The Social Responsibility of Business Is to Increase Its Profits," *New York Times Magazine,* September 13, 1970, pp. 122–126.
14. CED, *Social Responsibilities of Business Corporations,* p. 12.
15. Carroll, *Business and Society,* p. 19.
16. Tom Donaldson, *Corporations and Morality* (Englewood Cliffs, NJ: Prentice Hall, 1982); Tom Donaldson, *The Ethics of International Business* (New York: Oxford University Press, 1989).
17. Tom Donaldson and Tom Dunfee, "Toward a Unified Conception of Business Ethics: Integrative Social Contracts Theory," *Academy of Management Review,* Vol. 19 (1994), pp. 252–284.
18. John Rawls, *A Theory of Justice* (Cambridge, MA: Harvard University Press, 1971).
19. Paul Shrivastava, "Ecocentric Management for a Risk Society," *Academy of Management Review,* 20, no. 1 (January 1995), 188–120.
20. Ibid., pp. 120–121.
21. Ibid., pp. 121–127.
22. Ibid., pp. 129–134.
23. Ibid., pp. 127–129.
24. Mark Starik, "Should Trees Have Managerial Standing?: Toward Stakeholder Status for Non-Human Nature," *Journal of Business Ethics,* Vol. 14 (1995), pp. 207–208.

25. Ibid., pp. 208–209.
26. Ibid., pp. 209–213.
27. Ibid., pp. 214–215.
28. Thomas N. Gladwin, James J. Kennelly, and Tara-Shelomith Krause, "Shifting Paradigms for Sustainable Development: Implications for Management Theory and Research," *Academy of Management Review,* 20, no. 4 (October 1995), 875.
29. Ibid., p. 876.
30. Ibid., p. 878.
31. Ibid., p. 886.
32. Ibid., p. 889.
33. Ibid., pp. 890–894.
34. Ibid., pp. 896–897.
35. Mark Starik and Gordon P. Rands, "Weaving an Integrated Web: Multilevel and Multisystem Perspectives of Ecologically Sustainable Organizations," *Academy of Management Review,* 20, no. 4 (October 1995), 908–909.
36. Ibid., p. 909.
37. Ibid., p. 916.
38. Ibid., p. 930.
39. Stuart L. Hart, "A Natural-Resource-Based View of the Firm," *Academy of Management Review,* 20, no. 4 (October 1995), 986–991.
40. Ibid., p. 991.
41. Ibid., pp. 991–998.
42. Ibid., p. 998.
43. Ibid., pp. 998–1007.
44. Ibid., p. 1009.
45. Ronald E. Purser, Changkil Park, and Alfonso Montuori, "Limits to Anthropocentrism: Toward an Ecocentric Organization Paradigm?" *Academy of Management Review,* 20, no. 4 (October 1995), 1055.
46. Ibid., pp. 1055–1056.
47. Ibid., p. 1056.
48. Ibid., p. 1057.
49. Ibid.
50. Ibid., p. 1058.
51. Ibid., pp. 1058–1059.
52. Ibid., p. 1062.
53. Ibid., pp. 1063–1064.
54. Ibid., pp. 1067–1068.
55. Ibid., p. 1065.
56. Ibid., p. 1071.
57. Ibid., p. 1070–1071.
58. Ibid., p. 1071.
59. Ibid., p. 1073.
60. Ibid., pp. 1079–1080.
61. Ibid., pp. 1080–1082.

Suggested Readings

Alkhafaji, Abass F. *A Stakeholder Approach to Corporate Governance: Managing in a Dynamic Environment.* New York: Quorum Books, 1989.

Anshen, Melvin. *Managing the Socially Responsible Corporation.* New York: Macmillan, 1974.

Atchia, Michael, and Shawna Troop, eds. *Environmental Management: Issues and Solutions.* New York: Wiley, 1995.

Carroll, Archie B. *Business and Society: Ethics and Stakeholder Management,* 3rd ed. Cincinnati: Southwestern, 1996.

Chamberlain, Neil W. *The Limits of Corporate Responsibility.* New York: Basic Books, 1973.

Committee for Economic Development. *Social Responsibilities of Business Corporations.* New York: Committee for Economic Development, 1971.

Compton, Paul A., and Marton Pecsi. *Environmental Management.* London: Collets, 1994.

Donaldson, Tom. *Corporations and Morality.* Englewood Cliffs, NJ: Prentice Hall, 1982.

Etzioni, Amitai. *The Moral Dimension: Toward a New Economics.* New York: Free Press, 1988.

Fien, John, et al., eds. *Environmental Education: A Pathway to Sustainability.* New York: St. Martins, 1993.

Freeman, R. Edward. *Strategic Management: A Stakeholder Approach.* Boston: Pitman, 1984.

Hayner, Stephen, and Gordon Aeschliman, eds. *Environmental Stewardship.* New York: InterVarsity, 1990.

Karpoff, James A. *Environmental Issues: A Managerial Approach.* Upper Saddle River, NJ: Prentice Hall, 1995.

Schramm, Gunter, and Jeremy J. Warford. *Environmental Management and Economic Development.* Baltimore: Johns Hopkins, 1989.

Seldner, Betty J. *Environmental Decision Making for Engineering and Business Managers.* New York: McGraw-Hill, 1994.

Shrivastava, Paul. *Greening Business: Profiting the Corporation and the Environment.* Cincinnati: Thompson Executive Press, 1996.

Stead, W. Edward, and Jean Garner Stead. *Management for a Small Planet: Strategic Decision Making and the Environment.* Newbury Park, CA: Sage Publications, 1992.

Welford, Richard. *Environmental Strategy and Sustainable Development: The Corporate Challenge for the Twenty-First Century.* New York: Routledge, 1995.

Welford, Richard, and Andrew Gouldson. *Environment Management and Business Strategy.* London: Pitman Publishers, Ltd., 1993.

CHAPTER 13

Management Practice and the Environment

The environmental issues discussed in previous chapters have had an impact on business organizations that has both short- and long-run implications. In the short run, business organizations face a host of regulations designed to deal with these problems, which require new procedures and techniques for business to respond to environmental problems in an effective manner. Business also faces a different kind of environment out of which litigation may arise and thus must adopt new ways of protecting itself from problems of this type by doing environmental audits or investigations of potential environmental liability when buying land for expansion or considering a merger with another company. A whole new set of considerations has become important to management that must be factored into the decision-making process.

In the long run, business must consider new forms of organizational structures and processes that will enable it to respond to a changing environment in a evermore responsible fashion. There is no turning back to a simpler world where environmental impacts were limited to a few considerations of mostly economic nature. Business must now recognize the environment as being a major factor in decision making that has long-run implications for the way it does business and the relationships it develops with its stakeholders. Already in most companies, structures have changed to allow for environmental input at every level, and manufacturing processes have changed to minimize waste and allow for more recycling.

In a nutshell, business as a whole must take responsibility for the environment and manage its environmental responsibilities the way it manages other parts of the business. Managers must learn how take account of the impacts their operations have on ecosystems and habitats in order to promote sustainable operations that do not undermine the possibilities of economic growth for future generations. Business operations are at the heart of sustainable development, as no other organization in our society has such an impact on the environment as does business. Consequently, no other organization has the possibility of making such a positive contribution to sustainable development as does business.

THE GREENING OF MANUFACTURING

Such changes, however, call for new kinds of thinking and new strategies regarding manufacturing. While new technologies and products introduced by business have greatly enhanced living standards in a number of parts of the world, many adverse effects on the environment have also been created. Some of these have been brought under a certain degree of control, but as population around the world increases and people strive for a higher and higher standard of living, some of the old solutions to industrial pollution and the disposal of industrial wastes are no longer workable. New strategies have to be developed that are responsive to a changing environment.

Waste Minimization Programs

One of the key issues concerning our present system of industrial manufacturing is what to do with all the waste that is generated. Even if business is able to comply with the latest waste disposal regulations, and even if existing dump sites that pose a threat to human health and the environment are "cleaned up" using the best technology available, this waste material has to go somewhere. Who is to say that our present methods of waste disposal won't also cause problems for future generations? In some sense, we may just be moving the problem around by cleaning up existing dump sites and disposing of the waste in a so-called more appropriate manner.

In some sense, the disposal of waste has become more of a political than a technical problem. The driving force behind the cleanup of hazardous waste dumps and proper disposal of newly generated waste is a public concern. The public is afraid of dump sites and wants them cleaned up as soon as possible. The public agenda in these situations is often set by the news media as they begin to write stories about subjects such as hazardous waste dumps and improper disposal. Politicians then start to take notice and bills are passed that create new programs to deal with the problem. Agencies that implement these programs see an opportunity for more money and public visibility by keeping the issue in the headlines.

Companies themselves have an incentive to clean up existing dump sites where they are potentially responsible parties as quickly and efficiently as possible. In most if not all cases, they want to get these messy situations behind them and get on with the job of producing goods and services. Yet companies are often not in control of cleanup situations when they become highly politicized, and they are at a disadvantage in relation to a state or federal agency that is in a better position to play political gamesmanship. Companies often have very little control over a situation that may require them to spend millions of dollars.

The ultimate solution to this waste disposal problem, if there is one, seems to be one of not producing so much waste in the first place. This approach has been called waste minimization, and it involves a new way of thinking about the generation of waste material. This concept typically involves four stages: (1) prevention of waste generation at the source through redesign of products, (2) recycling or reuse of waste material to recover useful products, (3) treatment of wastes with methods such as incineration to reduce toxicity, and (4) disposal of remaining waste material in an appropriate manner. Many companies are focusing their efforts on the first two stages where the benefits can be substantial.

As the costs of waste disposal have increased because of government regulations, companies have found it profitable to develop such programs that focus on pollution prevention strategies to reduce these costs. The benefits of such programs include (1) lower equipment and raw material costs, (2) lower waste disposal and compliance expenses, (3) improvement of management and operating practices, (4) minimization of liabilities and cleanup costs, and (5) earning the trust of regulators and the community. Reducing the amount of waste a company produces would seem on the surface to be one of the best approaches to pollution that can be taken.

The 3M Corporation

In 1975, 3M Corporation unveiled a program called Pollution Prevention Pays (The 3Ps Program), one of the first of its kind in the country. The goal of the program was to eliminate or reduce sources of pollution in 3M products and processes, in other words, to stop creating pollution in the first place. The 3Ps Program was born out of the environmental awareness of the early 1970s, when new federal and state environmental laws and regulations emerged in abundance, restricting pollutant releases and tightening requirements for pollution monitoring and reporting. Companies like 3M took the necessary steps to comply, usually by installing conventional end-of-line controls.[1]

At 3M, top management responded to stricter regulation with a firm position, indicating that the firm would comply with existing regulations. Nevertheless, it understood the potentially high cost of end-of-line control and recognized a developing corporate problem in the strong, negative reactions of plant and operating division managers to increasing pollution control regulations. While these people agreed with the general intent of the regulations, they were quite concerned about investments with no return and encroachment on their management responsibilities.[2]

Thus top management was motivated to find a way to meet environmental requirements and keep 3M's products cost-competitive. In response to the CEO's request, the corporate environmental staff developed a positive, voluntary employee recognition program optimistically entitled Pollution Prevention Pays. Recognition was given to projects that (1) prevented pollution, (2) had a return consistent with corporate expectations, and (3) had some innovative content. The program emphasized prevention of pollution rather than control after it was generated.[3]

Under the 3Ps Program, technical innovation to prevent pollution and generation of waste at the source is encouraged through the following: (1) product reformulation involving the development of nonpolluting or less polluting products or processes by using different raw materials or feedstocks; (2) process modification involving changing manufacturing processes to control by-product formation or to incorporate nonpolluting or less polluting raw materials or feedstocks; (3) equipment redesign including modifying equipment to perform better under specific operating conditions or to make use of available resources; and (4) resource recovery including recycling by-products for sale or for use in other 3M products or processes.[4]

With the program having been firmly in place for more than two decades, the company claims that 4,100 ideas have been generated to reduce pollution, which has eliminated 1.3 billion pounds of pollutants and saved the company more than $710 million.

The program has been extended to 3M plants worldwide, even to countries where the environment is not such an important concern. More recently, 3M has begun to emphasize its environmental concern in its marketing pitch to consumers, and in the future, product developers will have to consider environmental standards as part of a broader "green marketing" strategy.[5]

These results amounted to a reduction of waste streams to all media (the program was multimedia) that are roughly one-half the size today they would have been without these projects. The program moved 3M up the waste-management hierarchy, and it helped fit environmental protection more neatly into the overall corporate strategy. And while many projects were not initiated because of their effect on waste streams, the fact that they did have an effect was not lost on the employees. After a time, they began looking for waste streams to profitably reduce.[6]

In summary, the 3Ps Program broke some new ground, substantially reduced the company's reliance on pollution controls, but was not a panacea for eliminating the continuing need for treatment and disposal. Substantial further waste reduction will require longer-term research projects and more economic justification. In 1988, 3M initiated a new phase of the 3Ps Program to reduce their annual hydrocarbon emissions by more than 55,000 tons over and above reductions achieved through the 3Ps Program. Called 3P+, the program called for a $150 million investment in air pollution control equipment that was entirely voluntary and reduced emissions beyond anything that was necessary to comply with state and federal environmental regulations.[7]

Dow Chemical

In 1986, the management of Dow Chemical U.S.A. decided to streamline and formalize its waste reduction efforts under a single program called Waste Reduction Always Pays (WRAP). Under this program, the company proposed to take a comprehensive approach to reducing the amount of waste material it dumped into the air, water, and land with the overall goal of reducing the amount of wastes produced by all its facilities. Waste reduction pays long-term dividends in two ways from the company's point of view: (1) When emissions are reduced, the company is making an investment in a clean and safe environment; and (2) waste reduction makes good sense from a business standpoint as where there is waste there is inefficiency, and where there is inefficiency, there is an opportunity to reduce costs.[8]

The company believes that by actively pursuing waste reduction opportunities, it can reduce its waste-management costs, improve the productivity of its operations, demonstrate to the public its commitment to environmental protection, and perhaps most importantly, show that a voluntary program of waste reduction can work without government regulation. The program has the following objectives: (1) reduction of waste and emissions to the environment, (2) provision of incentives for waste reduction projects, (3) provision of recognition for those who excel in waste reduction, and (4) reemphasis of the need for continuous improvement by recognizing opportunities in waste reduction.[9]

The reduction of waste and emissions to the environment involves elimination, reclamation, treatment, and destruction, and the development of secure landfills. The first effort is to avoid producing waste in the first place. Through research and development efforts, Dow strives to utilize production and operating processes that have the highest efficiencies and minimize waste. When the generation of some waste is unavoidable, it is recycled back into the production process or used as a raw material for another process wherever possible. When reclamation is not possible, the waste is treated or destroyed by means of water treatment and/or incineration to eliminate it from the environment. If neither treatment nor destruction is possible, the wastes are buried in a secure landfill.[10]

Providing incentives is based on the recognition that economic incentives are needed to cultivate a waste reduction mentality throughout the company. The WRAP Capital Projects Contest was started in 1988, and because of the company's decentralized approach to managing its production facilities, each division has a slightly different version of this program. In some divisions, for example, plants submit projects demonstrating waste reduction or yield improvement that contains a certain return on investment (ROI). Any project that demonstrates an improvement in waste reduction or yield improvement as measured by the ROI, receives capital dollars.

In 1988 and 1989, the company committed nearly $6 million to underwrite 42 projects aimed at reducing waste as part of the WRAP Capital Projects Contest. It estimates that these capital projects alone reduced waste streams by 88 million pounds annually. These capital funds were only a small part of overall waste reduction spending.[11] Such waste reduction is the cornerstone of Dow's waste-management policy and plays a crucial role in environmental protection and the long-term growth of the business. While waste is an inevitable part of the manufacturing process and can never be reduced entirely, making continued compliance with state and federal laws a top priority, continuous improvement in the reduction of waste can be accomplished.[12]

Regarding the future of the program, the following challenges were stated by Dow management: (1) how to build upon the initial success of the WRAP program, (2) how to extend WRAP to Dow's growing international operations, and (3) how to respond to continued pressure from environmental advocacy groups on numerous other issues. In response to the first challenge, Dow's environmental staff is considering the establishment of specific reduction goals to further institutionalize the Continuous Improvement Process in waste reduction. These goals may be difficult to implement given Dow's decentralized structure. With regard to the second challenge, Dow recognizes that it will have to work very hard to export a waste reduction mentality, particularly since significant regulatory pressures do not exist in many countries. The final challenge relates to the way in which Dow see its role in the public policy process. The company takes a long-range approach to public policy by working closely with other groups on the issues, which often leads to a healthy debate over implementation strategies.[13]

While programs such as these are now in place in many companies, particularly in firms in high-polluting industries such as the chemical industry, there are also many problems in implementing such programs within a given company. Many managers view waste minimization programs as merely an extension of existing regulatory programs that are regarded as costly and burdensome. Environmental accounting systems are inadequate to measure the costs and savings of such programs, and thus companies cannot often justify them on a cost basis. Waste minimization programs also involve changing production processes, which introduces risks that some plant managers are reluctant to take without appropriate incentives from top management. And finally, top managers of many companies are not involved enough in promoting the implementation of waste minimization programs.

From a broader perspective, experience indicates that in the early stages of such a program, easy and inexpensive changes can be made that result in large reductions in emissions relative to the costs involved. But as the firm's performance improves, further reductions in waste become progressively more difficult, often requiring significant changes in processes or even entirely new production technology. Diminishing returns set in, and as a company moves closer to "zero emissions," reductions become more capital intensive and may require broader changes in underlying product design and technology.[14]

Product Life-Cycle Analysis

A broader approach to dealing with environmental impacts in manufacturing is called life-cycle analysis. This approach recognizes that environmental concerns enter into every step of the process with respect to the manufacture of products, and thus examines environmental impacts of products at all stages of the product life cycle. This includes the stages of product design, development, manufacturing, packaging, distribution, usage, and disposal. Life-cycle analysis is concerned with reducing environmental impact at all these stages and looking at the total picture rather than just one stage of the production process. The process asks questions related to resource depletion, pollution, and health effects at all stages of a product life cycle.

Through utilizing this concept, firms are driven to minimize the life-cycle environmental costs of their total product systems. Through such analysis, firms can determine whether they can redesign existing products' systems to reduce their environmental impact, or develop new products with lower life-cycle costs. For new firms in particular, the life-cycle concept can be useful in making decisions about a product and how to produce it, as there are no preexisting commitments to products, facilities, or manufacturing processes.

Unanticipated trade-offs are a natural part of complex systems. Using less packaging may involve more spoilage. Designing a product to be less polluting may involve manufacturing delays. Life-cycle analysis is an effort to deal with this complexity. It monitors the streams of energy, materials, and information that feed into and emerge from a manufacturing process. It is thus a much more sophisticated approach to controlling pollution and reducing environmental impacts than waste minimization programs and involves many more people at various stages of the production process.

The typical life-cycle process is depicted in Figure 13.1, which shows the product life cycle of Bristol-Meyers Squibb Company. The company uses product life-cycle principles to integrate environmental concerns into every aspect of its business. The business units of the company include product life-cycle goals in their five-year strategic plans and are committed to conducting product life-cycle reviews of their existing and new products. The company claims that use of this process has shifted responsibility for environmental excellence from the exclusive domain of an environmental professional to a shared responsibility of management and employees.[15]

During a product life-cycle review, employees from marketing, research and development, manufacturing, packaging, finance, and other functional areas work together to inventory the environmental impacts of their products and processes. If the product is currently existing, the life-cycle review helps them identify where improve-

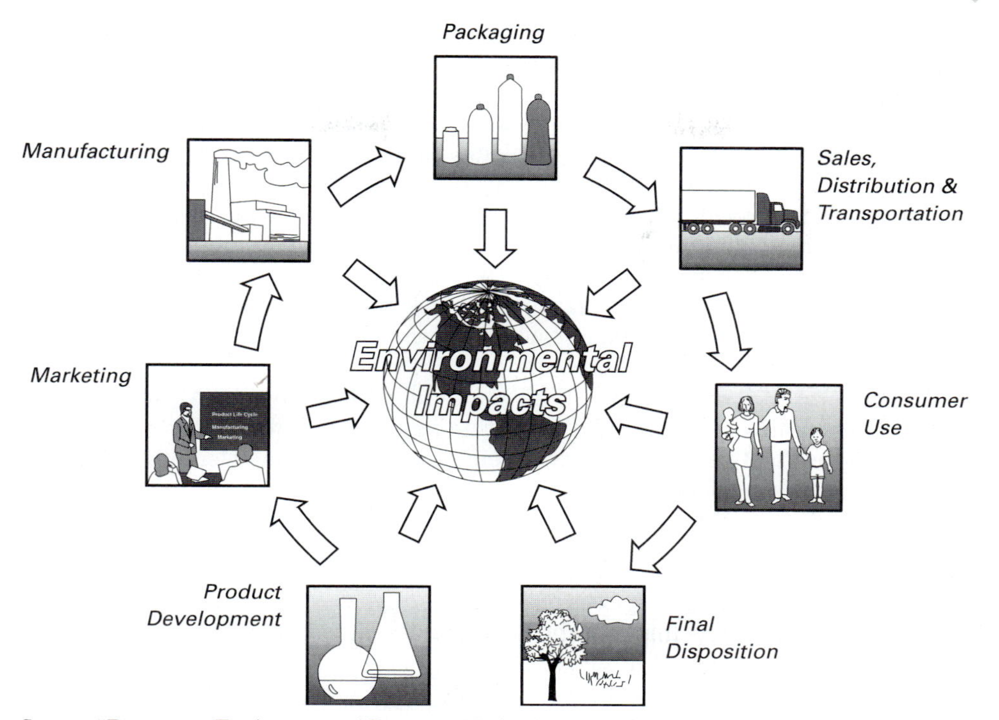

Source: "Report on Environmental Progress," Bristol-Meyers Squibb Company, May 1993, p. v.

FIGURE 13.1 Product Life Cycle

ments can be made. If the product is a new one under development, product life-cycle analysis helps the company to prevent the use of materials and practices that may be detrimental to the environment.[16]

Product life-cycle analysis can help to foster a cradle-to-grave environmental responsibility with respect to the environmental impact of the company's products. The objective of the program is to systematically identify opportunities to prevent pollution and reduce environmental impacts through all stages of a product's life cycle by highlighting how employees from many functional areas can prevent pollution and reduce environmental impacts of the business. In implementing this process, the company hopes to reduce its environmental liabilities, increase the cost-effectiveness of complying with environmental regulations, develop safer and more environmentally sound products, minimize waste of energy and raw materials, differentiate its products from the competition, and become recognized as a good environmental citizen.[17]

Design for Disassembly

Consistent with this emphasis on the product life-cycle concept where concern is centered on not only reducing pollution during the manufacturing of a product but also what happens to the product when it has reached the end of its life cycle, is a new concept called "design for disassembly." This concept generally means the simplification of parts and materials to make them easy and inexpensive to snap apart, sort, and recycle. Taking something apart may rapidly become as important as putting something together, and design for disassembly (DFD) focuses on taking things apart easily so that the various components of a product can be used again to make other products.[18]

BMW got an early start in implementing the concept by producing a new car called the Z1, which was said to be the first real DFD product designed to be disassembled and recycled. The car has an all-plastic skin that can be disassembled from its metal chassis in 20 minutes, and has doors, bumpers, and front, rear, and side panels made of recyclable thermoplastic. BMW also has a pilot disassembly plant where it chops apart five cars a day to learn new things about DFD, such as the use of pop-in, pop-out two-way fasteners which facilities disassembly whereas glue and screws are to be avoided.[19]

Several companies such as Whirlpool, Digital Equipment, 3M, and General Electric are beginning to incorporate DFD thinking in their products in order to make recycling cost-effective for all kinds of complex products. In the electronics and electrical equipment industry, computer and appliance manufacturers are among the leaders in design for disassembly. Several companies have recycling centers and take-back commitments. Some have set a goal of 90 percent recyclability by the year 2000 and have achieved significant reductions in assembly and disassembly time with their products.[20]

Composites, which combine glass, metals, plastics, and other fibers, make coding and separation of materials nearly impossible and are enemies of recycling. A good example of a composite product that made the news a few years ago is the palm-sized juice boxes that have become so popular. These boxes were safe to use and very good from a nutrition standpoint, but were impossible to recycle. The box is made of five layers of paper and plastic, and a sixth layer of aluminum foil, which prevents light and oxygen from entering and spoiling the food. It is said that milk can be safely kept in these packages for five months without refrigeration.[21]

These ultraslim layers would have to be separated, making recycling impossible to justify on a cost-effective basis. Because of this difficulty, the state of Maine banned use of the juice boxes to prevent them from crowding the state's landfills. Other states were considering similar bans, causing the maker of the boxes to launch a $3 million dollar lobbying campaign. It claimed that its boxes used less packaging material than larger sized glass and metal containers, and that they could be recycled without separating all the layers of material. In order to prove this latter claim, one of the manufacturers was financing a $1.6 million operation to recycle used plastics into wood substitutes.[22]

In order to increase the recyclability of plastics in cars and other products, designers and manufacturers have begun to work with suppliers to reduce the number of plastics used in products, to use and code plastics that recycle easily, and to create recycling relationships with other companies. A consortium of plastics and automobile companies in Europe is conducting a six-year study of automotive recycling processes and waste-management systems. The challenge for car makers is significant, with 12 million to 14 million cars a year being disposed of in Europe alone.[23]

There are several obstacles, however, to implementation of the concept on a broader scale. One is the development of an economically viable infrastructure for the collecting, refurbishing, and reprocessing of recycled materials. Another is the challenge of integrating design for disassembly initiatives into the product development process. And finally, there is an issue of product confidentiality. When products are dismantled on a large scale, companies fear the loss of control over the secrets of how they put products together and what materials are used in the products.[24]

THE GREENING OF MARKETING

The success of many of these manufacturing strategies depends on consumer demand for products that are environmentally sensitive and produced by companies that incorporate environmental concerns into their manufacturing and distribution processes. There has thus taken place in the past several years something called "green marketing," a response companies have made to an increasing demand for environmentally sensitive products. The products in this so-called green market include biodegradable products, recycled products, and more fuel-efficient cars and appliances. Consumers are increasingly becoming concerned about the environment, and this concern is showing up in the marketplace in the kind of products these consumers are willing to buy.

Wal-Mart stores was the first retailer to publicly call for more environmentally friendly products from vendors and tag shelves to highlight environmental improvements in products and packaging. The tag does not say a product is safe as far as environmental impacts are concerned, but only that improvements have been made in the product to minimize its environmental impact. The retailer works with vendors and typically asks them what improvements have been made in their product to make it more environmentally friendly. In a more recent effort, the company has also set up recycling centers in the parking lots of its 1,511 stores and is encouraging recycling in TV commercials.[25]

Many companies, like Procter & Gamble, have begun to advertise the environmentally sensitive features of their products. When first introduced in the 1960s, disposable diapers like the Pampers produced by P&G were hailed by many to be the product breakthrough of the decade. They were consistent with the throwaway society mentality of that era, and facilitated care of babies by parents who wanted to save time and effort. These diapers, however, became the focus of public concern about garbage, and even though they constituted only 1 to 2 percent of the total solid waste stream, they were singled out as representative of the problem with the throwaway society.[26]

The issue heated up in February of 1989, when the National Association of Diaper Services and Environmental Action, a consumer activist group, announced a $100,000 nationwide public education campaign to tell the country about the environmental cost of using disposable diapers. Disposable diapers were called "ecological booby traps that stuff the nation's landfills with an unhealthy legacy of virtually indestructible paper and plastic time capsules." The use of reusable cloth diapers was advocated as the solution to the problem, and would reduce the more than 18 billion diapers that have to be landfilled each year, according to the organization.[27]

In response, P&G discovered that some 80 percent of the diaper could be successfully composted, and is working to increase this percentage by replacing the plastic backsheet liner, wasteband, and tape tabs with other material. The biggest problem, however, is that most communities do not have composting facilities, so the company created a $20 million solid waste composting fund to help communities analyze solid waste problems and fund composting research. These composting initiatives are hoped to protect its leadership in the $3.5 billion disposable diaper industry, and counter claims other companies are making about the biodegradibility of their products.[28]

The biodegradibility issue surfaced as many companies began to advertise these features of their products in order to appeal to the emerging green consumer market. Most of these claims were false, as they failed to inform the consumer that biodegradable products need sunlight to decompose and, when buried other tons of other trash and dirt in a landfill, do not receive sunlight under normal conditions. Thus the consumer was being misled. Mobil Corporation was caught in this controversy when it launched its line of biodegradable Hefty trash bags to counter the competition, which had beaten it to the market with biodegradable products. Soon after this introduction, biodegradability became a dirty word and Mobil became caught in the middle.[29]

Green marketing, while initially appearing to be a new area for marketers to rush into in a competitive race, quickly became more complicated. Consumers wanted to express their concerns about the environment through marketplace behavior, but in the absence of knowledge about the environment, were easily exploitable and were left without any means to evaluate environmental claims being made by companies. The federal government and the states became increasingly concerned about these claims and began to investigate the advertising and promotional campaigns of several companies in order to prevent the process from getting out of hand. The attorney general of Minnesota sounded the alarm when he stated, "The selling of the environment could make the oat-bran craze look like a Sunday school picnic."[30]

Eventually, the environmental theme hit a sour note, as consumers became more wary about environmental claims and companies began to discard their environmental messages.[31] In late 1990, two reports were issued that indicated environmental claims made by manufacturers were confusing rather that helpful to consumers, and called for the development of national standards. One report urged business to adopt specific and substantive environmental claims backed up by reliable evidence. The report suggested that companies avoid vague phrases such as "environmentally friendly" and clarify whether environmental claims are being made for the package or the product. The term "recycling" also needed to be clarified, as any claims about recyclability were meaningless unless people have access to recycling facilities.[32]

The lack of industry-wide norms for "green" marketing led both industry groups and consumer groups to lobby Congress for federal standards. Before the federal government acted, state legislation began to appear. Several states enacted legislation governing the use of environmental terms. And state attorneys general circulated final recommendations for "responsible environmental advertising" in a separate effort. But their efforts were geared at least in part to prod the federal government into action, as they wanted to see the FTC or the EPA develop national green marketing standards. These standards are necessary, it was claimed, so that consumers will have the information they need to make purchasing decisions based on environmental considerations.[33]

The Federal Trade Commission (FTC) finally acted in mid-1992 by issuing guidelines to help reduce consumer confusion and prevent the false or misleading use of environmental terms such as "recyclable," "degradable," and "environmental friendly" in the advertising and labeling of products in the marketplace. The guidelines were also developed to reduce manufacturers' uncertainty about which claims might lead to FTC law-enforcement actions, thereby encouraging marketers to produce and promote

products less harmful to the environment. The EPA was fully supportive of the effort and cooperated in the guidelines project.[34]

The guides focus on how consumers are likely to interpret environmental claims and identify types of claims that should be explained or qualified to avoid deception. While the guides are not legally enforceable, they were meant to provide guidance to marketers in conforming with legal requirements. They apply to advertising, labeling, and other forms of marketing to consumers, and do not preempt state or local laws or regulations. In general, any time marketers make objective environmental claims they must be substantiated by competent and reliable evidence. Otherwise, broad environmental claims should be avoided or qualified. The guides outline four other general concerns that apply to all environmental claims.[35]

- Qualifications and disclosures should be sufficiently clear and prominent to prevent deception.
- Environmental claims should make clear whether they apply to the product, the package, or a component of either. Claims need not be qualified with regard to minor, incidental components of the product or package.
- Environmental claims should not overstate the environmental attribute or benefit. Marketers should avoid implying a significant environmental benefit where the benefit is, in fact, negligible.
- A claim comparing the environmental attributes of one product with those of another product should make the basis for the comparison sufficiently clear and should be substantiated.

The guides do not rigidly define environmental terms, but they do present a series of examples showing both acceptable and deceptive claims. The guides set out the different meanings that might be conveyed by the use or omission of particular language describing environmental features. The types of claims addressed by the guidelines include recyclable, degradable, compostable, recycled content, source reduction, refillable, and ozone safe (see box).

In 1995, the FTC began a review of these guidelines and issued a call for public comment to find out how well they had been working. The agency believed that the guidelines achieved their purpose of reducing consumer confusion about green claims and preventing deceptive environmental claims. Since the guidelines were issued, the FTC entered in 22 consent orders with companies and individuals concerning false and/or unsubstantiated environmental claims about products.

In addition to these efforts to protect consumers for misleading and deceptive claims about the environmental aspects of products, two private services have been developed to evaluate environmental claims and help consumers sort through all the hype by offering environmental seals of approval for products. Scientific Certification Systems (SCS) was founded in 1984 to address the growing confusion over environmental and food safety claims in the marketplace.[36]

The program was founded in the belief that the most enduring solutions for the environment will arise out of voluntary industry achievement. The goals of the programs are to (1) support manufacturer efforts to meet the highest environmental standards in product and packaging design, as well as in their production practices; (2) provide retailers with a means to distinguish and identify environmental product claims that are both accurate and significant; and (3) provide consumers with the added assurance of independent, third-party certification of environmental claims in the marketplace.[37]

DEGRADABLE, BIODEGRADABLE, AND PHOTODEGRADABLE

In general, unqualified degradability claims should be substantiated by evidence that the product will completely break down and return to nature, that is, decompose into elements found in nature within a reasonably short period of time after consumers dispose of it in the customary way. Such claims should be qualified to the extent necessary to avoid consumer deception about: (a) the product or package's ability to degrade in the environment where it is customarily disposed; and (b) the extent and rate of degradation.

COMPOSTABLE

In general, unqualified compostable claims should be substantiated by evidence that all the materials in the product or package will break down into, or otherwise become part of, usable compost (e.g., soil-conditioning material, mulch) in a safe and timely manner in an appropriate composting program or facility, or in a home compost pile or device. Compostable claims should be qualified to the extent necessary to avoid consumer deception: (1) if municipal composting facilities are not available to a substantial majority of consumers or communities where the product is sold; (2) if the claim misleads consumers about the environmental benefit provided when the product is disposed of in a landfill; or (3) if consumers misunderstand the claim to mean that the package can be safely composted in their home compost pile or device, when in fact it cannot.

RECYCLABLE

In general, a product or package should not be marketed as recyclable unless it can be collected, separated, or otherwise recovered from the solid waste stream for use in the form of raw materials in the manufacture or assembly of a new product or package. Unqualified recyclable claims may be made if the entire product or package, excluding incidental components, is recyclable. Claims about products with both recyclable and non-recyclable components should be adequately qualified. If incidental components significantly limit the ability to recycle a product, the claim would be deceptive. If, because of its size or shape, a product is not accepted in recycling programs, it should not be marketed as recyclable. Qualification may be necessary to avoid consumer deception about the limited availability of recycling programs and collection sites if recycling collection sites are not available to a substantial majority of consumers or communities.

RECYCLED CONTENT

In general, claims of recycled content should only be made for materials that have been recovered or diverted from the solid waste stream, either during the manufacturing process (pre-consumer) or after consumer waste (post-consumer). An advertiser should be able to substantiate that pre-consumer content would otherwise have entered the solid waste stream. Distinction made between pre- and post-consumer content should be substantiated. Unqualified claims may be made if the entire product or package, excluding minor, incidental components,

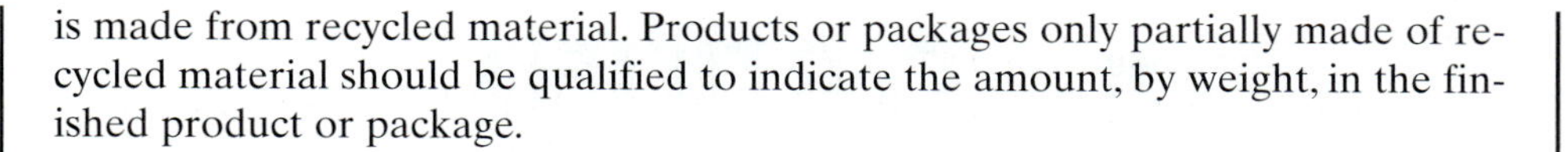

is made from recycled material. Products or packages only partially made of recycled material should be qualified to indicate the amount, by weight, in the finished product or package.

SOURCE REDUCTION

In general, claims that a product or package has been reduced or is lower in weight, volume, or toxicity should be qualified to the extent necessary to avoid consumer deception about the amount of reduction and the basis for any comparison asserted.

REFILLABLE

In general, an unqualified refillable claim should not be asserted unless a system is provided for: (1) the collection and return of the package for refill; or (2) the later refill of the package by consumers with product subsequently sold in another package. The claim should not be made if it is up to consumers to find ways to refill the package.

OZONE SAFE AND OZONE FRIENDLY

In general, a product should not be advertised as "ozone safe," "ozone friendly," or as not containing CFCs if the product contains any ozone-depleting chemical. Claims about the reduction of a product's ozone-depletion potential may be made if adequately substantiated.

Source: Federal Trade Commission, *Summary of FTC Environmental Marketing Guidelines* (Washington, D.C.: FTC, 1992), pp. 2–3.

Several levels of evaluation and certification are awarded. Through its Compliance Evaluation Program, SCS works with leading retailers to help them monitor the environmental claims made by their vendors and ensure that these claims meet responsible green marketing guidelines. Under the Environmental Claims Certification Program, SCS verifies the accuracy of environmental claims on products. Claims such as "recycled content" (paper, glass, steel, plastic products), "biodegradable" (cleaning products), and "energy efficient" (light bulbs) are among those being evaluated. The certification is accompanied by a specific statement of the achievement that has been documented (Figure 13.2).

In addition, SCS is a leading practitioner of "life-cycle" assessment, which involves an examination of the resources used, energy consumed, wastes produced, and emissions released as a result of the manufacture, distribution, use, and disposal of a product. Based on this "life-cycle" assessment, an environmental label is issued called a Certified Eco-Profile (Figure 13.3), the environmental equivalent of a nutritional label. This label provides crucial environmental facts about a product and helps manufacturers, consumers, retailers, procurement officials, and policy makers understand the array of environmental considerations and ramifications behind each product choice. It also helps put specific environmental achievements into perspective.

Courtesy of Scientific Certification Systems

FIGURE 13.2 Example of SCS Certification Emblem

FIGURE 13.3 Example of Certified Eco-Profile

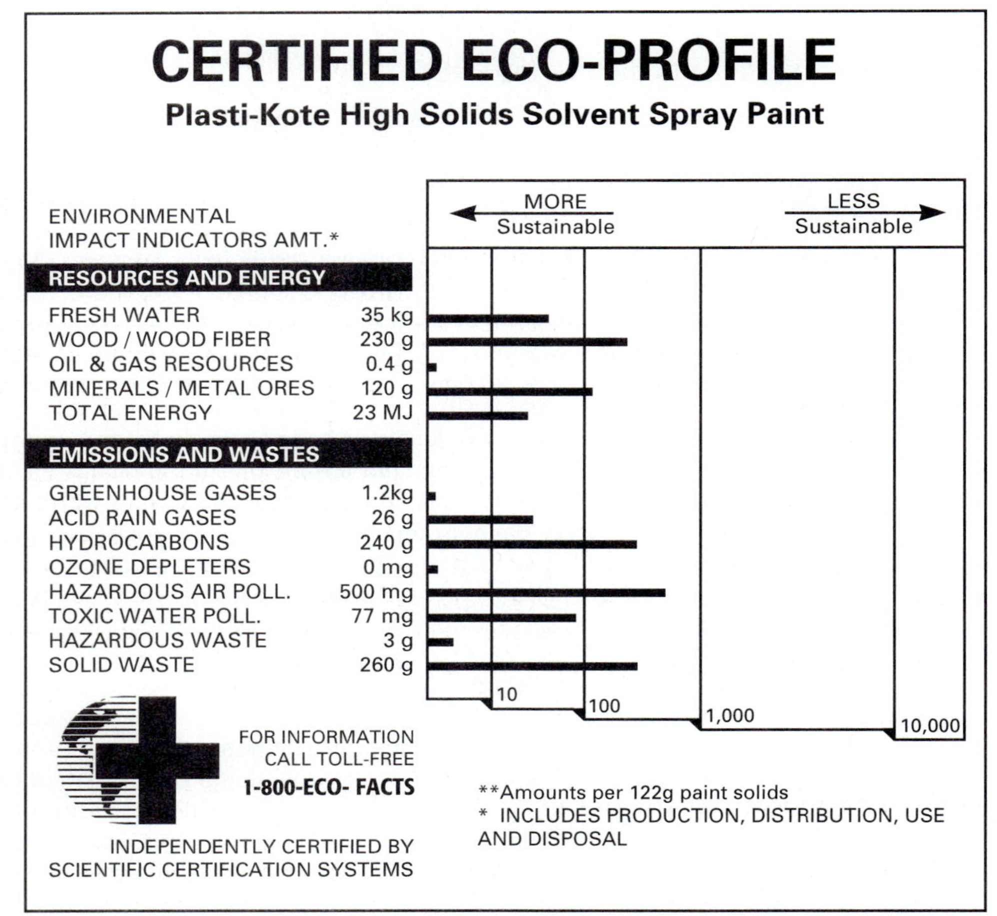

Courtesy of Scientific Certification Systems

The Green Seal program, started by Earth Day organizer Dennis Hayes, is an independent, nonprofit environmental labeling and consumer education organization. The program establishes product categories based on the significance of the environmental impact of products within each category. Standards set for these categories are publicly circulated for comment and revised every three years. Companies can then apply to have their products evaluated against these standards. Underwriters Laboratories Inc. (UL) does most of this evaluation work for Green Seal, and if the product meets the standards, the manufacturer is authorized to use the Green Seal Certification Mark on the product and in product advertising (see below).[38]

Courtesy of Green Seal, Inc.

With respect to wood and wood products, four organizations, including Scientific Certification Systems, are active internationally to promote sustainable forestry practices. Wood or wood products that are certified "green" are supposed to come from environmentally "well-managed" forests. The certifiers consider whether the products have been produced through a sustained-yield harvest, and whether consideration has been given to the entire forest ecosystem, which includes adequate wildlife habitat and watershed protection as well as the economic and cultural impact of the operation on local communities.[39]

The debate on national green marketing standards is thus enjoined. Marketers face the challenge of making truthful environmental claims for their own products while wondering if other companies will do the same or engage in questionable advertising practices to gain market share. This age-old problem for marketing appears in all markets but especially in newly developing ones like environmental marketing or the controversy over health claims that surfaced several years ago. Yet marketing offers great potential to deal with environmental problems through the market mechanism without resorting to further government regulation. But the consumer needs further education and companies need uniform guidelines so they are all playing on a level playing field.

THE GREENING OF STRATEGY

The concept of strategic management has evolved over the past decades as a means to integrate the various functional areas along with broader environmental considerations, where the environment refers to various influences external to the organization.

Strategic management involves the formulation of basic organizational missions, purposes, and objectives as well as policies and programs that determine the long-run performance of an organization. The concept of strategic management emphasizes the monitoring and evaluating of environmental opportunities and constraints in the light of a corporation's strengths and weaknesses.[40] Strategic management is concerned primarily with relating the organization to its environment, formulating strategies to adapt to that environment, and assuring that the implementation of strategies takes place in a manner consistent with organizational realities.[41]

The concept of strategic management enables a firm to both anticipate and create the future and prepare suitable guidelines for making better current decisions. To a large extent the success of a company will depend on how well it formulates its strategy in light of its evolving environment, how well it defines and articulates its mission and strategy, and how well it assures its implementation. Without this long-run perspective a company is forced to live from day-to-day and react to current events. The elements of the strategic planning process concern an understanding of the changing environment in which a company finds itself, the basic missions of the organization, basic company purposes, long-range planning objectives, and program/policies and strategies.[42]

The strategic planning process is concerned with defining the basic mission of the company, taking into account the company's products and/or businesses and the markets in which they are distributed. These basic purposes include factors such as product quality, customer service, response to community interests, and ethical conduct. The strategic planning process focuses on opportunities and threats in the environment. It provides a unified framework within which managers can deal with the opportunities that are there and assess the strengths of the company that can be utilized to take advantage of these opportunities, and identify weaknesses that must be addressed for the company to be successful.[43]

While there are many models of the strategic management process, the process typically begins with an environmental assessment that includes a look at the economic, legal, political, and social environments of relevance to the organization. The purpose of such an assessment is to get an idea of the opportunities and challenges present in these environments and the broader societal expectations that exist relative to the firm's purpose and behavior. The environment is the context in which the organization must survive and attempt to do its business.

Strategy formulation is the process of developing long-range and short-range plans to deal effectively with environmental opportunities and challenges in light of corporate strengths and weaknesses. Strategy formulation requires the manager to create a fit among the opportunities in the external environment, broader societal expectations of the corporation, the strengths and weaknesses of the firm, and the personal values of the key implementors. This stage involves a statement of the corporate mission, objectives, goals, and policies.

The corporate mission is the purpose or reason for the organization's existence. This mission defines the business the corporation is in and what it intends to accomplish as far as society and its stakeholders are concerned. These mission statements often involve a statement of the core values of the organization and the principles that guide its actions. Objectives are more specific statements of what the corporation hopes to achieve and can be stated in terms of product lines, geographic markets the corporation serves, market segments, and distributional channels along with other business factors. The goals state who is responsible for accomplishing these purposes and by what means and according to what time frame. The achievement of corporate goals should result in fulfillment of corporate objectives and its mission. Policies provide broad guidelines for making decisions. They flow from the goals and provide guidance for decision making throughout the organization so that the goals have some chance of being achieved.

The outcome of this formulation process is the development of some kind of strategic plan that will guide corporate efforts over some period of time. This plan basically will involve expansion into new markets or geographic areas of the country or world, acquisitions of new businesses or integration of existing businesses, joint ventures for endeavors that are extremely risky or expensive, or other strategies that involve growth and development. Other strategies may involve retrenchment by selling off some part of the business, cutting back on production and personnel, getting out of some product lines entirely, and other such strategies that involve refocusing or restructuring. Still other strategies could involve attempts to change the image of the company and present it as more socially or environmentally concerned.

Strategy implementation involves the process of putting this plan into action through the development of programs, budgets, and procedures. Implementation is concerned with day-to-day resource allocation problems. Programs are statements of activities or steps needed to accomplish the plan and make a strategy action-oriented. Budgets list the detailed costs of each program for planning and control purposes. Procedures are a system of steps or techniques that describe how to perform a particular task and typically detail the various activities that must be carried out to complete a corporation's program.[44] Activities at this stage mainly involve questions of structure and process that determine how the formulated strategy is going to be carried out and put into action within the organization.

Strategy evaluation involves the monitoring of corporate activities and measurement performance so that actual performance can be compared with desired performance. Although evaluation is the final stage of the strategic management process, it may also serve to stimulate the beginning of the entire process by highlighting weaknesses in previously formulated and implemented strategies. Thus the evaluation system serves as a feedback mechanism to provide necessary correctives to the other stages in the light of actual experience. The strategic planning system is thus self-correcting within the confines of the system itself.[45]

The natural environment has traditionally been excluded from explicit consideration in the strategic management process. It has been assumed that natural resources will continue to be available for the company to function and that waste will be able to be disposed of without major problems. The natural environment has been taken as a given, and even though the corporation is dependent on the natural environment for its very existence, the natural environment has not been a part of strategic thinking, reflecting the dualistic relationship with nature that exists in human consciousness. Given that these assumptions are now under question, the natural environment must be given consideration at all stages of the strategic management process.

While an environmental assessment has typically focused on the economic, legal, political, and social environments, the natural environment must also be given explicit consideration in order for business organizations to get a comprehensive picture of the external environment in which they function. The purpose of an environmental assessment process is the early identification of major issues confronting the business, giving it sufficient lead time for effective strategies to be developed. Increasingly, business organizations must be aware of issues arising out of the natural environment in order to develop strategies that take these issues into account.

An assessment must be made of natural resource availability and the impacts that obtaining these resources will make on the environment. If the resource base of the company is nonrenewable, strategies will have to be made to conserve these resources and search for substitutes as the resources begin to become exhausted. Strategies will have to be developed to extract these resources in an environmentally responsible manner or look for suppliers that have environmentally responsible practices. At the other end of the cycle, concern will have to be given to responsible disposal practices that take environmental impacts into consideration. Techniques such as product life-cycle analysis will be useful to make these kinds of assessments.

The insurance industry, for example, has had to give serious consideration to global climate change and its implications for the industry. Over the five-year period from 1990 to 1995, the industry paid out $48 billion on weather-related claims, over three times the amount paid out for the entire previous decade. Some analysts have stated that climate change could bankrupt the industry. Thus some companies have moved to limit exposure in areas that are particularly vulnerable, such as islands and coastal zones. Others have taken a more proactive stance by encouraging sound environmental practices among their clients, finding common cause with environmentalists. The natural environment poses a significant threat to this industry that companies are beginning to take into account.[46]

Strategy formulation will have to consider the opportunities and threats posed by the natural environment as well as the economic, legal, political, and social environments. The mission statement of the company should include environmental goals and objectives, for example, and state the company's commitment to be an environmentally responsible company. The objectives can include items such as how much recycled material the company plans to use in its production process, and how much recycled content it plans to use in paper products used in its offices. Expansion into new product lines or acquisition of other companies will have to take environmental considerations into account. Specific goals and policies should also reflect these environmental concerns.

The implementation process must also reflect environmental concerns. Programs, budgets, and procedures must reflect environmental considerations such as waste minimization, design for disassembly, green marketing strategies, and other such considerations. Finally, the evaluation process must measure how well the company is meeting its environmental commitments as reflected in its mission statement and how well it has responded to threats and opportunities in the natural environment. The evaluation process should show where adjustment in the company's strategic plan needs to be made to be more environmentally aware and responsible.

Strategy with regard to environmental issues needs to be developed at various levels that are analogous to the development of strategy for more traditional business concerns. These levels of strategy development include (1) the enterprise level, which encompasses the development and articulation of the role of the organization in society; (2) the corporate level, where product market decisions about entering, withdrawing, or remaining in an industry are considered; (3) the business level, where the focus of strategy is how to compete within an industry; and (4) the functional level, where operational decisions are of concern.[47] Strategy at all of these levels must be considered in developing an effective response to an environmental issue.

Enterprise-level strategy deals with the basic question of what the corporation stands for and involves an understanding of the role of the firm as a whole entity in society and its relationships to other social institutions. Enterprise-level strategy deals with such questions as the perception of the organization by major stakeholders, the principles or values the organization represents, the obligations the organization has to society at large, and what implications the current mix of business and allocation of resources has for the company's role in society. The point of enterprise-level strategy is that the firm needs to specifically and intentionally address these questions in its strategy formulation process.[48] Enterprise-level strategy, however, must also deal with issues in the natural environment and state the corporation's role in relating to this environment in a responsible and sustainable fashion.

At the *corporate level,* strategy has to do with mergers, acquisitions, diversification, divestment, and other corporate-wide decisions. With regard to environmental issues, for example, companies now have to be concerned about the potential liabilities they may incur should they acquire a company with major environmental problems that could involve the acquiring company in expensive and lengthy litigation. When building new plants to diversify or expand, companies need to be concerned that the land on which they are building was not previously used for dumping of hazardous materials for which they could incur liability.

At the *business level,* the concern is with the industry and competitor environment and how environmental issues are likely to affect this environment. Certain kinds of regulations, for example, may give competitive advantages to some companies and penalize others in the same industry. Many business scholars as well as company executives are beginning to look at environmental regulation as providing opportunities for innovation and gaining competitive advantages rather than as just a threat to profits and autonomy.[49] Some argue that regulation can enhance competitiveness through increased innovation and promotion of better business practices. The most successful companies are not necessarily those who generate the least pollution, but those who innovate and improve their practices. Such a response generates economic offsets that not only lower the net cost of compliance, but also lead to absolute advantages over firms not subject to similar regulations.

Regulations also create a market for environmental technology. The world environmental market has been estimated at $250 billion and will be worth $600 billion by the year 2000. To put these figures into perspective, the size of the aerospace industry is $180 billion, and the chemical industry $500 billion. Growth in this market is running at two to three times the growth of GNP in some countries.[50] Companies in countries like Germany that mandate clean technologies are poised to profit from this market.[51] This market is driven by environmental regulation that provides opportunities for new products and technologies to enhance the environment.

At the *functional level,* operational decisions are affected by new issues with respect to the environment as well as government regulations in response to environmental issues of concern to society. The New Clean Air Act, where many more chemicals emitted into the air have been regulated, has resulted in many changes at the operational level for the chemical companies involved. New operating procedures have had to be adopted by companies that service air-conditioners as well as auto service facilities to comply with regulations with respect to freon, which contains CFCs that have to be captured rather than released into the air as was done previously. A host of other issues of a similar nature have resulted in operational changes at the functional level.

THE GREENING OF COMMUNICATIONS

The area of environmental communications has become more and more important as environmental impacts are of more concern in society. Companies have problems communicating their position on environmental issues to a distrusting public that is concerned about company operations. After incidents like Bhopal, people want to know what risks plants pose for the community as well as what steps they are taking to reduce exposing people to environmental hazards. Communicating the actual risk involved is a problem, particularly when people are outraged over the problems and the company's response. When people are outraged, they tend to think a hazard is more serious than may actually be the case. Thus corporations need to learn how to calm the fear and anger the public has over environmental hazards.[52]

Many companies have established citizen action panels composed of local people who have an interest in what the company does and its potential impacts on the community. Such panels give local citizens a chance to voice their concerns to company personnel and develop some kind of relationship with company management. Some panels have the responsibility to monitor the company's performance with respect to environmental matters and make suggestions for improvement. For such panels to work, management must take them seriously and attend meetings regularly rather than send public affairs personnel to deal with the public. At their best, such panels can become a channel for two-way communication between the company and the public.

There are a whole range of community activities companies are involved in that include helping schools with environmental education, emergency planning networks, and citizen action panels. Many companies donate employee time and provide funds to help elementary and high schools in their local communities develop environmental education programs for teachers and students. Such programs can be very helpful in dealing with the general lack of knowledge related to environmental concerns that seems to exist in society.

Emergency planning networks help communities to develop evacuation plans in case a major problem develops at a local plant that involves release of toxic gases which could be fatal or harmful to health. Such evacuation plans can enable local citizens to get out of the area to protect themselves if some kind of industrial accident takes place. Such planning can also help firefighters know what kind of chemicals the company produces or stores so that the proper extinguishing materials can be used to put out fires should they develop.

Many companies have developed some form of accounting systems to communicate to the public concerning its environmental record. A concern with pollution accounting began with the Toxics Release Inventory, which is required of all companies discharging toxics into the environment. This inventory forced companies to pay more attention to their toxic discharges and account for them more accurately than they had previously. It also resulted in serious efforts to reduce toxic discharges as some estimates indicate that toxic emissions overall have been cut in half since implementation of the program. Many companies have continued this effort by developing systems for other emission calculations.

The starting point for most companies is to publish in an annual environmental report numerical data that's already public. The next step is to go beyond what the law requires and publish data that are not required by federal or state laws. Many such disclosures amount to no more than a laundry list of measures that reflect a mixed bag in terms of impacts on the environment. Some companies, however, go to great pains to create a single matrix that combines measures in a way that gives each its proper weight.[53] The effort to account for pollution is subject to all the problems that plagued the development of a social audit that was a concern in the United States some years ago and then lost favor because of all the complexities involved.

THE SUSTAINABLE CORPORATION

It should be clear to even the most casual observer that the present system of industrialization has to be radically changed to even begin to provide for the needs of people for higher living standards and at the same time reduce environmental impacts. The traditional model of industrial activity is one in which raw materials are used by manufacturing processes to produce products to be sold to enhance people's standard of living and waste material is generated to be disposed of in some fashion. This kind of industrial process exists in market economies as well as economies that are centrally planned. The major difference is that centrally planned economies are not as efficient or as productive as market economies, and thus do not produce as high a standard of living, nor do they, as we are now learning, mitigate their environmental problems as successfully as market economies.

But while market economies may be able to generate a higher standard of living for those people served by such economies, the industrial system that is at the heart of these economies must be changed. What some experts advocate is an industrial system that is more integrated, an industrial ecosystem if you will, where consumption of energy and raw materials is optimized, waste generation is minimized, and the effluents of one process serve as the raw material for another process. There must be more incentive for recycling, conservation, and a switch to alternative materials than the present system contains, and these incentives must lead to a different kind of manufacturing system than we know at present.[54]

It is recommended that we think of the industrial system as an analogue of biological ecosystems, where conservation and recycling are readily apparent. There is much that we could learn, for example, from the way the rain forests operate and the efficient manner in which nutrients are recycled and a rich growth is supported even though the soil underneath the rain forest is poor in nutrients. While such an ideal manufacturing ecosystem may never be attained in practice, it can serve as a model on which to base our strategies for the future. Manufacturers and consumers must change their habits to approach this ideal more closely, it is said, if the industrialized world is to maintain its present standard of living and if developing nations are to raise theirs to something approaching a similar level without seriously degrading the environment to a point where living standards of the kind we now enjoy are simply no longer sustainable.[55]

If changes of this sort are embraced by industrialized and developing nations, it may be possible to develop a more closed industrial ecosystem that is sustainable in light of decreasing supplies of raw materials and increasing problems with pollution and waste disposal. While industrial nations will have to make many changes in their current manufacturing practices, some of which will be major in nature, developing nations will have to leapfrog older technologies that are less ecologically sound and adopt new industrial methods that are more compatible with the ecosystem approach and more environmentally sustainable.[56]

Manufacturing processes simply transform stocks of materials from one shape to another. The circulating stock decreases when some material is unavoidably lost as waste material, and the stock increases to meet the needs of a growing population. An industrial ecosystem based on recycling and waste minimization still requires the use of energy and will still generate some waste material and harmful by-products, but at much lower and sustainable levels than are typical with today's system. Industrial processes are required that minimize the generation of unrecyclable wastes and minimize the permanent consumption of scarce material and energy resources. But industrial processes cannot be considered in isolation. They must be linked together and considered to be part of a whole system.[57]

The incentives for such changes are already beginning to appear. Waste minimization activities have been aided by regulations to control both solid and hazardous waste disposal. These regulations reflecting long-term environmental costs have increased disposal costs to the point where alternatives to disposal are economically feasible. Many companies are finding it profitable to sell their wastes as raw materials. Other companies are finding it profitable to reuse their own wastes by designing recycling loops in their own manufacturing processes and saving a great deal of money by minimizing the amount of final waste product they create. There is more of an emphasis on pollution prevention in both business and government and concepts such as dematerialization, which means doing more with less, are being talked about and implemented.[58]

The manufacturing process represents only the supply side of the industrial ecosystem where harmful by-products and effluents are created that have to be disposed of in some fashion. The consumer who buys products and throws things away is the demand side of the equation. Materials that are discarded by consumers are the raw materials for the next cycle of production. Changes in manufacturing must be matched by changes in demand patterns if the industrial ecosystem approach is to be fully implemented, and by changes in the treatment of materials once they have reached the end of their useful life. An effective infrastructure for the collection and segregation of various types of wastes can dramatically improve the efficiency of the industrial ecosystem. Since landfills for municipal wastes are becoming harder to find, consumers also face incentives for waste reduction.[59]

Creation of a sustainable industrial ecosystem is highly desirable from an environmental perspective and can also be profitable for companies and consumers. But corporate and public attitudes must change to favor this approach and government regulations must become more flexible so as to encourage recycling and strategies for waste minimization. Regulations, according to some experts, sometimes make waste minimization more difficult than waste disposal. Buying of hazardous wastes, for example, even where they could be used as raw materials, is difficult because of strict requirements for handling and documenting of these wastes, to say nothing of the potential liability the company may face if an accident happens. These regulations discourage innovative treatment of wastes, and many companies choose to avoid these risks and buy their materials through conventional channels.[60]

Regulations traditionally focused on end-of-pipe disposal practices and provided no advantages for manufacturers who capture and treat low-level effluents or who shift to production processes with more benign by-products. Thus companies have to meet regulatory requirements but have no incentive to develop more innovative practices that will reduce their waste material. This focus is beginning to change with the advent of the Pollution Prevention Act of 1990, which encourages companies to take steps to reduce the production of pollution.

Instead of rules regarding waste disposal, some economists advocate financial incentives to reduce pollution in the form of fees or taxes imposed on manufacturers according to the amount and nature of the hazardous materials they produce. These fees and taxes would give them an incentive to change their manufacturing processes to reduce hazardous waste production, and would make environmental costs internal so they can be taken into account when making production decisions.[61]

Such techniques might make it more feasible to solve environmental problems at the source where they are created, rather than to destroy or dispose of effluents once they have been created. Manufacturers would be able to share in the overall economic savings that accrued from reduced levels of hazardous materials and would harness strong economic incentives to reduce costs and gain competitive advantages in the marketplace. Manufacturers that ignore these incentives do not stay in business, and allocating the social costs of pollution would assure that only those environmentally sensitive manufacturers would stay in business.[62]

But such economic incentives may not be enough. Some kind of holistic approach will be required if a proper balance is to be maintained between narrowly defined economic benefits and environmental needs consistent with a sustainable society. The concepts of industrial ecology and system optimization must be taught more widely in engineering and business schools and must be recognized and valued by public officials and industrial leaders. Such values and approaches must be instilled into the society and reflected in the policies and procedures adopted by government and industry.[63]

Regulation must avoid counterproductive control measures such as appears to be the case with Superfund regulations. But a regulatory framework that is rational and efficient will be impossible to construct unless government, industry, and environmental groups learn to overcome their current adversarial relationships and work together to solve shared problems. This will come about only if there are shared values with respect to the environment, and shared agreement on strategies relative to what needs to be done to create a sustainable society. Education about ecology and environmental issues can help to create these common values and agreement on what strategies will work to the advantage of the entire society.[64]

The President's Council on Sustainable Development issued a report in 1996 that emphasized the integration of the economy, equity, and the environment in the development of national policy. This council was composed of leaders from government, business, environmental, civil rights, labor, and Native American organizations. It presented sustainable development as the framework that integrates economic, environmental, and social goals in discourse and policies that enhance the prospects of human aspirations. Thinking narrowly about jobs, energy, transportation, housing, or ecosystems, as if they were not connected, is no longer realistic, nor is seeing choices in terms of trade-offs and balance. These ways of thinking reflect a history of confrontational politics and pit vital necessities against each other in a false contest that inhibits exploration of solutions that build common purpose from shared goals.[65]

A regulatory framework based upon such shared values may be emerging in the form of the ISO 14000 standards devised by the Geneva-based International Organization for Standardization (ISO), whose members include business and government groups that deal with standards. These standards are entirely voluntary, and rather than replace state and national environmental laws, they aim to make environmental management a top priority in participating companies.[66] They will direct companies to manage, measure, improve, and communicate the environmental aspects of their operations in a systematic manner. The standards cover the design, manufacture, and marketing of products; the selection of raw materials; and the type of environmental information that is collected and how this information is communicated internally, to governments, and to the public. More specifically, the standards cover environmental management systems, environmental performance evaluation, environmental labeling, and life-cycle assessment.[67]

Companies that are certified under the standards may win a partial respite from government policing of their operations at the same time they better protect the environment and maybe even save some money through reducing waste and adopting more efficient practices. To stay certified, companies must be inspected regularly by outside auditors. This practice would free agencies such as the EPA from regular inspections to devote time to other activities. These standards represent a move away from command-and-control regulation to voluntary compliance, with government intervening only when something goes wrong.[68] It remains to be seen exactly how ISO 14000 will fit into the entire scheme of regulatory compliance, enforcement, and penalties, but the future of the standards looks promising.

Creation of an ideal industrial ecosystem, where use of energy and materials is optimized, wastes and pollution are minimized, and where there is an economically viable role for every product of a manufacturing process, will not be easy to attain. But the incentives are beginning to change and point in this direction. Companies will be able to minimize costs and stay competitive through adhering to an ecological approach to manufacturing where ecological costs and benefits are taken into account. It is also becoming clear that societies will have a chance to raise their living standards only if they take environmental sustainability into account in the policies and consumption patterns they adopt. Coming to recognize that people and their technologies are a part of the natural world may make it possible to imitate the best workings of biological ecosystems and construct artificial ones that are sustainable.[69]

Questions for Discussion

1. What impacts has the environment made on business? How is business responding to these impacts? What more will it have to do in the future to respond to people's material needs and at the same time be environmentally sensitive?
2. What new kinds of thinking and strategies seem to be necessary? How should the industrial system be redesigned to meet the needs of the future? How could such redesign take place? What changes will consumers have to make in their practices?

3. What are some of the problems with the current system of regulatory controls? What changes need to be made to enable corporations to restructure their operations? What incentives need to be changed to enable corporations to be more environmentally sensitive?
4. What does waste minimization mean in theory and practice? What problems exist with trying to clean up existing waste dumps? Is it possible to cut down on the amount of waste generated to avoid cleanups in the future? How can this be done? Give some examples.
5. What problems exist in the life-cycle approach to dealing with environmental problems? Is this a comprehensive manner of looking at the environmental impacts of a company? What elements are important to the success of this approach?
6. Describe the design for disassembly concept. How does this concept relate to recycling? What are some companies doing to promote this concept? Do you believe it will catch hold in more companies?
7. Where do marketing strategies fit into the environmental picture? What is currently being done by some companies in this regard? What pitfalls exist with respect to so-called "green" marketing strategies?
8. Is the development of federal standards with respect to "green" products the answer? What problems are these standards likely to cause? Will the Green Cross and Green Seal programs work very well in dealing with this problem?

Endnotes

1. Robert P. Bringer, "Pollution Prevention Plus," *Pollution Engineering,* XX, no. 10 (October 1988), 84–89.
2. Robert P. Bringer, "The Prevention of Hazardous Waste Generation: An Idea Whose Time Has Come," *International Environment Reporter,* 11, no. 8 (August 10, 1988), 452–454.
3. Ibid.
4. Bringer, "Pollution Prevention Plus," pp. 84–85.
5. Kevin Kelly, "It Really Can Pay to Clean Up Your Act," *Business Week,* November 7, 1994, p. 141.
6. Bringer, "The Prevention of Hazardous Waste Generation," p. 452.
7. Bringer, "Pollution Prevention Plus," p. 84.
8. Dow Chemical Company, *Waste Reduction Always Pays,* undated, pp. 1–2.
9. Dow Chemical Company, *Managing Environmental Issues,* pp. 9–10.
10. Ibid., p. 10.
11. Dow, *Waste Reduction Always Pays,* p. 2.
12. Ibid., p. 5.
13. Dow, *Managing Environmental Issues,* p. 15.
14. Stuart L. Hart, "A Natural-Resource-Based View of the Firm," *Academy of Management Review,* 20, no. 4 (October 1995), 993.
15. Bristol-Myers Squibb Company, "Report on Environmental Progress," May 1993, p. iv.
16. Ibid.
17. Bristol-Myers Squibb Company, "Environment 2000: Pollution Prevention Throughout the Product Life Cycle," 1992, p. iii.
18. Bruce Nussbaum, "Built to Last—Until It's Time to Take It Apart," *Business Week,* September 17, 1990, p. 102.
19. Ibid.
20. Karen Blumenfeld, "Design for Disassembly," *Tomorrow,* V, no. 4 (December 1995), 41.
21. Gary McWilliams, "The Big Brouhaha Over the Little Juice Box," *Business Week,* September 17, 1990, p. 36.
22. Ibid.

23. Blumenfeld, "Design for Disassembly," p. 41.
24. Ibid.
25. "Tending Wal-Mart's Green Policy," *Advertising Age,* January 29, 1991, p. 20.
26. Laurie Freeman, "Procter & Gamble," *Advertising Age,* January 29, 1991, p. 16.
27. Ibid.
28. Ibid., pp. 16, 34.
29. Jennifer Lawrence, "Mobil," *Advertising Age,* January 29, 1991, p. 12–13.
30. Randolph B. Smith, "Environmentalists, State Officers See Red as Firms Rush to Market Green Products," *Wall Street Journal,* March 13, 1990, p. B1.
31. Randolph B. Smith, "Plastic Bag Makers Discarding Environmental Claims," *Wall Street Journal,* March 30, 1990, p. B1.
32. Joann S. Lubin, "Environment Claims Are Sowing More Confusion, 2 Reports Say," *Wall Street Journal,* November 8, 1990, p. B8.
33. Jennifer Lawrence and Steven W. Colford, "Green Guidelines are the Next Step," *Advertising Age,* January 29, 1991, p. 28.
34. *FTC News,* July 28, 1992, p. 1.
35. Federal Trade Commission, *Summary of FTC Environmental Marketing Guidelines* (Washington, D.C.: FTC, 1992), p. 1.
36. Letter to the author from Mitchell Friedman, Public Relations Manager, Green Cross Certification Company, September 30, 1991.
37. Green Cross Certification System, "The Informed Choice," undated, p. 3; Scientific Certification Systems, "Harnessing the Power of Science and the Marketplace for a More Sustainable Future," undated.
38. Laurie Freeman, "Ecology Seals Vie for Approval," *Advertising Age,* January 29, 1991, p. 30; Green Seal, "Now You Can Give It to Them," 1994.
39. Sheila Polson, "Cutting with Conscience: Sustainable Harvested 'Certified' Wood Is Gaining Popularity," *E Magazine,* May/June 1996, pp. 42–43.
40. Thomas L. Wheelen and J. David Hunger, *Strategic Management* (Reading, MA: Addison-Wesley, 1984), pp. 3–4.
41. George A. Steiner and John B. Miner, *Management Policy and Strategy,* 2nd ed. (New York: Macmillan, 1982), p. 30.
42. Ibid., p. 29.
43. Ibid., pp. 29–30.
44. Wheelen and Hunger, *Strategic Management,* pp. 10–11.
45. Ibid., pp. 11–12.
46. Peter Knight, "What Price Natural Disasters?" *Tomorrow,* 5, no. 2 (March–April, 1995), 48–50; Martin Wright, "God, Mammon and the Markets," *Tomorrow,* VI, no. 3 (May–June 1996), 10–11.
47. Liam Fahey and V. K. Narayanan, *Macroenvironmental Analysis for Strategic Management* (St. Paul, MN: West, 1986), p. 189.
48. Archie B. Carroll, *Business and Society: Ethics and Stakeholder Management* (Cincinnati: Southwestern, 1989), p. 453.
49. Stuart L. Hart, "A Natural-Resource-Based View of the Firm," *Academy of Management Review,* 20, no. 4 (October 1995), 986–1014.
50. Alex Goldsmith, "Legislation Still the Market Motor," *Tomorrow,* June 1995, p. 30.
51. Curtis Moore, "Green Revolution in the Making," *Sierra,* January–February 1995, pp. 50–52. In late 1994, the Clinton administration launched an effort to reduce regulatory and statutory barriers to U.S. environmental technology companies. The program was to provide greater assistance to companies that want to market their technology here and abroad. Without such assistance, it was believed that the industry would fall behind Germany and Japan, which are aggressively promoting their environmental-technology sectors. Timothy Noah,

"Clinton Administration to Launch Effort to Promote Environmental Technology," *Wall Street Journal,* December 12, 1994, p. C15.
52. Dwight Holing, "It's the Outrage, Stupid," *Tomorrow,* V, no. 3 (May–June 1995), 20–21.
53. Bill Birchard, "By the Numbers," *Tomorrow,* June 1995, pp. 52–53. See also Jonathan S. Naimon, "Corporate Reporting Picks Up Speed," *Tomorrow,* June 1995, pp. 62–66.
54. Robert A. Forsch and Nicholas E. Gallopoulos, "Strategies for Manufacturing," *Scientific American,* 261, no. 3 (September, 1990), 144.
55. Ibid.
56. Ibid., pp. 144–145.
57. Ibid., pp. 146, 149.
58. Ibid., p. 150. See also Cutter Information Corporation, *Business and the Environment,* VI, no. 8 (August 1995), 1–6.
59. Ibid., p. 151.
60. Ibid.
61. Ibid., p. 152.
62. Ibid.
63. Ibid.
64. Ibid.
65. The President's Council on Sustainable Development, *Sustainable America: A New Consensus* (Washington, D.C.: U.S. Government Printing Office, 1996).
66. Lisa Sanders, "Going Green with Less Red Tape," *Business Week,* September 23, 1996, pp. 75–76.
67. "Business and the Environment," November 1995, pp. 1–2.
68. Sanders, "Going Green," pp. 75–76.
69. Forsch and Gallopoulos, "Strategies for Manufacturing," p. 152.

Suggested Reading

Cameron, Donald R. *Environmental Concerns in Business Transactions: Avoiding the Risks.* New York: Butterworth Legal Publications, 1993.

Coddington, Walter, and Peter Florian. *Environmental Marketing: Positive Strategies for Reaching the Green Consumer.* New York: McGraw-Hill, 1993.

Dennison, Mark. *Environmental Reporting, Recordkeeping, and Inspections: A Compliance Guide for Business and Industry.* New York: Van Nos Reinhold, 1994.

Dominguez, George S. *Marketing in a Regulated Environment.* New York: Wiley, 1978.

Dow Chemical Company, "Recycling Plastics: Great Things Are Worth Repeating." Special Reprint from 1989 Annual Report, 1990.

Edwards, Felicity N., ed. *Environmental Auditing: The Challenge of the 1990s.* Calgary: Paul and Company, 1993.

Fisher, Kurt, and Johan Schot, eds. *Environmental Strategies for Industry: International Perspectives on Research Needs and Policy Implications.* Washington, D.C.: Island Press, 1993.

Fitzgerald, Chris. *Environmental Management Information Systems.* New York: McGraw-Hill, 1995.

Forsch, Robert A., and Nicholas E. Gallopoulos. "Strategies for Manufacturing," *Scientific American,* 261, no. 3 (September 1990).

Frause, Bob, and Julie Colehour. *Environmental Marketing Imperative: Strategies for Transforming Environmental Commitment.* New York: Probus Publishing Co., 1993.

Friedman, Frank B. *Practical Guide to Environmental Management.* Washington, D.C.: Environmental Law Institute, 1988.

Gilpin, Alan. *Environmental Impact Assessment: Cutting Edge for the 21st Century.* Cambridge: Cambridge University Press, 1994.

Greenwood, Brian, and Anna Marshall. *Environmental Law in Corporate Transactions.* New York: Butterworth Legal Publications, 1994.

Harrison, E. Bruce. *Environmental Communications and Public Relations Handbook,* 2nd ed. Washington, D.C.: Government Interests, 1992.

Hopfenbeck, Waldemar. *Environmental Management and Marketing.* Upper Saddle River, NJ: Prentice Hall, 1993.

International Labor Conference. *Environment and the World of Work.* Geneva: International Labor Organization, 1990.

Jain, R. K., et al. *Environmental Assessment.* New York: McGraw-Hill, 1993.

Kolluru, Rao V. *Environmental Strategies Handbook: A Guide to Effective Policies and Practices.* New York: McGraw-Hill, 1993.

Ledgerwood, Grant, et al., eds. *Environmental Audit and Business Strategy: A Total Quality Approach.* London: Pitman Publishing, Ltd., 1992.

McLoughlin, J., ed. *Environmental Pollution Control: An Introduction to Principles and Practice of Administration.* Netherlands: Kluwer Academic Press, 1993.

Petulla, Joseph M. *Environmental Protection in the United States: Industry, Agencies, Environmentalists.* San Francisco: San Francisco Study Center, 1987.

Polonsky, Michael J., and Alma T. Mintu-Wimatt. *Environmental Marketing Strategies, Practice, Theory, and Research.* New York: Harworth Press, 1995.

Smith, Ann C., and William A. Yodis. *Environmental Auditing Quality Management.* New York: Wiley, 1994.

Smith, Owen T. *Environmental Lender Liability.* New York: Wiley, 1991.

Tokar, Michael. *The Green Alternative: Creating an Alternative Future.* San Pedro, CA: R. & E. Miles, 1988.

Winter, George. *Business and the Environment.* New York: McGraw-Hill, 1988.

CHAPTER 14

Toward a Sustainable Society

Dealing at the firm level is something of a micro approach to environmental problems in analyzing what corporations either individually or as a whole can do to make themselves more sustainable or ecocentrically managed. This level also focuses on the production side of the picture and does not deal with issues related to continued consumption of goods and services. But a more macro approach must also be taken and some discussion devoted to what society as a whole needs to do to become more sustainable. Questions need to be raised about the nature of an industrial society that is oriented toward continued consumption of goods and services and understands growth in economic and materialistic terms.

Despite its shortcomings, the concept of sustainability seems to be a viable way to think about these issues. Many experts believe that our present industrial societies are not sustainable in that they use too many virgin resources and degrade the environment in too many ways, and that such practices cannot continue much longer. If people all over the world want to increase their standard of living on a par with advanced industrial nations like the United States, resources will be used even faster and environmental degradation will increase. The world simply cannot sustain such activities, it is believed, and aspirations of this sort will exceed the earth's carrying capacity with respect to resources and all aspects of the environment.

According to some estimates, since 1900, the number of people inhabiting the earth has multiplied more than three times, and the world economy has expanded more than 20 times during the same time period. The consumption of fossil fuels has grown by a factor of 30, and industrial production has increased by a factor of 50, with four fifths of that increase occurring since 1950 alone. While there have been great gains in human welfare because of these developments and the potential for future gains is even more promising, development at this pace has also produced environmental destruction on a scale never before imagined and is undermining prospects for future economic development as well as threatening the very survival of the earth's inhabitants.[1]

By the year 2030 there will be approximately 10 billion people on this planet, all of whom would like to enjoy a living standard roughly equivalent to the advanced industrial democracies. But if all these people were to consume critical natural resources such as copper, cobalt, petroleum, and nickel at current U.S. rates, and if new resources are not discovered or substitutes developed, such living standards could be supported by existing resources for only a decade or less. On the other side of the ledger, with regard to waste disposal, at current U.S. rates, 10 billion people would generate 400 billion tons of solid waste every year—enough to bury greater Los Angeles under 100 meters of such waste material.[2]

Changes taking place in the world have profound implications for resource usage and waste disposal problems. The citizens of Eastern Europe, who have recently been freed from decades of Communist rule, are attempting to establish democracies and some kind of market systems to provide more goods and services for their people. Changes in this part of the world alone have profound implications for resource usage and waste disposal. If changes of a similar nature take place in other countries such as Russia and China and more consumer goods and services are produced for the people of these nations, further demands will be placed on the environment. And there are many Third World nations who have aspirations for a better life for themselves.

A fivefold to tenfold increase in economic activity translates into a greatly increased burden on the ecosphere. Such an increase is not unrealistic as it represents annual growth rates of only between 3.2 and 4.7 percent, well within the aspiration levels of many countries of the world. Such growth has severe implications for investment in housing, transportation, agriculture, and industry. Energy use would have to increase by a factor of five just to bring developing countries, given their present populations, up to the levels of consumption now existing in the industrialized world. Similar increases could be projected for food, water, shelter, and other things that are essential to human existence.[3]

The question being asked ever more frequently by commissions and policy makers all over the world is "can growth on the scale projected over the next one to five decades be managed on a basis that is sustainable, both economically and ecologically? The answer to this question is not evident to some experts, as the obstacles to sustainability are mainly social, institutional, and political. It is believed that economic and ecological sustainability are usually dealt with as two separate questions by all governments and international organizations, when they are, in fact, interrelated. Economic growth cannot continue if such growth undermines the ecological conditions that support continued growth.[4]

If we have not attained sustainability within the next 40 years, say some experts, environmental deterioration and economic decline are likely to be feeding on each other, pulling us into a downward spiral of social and economic disintegration.[5] The foundations for further economic growth will be eroded and social upheaval will take place throughout the world on an unprecedented scale. Progress toward sustainability hinges on a collective sense of responsibility for the earth and to future generations, as there are difficult questions regarding national sovereignty and individual rights and responsibilities that need to be considered. The capacity of national leaders and of international institutions will be severely tested in the effort to put the world on a firm ecological and economic footing.[6]

Sustainable growth has implications for the distribution of economic wealth and income throughout the world, and raises questions about intergenerational equity as well as equity among the peoples of the world as developing nations strive to better themselves with shrinking resources. Such growth would require a minimum of 3 percent annual growth in per capita income in developing countries, and the need for policies to achieve greater equity within these countries. Greater equity must also be achieved between the industrialized world and developing countries, as the former consumes about 80 percent of the world's goods and has only one quarter of the world's population. With three quarters of the world's population, developing countries command less than one quarter of the world's wealth. This imbalance is getting worse and cannot be continued if sustainable growth is to become a reality throughout the world.[7]

Many developing countries, as well as large parts of many developed countries, are resource based, in the sense that their economic capital largely consists of stocks of environmental resources such as soil, forests, fisheries, and other such natural resources. Their continued development depends on maintaining and perhaps increasing these stocks of resources to support agriculture, fishing, and mining for local use and for export purposes. But during the past two decades, the poorer countries of the developing world have experienced a massive depletion of this capital. Environmental and renewable resources are being used up faster than they can be restored or replaced, and some developing countries have depleted virtually all of the ecological capital and are on the brink of environmental bankruptcy.[8]

According to the World Commission of Environment and Development, sustainable growth is based on forms and processes of development that do not undermine the integrity of the environment on which they depend. But modern civilizations have been characterized by unsustainable development, utilizing forms of decision making that do no take the future into account. They have ignored the long-term ecological costs of development and these costs are now coming due in economies all over the world. Yet many governments refuse to change their policies to correspond with this emerging reality and continue to act as if environmental conditions can be ignored and that nature will take care of itself.[9]

Conditions necessary to make development sustainable include the following: (1) reviving growth particularly in developing nations; (2) addressing equity issues between generations and between countries of the world; (3) meeting the essential needs and aspirations of people all over the world; (4) reducing rates of population growth recognizing that development is the best means of population control; (5) adopting policies that do not deplete the basic stock of ecological capital over time; (6) making a significant and rapid reduction in the energy and raw-material content of every unit of production; and (7) merging environmental and economic concerns into decision-making processes in both the public and private sectors.[10]

While economic and ecological systems have become totally interlocked in the real world, they remain totally divorced in many of our institutions. Environmental protection agencies have been set up by governments around the world, but they have been hamstrung by limited mandates and budgets. Meanwhile, many agencies created to promote economic development have not been made to take responsibility for the environmental implications of their policies and expenditures. The resulting imbalance results in promotion of economic growth at the expense of the environment.[11]

> Environmental agencies must be given more capacity and more power to cope with the effects of unsustainable development policies. More important, governments must make their central economic, trade and sectoral agencies directly responsible and accountable for formulating policies and budgets to encourage development that is sustainable.

> Only then will the ecological dimensions of policy be considered at the same time as the economic, trade, energy, agricultural and other dimensions—on the same agendas and in the same national and international institutions.[12]

The market is one area where the merging of environmental considerations with economic decision making could have a major impact. The market is the most powerful instrument for driving development that societies have available, but whether or not the market supports sustainable or unsustainable forms of development is largely a function of public policy on the part of government structures. The market does not take into account the external environmental costs associated with producing, consuming, and disposing of goods and services. But policy makers often do not take these external costs into account when considering fiscal and monetary policies or trade policies, all of which have environmental implications. When these externals are taken into account, the assumption is often made that resources are inexhaustible or that substitutes will be found before we run out of some resources, or that the environment should subsidize the market.[13]

Ignoring the environmental implications of policy means that the costs of environmental degradation are borne by the larger community in the form of air, water, land, and noise pollution and of resource depletion. Or, these costs are transferred to future generations who are stuck with a degraded environment that will no longer support growth rates that have been attained in the past. Internalizing these costs requires some acceptable means of determining the costs of this degradation and then finding the political will to impose these costs on marketplace transactions, so the cost of goods and services will reflect the environmental impacts of production and consumption activities. Environmental degradation should also be integrated into resource accounts in national economic systems, so policy makers would have a more accurate picture of the way certain economic policies are affecting ecological systems and stocks of resources.[14]

Creating a sustainable society calls for major shifts on several fronts simultaneously to restore some kind of equilibrium that will continue to make the planet hospitable. There needs to be a global balance between births and deaths, carbon emissions and carbon fixation, soil erosion and soil formation, and tree cutting and tree planting. The people of the earth will either mobilize to reestablish a stable relationship with the earth's natural support systems or continue down the path of environmental deterioration. At some point, experts say that a continuing preoccupation with the unstable present will begin to obliterate hopes for reclaiming the future. Environmental deterioration is gradual and difficult to arrest, let alone turn around, once it has reached a certain point. The adjustments needed in economic and social systems around the world are permanent for they are the prerequisites for long-term survival.[15]

The move toward sustainable development could be compared to two other major changes in societies in past centuries—the agricultural revolution and the industrial revolution. Such a change will have far-reaching implications for societies all over the world, as they begin to see that environmental protection and economic development are complementary rather than antagonistic processes. The old notion of trade-offs between economic growth and economic development is no longer viable. Economic growth and development must take place and be maintained over time within the limits set by ecology—by the interrelations of human beings and their works, the biosphere, and the physical and chemical laws that govern our world.[16]

CHANGING CONSCIOUSNESS

The problems we face with respect to the environment and the move toward sustainability require another major change in human consciousness regarding the place of human beings in the larger scale of the universe. There have been several such shifts in thinking throughout history as people have had to come to grips with new ways of thinking about and looking at the world (Exhibit 14.1). Adoption of these changes has never come easily as people are reluctant to change perceptions unless forced to do so by overwhelming evidence or by the magnitude of problems that have to be solved.

The first such change in human consciousness was stimulated by the development of science and the scientific method as it was applied to the physical world in which we exist, particularly to the nature and composition of the universe. Prior to that time it was accepted as an article of faith that the sun and the other planets revolved around the earth, that human beings were the center of the universe. When observations began to be made regarding movement of the planets and the sun, early scientists developed elaborate theories to explain the irregularities of their rotation based on this assumption. But none of these explanations proved satisfactory.

Finally, the only thing that made sense was to abandon this assumption and recognize that the earth and the other planets revolved around the sun instead. The earth was not the center of the universe or even of our own solar system. But the early scientists who developed these new theories did so at considerable personal cost as they received formidable opposition from the church. The result of this change of perception was a humbling experience for humans, as they were removed from the center of the universe. We now know that we live on this rather small, inconspicuous planet that is only one of billions of such bodies in the vast universe with which we are surrounded.

The second such change in human consciousness was presented by theories of evolution, theories that have not been accepted by some religions even today. The theory of evolution challenged the notion of creationism, as it is now called, namely, that human beings were created directly by God after the world had been created, and that they have a special status with respect to the rest of creation. The theory of evolution is again a humbling experience as it places human beings in an evolutionary process, where the creation of plant and animal life is the result of a lengthy process of natural selection that has gone on for centuries before humans existed. Many people have not accepted this perception because they believe it degrades humans to mere animals in a long chain of evolution. Battles between creationists and evolutionists continue.

EXHIBIT 14.1 Changes in Human Consciousness

Theories About the Universe	The earth is not at the center of the solar system let alone the universe
Theories of Evolution	Human life is the result of an evolutionary process that incorporates the principle of natural selection
Theories of the Unconscious	Human freedom is circumscribed by unconscious wishes and desires
Theories About Nature	Humans are a part of nature and must see them as but one part of a vast and interdependent ecosystem

The third such intellectual challenge was posed by psychology and the discovery of the unconscious. Human freedom has been of particular importance in American culture and in the Protestant religion, which emphasizes free will and the importance of choice. Yet psychology presents a different perception of choice and places limits on free will with its notion of the unconscious. Many of our so-called choices are not really choices at all, in the sense in which we usually think about choice. Many of our decisions are based on unconscious wishes or desires, so we are told, and are not really free at all in the sense we would like to believe. It is only as these unconscious wishes or desires are brought to consciousness in psychotherapy or psychoanalysis that we can expand our freedom of choice and free ourselves from these unconscious motives and fears.

Environmentalism can be seen as posing yet another challenge to human self-understanding and providing another humbling experience. The traditional view of humans and their relationship to nature is dualistic, that humans stand over against nature and are somehow apart from nature. The task of humans is to conquer nature, to take dominion over the animals and the natural world as some Christian doctrine has emphasized. This view led to an objectification of nature and allowed us to manipulate nature to our advantage and exploit it for our own purposes. This view is now being challenged by those who advocate that humans must instead see themselves as a part of nature, and through education about ecology must come to see themselves as but one link in the great chain of being. Only by adopting this perspective, it is argued, can humans see nature properly and understand what must be done to promote survival of the planet and the human race.

The earth and its creatures are considered the property of humankind, to be dominated, manipulated, and controlled for purposes of enhancing the welfare of humans. But humans believe they somehow stand outside of nature and are not subject to the laws of ecology. Advanced technology gives impetus to the basic assumption that there is essentially no limit to the power humans have over nature. These unconscious assumptions give rise to unsustainable practices. In the developing world, it takes the form of development at any cost, even if it means the wholesale destruction of forests and the creation of industrial centers that are sources of severe environmental pollution. In the developed world, unsustainable development has generated wealth and comfort for about one fifth of humankind, and while environmental protection activities have developed in these countries, they have been ameliorative and corrective, not a force for serious restructuring. These activities have been encompassed within the consciousness of unsustainability.[17]

Creation of a sustainability consciousness includes the following elements: (1) The human species is part of nature, its existence depends on its ability to draw sustenance from a finite natural world, and its continuance depends on its ability to abstain from destroying the natural systems that regenerate this world; (2) economic activity must account for the environmental costs of production; and (3) the maintenance of a livable global environment depends on the sustainable development of the entire human family.[18]

According to William Ruckelshaus, a former head of the EPA, this change in consciousness will come about when it is in the interests of individuals and organizations to change, either because they benefit from changing or because they incur severe sanctions from not changing. Changing interests with regard to the environment require three things: (1) A clear set of values consistent with the consciousness of sustainability must be articulated by leaders in both the public and the private sector; (2) motivations need to be established that will support the values; and (3) institutions must be developed that will effectively apply the motivations.[19]

The first of these has happened as leaders around the world have enunciated sustainable development as a desirable goal for their countries. The second and third steps are more difficult as the appropriate motivations and institutional structures are inadequate to the task or nonexistent. The difficulty of moving from stated values to actual practice stems from certain basic characteristics of modern industrialized nations. The first problem is the economic one of externalities, that the environmental cost of producing a good or a service is not accounted for in its price, thereby producing the tragedy of the commons mentioned by Garrett Hardin in an earlier chapter. Industrialized societies refuse to treat environmental resources as capital and spend them as income, which leads to overspending.[20]

The second problem is the political one of motivating people to act in a democracy. Modifying the market to reflect environmental costs is by necessity a function of government, and those adversely affected by such changes often have disproportionate influence on public policy to prevent such modifications from taking place or to mitigate their effects. The critical question for industrial democracies is whether they will be able to overcome political constraints on shaping the market system to promote long-term sustainability.[21]

Well-designed incentive programs are cost-effective and can work to reward people for implementing ecologically sound practices. Incentives could be provided for efficiency improvements for utilities and other companies, as well as consumers. The tax base could be restructured to favor environmentally friendly investments, and because taxes adjust prices, they can be used to help meet many environmental goals such as reducing the possibility of global warming and promoting the use of recycled materials. Fees could be set on carbon emissions from the burning of fossil fuels and thereby slow global warming, and on the use of virgin materials, thus encouraging recycling and reuse. Determining tax levels that reduce harm to human health and the environment without damaging the environment is a complicated task, but it could be done if the political will were available.[22]

Sustainable development requires a multifaceted approach, as there may be as many responses as there are ecological circumstances, economic factors, and cultural constraints. There is no single cookbook-style recipe for sustainable development, and no manual can ever set out the principles and practices of such a broad concept.[23] It is as much a state of mind as anything else, a new kind of consciousness that looks at economic growth in a different perspective. It involves a redefinition of what is politically and economically possible and what has to be done to sustain growth into the future.

TOWARD AN ETHICS OF CONSUMPTION

Sustainable growth has implications for resource availability and limits relative to the ability of the planet to provide everyone with an improved material standard of living. While an assumption behind the concept may be that continued growth is possible, as was mentioned in an earlier chapter, sustainable growth may take different forms and directions than has been true of growth in past decades. Continued growth in consumption of goods and services and the continued development of a materialistic lifestyle may not be possible under conditions of sustainable growth. The world is already overconsuming resources, witness the depletion of fish stocks around the world, and cannot continue on the path of more and more production.

Such concerns have profound implications for corporate activity based on the promotion of consumption and an ever-increasing material standard of living. The corporation is the primary instrument of economic growth in industrialized societies, and one of its primary reasons for existing is producing goods and services to enhance people's material well-being. If this activity is to be curbed in the interests of conserving resources and reducing pollution, what will become of the corporation and the continued growth on which its legitimacy is predicated? Thus the challenge of the environment is an important one, and whether overconsumption is a legitimate problem and changing patterns of consumption are necessary are questions that need examination.

The Protestant Ethic

The best place to begin such a discussion is to examine the moral system that existed during the development of market systems and provided a legitimacy for their existence. The primary ethical emphasis behind the development of market systems in Western Europe and the United States has been called the Protestant Ethic, because this ethic had religious origins in the newly emerging industrial societies that developed after the Reformation period. The Protestant Ethic helped to legitimize the capitalist system by providing a moral justification for the pursuit of wealth and the distribution of income that were a result of economic activity within this system.[24]

The self-discipline and moral sense of duty and calling at the heart of this ethic were vital to the kind of rational economic behavior that early capitalism demanded (calculation, punctuality, productivity). The Protestant Ethic thus contributed to the spirit of capitalism—what might now be called cultural values and attitudes—a spirit supportive of individual human enterprise and accumulation of wealth necessary for the development of capitalism. Within this climate, people were motivated to behave in a manner that proved conducive to rapid economic growth of the capitalistic variety and shared values consistent with this kind of development.[25]

This ethic contained two major elements: (1) an insistence on the importance of a person's calling, which meant that one's primary responsibility was to do one's best at whatever station God had assigned him or her in life, rather than to withdraw from the world and devote oneself entirely to God, as the Catholic Church had taught as a counsel of perfection; and (2) the rationalization of all of life as introduced by Calvin's notion of predestination whereby work became a means of dispersing religious doubt by demonstrating to oneself and others that he or she was one of the elect.[26]

Thus one was to work hard, be productive, and accumulate wealth. But that wealth was not to be pursued for its own sake or enjoyed in lavish consumption, because the world existed to serve the glorification of God and for that purpose alone. The more possessions one had, the greater was the obligation to be an obedient steward and hold these possessions undiminished for the glory of God by increasing them through relentless effort. A worldly asceticism was at the heart of this ethic, which gave a religious sanction to acquisition and rational use of wealth to create more wealth.

> The upshot of it all, was that for the first time in history the two capital producing prescriptions, maximization of production and minimization of consumption, became components of the same ethical matrix. As different from medieval or communist culture these norms were not reserved for or restricted to specific individuals or groups. Everyone hypothetically belonged to that universe from which the deity had drawn the salvation sample, without disclosing its size or composition. The sampling universe had no known restriction of biological or social background, aptitude, or occupational specialization. Nobody could opt out from the sampling process, indeed, everyone had to act as if indeed he had been selected.[27]

Thus the pursuit of material wealth was given a moral justification in that wealth, which was believed to be the fruits of hard work, was a sign of election—as sure a way as was available to dispose of the fear of damnation. But one was not to rest on his or her laurels or enjoy the fruits of his or her own labor. Whatever wealth was earned must be reinvested to accumulate more wealth in order to please God and as a further manifestation of one's own election. This represented a new approach to acquisitiveness and the pursuit of profit over earlier periods. What had been formally regarded at best as something of a personal inclination and choice, had now also become something of a moral duty.[28]

The Protestant Ethic proved to be consistent with the need for the accumulation of capital that is necessary during the early stages of industrial development. Money was saved and reinvested to build up a capital base. Consumption was curtailed in the interests of creating capital wealth. People dedicated themselves to hard work at disagreeable tasks and justified the rationalization of life that capitalism required. All of this was a major change from the way people behaved in medieval agrarian society.

Thus the Protestant Ethic was a social and moral counterpart to the early stages of capitalism, which emphasized both the human and capital sources of productivity and growth and in this sense was the first supply-side theory. It emphasized the human side of production through hard work and the aspect of the calling. But it also advocated not only that people should work hard but also that the money they earn in the process should also be put to work and not spent on lavish consumption. Inequality was thus morally justified if the money earned on capital was reinvested in further capital accumulation, which would benefit society as a whole by increasing production and creating more economic wealth.

This notion of the Protestant Ethic became of particular importance in American society as capitalism developed and economic wealth was created. It stood as one of the most important underpinnings of American culture, and thrift and industry were believed to hold the key to material success and spiritual fulfillment. Eventually, the Protestant Ethic was stripped of its religious trappings, but the basic assumptions about work and its importance remained the same. The Protestant Ethic thus became known as the work ethic and is now almost exclusively discussed in secular terms with very little reference made to its religious origins except in certain scholarly and religious circles.

One topic of interest and concern that appeared frequently in both popular and professional literature during the 1970s, was the weakening or disappearance of the Protestant Ethic or work ethic from the American scene. A good deal of evidence suggested that the traditional values regarding work and the acquisition of wealth as expressed by the Protestant Ethic were changing in some fashion. Many articles indicated that young adults, in particular, had little interest in the grinding routine of the assembly line or in automated clerical tasks. They were turning away, it was suggested, from their parent's dedication to work for the sake of success and were more concerned about finding meaningful work—something that was satisfying and personally rewarding in terms other than money. Young people were seeking to change existing industrial arrangements to allow these intangible goals to be pursued.[29]

This change in values was already noted as early as 1957 by Clyde Kluckhohn, who did an extensive survey of the then available professional literature to determine if there had been any discernible shifts in American values during the past generation. As a result of this survey, he discovered that one value change that could be supported by empirical data was a decline of the Protestant Ethic as the core of the dominant middle-class value system.[30] Then in 1976, Daniel Bell argued that the Protestant Ethic had been replaced by hedonism in contemporary society—the idea of pleasure as a way of life. During the 1950s, according to Bell, American culture had become primarily hedonistic, concerned with fun, play, display, and pleasure. The culture was no longer concerned with how to work and achieve, but with how to spend and enjoy.[31]

Consistent with these views, Christopher Lasch argued that a new ethic of self-preservation has taken hold in American society. The work ethic has been gradually transformed into an ethic of personal survival. The Puritans believed that a godly man worked diligently at his calling not so much in order to accumulate personal wealth as to add to the wealth of the community. But, the pursuit of self-interest, which was formerly identified with the rational pursuit of gain and the accumulation of wealth, has become a search for pleasure and psychic survival. The cult of consumption with its emphasis on immediate gratification has created the narcissistic man of modern society. Such a culture lives for the present and does not save for the future, because it believes there may not be a future to worry about.[32]

Perhaps the crowning blow to the work ethic was provided by Daniel Yankelovich in a more recent publication. Yankelovich states that traditionally, Americans have been a thrifty and productive people adhering to the major tenets of the Protestant Ethic, and in the process helping to create an abundant and expanding economy. But in the past two decades, Americans have loosened their attachment to this ethic of self-denial and deferred gratification and are committed in one way or another to the search for self-fulfillment.[33]

Behavioral Changes in American Society

The weakening of the traditional work ethic with its inherent restriction on consumption is consistent with a behavioral change in American society. Prior to World War II, people by and large were savings-oriented and lived by the ethic of deferred gratification. They would not buy houses with large mortgages and run up huge credit card balances, but would save their money until they could buy things outright. Gratification of their desires was deferred until they could afford to satisfy them, and then, and only then, was it proper to buy things to enjoy. In other words, most people lived within their immediate means and did not borrow for purposes of increased consumption.

During the 1950s, however, this ethic changed into one of instant gratification, as a consumption society was created where people were encouraged to satisfy their desires now rather than wait until they had the money in hand. Buying on credit was encouraged and long-term mortgages became the order of the day with regard to housing. Why defer gratification when one could buy and enjoy things immediately and pay for them in the future? Companies helped to create this kind of society by making credit easy to obtain through the use of credit cards and by using more sophisticated forms of advertising to increase consumption of their products. In fact, some theories even advocated that companies controlled not only the supply of products but also the demand function through manipulation of consumer desires.[34]

These were the days when the throwaway society was created and obsolesence was built into products so that people could buy newer products faster. Packaging was improved so that products looked more attractive and could be purchased easier. This meant that the amount of stuff to be disposed of increased dramatically as products that had outlived their usefulness had to be discarded along with all the packaging materials used to encase new products. The United States became a society where consumption was emphasized and money was made available so people could buy on credit more easily and pay their debts some time in the future.

This change in ethic was also a change in the culture. Television fed this change with sitcoms that portrayed the typical American family as one that lived in a nice house in the suburbs with two cars and all the latest kitchen appliances and electronic gear in the rest of the house. Advertising on television also became more sophisticated to stimulate demand for products. Companies fed the consumption binge with a proliferation of products that appealed to every taste that could be imagined, which encouraged people to go into debt to enjoy the pleasures these products could bring immediately rather than in some future time period.

Government contributed to the development of this culture with the notion of entitlements and the development of programs based on the idea that people in American society were entitled to certain amenities whether or not they earned them in the traditional sense. Social security was provided to assure that people could retire with a certain amount of income. Welfare programs were established to provide a minimum level of goods and services to those who were not working. Medicare and Medicaid programs provided medical services to older people and those in poverty. All of these entitlement programs came at great cost and involved the government itself going into greater and greater debt to pay for them.

Perhaps the development of the atomic bomb also had an impact on generations growing up after World War II, because the future has never been as certain since that time, and we all have had to live with the knowledge that humans have the ability to destroy the planet. Thus one might as well live as well as one possibly can now rather than defer gratification for some future time that may not be there. Changes in religious beliefs and the increasing secularization of society may also have weakened belief in an afterlife, and more people came to hold the belief that you only go around once in life and might as well enjoy it to the fullest extent possible.

There were thus many factors behind this change in behavior of the American people, and no one in particular was responsible for this change but everyone in general helped to create a new approach to consumption where instant gratification became a cultural trait in contrast to earlier times when saving was emphasized. The implications of this change were profound for lifestyles and habits of people, as society became more wealthy and prosperous. Many people lived more interesting lives and had more diversity available to them as never before. They traveled more miles, wore more and different clothes, drove more expensive and sophisticated cars, and in general, enjoyed rising standards of living that involved consumption of the latest products. By inference, these same changes took place in industrial societies in Western Europe where the Protestant Ethic was operative.

Implications of Instant Gratification

But there were adverse implications to this change as well, particularly as far as resource usage and environmental impacts were concerned. In the 1960s, concern about the natural environment began to emerge, and a great deal of legislation related to the environment was passed. These environmental concerns about pollution and resource usage run headlong into cultural values related to increased consumption and immediate gratification. The question now being asked increasingly is whether advanced industrial societies like the United States are sustainable from an environmental point of view, and whether they are just in relation to developing countries from a moral point of view. Questions are being raised about the feasibility and morality of our society hooked on an ever-increasing standard of living, using up more and more of the world's resources, and causing more and more pollution of the environment.

Do the United States and other advanced industrial societies need to cut back on consumption and share some of their largess with developing nations? Do developed societies need to save something for future generations if they take the concept of sustainability seriously? There are moral questions thus related to intragenerational and intergenerational equity. Is there a need for some new kind of ethic, perhaps an environmental ethic, that would essentially function like the work ethic did in terms of providing moral limits on consumption? These are critical questions that need to be raised as more and more nations around the world develop some form of market economies and economic growth is promoted.

Alan Durning has written a book appropriately entitled *How Much Is Enough?*, in which he argues for the creation of what he calls the culture of permanence—a society that lives within its means by drawing on the interest provided by the earth's resources rather than its principal, a society that seeks fulfillment in a web of friendship, family, and meaningful work. Yet he recognizes the difficulty of transforming consumption-oriented societies into sustainable ones and the problem that the material cravings of developing societies pose for resource usage. These forces cause what he calls a conundrum that is described as follows.

> We may be, therefore, in a conundrum—a problem admitting of no satisfactory solution. Limiting the consumer life-style to those who have already attained it is not politically possible, morally defensible, or ecologically sufficient. And extending the life-style to all would simply hasten the ruin of the bioshpere. The global environment cannot support 1.1 billion of us living like American consumers, much less 5.5 billion people, or a future population of at least 8 billion. On the other hand, reducing the consumption levels of the consumer society, and tempering material aspirations elsewhere, though morally acceptable, is a quixotic proposal. It bucks the trend of centuries. Yet it may be the only option.[35]

For the past 40 years, with some exceptions in the form of communal living and religious groups such as the Amish, the overriding goal of people in advanced industrial societies has been one of buying more goods, acquiring more things, increasing their stock of material wealth. Companies have profited from this consumer culture by catering to the consumer, making goods more convenient to buy, bombarding them with advertising—in general, promoting a consumer society by creating a certain materialistic conception of the good life. Because of this trend, the world's people have consumed as many goods and services since 1950 as all previous generations put together. Since 1940, according to Durning, the United States alone has used up as large a share of the earth's mineral resources as did everyone before them combined.[36]

Aside from the question as to whether all this consumption has really made people happier and more fulfilled, the environmental impacts have been severe as more and more resources have become depleted, and it becomes more and more difficult to dispose of waste material. Reduction of consumption in industrial societies, however, can have severe repercussions. Since about two thirds of the gross national product or its equivalent in developed countries consists of consumer purchases, it seems obvious that any severe reduction of consumer expenditures would have serious implications for employment, income, investment, and everything else tied into economic growth. Lowering consumption could be self-destructive to advanced industrial societies. Yet if such measures aren't taken, Durning warns, ecological forces may eventually dismantle advanced societies anyhow, in ways that we can't control and that would be even more destructive.[37]

Toward a New Understanding of Growth

Is there any way out of this dilemma? Several things suggest themselves. Corporations could be more responsible in their advertising and promotion activities and consumers in their consumption activities by promoting and buying products that have less adverse impacts on the environment. This was supposed to be the goal of "green marketing," but because corporations were more concerned with exploiting a trend to increase market share than they were to promote more responsible consumption, the effort has not been able to realize its potential. If an environmental consciousness were to be expanded throughout society through responsible green marketing, much could be done to mitigate environmental impacts by promoting more ecologically sound products and packaging.

However, the goal of green marketing is to change consumption patterns, not necessarily limit consumption. The effort still sends the message that material acquisition can continue unabated and reinforces the image of self-interested human beings whose mission in life is centered upon the consumption and acquisition of material comforts. It focuses on changing personal levels of consumption rather than reducing those levels, particularly in industrialized countries.[38] Thus it does not seem to offer a long-range solution to the problem of overconsumption and continuation of a materialistic lifestyle.

If consumption does need to be limited in order to move toward a sustainable society, the adverse impacts on employment and other aspects of a growth-oriented economy could be mitigated by promoting more employment and investment in companies that produce goods and services that directly enhance the environment. In other words, more people would be employed and more investment made in an environmental sector, where technologies are developed to deal with environmental problems related to pollution and waste disposal, and services are provided related to recycling and restoration of the environment. Growth could still increase under this scenario, but people would be employed and profits made in a different manner by producing goods and providing services that directly enhance the environment rather than devoting so many of our economic resources to producing and providing consumer goods and services that destroy it in the interests of more and more consumption.

Over the past decade, there has been a good deal of growth in what is called an environmental industry (Table 14.1). In 1994, there were approximately 59,000 private and public companies engaged in environmental activities, earning revenues of $161.5 billion, an increase of over 200 percent from 1980 revenues. Employment jumped from 463,000 in 1980 to 1,262,000 in 1994, an increase of almost 175 percent. Industry segments include solid waste management, which is the largest revenue producer, remediation/industrial services, air pollution control equipment, resource recovery, and process and prevention technology, which produced the smallest revenues.

TABLE 14.1 Environmental Industry Revenues and Employment by Industry Segment 1980–1994

[Covers approximately 59,000 private and public companies engaged in environmental activities]

	Revenue (bil. dol.)					***Employment (1,000)***				
Industry Segment	*1980*	*1990*	*1992*	*1993*	*1994*	*1980*	*1990*	*1992*	*1993*	*1994*
Industry total	**52.0**	**137.7**	**147.0**	**152.8**	**161.5**	**463**	**1,174**	**1,204**	**1,223**	**1,262**
Analytical services[1]	0.4	1.5	1.6	1.6	1.6	6	20	20	20	20
Water treatment works	9.2	19.8	21.7	23.4	25.7	54	95	100	106	114
Solid waste management[2]	8.5	26.1	28.2	29.4	31.0	83	210	218	222	230
Hazardous waste management	0.6	6.3	6.6	6.5	6.4	7	57	58	55	53
Remediation/Industrial services	0.4	8.5	8.2	8.4	8.6	7	107	99	100	100
Consulting & engineering	1.5	12.5	14.3	14.6	15.3	21	144	158	158	163
Water equipment and chemicals	6.3	12.1	13.0	13.2	13.5	62	98	101	101	101
Instrument manufacturing	0.2	1.6	1.8	1.8	1.9	3	19	23	24	25
Air pollution control equipment	3.0	3.7	3.8	3.8	3.7	28	83	83	84	83
Waste management equipment	4.0	10.4	11.1	10.9	11.2	42	89	91	88	88
Process and prevention technology	0.1	0.4	0.6	0.7	0.8	2	9	13	14	15
Water utilities	11.9	19.8	21.9	23.1	24.2	77	105	111	115	118
Resource recovery	4.4	13.1	12.2	13.3	15.4	49	118	106	113	128
Environmental energy sources	1.5	1.8	2.0	2.1	2.2	22	21	23	23	24

[1]Covers environmental laboratory testing and services. [2]Covers such activities as collection, transportation, transfer stations, disposal, landfill ownership, and management for solid waste.

Source: U.S. Department of Commerce, Bureau of the Census, *Statistical Abstract of the United States 1995,* 115th ed. (Washington, D.C.: U.S. Government Printing Office, 1995), p. 237.

Thus an environmental sector is developing that indicates a need for and the development of something called an environmental ethic consistent with this development (Exhibit 14.2). While the Protestant Ethic served an important function in limiting consumption to build up a productive base, once that base was established people needed to consume more in order to keep the system going. Such increased consumption, however, has precipitated environmental problems related to pollution and resource usage, so that there is a need for a new ethic that again provides moral limits to consumption and increased investment in environmental technologies and services as well as promoting alternative meanings of growth.

Such alternative meanings of growth involve a reeducation of the American consumer to realize that growth can mean more than mere accumulation of things and that wealth can mean more than just material wealth. Perhaps Durning is right in suggesting that consumers in industrial societies can curtail the use of those things that are ecologically destructive and cultivate the deeper nonmaterial sources of fulfillment that he claims are the main psychological determinants of happiness—things like family, social relationships, meaningful work, and leisure.

If this is so, then a new philosophy is needed that emphasizes that growth does not have to mean mere accumulation nor mere economic development, but rather can involve the integrative expansion of both the individual and the community through ongoing dynamic interaction. Growth could be seen to involve constructive reintegration of problematic situations in ways that lead to widening horizons of self and community and understood as an increase in the moral-esthetic richness of experience, an increased infusion of experience with meaningfulness and expansion of value.

In the final analysis, the cause of our environmental problems is rooted in the culture in which we live, a culture based on certain ethical perspectives with regard to economics and the environment and central values that guide decision making in both the public and private sectors. Economic values are dominant in our culture and provide the basis for decisions as to the uses and abuses of the environment. We do not have an ecological perspective that informs our decisions and guides our actions. Our approach to nature is dualistic, our vision of reality is fragmented, and we view resources and technology as infinite in some sense.

In order to deal adequately with our environmental problems something of a cultural revolution may have to take place. Our culture needs a new kind of holistic belief system that has a worldview that is ecocentric in nature. We need a new order that is based on different cultural values and a new kind of environmental ethic that guides our policies. The economic system must be made to serve environmental values so that sustainable growth becomes a possibility. Otherwise the conditions for further economic

EXHIBIT 14.1 Changing Ethical Concerns

Protestant Ethic	*Consumption Orientation*	*Environmental Ethic*
Limited Consumption	Limited Saving	Limit Consumption
Promoted Investment in Productive Capacity	Promoted Consumption	Promote Investment in Environment
		Alternative Meanings of Growth

growth of any kind will be destroyed. This kind of change will not be easy, as cultures such as ours that have been successful tend to do things the same way as in the past. But change must come because of environmental pressures that can no longer be ignored.

Questions for Discussion

1. What is a sustainable society? In your opinion, is this a feasible option for industrial societies? Is it necessary? What path is sustainable growth most likely to take in Third World countries?
2. What is intergenerational and intragenerational equity? Does sustainable growth require that these issues be addressed? Why or why not? What implications does sustainable growth have for the distribution of income and wealth throughout the world?
3. What conditions are necessary to promote sustainable development? How can these conditions be created? What changes would have to be made in governmental policies to bring these conditions about? What changes are necessary in the industrial system?
4. What issues have to be addressed in a sustainable society? Discuss each of these issues and think of policies that could be justified on sustainable grounds. Are these policies realistic for industrial countries? Third World countries?
5. Is a new consciousness necessary for the future? What elements will this new consciousness include? Do you see these elements emerging? What can be done to hasten this development?
6. What institutional changes are necessary to promote sustainability? Are these changes taking place? What kind of systems are most effective for motivating people to respond to environmental problems? Are these in evidence in our society?
7. What is the Protestant Ethic? How did the discovery of this ethic come about? What evidence supports its existence? What was the nature of the Protestant Ethic? How was it related to the development of capitalism? What kind of a person did such an ethic produce?
8. What changes took place in American society during the 1950s in particular? What kind of society was created? What were some of the social forces behind these changes? What environmental impacts did these changes produce? What issues brought about by environmental problems began to be discussed in society?
9. How does the concern about limits to growth differ from the concern about sustainable growth? Do advanced industrial nations need to cut back on their consumption activities? How much is enough? Is there a need for a new environmental ethic? What would it look like? How would lifestyles be changed?
10. What are the implications of the issues discussed in this chapter for business organizations? What role could business adopt in society that would be more consistent with changes in the ability of the natural environment to support an ever-increasing standard of living? What are the ethical issues involved in this kind of change?

Endnotes

1. Jim McNeill, "Strategies for Sustainable Economic Development," *Scientific American,* 261, no. 3 (September 1989), 155.
2. Robert A. Forsch and Nicholas E. Gallopoulos, "Strategies for Manufacturing," *Scientific American,* 261, no. 3 (September 1990), 144.
3. McNeill, "Strategies for Sustainable Development," p. 156.
4. Ibid., pp. 155–156.
5. Lester R. Brown, Christopher Flavin, and Sandra Postel, "Picturing a Sustainable Society," *State of the World 1990* (New York: W. W. Norton & Company, 1990), pp. 173–174.
6. Lester R. Brown, Christopher Flavin, and Sandra Postel, "Outlining a Global Action Plan," *State of the World 1989* (New York: W. W. Norton & Company, 1989), pp. 174–175.
7. McNeill, "Strategies for Sustainable Development," p. 156.

8. Ibid., p. 157.
9. Ibid.
10. Ibid., pp. 158–163.
11. Ibid., pp. 162–163.
12. Ibid., p. 163.
13. Ibid.
14. Ibid., pp. 163–164.
15. Brown, Flavin, and Postel, "Outlining a Global Action Plan," pp. 192–93.
16. William D. Ruckelshaus, "Toward a Sustainable World," *Scientific American,* 261, no. 3 (September 1989), 167.
17. Ibid., p. 168.
18. Ibid.
19. Ibid., pp. 168–169.
20. Ibid., p. 169.
21. Ibid., pp. 169–172.
22. Sandra Postel, "Accounting for Nature," *World Watch,* 4, no. 2 (March–April 1991), 30–32.
23. Norman Myers, "Making the World Work for People," *International Wildlife,* 19, no. 6 (November–December, 1989), 14.
24. Max Weber, *The Protestant Ethic and the Spirit of Capitalism* (New York: Scribner, 1958).
25. Richard LaPiere, *The Freudian Ethic* (New York: Duell Sloan, and Pearce, 1959), p. 16.
26. David C. McClelland, *The Achieving Society* (New York: Free Press, 1961), p. 48.
27. Gerhard W. Ditz, "The Protestant Ethic and the Market Economy," *Kyklos,* 33, no. 4 (1980), 626–627.
28. Ralph Barton Perry, *Puritanism and Democracy* (New York: Vanguard Press, 1944), p. 302.
29. See *Editorial Research Reports on the American Work Ethic* (Washington, D.C.: Congressional Quarterly, 1973); Harold L. Sheppard and Neal Q. Herrick, *Where Have All the Robots Gone* (New York: Free Press, 1972); Special Task Force to the Secretary of Health, Education, and Welfare, *Work in America* (Cambridge: MIT Press, 1973); and Judson Gooding, *The Job Revolution* (New York: Walker & Co., 1972).
30. Clyde Kluckhohn, "Have There Been Discernible Shifts in American Values During the Past Generation?" *The American Style: Essays in Value and Performance,* Elting E. Morrison, ed. (New York: Harper & Bros., 1958), p. 207.
31. Daniel Bell, *The Cultural Contradictions of Capitalism* (New York: Basic Books, 1976), p. 70.
32. Christopher Lasch, *The Culture of Narcissism: American Life in an Age of Diminishing Expectations* (New York: W. W. Norton, 1978), pp. 52–53.
33. Daniel Yankelovich, *New Rules: The Search for Self-Fulfillment in a World Turned Upside Down* (New York: Random House, 1981).
34. See John Kenneth Galbraith, *The New Industrial State* (Boston: Houghton Mifflin, 1967).
35. Alan Durning, *How Much Is Enough?* (New York: W. W. Norton, 1992), p. 25.
36. Ibid., p. 38.
37. Ibid., p. 107.
38. Ronald E. Purser, Changkil Park, and Alfonso Montuori, "Limits to Anthropocentrism: Toward an Ecocentric Organization Paradigm?" *Academy of Management Review,* 20, no. 4 (October 1995), 1075–1076.

Suggested Reading

Angeli, D. J., et al., eds. *Sustaining Earth: Responses to the Environmental Threat.* New York: St. Martins, 1991.

Bertalmus, Peter. *Environment, Growth and Development: The Concepts and Strategies of Sustainability.* New York: Routledge, 1994.

Borelli, Peter, ed. *Crossroads: Environmental Priorities for the Future.* Covelo, CA: Island Press, 1988.

Brown, Lester R. "The Illusion of Progress." *State of the World 1990.* New York: W. W. Norton & Company, 1990, p. 8.

Brown, Lester R., Christopher Flavin, and Sandra Postel. "Outlining a Global Action Plan." *State of the World 1989.* New York: W. W. Norton & Company, 1989, pp. 174–175.

Brown, Lester R., Christopher Flavin, and Sandra Postel. "Picturing a Sustainable Society." *State of the World 1990.* New York: W. W. Norton & Company, 1990, pp. 173–174.

Campiglio, Luigi, ed. *Environment After RIO: International Law and Economics.* Netherlands: Kluwer Academic Press, 1994.

Carlin, Alan. *Environmental Investments: The Cost of a Clean Environment.* New York: Diane Publishers, 1994.

Clark, W. C. *Sustainable Development of the Biosphere.* New York: Cambridge University Press, 1986.

Collard, David, et al. *Economics, Growth, and Sustainable Environments.* New York: St. Martin's Press, 1988.

Costanza, Robest, ed. *Ecological Economics: The Science and Management of Sustainability.* New York: Columbia University Press, 1992.

Cummings, Charles. *Eco-Spirituality: Toward a Reverent Life.* New York: Paulist Press, 1991.

Daly, Herman E., and Clifford W. Cobb. *For the Common Good: Redirecting the Economy Toward Community.* Boston: Beacon Press, 1989.

Doob, Leonard W. *Sustainers and Sustainability: Attitudes, Attributes, and Actions for Global Survival.* New York: Greenwood, 1995.

Hawken, Paul. *Ecology of Commerce: A Declaration of Sustainability.* New York: Harper, 1994.

Henning, Daniel H., and William R. Manguin. *Managing the Environmental Crisis.* Durham, NC: Duke University Press, 1989.

Hickey, James E., and Linda A. Longmire. *Environment: Global Problems, Local Solutions.* New York: Greenwood, 1994.

Kristensen, Thorkil, and Johan Peter Paludan. *The Earth's Fragile Systems: Perspectives on Global Change.* Boulder, CO: Westview Press, 1988.

Miller, G. Tyler, Jr. *Environment: Problems and Solutions.* Cincinnati: Thomson International Press, 1994.

Miller, G. Tyler, Jr. *Sustaining the Earth: An Integrated Approach.* Cincinnati: Thomson International, 1994.

Pearce, David W., et al. *Sustainable Development: Economics and Environment in the Third World.* Brookfield, VT: Gower Publishing Company, 1990.

Postel Sandra. "Accounting for Nature." *World Watch.* 4, no. 2 (March–April 1991), 30–32.

Princen, Finger. *Environmental NGOs in World Politics: Linking the Global and the Local.* New York: Routledge, 1994.

Przeworski, Adam. *Sustainable Democracy.* Cambridge: Cambridge University Press, 1995.

Redclift, Michael. *Sustainable Development: Exploring the Contradictions.* New York: Routledge, 1989.

Timberlake, Lloyd. *Only One Earth: Living for the Future.* New York: Sterling, 1987.

Tisdell, Clem. *Environmental Economics: Policies for Environmental Management and Sustainable Development.* London: Ashgate Publishing Co., 1993.

Turner, R. Kerry, ed. *Sustainable Environmental Economics and Management: Principles and Practices.* New York: Wiley, 1993.

United States Environmental Protection Agency. *Reducing Risk: Setting Priorities and Strategies for Environmental Protection.* Washington D.C.: U.S. Government Printing Office, 1990.

Uno, Kimo. *Environmental Options: Accounting for Sustainability.* Netherlands: Kluwer Academic Press, 1995.

Young, John. *Sustaining the Earth.* Cambridge: Harvard University Press, 1992.

Chlorofluorocarbons

Chlorofluorocarbons (CFCs) were invented in the 1930s and became widely used in a variety of products because they were chemically stable, low in toxicity, and nonflammable. They were, in many ways, an ideal chemical that had many uses. They became the leading heat transfer agent in refrigeration equipment and air-conditioning systems for buildings and vehicles. CFCs were used in the manufacture of various kinds of foam, including building insulation, and as solvents and cleaning agents in semiconductor manufacturing and other businesses. They were once widely used as propellants in aerosol containers, although they were banned for this use in the United States several years ago, while in Europe and Japan they continued to be used for this purpose.

Du Pont became the world's largest manufacturer of CFCs, receiving $600 million in 1987 in revenues from this business. Du Pont was the only firm that produced CFCs in all three major markets that included the United States, Europe, and Japan. In 1985, Du Pont produced 882 million pounds of CFCs in these three markets, with 706 million pounds being produced in the United States alone. U.S. sales of CFCs peaked in 1973, when concern began to develop about depletion of the ozone layer. Aerosol uses accounted for about half of CFC consumption in the United States, and when this use was banned in 1978 because it was a nonessential use of the product, consumption of CFCs fell by 50 percent, but began climbing slowly back toward mid-1970 levels as more uses were found and demand in nonaerosol sectors grew.

As concern about CFCs and their role in depleting the ozone layer mounted, Du Pont led producers and users in opposing CFC regulation, citing scientific uncertainty as the primary reason for this opposition. In the absence of regulation, CFC use was expected to grow at about the level of GNP in the industrialized world, and somewhat faster in developing countries. All of these forecasts changed with the signing of the Montreal Protocol in 1987, where virtually all of the world's industrial nations agreed to cap production. The countries who signed the Protocol could decide for themselves how to allocate production among CFC compounds and among producers, subject to an overall ceiling for each country that was expressed in terms of ozone depletion potential.

While opposing regulation because of scientific uncertainty, Du Pont also made a public promise to change its position if the scientific case against CFCs should solidify. Subsequent findings led Du Pont to change its position. The company stated that "it would be prudent to limit worldwide emissions of CFCs while science continues to work to provide better guidance to policymakers." But only international action would be effective, the company argued, because unilateral action by the United States would provide an excuse for other nations to delay regulating their own producers. Thus Du Pont supported ratification of the Montreal Protocol.

The issuance of the NASA Executive Summary Report on ozone depletion caused the company to reassess its position again. This report described a fundamental change in the scientific understanding of the CFC-ozone connection. It presented hard evidence of reductions in stratospheric ozone concentrations over temperate populated regions as well as Antarctica, firmly established the link between CFCs and ozone depletion, and suggested that implementation of the Montreal Protocol would result in little net depletion of ozone and that continuing ozone decreases were expected even if the Protocol were implemented.

Soon after this report was released, Du Pont decided to stop production of CFCs altogether. Because of difficulties in developing substitutes and obtaining regulatory approval to produce them, this exit would be phased in over a ten-year period. But by 1999, when the Protocol would require a cutback to 50 percent of 1986 levels, Du Pont planned to stop manufacturing regulated CFC compounds entirely. The company received widespread accolades for this announcement, while competitors either kept silent or commented that they were still reviewing the evidence and were not at all sure that such drastic action was warranted.

1. Did Du Pont do the right thing with regard to this controversy? Did it act responsibly to shareholders? Why did Du Pont decide to do more than the Protocol actually required? What was its decision based on?
2. How shall companies deal with controversies of this nature? Are international solutions necessary for these kinds of global environmental problems? What role can companies play in bringing about such international agreements?

SOURCE: Rogene A. Buchholz, *Business Environment and Public Policy: Implications for Management,* 5th ed. (Englewood Cliffs, NJ: Prentice Hall, 1995), pp. 646–647.

Ethylene Dibromide

A major scandal over ethylene dibromide (EDB) erupted in late 1983. The chemical was widely used as a soil fumigant for crops and as a fumigant for citrus and other fruits and vegetables after harvest as well as an antiknock agent in gasoline. It was also used to protect stored wheat, corn, and other grains against destruction by insects and contamination by molds and fungi, and as a fumigant to keep milling machinery free of insects. EDB proved useful to keep grain supplies free from insect fragments, insect excrement, and mold toxins and thus made a major contribution to the safe storage of grain products over extended periods.

Ethylene dibromide had been produced in the United States since the 1920s. It was used primarily as a lead scavenger in leaded antiknock gasoline additives. During the 1950s and 1960s, federal pesticide registrations for uses of EDB as a fumigant were granted under the Federal Insecticide, Fungicide and Rodenticide Act (FIFRA). EDB has been registered as a pesticide since 1948, and thus became subject to EPA regulations after the agency was formed and given responsibility for pesticide control.

Eventually the chemical was tested to be an animal carcinogen, linked to cancer and reproductive disorders. Many scientists, however, were confident that the minute levels of EDB found in grain-based products posed no imminent health hazard. Roughly 350 million pounds of EDB were produced annually in the United States of which about 245 million pounds were used domestically. About 230 million pounds (about 93 percent) of this amount was used as an additive in leaded gasoline. The remaining 7 percent, which was about 15 million pounds annually, was used for various agricultural activities.

On September 30, 1983, the EPA ordered immediate suspension of the use of EDB for soil fumigation. Soil fumigation accounted for over 92 percent of the more than 20 million pounds of EDB used annually as an agricultural pesticide. Appeals to this suspension could be made, but use of the product ceased during the appeal process. The September 30 decision also proposed immediate cancellation of EDB fumigation of stored grain, grain milling machinery, felled logs, and proposed cancellation of quarantine use on citrus and other tropical fruits to control the fruit fly effective September 1984. Minor uses of EDB were retained with added safety precautions and the requirement that applicators be trained and certified. Appeals were made to have administrative hearings for each use the EPA proposed to cancel.

Shortly after the EPA announced its decision to ban EDB, the state of Florida began testing food products for EDB residues. Dr. Stephen King, the state's health officer, decided that if EDB could be detected at a level of one part per billion, the food should not be sold. As a result of this discovery, on February 3, 1984, the EPA Administrator, William D. Ruckelshaus, announced the immediate emergency suspension of EDB for use as a fumigant for treating stored grain and milling machinery and recommended residue levels for grain and grain-related products already in the "pipeline" to protect the public from EDB contamination.[1]

Ruckelshaus said that the emergency suspension of EDB use as a grain fumigant will result in the "clearing of EDB from the food pipeline in this country. I firmly believe that the guidelines we are recommending today are fully protective of public health. I expect the residue levels on all grain products will begin to decline almost immediately as a result of the actions we are announcing today. In fact, in the very near future, that rate of decline should become quite pronounced."[2]

The EPA recognized its responsibility to take action at the federal level to deal with EDB residues in a way that would protect public health. The balance between the risks and benefits of EDB usage changed in recent years due to increased concern about potential health effects and the development of alternatives for most major uses of EDB. Based on a detailed evaluation of health effects and economic benefits, the EPA concluded that for all the major uses of EDB as a pesticide, the risks of continued usage outweighed the benefits and thus registrations of EDB for use as a pesticide should be canceled.[3]

1. How could the public be affected by the severe contradictions among the opinions on the health effects of EDB of a large number of respectable scientists and organizations in charge of protecting public health? What might be the consequences of these differences in terms of public trust in scientists and federal agencies?
2. Given the fact that the public has a right to know the truth about the safety of products that it is consuming, particularly if a cancer issue is involved, how much information and what type of information should be disclosed to the public? Are there indisputable facts in a highly complicated controversy such as the EDB situation? Where do values enter into the discussion? To whom should the public listen?

3. Since evidence about EDB was not conclusive and the issue was not clear-cut, could government agencies be blamed for being conservative and being on the safe side? Is it acceptable to say that safety comes first when there is a reasonable doubt about the safety of a product?
4. How should trade associations, the industry in general, and individual companies conduct themselves in this kind of situation? Should there be any self-imposed constraints on promoting and marketing such a substance? What precautions, if any, should be taken? How should responsibilities to stockholders and consumers be weighed?

Case Notes

1. Office of Public Affairs, United States Environmental Protection Agency, "EDB Facts," Washington, D.C., February 3, 1984, pp. 2–3.
2. Ibid., p. 3.
3. Testimony of Edwin L. Johnson before the Senate Committee on Agriculture, Nutrition, and Forestry, Orlando, Florida, January 23, 1984.

SOURCE: Rogene A. Buchholz, *Business Environment and Public Policy: Implications for Management,* 5th ed. (Englewood Cliffs, NJ: Prentice Hall, 1995), pp. 45–46.

The Forgotten Dumps

The Pinelands National Reserve covers parts of seven counties in southern New Jersey. The reserve is a million-acre wilderness of prime forests, cedar swamps, tidal creeks, and cranberry bogs.[1] Beneath the reserve lies the pristine waters of the Cohansey Aquifer, said to be the largest on the East Coast. The aquifer contains about 17 trillion gallons of water, the equivalent of a 2,000-square-mile lake with a uniform depth of 37 feet.[2] This aquifer supplies drinking water for much of South Jersey, as well as supporting the region's fragile wetlands ecology, its blueberry and cranberry industries, and its costal estuaries.[3]

Because it was remote and sparsely populated, the Pinelands was also popular as a dumping ground for toxic wastes. In 1984, the region contained at least 7 industrial lagoons and 43 known landfills, 17 of which had been closed since 1980 by order of the Pinelands Commission or the State Department of Environmental Protection. From 1976 to 1979, 60,000 gallons of hazardous chemicals were spilled in the Pinelands—either deliberately or accidentally—in 41 separate incidents. Eventually, these wastes began to cause serious groundwater contamination problems and threatened the pristine quality of the water.[4]

Tests conducted through 1984 concluded that groundwater at more than a dozen sites in the Pinelands contained contaminants, but that pollutants had not spread beyond the immediate areas of those dumps. But if the dumps were not cleaned up soon, the pollutants were expected to spread to other areas, particularly if large amounts of water were pumped from the aquifer to supply other cities. The groundwater movements caused by this pumping could disperse the contaminants now concentrated near the dump sites.[5]

Two abandoned chemical dumping sites surfaced during 1979 in the center of Pinelands within the headwaters of the Wading River used to irrigate the many cranberry bogs in the Pinelands. Piles of rusted out drums were strewn about each of the sites, and puddles of colored liquids were visible. The dumps were closed in the early 1960s, but odors of chemicals lingered in many sections of the dumps that were potent enough to give a visitor a slight headache.[6] The nearest home was more than mile away, so there was little fear of a public health danger. The most severe threat was initially believed to be to cranberry production, which was the biggest industry in the area. Cranberries are not known to be particularly sensitive to water quality, and the impact on the cranberry bogs was not immediately known.

Samples of the chemicals found at the site were sent to the State Department of Environmental Protection for analysis. Scattered throughout one of the sites were patches of an asphalt-like substance found to contain carcinogens such as benzene, dimethylphenol, and other similar substances. More than 30 other chemicals were also found in various areas of the site. Groundwater tests indicated that if pollution had entered the ground, it had not yet spread to adjacent areas. At the other site, there were standing trenches of drum-laden water that were oddly colored and thickly coated with sticky tar along the bottom. One of the wells at this site was found to contain concentrations of several different agents, some of which were known toxins, and many of which were suspected of being carcinogenic. Thus groundwater below the site was believed to be highly contaminated, but the contaminant plume was not detected in off-site wells. However, the Cohansey Aquifer is only 30 to 40 feet below the surface in that area and runs very deep.[7]

The chemicals were believed to have originated with the 3M Company and Rohm and Haas as well as the Hercules Company. None of these companies owned or operated the sites, but it was believed that waste material from the 3M plant in Bristol Township and a Rohm and Haas plant in Bristol, Pennsylvania, were dumped at the sites.[8] These firms indicated that Mr. Rudy Kraus of Industrial Trucking in Bucks County was hauler and operator of the sites. Mr. Kraus hauled wastes from their plants over a ten-year period beginning in 1950 and ending in the 1960s. Identification of the companies involved in dumping at the sites was important to the Department of Environmental Protection's investigation because these companies would be in the best position to know what was buried there and have the expertise to help neutralize whatever dangers exist.[9] Thus the first step in cleaning up these dumps was to notify the potentially responsible parties to request information on the types and quantities of wastes sent to the Woodland sites.

The 3M Corporation received a letter dated July 19, 1983, from Mr. John F. Dickinson with the Office of Regulatory Services in the Department of Environmental Protection of the State of New Jersey. The letter was addressed to Mr. Lewis W. Lehr, chairman of the board and chief executive officer of 3M Corporation. The letter indicated that the company had utilized the services of Rudolph Kraus to haul waste material from its plant in Bristol Township in Pennsylvania to the routes 72 and 532 Woodland Township dump sites. The letter stated that testing of the sites had disclosed that the soil and groundwater had been substantially contaminated by hazardous chemical compounds leaching from the disposed material.[10]

The letter mentioned that the sites sit directly on top of the Cohansey Aquifer and at the headwaters of the Wading River, which was stated to be the source of irrigation waters for the numerous cranberry farms in the area. In order to undertake remedial action at the sites, Mr. Dickinson asked 3M to supply the state of New Jersey with information about its disposal operations, including the types and quantities of waste placed with Mr. Kraus for disposal. The company replied to this letter and stated that a review had been made of files and records regarding operations at the 3M plant in Bristol, but that no records had been found pertaining to the disposal sites in Woodlands Township. Nonetheless, the company did put together some information based on the recollections of former 3M employees. The types and quantities of waste were estimated based upon the product mix manufactured during the time Mr. Kraus would have hauled wastes from the Bristol plant.[11]

The letter mentioned that the company understood that the organic liquid waste was burned to facilitate recovery of the drums. These drums were sold to junkyards for scrap. The letter mentioned that the company would cooperate with the state of New Jersey regarding this matter. A copy of the letter went to Mr. Brian Davis, senior attorney with the Office of General Counsel for 3M Corporation. It mentioned a meeting that would be held with Mr. Dickinson on September 20 of that year.

The New Jersey Department of Environmental Protection (NJDEP) did call a meeting in Trenton for all the potentially responsible parties (PRPs), which at that time included three major dischargers, two minor dischargers, one transporter/operator, and two landowners. The NJDEP stated its concern about the potential environmental effects of waste material discharged at the sites, and called for the PRPs to conduct an investigation and begin remedial action to clean up the sites. If the PRPs did not do so, the NJDEP would carry out this action itself and seek treble damages from the PRPs under the provisions of the Spill Compensation and Control Act. The PRPs agreed to cooperate with the state and requested time to meet and prepare a response to the state's request.

1. What threats do toxic waste dumps pose for the environment and human health? Why were these threats not apparent when the dumps were first created? What happened to change perceptions regarding the disposal of toxic waste material?
2. Why didn't companies keep records of what waste material they disposed of and in what quantities? What ethical responsibilities did they have to dispose of their waste material safely? Were there any market incentives for them to pay attention to waste disposal?

Case Notes

1. Marc Duvoisin, "The Tainting of the Pinelands' Most Precious Resource," *The Philadelphia Inquirer,* January 16, 1984, p. 4B.
2. Brett Skakun, "Chemical Dumps Found in Pines," *Atlantic City News,* April 29, 1979, p. 1.
3. Duvoisin, "Tainting of the Pinelands," January 16, 1984, p. 4B.
4. Ibid.
5. Ibid.
6. Skakun, "Chemical Dumps," p. 1.
7. Ralph Siegel, "Two Dumps in Woodland Pushed for Superfund Aid," *Burlington County Times,* November 19, 1982, p. 1.
8. Ibid.
9. Tony Muldoon, "Pine Barrens' Breeze Could Be Lethal," *Camden Courier-Post,* April 27, 1979, p. 3.

10. Letter from John F. Dickinson, Office of Regulatory Services, Department of Environmental Protection, State of New Jersey, July 19, 1983.
11. Letter from Russell H. Susag, Director, Environmental Regulatory Affairs, 3M Corporation, St. Paul, Minnesota, September 1, 1983.

SOURCE: Adapted from Rogene A. Buchholz et al., *Managing Environmental Issues: A Casebook* (Englewood Cliffs, NJ: Prentice Hall, 1992), pp. 130–40.

Bhopal

During the evening of Sunday, December 2, 1984, an incident happened in Bhopal, India, that has been called the worst industrial accident in history. The first sign that something was wrong came shortly before midnight, when a worker at the Union Carbide pesticide plant on the outskirts of Bhopal (pop. 672,000) noticed that pressure was building up in a tank that contained 45 tons of methyl isocyanate (MIC), a deadly chemical used to make pesticides. Pressure in the tank continued to build until sometime after midnight when the highly volatile and highly toxic MIC began to escape from the tank into the surrounding atmosphere. The escaping gas overwhelmed inadequate and reportedly out-of-commission safety backup systems and spread in a foglike cloud over a large and highly populated area close to the plant.[1]

Early reports indicated that 2,500 people had died and at least another 1,000 were expected to die within a two-week period. Some 150,000 people were said to have been treated at hospitals and clinics in Bhopal and surrounding communities. Most of the deaths were caused by the lungs filling up with fluid, causing the equivalent of death by drowning. Other people suffered heart attacks. Some of the survivors were permanently blinded, others suffered serious lesions in their nasal and bronchial passages. Doctors also noticed concussions, paralysis, and signs of epilepsy. Six days after the accident, it was reported that patients were still arriving at Hamidia Hospital in Bhopal at the rate of one a minute, many of them doubling over with racking coughs, gasping for breath or convulsed with violent spasms.[2] While there was some dispute over the exact number of victims that were killed and injured by the cloud, it is clear that the disaster was of significant proportions.[3]

Methyl isocyanate is a colorless chemical compound used by Union Carbide as an ingredient in producing relatively toxic pesticides known as Sevin and Temik. Isocyanates in general are reactive and resemble aldehydes and ketones in their propensity to undergo additional reactions with a variety of compounds containing active hydrogen atoms. This reactivity makes them useful as chemical intermediates but also makes them tricky to handle.[4] MIC in particular will react with many compounds including water. At room temperature, the MIC water reaction starts slowly, but the reaction produces heat, and if the reaction continues long enough, the MIC will start to boil violently and build up pressure.[5]

Because of its hazardous properties, great precautions were taken in handling and storing the chemical. At the Bhopal facility, MIC was stored in three double-walled stainless steel tanks, buried mostly underground to limit leakage and shield them from outside air temperatures. These tanks were refrigerated to keep the highly volatile gas in liquid form, and also equipped with thermostats, valves, and other devices to warn when the temperature of the chemical exceeded the boiling point. The Bhopal plant had two safety devices that were supposed to operate automatically in case a tank ruptured and the gas started escaping. The first was a scrubber that would neutralize the highly reactive gas by treating it with caustic soda. If the scrubber failed to do its job sufficiently, another mechanism was supposed to ignite the gas and burn it off in the air harmlessly before it could do much damage.[6]

Union Carbide was first incorporated in India some 50 years ago when it began manufacturing batteries. The Indian subsidiary was allowed to stay on after India won independence from Britain and is one of the few firms in India in which the parent company is permitted to hold a majority interest. Union Carbide owned 50.9 percent of the Bhopal facility. The Indian government has long favored Union Carbide because of its interests in developing sophisticated industry and in promoting the "Green Revolution" in agriculture. Pesticides are an important ingredient in this revolution, which is important to India because of its huge population, much of which is very poor by U.S. standards.[7]

The Bhopal plant was built in 1969 with approval from the local authorities and the blessing of the national government. When the plant was first built, it was located just outside the city limits in an open area, but by the time an expansion program got underway six years later, the area between the town and the plant had begun to be settled by squatters. Many of them were attracted by the roads and water lines that accompanied the plant. In 1975, the administrator of the municipal corporation asked that the plant be removed because of potential dangers to the people living nearby. Instead, the administrator was removed from office and the plant remained.[8]

The deadly cloud had hardly dissipated before lawyers became involved. Five American attorneys, including Melvin M. Belli, filed a class action suit against Union Carbide on behalf of the victims seeking $15 billion in damages. The suit sought to represent all those who were injured or who lost relatives as a result of the disaster and claimed that the corporation was negligent in designing the Bhopal plant and that it failed to warn the area's residents about the dangers presented by the stored chemical. The suit charged that Union Carbide acted "willfully and wantonly" with utter disregard for the safety of Bhopal residents. This charge was partly based on the allegation that the Bhopal plant lacked a computerized early-warning system that had been installed in the company's plant in Institute, West Virginia, which was supposedly identical to the Bhopal facility.[9]

The Indian government itself got into the legal picture and further complicated matters. Carbide had been trying to work out a settlement with the government and according to some sources had offered the government an immediate $60 million and further $180 million over the next 30 years as compensation for the victims of the disaster.[10] Whatever the amount was, it was rejected by the Indian government when it filed suit on April 8 in New York on its own behalf. The suit charged that Union Carbide was liable for any and all damages arising from the poison gas leak, but because of the enormity of the disaster, the government wasn't able to allege with particularity the amount of compensatory damages being sought. The suit also sought punitive damages "in an amount sufficient to deter Union Carbide and any other multinational corporation from the willful, malicious and wanton disregard of the rights and safety of the citizens of those countries in which they do business."[11]

The suit claimed that Union Carbide was negligent in designing and maintaining the Bhopal plant and that the company made false representations to the government about the plant's safety. The company is alleged to have encouraged the storing of MIC in "dangerously large quantities," failed to equip the storage tanks with alarm devices and temperature indicators, and didn't provide "even basic information" about appropriate medical treatment for exposure to the chemical.[12]

Thus the legal problems for Carbide mounted and it faced a great deal of uncertainty regarding the eventual outcome of the situation. The greatest uncertainty, of course, involved the amount of eventual compensation awarded to the victims of the tragedy. With a reported $200 million in insurance coverage, Carbide could probably weather a settlement in the range of $250 million to $300 million, but anything near $500 million would hurt the company's performance. If the final settlement were to be paid out over a period of time, this would substantially reduce the immediate burden for the company.[13]

John Tollefson, dean of the School of Business at the University of Kansas, stated that three things are crucial to successful management of such a crisis situation as the tragedy at Bhopal: (1) Executives must give long-range considerations priority over short-term costs and benefits, (2) action must be taken immediately, and (3) truthful information must be provided to the public from the beginning. The worst thing that could happen in a situation like Bhopal, added John D. Aram, professor of Management at Case Western Reserve University, "is that (executives) get into a bunker mentality where assumptions get frozen and alternatives get closed down instead of opened up."[14]

While Union Carbide was conducting its own investigation of the incident, an inquiry was conducted by the government of India that identified a number of design flaws, operating errors, and management mistakes that helped cause the accident. Plant safety procedures were said to be inadequate to deal with a large-scale leak of the deadly MIC, despite the fact that the dangers such a leak would pose were well known. Nor had any precautions been taken to protect people living near the plant. No procedures were developed for alerting or evacuating the population that would be affected by an accident.

In addition, some important safety systems were not working at the time of the accident. Refrigeration units designed to keep MIC cool so that it could not vaporize had been shut down before the accident. Other equipment, including devices designed to vent and burn off excess gases, was so inadequate that it would have been ineffective even if it had been operating at the time of the accident. Finally, plant workers failed to grasp the gravity of the situation as it developed, allowing the leak to go unattended for about an hour. Brief and frantic efforts to check the leak failed. As the situation deteriorated, the workers panicked and fled the plant.[15]

Finally, the company issued its own report about what caused the disaster. The company report stated that it believed that the accident resulted from a large amount of water entering a storage tank and triggering a chemical chain reaction. The report stated that the water was put in the tank either "inadvertently or deliberately," but the chairman of the company couldn't say that it was an act of sabotage. One possible source of the water was a utility station where a pipe marked water was located next to one marked nitrogen that was used to pressurize the tank. Quite possibly, someone connected the wrong pipe to the tank, allowing as much as 240 gallons of water to mix with the MIC in the tank. The report covered an investigation of nearly three months by a team of company scientists and engineers who conducted about 500 experiments to determine the technical aspects of the accident.[16]

The report also showed that the Bhopal plant was ill run, violated a number of standard operating procedures, and failed to maintain safety devices. Conditions were so poor at the time of the disaster, that the plant "shouldn't have been operating," according to Warren M. Anderson, the chairman of Union Carbide. The report confirmed that the scrubber unit intended to neutralize the escaping gas wasn't operating prior to the accident, and that another safety device, a flare tower, also wasn't operating because it had been shut down for maintenance.[17]

1. Is there anything a company can do about citizens who choose to live near its facilities? Should companies be required to provide a "buffer zone" between their plants and the nearest residents? Should companies actively participate in the preparation of an evacuation plan for communities in the event a disaster like the one at Bhopal occurs? What moral and legal implications do these questions involve?
2. Are there some products that are simply too toxic and hazardous to handle and thus shouldn't be produced? Can the EPA administrator ban such substances from being produced in this country? If so, on what grounds? What about foreign nations? Did the safety systems at the Bhopal plant appear to be adequate? Can safety systems be designed and operated to prevent such accidents from happening?
3. How much is a life worth? What is adequate compensation for an ongoing illness that was the result of the Bhopal accident? How do courts usually deal with these issues? Is the present value of future earnings an adequate way to figure compensation for the family of a deceased wage earner?
4. Was Union Carbide, the parent company, responsible and liable for the actions of its Bhopal subsidiary? What criteria would you use to answer this question? What are the implications of your answer for multinationals in general? Under what circumstances should they continue to be shielded from the actions of their foreign subsidiaries?
5. How can technology be better managed so that safety systems aren't shut down or not working as apparently was the case at the Bhopal facility? What kind of controls can management institute so that it can be assured on a day-to-day basis that all safety systems are up and operating properly? Is such a goal realistic? Is safety a management problem or a technical problem for the engineers?
6. Is more regulation the answer to controlling the use of chemicals and other hazardous substances? Do communities have the right to know what is being produced in their back yard so they can take appropriate action to protect themselves in the event of an accident? Should such a national right-to-know be passed or can corporations be relied on to do the "right" thing in this regard?

Case Notes

1. "India's Night of Death," *Time,* December 17, 1984, p. 22.
2. Ibid., pp. 22–23.
3. "Carbide Lawyer Says Number of Bhopal Victims Overstated," *Dallas Times Herald,* April 18, 1985, p. 28A.
4. Ward Worthy, "Methyl Isocyanate: The Chemistry of a Hazard," *Chemical and Engineering News,* 63, no. 6 (February 11, 1985), 27.
5. Ibid., pp. 27–28.

6. "India's Night of Death," p. 25.
7. Ibid., p. 26.
8. Ibid.
9. "Union Carbide Fights for Its Life," *Business Week,* December 24, 1984, pp. 53–56.
10. "India's Bhopal Suit Could Change All the Rules," *Business Week,* April 22, 1985, p. 38.
11. Roger Friedman and Matt Miller, "Union Carbide Is Sued by India in U.S. Court," *Wall Street Journal,* April 9, 1985, p. 14.
12. Ibid.
13. Matt Miller, "India Lifts Ban on Carbide Plan for Asset Sales," *Wall Street Journal,* December 1, 1986, p. 2.
14. Ron Winslow, "Union Carbide Mobilizes Resources to Control Damage From Gas Leak," *Wall Street Journal,* December 10, 1984, p. 29.
15. "Frightening Findings at Bhopal," *Time,* February 18, 1985, p. 78.
16. Barry Meier, "Union Carbide Says Facility Should Have Been Shut Before Accident," *Wall Street Journal,* March 21, 1985, p. 3. Union Carbide raised the possibility of sabotage again later in an apparent effort to lower expectations about the size of a final settlement. See Barry Meier, "Carbide's Bhopal Sabotage Claim Seen by Some as Effort to Shape Settlement," *Wall Street Journal,* August 15, 1986, p. 5.
17. Ibid.

SOURCE: Adapted from Rogene A. Buchholz et al. *Management Responses to Public Issues,* 3rd ed. (Englewood Cliffs, NJ: Prentice Hall, 1994), pp. 301–322.

The Exxon Valdez

The Exxon supertanker *Valdez* entered the port of Valdez on March 22, 1989, riding high in the water because its huge cargo chambers were empty. Tugs guided it into the dock at Berth 5 at the Alyeska oil terminal. Alyeska was the name given to a consortium of oil companies that had been formed to operate the terminal. The tanker, only two years old and built in the San Diego shipyards, cost $125 million. It was one of the best equipped vessels that hauled oil from the port of Valdez, having collision avoidance radar, satellite navigational aids, and depth finders.[1]

The commander of the *Valdez* was Captain Joseph Hazelwood of Huntington, New York, a 20-year veteran of Exxon and commander of the supertanker for 20 months. In 1985, he had been convicted of drunken driving in Long Island, New York, and was again found guilty of driving while intoxicated in September of 1988 in New Hampshire. In the span of five years, his automobile driver's license was revoked three times. He informed the company about his drinking problem in 1985, and Exxon immediately sent him to an alcohol rehabilitation program. The company claimed that it was not aware that his drinking problem persisted after he left the treatment program, however, at the time of the incident, Hazelwood apparently was still not permitted to drive a car but retained his license to command a supertanker.[2]

On Thursday, March 23, the ship was eased out into the harbor by the port pilot, a customary practice in most shipping facilities. The pilot apparently noticed alcohol on Hazelwood's breath, but noticed no impairment of the captain's judgment or faculties.

Thus he turned over command of the ship to Hazelwood and descended over the side of the tanker to a waiting pilot boat. The tanker increased its speed to 12 knots and entered the more open water of Prince William Sound. It was the 8,549th tanker to safely negotiate the Valdez narrows since the first tanker left Valdez fully loaded in August 1977. No serious accidents had happened during that time.[3]

There were icebergs, however, in the outgoing lane, and the ship radioed the Coast Guard for permission to steer a course down the empty incoming lane to avoid the icebergs. This permission was granted and the *Valdez* altered course. At some point after the pilot left the ship, Hazelwood left the command post and went below to his cabin, in violation of company policy, which requires the captain to stay in command of the ship until it is in open water. Third Officer Gregory Cousins was left in charge of the ship, even though he lacked Coast Guard certification to pilot a tanker in Alaskan coastal waters. The ship was in trouble almost immediately, as it had set out on a course that would take it due south on a potential collision course with Busby Island, five miles away.[4]

The Coast Guard station on Potato Point had been tracking the ship, but had not noticed a potential problem because the ship disappeared from the screen for a while. Apparently the Coast Guard had replaced its radar unit two years previously with a less powerful unit that was unable to maintain contact with the ship and warn it of potential danger. During this time, the ship rode over submerged rocks off Busby Island and minutes later plowed into Bligh Reef and began dumping its cargo. The reef had torn 11 holes in the ship's bottom, some as large as 6 by 20 feet. Eight cargo holds that were big enough to swallow 15-story buildings were ruptured. While a command had been given to change the course of the ship to avoid disaster, it came too late to have effect. At 12 knots, it took about half a mile for any rudder change to alter the course of the 987-foot ship substantially.[5]

The Coast Guard station in Valdez was notified of a vessel run aground about 12:28 A.M. on Friday. About 1 A.M. a Coast Guard pilot boat headed for the accident site, following a tugboat that had already been dispatched. At 3:23 A.M. they arrived on the site and saw that the ship was losing oil at a rate that was later reported to be 1.5 million gallons an hour. At 5:40 A.M. it was reported that the *Valdez* had lost 210,000 barrels of oil, or more than 8.8 million gallons. There are 42 gallons in a barrel of oil, which is the standard industry measure. Spotters aboard an Alyeska plane reported at 7:27 A.M. that the oil slick was 1,000 feet wide and 5 miles long and was spreading. Earlier a passing boat had reported encountering an oil slick about half a mile south of Bligh Reef. Later it was estimated that the *Valdez* had released about 240,000 barrels of oil equivalent to 10.1 million gallons into the sound.[6]

After Hazelwood's blood was tested fully nine hours after the ship ran aground, he still had a blood-alcohol level of .06, which is higher than the .04 the Coast Guard considers acceptable for captains. It was estimated that his blood-alcohol level at the time of the accident was about .19, assuming that he had not had anything more to drink after the accident and that his body metabolized at the normal rate. This level of .19 is almost double the amount at which most states consider a motorist to be legally drunk. After it learned of these test results, Exxon fired Hazelwood and the state filed criminal charges against him for operating a ship while under the influence of alcohol, reckless endangerment, and criminally negligent discharge of oil. The maximum penalty for the combined charges was 27 months in jail and a $10,000 fine. The state also issued a warrant for his arrest.[7]

After one day, the slick was eight miles long and four miles wide, and was clearly the worst spill in U.S. history. By the end of the week the slick covered almost 900 square miles to the southwest of Valdez, threatening the marine and bird life in the sound and spreading to the Chugach National Forest. On Thursday afternoon, the slick began taking its greatest toll of wildlife when oil began washing up on the beaches on Knight and Green islands. Scientists found many blackened animals huddled or dead on the beaches. Scores of mormorants and other birds were barely distinguishable from the oil-covered sand and gravel.[8]

The oil slick continued to spread, covering more than 1,000 square miles and hitting hundreds of miles of inaccessible beaches and drifting into the Gulf of Alaska where it threatened the port of Seward and the delicate shoreline of Kenai Fjords National Park. The area covered by the spill was said to be larger than the state of Rhode Island. Eventually the slick spread 100 miles out into the Gulf of Alaska, forcing federal officials to open a second front in their battle to contain its advancement. Scientists estimated that about half the oil lost by the *Valdez* had left Prince William Sound and had entered the gulf creeping south at about 15 to 20 miles a day.[9]

One of the first effects of the accident was to postpone indefinitely the fishing season for shrimp and sablefish, to which fishermen reacted bitterly. There was also some question as to whether the season for herring roe would also have to be canceled because of the spill. Herring roe, which are really eggs, are considered a delicacy in Japan and bring high prices that gives fishermen an economic boost to carry them through to the summer salmon season.[10] The herring lay their eggs on floating kelp beds that fishermen feared would be smothered by the oil slick. Millions of salmon fingerlings from the hatcheries were scheduled to be released into the sound's inlets to begin a two-year migration cycle. These fingerlings feed on plankton that may be poisoned by the oil thus beginning a contamination that would continue up the food chain. Clams and mussels were expected to survive, but hydrocarbons will probably accumulate in their body tissues, which would endanger any species that feeds on them.[11]

Before long, waterfowl by the tens of thousands would finish their northward migrations and settle in summer nesting colonies in the sound. More than 200 different species of birds were reported to be in the sound, and some 111 of them are water related. The Copper River delta, which is at the east end of the sound, is home to an estimated 20 million migratory birds, including one fifth of the world's trumpeter swans. It was later estimated that thousands of sea birds such as cormorants and loons died either because oil destroyed their buoyancy or because they were simply poisoned.[12]

Emergency teams sent out to clean up the oil found ducks coated with crude and sea lions with their flippers drenched with oil clinging to a buoy located near the damaged tanker. Environmentalists feared that a significant part of the sound's sea otter population of 12,000 would be totally wiped out by the spill. Sea otters die of hypothermia when their fur becomes coated with oil. They may also sink under the surface of the water and drown. Thus many different kinds of animals were placed under threat by the spreading oil slick.[13]

The long-term effect of the spill could be to change the balance of power between the oil industry and the environmentalists. The latter lost no time in getting their message across regarding oil and gas exploration on the North Slope. While they were unable to prevent development of the North Slope fields, the *Valdez* disaster gave them new ammunition in the fight against opening up the Arctic National Wildlife Refuge (ANWR) that lies between Prudhoe Bay and the Canadian border. As the name implies, the area teems with wildlife of all kinds, and the environmentalists want to keep the oil industry out of this preserve.[14]

Industry has lost the trust of the Alaskan people, who felt betrayed by believing the claims of the oil companies that they could protect the environment. People will be less likely to believe that the oil industry can develop the Arctic in a responsible manner. State lawmakers want assurances that current operations will not further harm the environment but are less likely to trust the oil companies to do this on their own without state regulation.[15] Federal officials began talking about stricter enforcement of existing laws as well as new requirements that tankers be equipped with double hulls for added protection. Other suggestions had to do with tougher personnel rules that would ban drunken drivers from commanding tankers, and proposals for updating the training standards for crews of tankers. Perhaps one of the most controversial proposals had to do with testing of employees for drug and alcohol abuse.[16]

1. Why was Alyeska not better prepared to deal with an oil spill of this nature? What happened over the years to create an attitude of complacency? How can companies guard against this attitude and keep themselves alert to potential accidents?
2. Did Exxon accept responsibility for the spill? In what ways? Was the bad coverage it received from the media deserved? How could Exxon have responded differently? How much should the spill be blamed on Hazelwood himself?
3. Is it possible for there to be a balanced approach to industrial development and preservation of the environment? Can industry and the environment live with each other in harmony? How can such a balance be achieved?
4. Who are the stakeholders in this incident? What rights do they have? What did Exxon owe each of them? How shall the rights of these stakeholders be balanced against each other? Do the animals whose lives were threatened have any rights?

Case Notes

1. "Disaster at Valdez: Promises Unkept," *Los Angeles Times,* April 2, 1989, p. I20.
2. "The Big Spill," *Time,* April 10, 1989, p. 39.
3. "Disaster at Valdez," p. I20.
4. Ibid.
5. Ibid., p. I22.
6. Ibid., p. I21.
7. "The Big Spill," p. 40.
8. Mark Stein, "FBI Starts Probe of Valdez Spill as Toll Mounts," *Los Angeles Times,* April 1, 1989, p. I1.
9. Larry B. Stammer and Mark A. Stein, "New Front Opened in Oil Spill Battle," *Los Angeles Times,* April 8, 1989, p. I23.
10. Mark A. Stein, "Arrest of Missing Tanker Captain Sought by Alaska," *Los Angeles Times,* April 2, 1989, p. I1.
11. Ken Wells and Marilyn Chase, "Paradise Lost: Heartbreaking Scenes of Beauty Disfigured Follow Alaska Oil Spill," *Wall Street Journal,* March 31, 1989, p. A1.
12. Ibid.
13. "Smothering the Waters," *Newsweek,* April 10, 1989, p. 57.

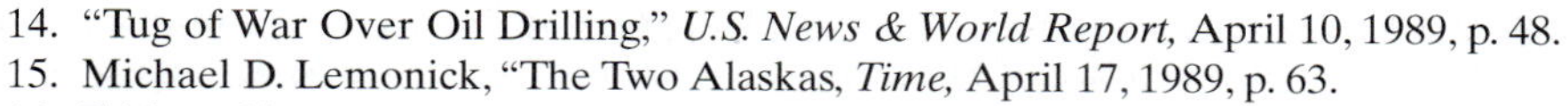

14. "Tug of War Over Oil Drilling," *U.S. News & World Report,* April 10, 1989, p. 48.
15. Michael D. Lemonick, "The Two Alaskas, *Time,* April 17, 1989, p. 63.
16. Ibid., p. 66.

SOURCE: Rogene A. Buchholz, *Public Policy Issues for Management,* 2nd ed. (Englewood Cliffs, NJ: Prentice Hall, 1992), pp. 292–94.

Oakdale

During the 1940s and 1950s, several companies, one of which was 3M, disposed of their wastes using the accepted practice of the day of hiring a contractor to dispose of waste material generated by its manufacturing plants. The contractor in this case disposed of the material in a 60-acre disposal site in an undeveloped, lowland area about 10 miles from St. Paul, Minnesota. The site consisted of an area for burning material, a drum recycling operation, and several waste disposal areas for both organic and inorganic residues.

As the amount and volume of waste increased and open burning was outlawed, the contractor dug trenches to dispose of the waste, and used other low areas in the site to try and keep up with the waste disposal activity. However, by 1960, most of the available disposal areas on the site were filled and the contractor ceased operations at this location. After the site was closed, it became covered with foliage and the property changed hands, which helped to obscure past usage. The state pollution control agency, however, learned about this site from some of the area residents in 1978, and conducted a preliminary investigation. The agency subsequently notified the potentially responsible parties in early 1980, including present and past landowners, transporters, and waste generators.

Later in that year, the potentially responsible parties, including 3M, were called to a meeting at the state pollution control agency to discuss a course of action. When no specific proposals were forthcoming, 3M pointed out the need for an immediate investigation of the site to determine the scope of the problem. The company offered to fund the investigation, and with the approval of the state pollution agency, hired Barr Engineering, a Minneapolis-based independent consulting firm, to conduct the investigation.

The results of the investigation showed that there were three separate disposal areas. These three sites were named the Abresch, Brockman, and Eberle Sites after the owners of the sites at the time of disposal activity. Based on the data collected during these investigations, Barr Engineering recommended a remedial action plan consisting of four major programs. The main objective of the remedial action plan was to remove the major sources of contaminants and prevent any future movement of contaminants into the deeper aquifers.

In July 1983, a detailed remedial action plan consisting of the four recommended programs of Barr Engineering was approved by 3M, the state pollution control agency, and the EPA. The key ingredient in the development and acceptance of the remedial action plan was the inclusion of provisions in the agreement that allowed the government agencies involved some degree of flexibility in modifying portions of the program as work progressed and more knowledge of the site and waste encountered were acquired. The basic philosophy of 3M in the agreement was to cooperate with the regulatory agencies and local government units and resolve issues on the basis of actual data and the operating history of the system. The company tried to keep politics to a minimum.

The entire cleanup effort was completed in 1985, which meant that only five years elapsed from the first notification to completion of the effort. It is useful to reflect on this experience with the Oakdale project to determine if there are any lessons to be learned and generalizations that can be developed that might apply to other situations. The success of this cleanup effort certainly indicates that under certain conditions, a waste dump can be cleaned up effectively and rather quickly compared with other cleanup efforts.

1. What was the accepted practice in the 1940s and 1950s regarding disposal of hazardous waste material? What problems has this practice caused? Is there any way these problems could have been avoided given the state of our knowledge at the time? Was it possible at that time to take more precautions?
2. What management factors made this cleanup a success story? Can the lessons learned from this situation be transferred to other waste sites? Are there some unique factors about Oakdale that make any generalizations suspect?

SOURCE: Adapted from Rogene A. Buchholz et al., *Managing Environmental Issues: A Casebook* (Englewood Cliffs, NJ: Prentice Hall, 1992), pp. 170–175.

Benzene

Benzene is a colorless, sweet-smelling gas with a high vapor pressure that causes rapid evaporation under ordinary atmospheric conditions. The gas has long been recognized as a potentially dangerous substance capable of causing toxic effects and diseases. Over 70 percent of the benzene manufactured in the United States is eventually released into the atmosphere. It is used in the processing and manufacturing of tires, detergents, paints, pesticides, and petroleum products. Ninety-eight percent of all benzene produced is used in the petrochemical and petroleum refining industries. Over 270,000 employees are exposed to benzene in product-related industries.[1]

In May 1977, OSHA issued an emergency temporary standard (ETS) ordering that worker exposure to benzene be reduced from the regulated level of 10 parts per million (ppm) to 1 ppm, and proposed to make this standard permanent in all industries except gasoline distribution and sales pending a public hearing as required in its enabling legislation. At the time the ETS was issued, little medical evidence existed showing a relationship between benzene and cancer at any level found in the industry. Nonetheless, OSHA decided to drop the exposure level because (1) adverse health effects were evident at certain exposure levels and thus a reduced exposure was necessary to maintain a customary factor of safety, and (2) there is no safe level of exposure to carcinogens, suggesting that exposure should be reduced to the lowest level that can be easily monitored.

Industry pressed for the use of benefit-cost analysis in setting such standards, arguing that the law defines a standard as a regulation "reasonably necessary or appropriate" to protect worker's health and safety. A standard that flunks a benefit-cost test is not reasonably necessary or appropriate, industry maintained. Based on this argument, a federal appeals court in New Orleans set aside the ordered reduction, ruling that OSHA cannot legally regulate occupational health hazards without using a benefit-cost analysis to "determine whether the benefits expected from the standard bear a reasonable relationship to the costs imposed." The Supreme Court, in a 5-to-4 decision, upheld the lower court ruling, but for mostly different reasons.[2]

The court held that OSHA must make a "threshold determination" of significant risk at the standard's present level before moving to a lower level. To the court a safe work environment is not necessarily a risk-free work environment, and OSHA must adequately prove that there will be significant health benefits at the lower level. The ruling stated that OSHA had established a risk at levels above 10 ppm, but that the agency failed to prove that any substantial benefit would result from dropping the level. While the court refused to uphold the 1 ppm standard, it in no way rejected the agency's right to regulate industry on a less strict standard.

Under pressure from labor unions and public advocacy groups, OSHA proposed a new rule for occupational exposure to benzene in December 1985, which again dropped exposure to 1 ppm. In proposing this rule, OSHA adhered to the standards set by the Supreme Court ruling. Quantitative risk studies were done to assess whether benzene exposure posed a threat to workers. OSHA then assessed whether the new standard would provide significant benefits to exposed workers. Finally, the agency looked at available data to set an exposure limit that was technologically and economically feasible.

1. Is the use of benefit-cost analysis appropriate when dealing with regulation of substances like benzene? What ethical theory is benefit-cost analysis based on? What are the problems in applying this theory to making decisions about appropriate regulations?
2. Is it possible to create a risk-free work environment? If not, what level of risk is it appropriate and ethical to ask workers to accept? Should they be in on the decision about the risks they face in the workplace? What is a fair way to handle these kinds of issues?

Case Notes

1. Unless otherwise noted, material for this case comes from Tom L. Beauchamp, *Case Studies in Business, Society, and Ethics,* 2nd ed. (Englewood Cliffs, NJ: Prentice Hall, 1989), pp. 203–211.
2. *Marshall v. American Petroleum Institute,* 581 F.2d 493 (5th Cir. 1978). See also "Court Battle over Benzene Safety Raises Issues of Weighing Agency Rules, Cost against Benefit," *Wall Street Journal,* January 9, 1980, p. 44; and "Regulations Limiting Worker Exposure to Benzene Are Voided by High Court," *Wall Street Journal,* July 3, 1980, p. 6.

SOURCE: Rogene A. Buchholz, *Business Environment and Public Policy: Implications for Management,* 5th ed. (Englewood Cliffs, NJ: Prentice Hall, 1995), pp. 347–348.

Mobil Oil Corporation

Several years ago, Mobil began to feel the pressure from environmentalists as one of the largest producers of all-plastic packaging. Degradability was then viewed as the ecological cure for the plastics industry by consumer and environmental groups, and marketers began making a push to make degradibility a theme in their advertising and promotion campaigns. Mobil reformulated its Hefty line of bags by using an additive that hastened degradation by 25 to 40 percent under the right conditions, and introduced this reformulated product in June 1989 in order to survive in the marketplace. While Mobil didn't extensively advertise this feature of its bags, partly because it didn't want to hype degradibility as a solution to the solid waste problem, the biodegradibility feature was prominently displayed on the package along with a disclaimer that degradability was activated by exposure to the elements.[1]

Mobil had earlier concluded that biodegradable plastics would not help solve the solid waste problem, but believed it had to introduce such a product in order to meet competitive pressures. The company favored source reduction, recycling, incineration, and selective landfilling, but recognized that there were some short-term gains to be made in switching to a biodegradable product, and the public-relations value of this move had to be considered as opposed to real solutions to the problem. This position came back to haunt the company when state attorneys general began investigating green claims. While its market share did not change either before or after the product was introduced, the company believed that sales of the Hefty line would have suffered because it would have lost shelf space to its competitors.[2]

The problem is that after the reformulated Hefty line was introduced, the public did a 180 degree turn on its thinking about degradability, and environmental groups, like the Environmental Defense Fund, called for a consumer boycott of many products that carried such claims. People began to realize that most garbage is sent to covered landfills where the degradation process in limited at best, or at worst, does not take place at all because of being buried. What Mobil believed was a marketing response turned out to be a marketing nightmare, as the Federal Trade Commission and state agencies began investigating Mobil and other companies for their degradability claims. By March of 1990, the pressures were so great from the other direction, that Mobil announced that it would voluntarily remove degradability claims from all Hefty packaging.[3]

This action, however, didn't protect Mobil from further controversy, as in June of 1990, seven state attorneys general, from the states of California, Massachusetts, New York, Texas, Minnesota, Washington, and Wisconsin, sued Mobil in seven separate lawsuits. These suits charged Mobil with deceptive advertising and consumer fraud for its degradability claims. The attorney general of Minnesota, for example, stated that "Mobil's advertising claims break down faster than their garbage bags." Officials said that Mobil was targeted partly because of its public position against degradability benefits, which made it look like it was simply trying to profit from consumer trends rather than be environmentally responsible. This incident has made the company learn that green marketing can be extremely changeable, and that consumer education is needed on a larger scale.[4]

1. Did Mobil have to make biodegradability claims about its Hefty product line to remain competitive? Does this action suggest anything about the nature of the marketplace and its ability to respond to environmental problems? What could Mobil have done to be more sensitive to consumer concerns about the environment?
2. What does this case suggest about the whole green marketing effort? Is some kind of intervention needed to make advertisers more honest about the environmental characteristics of their products? Will the government guidelines work to mitigate these misleading claims? What do you think about the efforts of the private groups that are certifying environmental claims?

Case Notes

1. Jennifer Lawrence, "Mobil," *Advertising Age,* January 29, 1991, pp. 12–13.
2. Ibid.
3. Ibid., p. 13.
4. Ibid.

Lead Contamination

Already in the late 1960s, there was concern in South Dallas about lead poisoning and its effects on human health. The lead came from three lead smelters that regularly spewed bluish-black clouds of lead-bearing smoke from their stacks that settled on the homes of the surrounding residential areas. In the case of Dixie Metals Company, the pollution was the byproduct of a process used to recover lead from used automobile batteries. The smoke was so dense at times that it obstructed visibility and brought tears to the eyes of residents. Complaints about headaches, stomach cramps, and swollen joints were commonplace.[1]

The surrounding communities and the smelters sort of grew up together when little was known or documented about the health effects of lead or the pathways that air and water emissions take upon leaving the boundaries of a lead plant. However, the city of Dallas did enact a local lead ordinance for air emissions in 1968 that set standards for lead emissions from a plant at 5 micrograms per cubic meter. This was 10 years before the federal government would set its own standard for lead based on mathematical modeling and preliminary scientific studies.

More than a decade after the residents of the areas first complained, the problem still remained. While the smelters did make some progress in reducing lead emissions over a 15-year period, the companies were generally slow and reluctant to respond. City records indicated that the city of Dallas repeatedly failed to enforce the city ordinance regulating lead emissions and failed to come up with a clean-up plan, even though studies showed a public health problem.

As early as 1967, city air pollution officials measured emissions of 23 times the state established pollution standard at the Dixie smelter and 7 times the standard at the National Lead smelter. In 1970, city officials measured lead emissions from Dixie at 120 times the city's lead standard and 18 times the standard at National Lead. Despite these violations, neither the city of Dallas or the state of Texas took any enforcement action, with the exception of issuing periodic notices of violation and municipal court citations, against the companies to force them to reduce emissions.[2]

Finally, in 1974, the city took legal action against the companies. The suits were settled out of court a few months later in exchange for the promise of additional pollution controls within two years and payment of $35,000 by each of the three lead companies. The settlement did not require the companies to remove tons of lead-contaminated soil, but only to reduce pollution emissions. RSR and Dixie did not fully comply with the settlement until three years beyond the deadline and National Lead closed in 1979 rather than comply. But during this period, the city did not take the smelters back to court for violation of the consent decrees.[3]

Eventually the federal government came into the picture. EPA Region VI, which includes Bartlesville, Oklahoma, and Dallas, El Paso, and Corpus Christi, Texas, established a Toxic Substances Coordinating Committee (TSCC) to ensure a multimedia and interdivisional approach to environmental problems. The committee selected lead and cadmium as the two top priority regional chemicals because of availability of vast amounts of data that showed widespread environmental contamination in Region VI and exposure of children residing in contaminated areas.

Lead is taken into the body by breathing contaminated air or absorbing it through the skin into the bloodstream. It also enters the bodies of children when they eat dirt or put dirty hands or toys into their mouths. Symptoms of lead poisoning include anemia, intestinal cramps, nerve paralysis—particularly in the arms and legs—fatigue, and loss of appetite. Children and adults with high lead levels frequently suffer from nausea and constipation. Lead contamination of the air and soil occurs when particles of the metal escape during smelting. The airborne particles are blown only a short distance from a smelter before settling to the ground.[4]

The EPA eventually awarded a contract for $9,999 to the University of Texas at Arlington (UTA) to study the problem. The results of the study showed that that cadmium soil levels found throughout Dallas were not high enough to be of major concern. However, lead levels in the soil were elevated around all of the facilities studied.[5] Based on these UTA figures, concern was expressed that these levels might be associated with potential lead poisoning in the population, especially children, living in the area. The EPA formed a study group in October of 1981 to evaluate the literature of lead and evaluate the health and environmental significance of the UTA data and data from other smelter sites. On November 2, 1981, the group reported publicly its findings on the relationship between lead in soil and lead in blood, concluding that no quantitative relationship can be drawn between soil lead and blood lead, but that each case had to be addressed individually.

At about the same time as the UTA study, the city of Dallas launched a voluntary blood-lead sampling program in which nearly 12,000 individuals participated. The EPA volunteered to assess the quality of the samples for the city and to evaluate the significance of the data. The EPA sought and obtained the assistance of the Department of Health and Human Services and the Centers for Disease Control in conjunction with the Texas Department of Health and the Dallas Department of Health and Human Services Health Division to assess the health implications of the Dallas blood sampling program results.

Based on 1980 census data, it was estimated that of all those people living within 2 miles of either site, 7.5 percent participated and of all those living within a half-mile of either site, 22.3 percent of the population participated. These participation rates were less than the 75 percent participation rate adopted by the Office of Management and Budget (OMB) as a quality requirement for federally funded surveys to be considered representative. Because of this low participation rate and since the individuals who participated in the screening program were self-selected, the potential for a sample bias existed.[6]

The EPA published another study on February 1, 1983, which found that 5.6 percent of the children living within one-half mile of the RSR smelter suffered from lead toxicity. When these results were publicized, they produced a very strong reaction from the residents of the contaminated areas for remedial action and the closing of the RSR smelter. A number of lawsuits were initiated against RSR on grounds of physiological and psychological impairment due to undue lead exposure. Some suits claimed damages by smelter emissions to homes caused a reduction in property value.[7]

The total costs for removal, replacement, and transportation of contaminated soil to a disposal site would be approximately $850 per lot. The disposal fee charged by the landfill accepting the soil was not included in the estimate. This charge would largely be based on the quantity of the soil to be disposed of and its characteristics (whether or not it met the Resource Conservation and Recovery Act (RCRA) criteria for classification as a hazardous waste). There might be a possibility of negotiating a one-time disposal of the soil in a local landfill if it could be demonstrated that there would be no hazard created and the EPA and state agencies agreed to such an arrangement. If the soil was not a hazardous waste, it could be disposed in a local sanitary landfill. In either case, the disposal fee would probably be approximately $40–60 per lot.[8]

Further soil studies were conducted by both the EPA and the city of Dallas to alleviate public concerns and to gather sound scientific data. While the studies were inconclusive about the connection between soil-lead and blood-lead levels, the damage had been done. Furthermore, RSR continued to violate EPA standards for lead emissions, and, in the summer of 1983, the state of Texas and the city of Dallas took RSR to court to force compliance with the air standards and to clean up the surrounding area. In October of 1983, an out-of-court agreement was formalized.

The efforts of the soil cleanup and remedial program (SCRP) were directed at property located within a half-mile of the smelter facility. Soil tests were used by the EPA to establish two zones related to lead concentrations. The soil in Zone A posed the greatest threat to children, and the cleanup procedures consisted of (1) removal of the top six inches of soil and replacement with six inches of tested, uncontaminated soil, (2) planting and maintenance of new sod, (3) removal and replacement of soil around trees and shrubs, (4) replacement. of soil in unpaved driveways with three inches of limestone, and (5) special efforts to minimize the air lead levels during cleanup activities.[9] Property in Zone B posed a lesser threat to health, and required only (1) establishment of a vegetative cover in areas with bare spots, and (2) removal of six inches of soil from vegetable gardens and replacement with six inches of tested, uncontaminated soil. Certain areas in both zones were granted special provisions, some requiring twelve inches of soil removal.[10]

1. What measures can be taken, both positive and punitive, to make corporations and their executives more responsive to the public and more accountable for their actions when they and their corporations fail to comply with the laws and thereby endanger public health and safety? Positive measures could take the form of tax and other financial incentives; punitive measures could be in the form of increased outside regulation or stiffer civil and criminal penalties including incarceration.

2. What methods are available (or should be available) to the public to assure that there is no significant health hazard in the environment, or, if one does exist, that it is within an acceptable range? What role should be played by scientific institutions and universities in providing information to the public?
3. To what extent and in what ways did major government agencies (federal, state, and city) involved in the case fail or succeed in performing their assigned functions to protect the public's health? What additional measures should be taken to ensure that these public agencies have the necessary legal power, adequate financial and human resources, and the accountability to perform their functions effectively?
4. What role should the news media play in this kind of situation? Can they be trusted to inform the public accurately and provide all the information that the public should know about? or is the press biased and likely to present only one side of the story? Can the press be an adequate watchdog in protecting public health?

Case Notes

1. D. W. Nauss, "The Danger: Despite Health Threat, City Lax With Law for over a Decade," *The Dallas Morning News,* June 26, 1983, p. M1.
2. Ibid.
3. Ibid., p. M16.
4. Chuck Cook and Bill Lodge, "City Keeps Hands Off Plants Despite Lead Contamination," *The Dallas Morning News,* July 26, 1981, p. A22.
5. United States Environmental Protection Agency, Region VI, *Status Report on the Lead/Cadmium Study Conducted in Dallas and a Scenario of Possible Actions,* June 29, 1981, p. 5.
6. Lead Smelters Study Group, Office of Toxics Integration, U.S. Environmental Protection Agency, *Report on Dallas, Texas Lead Study,* April 16, 1982, p. 2.
7. Milan S. Yancey, *A Case Study of an Environmental Issue: The West Dallas Lead Smelter,* unpublished Master's Thesis, The University of Texas at Dallas, May 1985, p. 3. See also D. W. Nauss, "City, Smelter Conspiracy Over Pollution Alleged," *Dallas Times Herald,* April 24, 1983, p. B4.
8. Yancey, Ibid., p. 16.
9. Yancey, Ibid., p. 68.
10. Yancey, Ibid., p. 70.

SOURCE: Adapted from Rogene A. Buchholz et al., *Management Response to Public Issues,* 2nd ed. (Upper Saddle River, NJ: Prentice Hall, 1989), pp. 322–344.

Save the Turtles

Since 1978, all six species of sea turtles found in waters of the United States have been listed as endangered or threatened under the Endangered Species Act of 1973 and have thus been protected from further decimination. But populations of these turtles in North America and all over the world have declined because of development on beaches where they breed; butchery of nesting females and theft of eggs from their nests; oil slicks; turtles eating plastic garbage; and death in fishermen's nets used to catch other fish and shellfish.[1]

The Kemp Ridley sea turtles are most threatened, since they nest on only one beach—near Rancho Nuevo, Mexico—where they have declined from 40,000 nesting females a day in the late 1940s, to 10,000 in 1960, and to little more than 500 in the 80s. According to some estimates, approximately 48,000 sea turtles are caught each year on shrimp trawlers in the Southeast, and about 11,000 of these turtles die because of drowning, since they must come to the surface every hour or so to breathe. About 10,000 of those turtles that have died are loggerheads, and 750 are Kemp Ridleys.[2]

The shrimping industry disputes government figures, which show a close correlation between the number of dead turtles on beaches and the number of trawlers working in the vicinity. While any one boat may not catch many turtles, the cumulative impact of approximately 7,000 offshore commercial vessels towing for four to five million hours per year can be serious. Shrimpers claim dead turtles are mostly victims of pollution or disease rather than of shrimpers' nets. But for many people there seems to be no doubt that shrimpers are killing a significant number of turtles along with other nonshrimp organisms. For every pound of shrimp caught, nine pounds of other fish such as juvenile trout, redfish, whiting, and flounder are dumped dead over the side of the boat in what is called the by-catch.[3]

The shrimping industry provides jobs for many people in the Southeastern part of the country. More than 30,000 commercial fishermen and their families rely on shrimp for their livelihood, and many more work in shoreside processing plants.[4] Many shrimpers are second and third generation, following in the paths of their fathers and grandfathers. Thus the industry has great social as well as economic value, and any threat to the industry is likely to be met with great resistance.[5]

Such a threat appeared in the form of turtle excluder devices (TEDs), which were developed to save turtles and other fish from being killed in the fishing nets of the shrimpers. The TED is a panel of large-mesh webbing or a metal grid inserted into the funnel-shaped nets of the shrimpers. When these nets are dragged along the bottom of the ocean, shrimp and other small animals pass through the TED and into the narrow bag at the end of the funnel where the catch is collected. Sea turtles, sharks, and other fish too large to get through the panel are deflected out an escape hatch. The problem is that some of the shrimp escape as well, as much as 20 percent or more of the catch, according to some estimates.[6]

These devices were actually developed to save the industry. Since the law requires that endangered species in the public domain be protected regardless of the cost, the industry was in danger of being totally shut down if some environmental groups were to sue the industry or the federal government. Thus the National Marine Fisheries Service (NMFS) tried to find a technological solution. They spent $3.4 million between 1978 and 1984 developing and testing the device. By 1981, the agency was promoting voluntary use of TEDs and began passing out free TEDs in 1983 to encourage shrimpers to use them. Shrimpers objected because they were difficult to use and lost a significant percentage of the shrimp catch.[7]

As more and more dead turtles washed ashore, environmental groups like Greenpeace and the Center for Marine Conservation demanded an end to the killing. Since the voluntary approach had failed, the U.S. Fish and Wildlife Service demanded mandatory use of TEDS, and the Center for Marine Conservation threatened to sue the NMFS and close down the industry completely. Industry representatives then agreed to phase in use of TEDS, but rank-and-file fishermen rose up in rebellion. They vowed civil disobedience against what they saw as a threat to their survival, and filed lawsuit after lawsuit, each of which was lost in court.[8]

The fight then moved to Congress where the Endangered Species Act was up for renewal. After prolonged debate, amendments passed in early fall 1988 made TEDs mandatory by May 1, 1989, but only in offshore waters, with the exception that regulations already in effect in the Canaveral area of Florida would remain in effect. Regulations for in-shore areas were to go into effect by May 1, 1990, unless the Secretary of Commerce determined that other conservation measures were proving equally effective in reducing sea turtle mortality by shrimp trawling. Further testing was to be done on TEDs under in-shore conditions, but until then, in-shore turtles would have virtually no protection.[9]

As the May 1 date for implementing TEDs according to the federal timetable drew closer, fishermen in Louisiana rallied to oppose installation of the device. Officials from across the South attended the rallies and pledged to help stop TED legislation from being implemented. Governor Roemer of Louisiana said that state wildlife agents should boycott TED laws until studies showed conclusively that the devices worked.[10] Congressional representatives from Louisiana persuaded the Secretary of Commerce, who was responsible for implementing the TED regulations, to delay their implementation still further to give shrimpers time to buy and install the devices. Only warnings would be issued through the end of June while a National Academy of Sciences committee studied the issue.[11]

When the warning period ended, penalties went into effect that could go as high as $10,000 and confiscation of catch. Criminal violators, those who repeatedly thumbed their noses at the law, could be convicted of a felony and be fined $20,000, in addition to losing their catch. Emotions ran high in some Louisiana communities, and many shrimpers vowed to break the law by not pulling TEDS, and dared officials to haul them off to jail. Some vowed to shoot the man that tried to take away their living.[12]

In view of these developments, many shrimpers installed and tried to use the device in order to comply with the regulations. Then nature struck with the largest bloom of seaweed in several years; the seaweed clogged the excluder panels and prevented much of the shrimp catch from being taken. Shrimpers who had installed TEDs cut them out of their nets, and the Coast Guard temporarily suspended the regulations for Louisiana coastal waters. Representatives from the state hoped that the Secretary of Commerce would make the suspension permanent.[13]

Then the Secretary of Commerce, after initially telling the Coast Guard not to enforce the law, reversed himself. When the shrimpers heard this, they streamed into port to protest, blocking shipping channels in Galveston and Corpus Christi, Texas, as well as several Louisiana locations. The blockade in Galveston halted all ship and ferry traffic, although by midafternoon the shrimpers agreed to let ferries pass through the blockade. The blockade threatened to shut down Houston's oil refineries. There was some violence as an Alabama man was arrested after he fired a semiautomatic rifle from his boat in Galveston, and two men were arrested in Corpus Christi for throwing an object through a window of a 41-foot Coast Guard patrol boat. Angry fishermen set fire to a huge pile of TEDs on shore.[14]

Experts on the use of TEDs defended their results and accused the shrimpers of refusing to learn how to use the devices correctly. If installed properly, the catch could even be increased. But the fishermen were said to believe that they were fighting for their lives in a depressed economy that resulted in an increase in the number of competing boats and kept prices stagnant. The TEDs were seen as the death blow to a dying industry, and research data with respect to use of the TEDs was rejected. With such a hardened position, nothing short of a court-ordered settlement seemed likely to resolve the issue.[15]

In February 1990, the shrimpers sued the federal government again, saying the laws requiring TEDs placed an unconstitutional burden on their businesses. The suit was filed in federal court in Corpus Christi, Texas, and sought immediate suspension of the regulations requiring the use of TEDs for offshore shrimpers. Attorney Robert Ketchand who filed the suit on behalf of the Concerned Shrimpers of America called the TEDs laws "regulatory taking" of shrimpers profits.[16] Thus the controversy seemed to have come full circle, with the shrimpers pursuing the cause through the courts as they did before the amendments to the Endangered Species were passed.

1. Is there a technological solution to this problem, or is the nature of the controversy so political at this point that the parties to the controversy have ceased to believe a technological solution exists? If so, what kind of a political solution will work to resolve the controversy?
2. Should the fishermen be paid compensation for the losses they claim because of using TEDS? How should these losses be determined? Who should pay for the protection of endangered species? What is a fair resolution of this issue?
3. Is the on-again, off-again nature of the regulations a serious problem? Was the Secretary of Commerce right in suspending TED regulations when shrimpers blockaded ports along the Gulf Coast? What else could have been done at this point?
4. What would you suggest be done now? Is our political/legal system structured in such a way that it can resolve conflicts of this nature? What makes this conflict different from other that seem to get resolved without resort to violence or stonewalling tactics that drag on forever?

Case Notes

1. Jack and Anne Rudloe, "Shrimpers and Lawmakers Collide over a Move to Save the Sea Turtles," *Smithsonian,* Vol. 20, No. 9 (December 1989), p. 47.
2. Ibid.
3. Ibid., p. 49.
4. Ibid., p. 47.
5. Ibid., p. 49.
6. Ibid., p. 45.
7. Ibid., p. 50.
8. Ibid.
9. Endangered Species Act Amendments of 1988, Conference Report 100–928 to Accompany H.R. 1467, House of Representatives, 100th Congress, 2d Session, p. 5.
10. Christopher Cooper, "Shrimpers Vow to Defy Law on TEDs," *The Times-Picayune,* April 9, 1989, p. B1.
11. Rudloe, "Shrimpers and Lawmakers Collide," pp. 52–53.
12. Christopher Cooper, "Furious Shrimpers Flouting TEDs Law," *The Times-Picayune,* July 9, 1989, p. B1.

13. Christopher Cooper, "La. Shrimpers Get Break on TEDs," *The Times-Picayune,* July 11, 1989, p. B1.
14. Christopher Cooper, "Shrimpers' TEDs Protest Turns Violent," *The Times-Picayune,* July 23, 1989, p. A1.
15. James O'Byrne, "Research Disputes Shrimpers' Claims," *The Times-Picayune,* July 27, 1989, p. A1.
16. Christopher Cooper, "Shrimpers File Federal Suit Against TEDs," *The Times-Picayune,* February 22, 1990, p. B1.

SOURCE: Adapted from Rogene A. Buchholz, et al., *Managing Environmental Issues: A Casebook* (Upper Saddle River, NJ: Prentice Hall, 1992), pp. 61–70.

Index

A

B

C

D

E

F

G

H

I

J

K

O

P

R

S

T

U